CONTEMPORARY CONCRETE STRUCTURES

AUGUST E. KOMENDANT, D.E., P.E.
Consulting Engineer

McGRAW-HILL BOOK COMPANY
New York   St. Louis   San Francisco   Düsseldorf   Johannesburg
Kuala Lumpur   London   Mexico   Montreal   New Delhi
Panama   Rio de Janeiro   Singapore   Sydney   Toronto

Library of Congress Cataloging in Publication Data

Komendant, August E.
  Contemporary concrete structures.

  1. Concrete construction. 2. Reinforced concrete construction. I. Title.
  TA681.K66   624'.1834   72-3296
  ISBN 0-07-035328-X

Copyright ©1972 by McGraw-Hill, Inc. All Rights Reserved. Printed in the United States of America. No part of this publication may be reproduced, stored in a retrieval system, or transmitted, in any form or by any means, electronic, mechanical, photocopying, recording, or otherwise, without the prior written permission of the publisher.

1234567890 KPKP 765432

The editors for this book were William G. Salo, Jr., Don A. Douglas, and Jean Ely, the designer was Naomi Auerbach, and its production was supervised by George E. Oechsner. It was set in Press Roman by Scripta Technica, Inc.

It was printed and bound by The Kingsport Press.

This book is dedicated to the memory of the great structural engineers

Prof. Dr. Eng. KURT BEYER
Prof. Dr. Eng. FRANZ DISCHINGER
Dr. Eng. EUGÈNE FREYSSINET

whose works have most extensively contributed to the present level of structural engineering in the advanced field of reinforced and prestressed concrete.

CONTENTS

Preface xi
Notations xv
Indices and Sign Conventions xvii

Part 1. THEORY OF STRUCTURES

1. **Design and Its Critical Evaluation** 3
   1. Approach to a Design 3
   2. The Means of Realization 5
   3. Critical Evaluation 6

   CARRYING ACTIONS

2. **Beam Action** 17
   4. Beam Theory 17
      Loadings; Internal forces, simple and combined bending; Force distribution between concrete and steel; Statically determinate and indeterminate beams
   5. Two-way Beam Action 30
      Simple and continuous slab
   6. Beam-Grid Systems 40
   7. Torsion 49
   8. Stress Analysis 53
      Noncracked and cracked sections; Shear, bond, and principal stresses; Prestressing stresses
   9. Plastic Flow and Shrinkage 66
      Stress transfer from concrete to steel; Influence of plain reinforcing upon shrinkage and plastic flow; Loss in prestress; Change in redundants
   10. Ultimate Carrying Capacity and Factor of Safety 84
       Phases of equilibrium condition; Local concentrated stresses
   11. Deformations 99
       Deflections of statically determinate and indeterminate beams; Beams on elastic soil; Time deflection due to plastic flow and shrinkage

3. **Arch Action** 111

   12. Three-hinged Arches............................... 115
       Analysis; Temperature change; Elastic and plastic deformations

13. Two-hinged Arches 120
    Theory of analysis; Formulas for common loadings; Temperature change
14. Fixed Arches 126
    Theory of analysis; Formulas for common loadings; Temperature change; Elastic and plastic deformations of two-hinged and fixed arches

4. **Suspension Action** 133

    Cable force; Deflections

5. **Combined Carrying Actions** 137

    15. Simple Frames...................................... 137
        Two-hinged frames; Theory and formulas for common loadings; Temperature change; Fixed frames; Theory and formulas for common loadings; Temperature change
    16. Multibay and Multistory Frames....................... 153
        Theory, symmetric and antisymmetric loadings; Wind loading; Temperature change, shrinkage
    17. Flat Slabs 170
        Theory, symmetric and antisymmetric loadings; Temperature change
    18. Arch Beams 177
        Analysis; Deformations
    19. Beam-Suspension Action 179
        Analysis; Prestressing
    20. Trusses (Arch-Suspension Action) 181
        Types, analysis; Simple and continuous trusses; Secondary stresses, influence of plastic flow
    21. Vierendeel Girders (Beam-Arch Suspension Action) 186
        Theory of analysis; Simple and continuous spans; Deflections
    22. Beam-Shear Action 204
        Theory; Rigidity; Deflections

6. **Circular Beams (Beam-Torsion Action)** 213

    Theory, statically determinate and indeterminate circular beams; Shear forces; Flexural and torsional moments for typical loadings; Prestressing

7. **Space Structures** 235

    23. Spherical Shells 239
        Theory, deformations; Particular integrals for common loadings; Temperature change; Shrinkage; Boundary conditions; Homogeneous solution; Marginal members

24. Polygonal Domes 255
    Theory, membrane condition; Ridge loading, normal forces
25. Rotationally Symmetric Cylindrical Shells 262
    Theory, particular integrals for common loadings; Temperature
    change; Shrinkage; Boundary conditions; Homogeneous solution
26. Segmental Cylindrical Shells 270
    Theory; Deformations; Boundary conditions
27. Polygonlinear Shells (Folded Plates)......................... 273
    Membrane and bending theories; Deflections, ridge and panel loads;
    Continuous polygonlinear shells; Prestressing
28. Curvilinear Cylindrical Shells 294
    Membrane theory, particular integrals for common loadings;
    Ellipse, circle, cycloid; Prestressing; Bending theory, prestressing,
    deformations; End supports
29. Pipes 325
    Particular integrals for common loadings; Prestressing
30. Hyperbolic Paraboloids 327
    Membrane theory; Ridge loads
31. Elliptic Paraboloids 331
    Membrane theory; Boundary conditions; Normal forces and shear
32. Stability of Shells.. 340

Part 2. MATERIALS 347

8. Concrete... 349

33. General Characteristics 349
    Cement; Portland cement minerals; Hydration process
34. Aggregates 357
    Fine aggregates; Coarse aggregates; Lightweight aggregates
35. Admixtures... 362
36. Mix Design 366
    High-strength concrete; Pumpcrete and shotcrete; Grout mix and
    grouting operation
37. Placing and Consolidation 378
    Vibration; Water movement in concrete during consolidation
38. Curing of Concrete 381
    Air curing; High-pressure steam curing; Duration of curing;
    Internal stresses during hardening
39. Surface Treatment 385
    Form surface; Sandblasting; Acid treatment
40. Physical Properties of Concrete 388
    Compressive, tensile, and shear strength; Deformations; Elastic
    strain and initial set; Modulus of elasticity; High-strength and
    lightweight concrete; Thermal properties
41. Shrinkage and Swell 393
    Capillary action, internal stresses; Influence of relative humidity
    and curing

viii Contents

    42. Plastic Flow 397
        Relationship between shrinkage and plastic flow; Factors
        influencing plastic flow

9. Steel 399

    43. Reinforcing Steel 399
        Grades and properties of reinforcing steel and welded wire fabric
    44. Prestressing Steel 401
        Physical properties of high-strength steel wires, strands, and rods
    45. Corrosion of Steel 405

10. Bearing Neoprene 408

    46. Physical Characteristics 408
        Compressive strength; Shear strength; Deformations; Allowable
        stresses

Part 3. STRUCTURAL SYSTEMS AND METHODS OF CONSTRUCTION ... 415

11. Basic Structural Elements 417

    47. Prefabricated Slabs and Planks 418
    48. Slab Beams 423
    49. Precast Beams and Girders 428
    50. Columns 431
    51. Wall Panels 439

12. Structural Systems 442

    52. One-way Structural Systems 446
        Narrow-spaced systems; Folded-plate systems; Circular one-way
        systems
    53. Vierendeel Systems 464
    54. Frame and Arched Systems 471
        Frame-suspension system; Channel frames; Vierendeel arches;
        Circular cantilevered frames; Prefabrication
    55. Two-way Structural Systems 479
        Wide- and narrow-grid systems; Vierendeel wide-grid systems

13. Space Systems 491

    56. Prismatic Shells 491
    57. Domes, Long-Span Arched Systems 496
        Folded plates, hyperbolic paraboloids; Cylindrical shells;
        Poured-in-place and precast

58. Curvilinear Cylindrical Shells 504
   Prestressed, prefabricated, and poured in place

14. Suspension Systems **509**

59. Circular One- and Two-way Systems 509
   Arrangement of cables; Marginal and supporting members
60. Rectangular Long-span One- and Two-way Systems 516
61. Drainage, Cable Anchorages, and Decking 520

15. Prefabricated Housing **524**

62. Panel Systems 525
   Design, manufacturing, erection, evaluation
63. Modular-box Systems 527
   Box and form types; Manufacturing, erection, evaluation
64. Housing Complexes 532
   Layouts, arrangement of units; Stair-elevator towers and public areas; Habitat '67

16. High-rises **544**

65. Medium-height High-rises 544
   Framing, floor systems; Construction methods
66. High-rises—Suspended System 548
   Design and practical considerations; Column stations, suspension and floor systems; Construction methods

17. Bridges **558**

67. Bridges for Grade Separation 560
   Structural systems related to terrain and span; Construction
68. Bridges for Pedestrian Crossings 570
69. Bridges for Intersecting Superhighways 572
   Principle of design; Type of superstructure; Columns; Construction methods
70. Medium-span Bridges 575
71. Long-span River and Valley Bridges 579
   Box and semibox girder bridges; Construction methods; Free cantilevering; Construction by use of launching girder and form carriers; Diagonal-cable bridges; Piers, superstructure; Construction methods
72. Suspension-Beam Bridges 596
   Triangular box-girder type; Stress-ribbon bridges
73. Arch and Arch-Beam Bridges 598
   Conventional arch bridges; Maillart-type arch-beam bridges; Prestressed arch-beam bridges; Construction methods

18. Miscellaneous Structures 605

74. Water Towers 605
    Design and construction methods
75. Surface Tanks 611
    Water tanks; Gasoline tanks
76. Silos 614
    Silo pressure; Grain silos; Cement silos; Raw-sugar silos; Silos for hygroscopic materials; Design and construction
77. Observation and Transmission Towers 623
    Design and construction
78. Power-line Towers, Telephone and Lighting Poles 629
    Design, construction, and prestressing
79. Wind and Earthquake Forces 631
    General theory; Wind loading; Earthquake loading

19. Why Structures Collapse or Fail 645

80. Design 646
81. Construction 649
82. Supervision and Inspection 651
83. Responsibilities 651

Appendix 655
Index 666

PREFACE

Progressive engineers are always seeking new ways to extend the boundaries of knowledge. Therefore, from day to day, what one "knows" changes and gives way to new principles, theories, and achievements. The progress is extensive and seems limitless. In these changes and new forms, the face of contemporary engineering and architecture appears.

The present principles and trend in structural engineering and architecture are simplicity, well-balanced masses and areas, and harmony of the structure with its surroundings. The structural system should agree with the particular material it is made of. The transmission of the loads and forces to which the structural element is subjected should be clear and visible. The construction must be simple, rapid, and economical. The limits of these aims are controlled only by safety of the structure, special requirements, and site conditions.

To satisfy all these concepts and requirements is possible but the technical and aesthetic problems involved are rather difficult and complex. In addition to this, economic necessities, arising from strong competititon, must be satisfied. It requires the teamwork of many specialists with high professional responsibility, from various fields of engineering, science, architecture, construction, and laboratory technique, to obtain maximum results at minimum costs.

Since advanced structural theories are extensive, very often rather complex, highly abstract, and presented mainly in special papers, proceedings, and technical periodicals, they are out of easy reach for professors, practicing engineers, students, and especially architects. This is the reason for this book, in which the required, most applicable structural theories, material characteristics, and construction methods are discussed for practical purposes. Most of the materials treated have been used by the author in his 35 years of combined experience as consulting engineer, contractor, and professor in the field of advanced structures. The accuracy of the theories and methods, selected from the vast amount of studies on the subject, has been proved by actual testing on structures in the field. The deviations of the computed deformations from measured ones have been values only of the second order. Thus, they are adequately accurate to describe the stress condition of the structure concerned.

The topics in this book are presented in three parts. The first two parts deal with the philosophical approach to and critical evaluation of design, statics, and materials, and the third part is concerned with the design! A knowledge of the elementary principles of structures and materials, usually covered in college and university courses has been taken for granted; these principles, therefore, are discussed only to the extent to obtain continuity.

In the first part, although only the most applicable and simplest advanced theories are presented, the discussion is highly mathematical. To keep the use of higher mathematics to a minimum, equations are developed for general cases and are presented so that they are direct, readable, and understandable for their meaning and application. Also, integration tables, to solve the integrals, have been added. Empirical formulas are used only where valid advanced theories are not available or are too complex for practical use. Prestressing, which is actually nothing more than an additional external loading, is not discussed separately but is incorporated directly into general structural analysis. For further clarification, the theories and analyses are illustrated through their numerical applications (slide rule accuracy) to actual structural situations. In order to reduce the static analyses, where possible, closed formulas or tables for reactions, shear, and moments of indeterminate systems subjected to common loadings are presented. For clarity, and to reduce the number of notations, indices are used.

For this reason, in spite of the higher mathematics involved, practicing engineers and students should experience no difficulty in understanding and using the theories and methods, presented in the first part. The architect's interest in this material would be to widen his knowledge about what is structurally available and required to avoid failures and to secure success. To facilitate the use of the numerous illustrations relative to countries in which the metric system is used, a conversion table from English to metric units is included in the Appendix.

The second part deals mainly with concrete and gives basic information about concrete technology, as we understand it at present. Without such knowledge, no success is granted in contemporary advanced concrete structures. Information and data about reinforcing and high-carbon prestressing steel and anchorage devices are readily available from manufacturers; therefore, only special characteristics of steel—its behavior and durability under high stresses and special conditions—are discussed. Also, data about neoprene, used for bearings, are given in the extent necessary for design.

The third part deals with structural systems and related construction methods for utility and public buildings: high-rises and mass-produced prefabricated houses and housing complexes, bridges, and miscellaneous structures such as towers and silos. At the end of this part the main reasons that structures collapse and fail are discussed.

It is believed that the second and third parts will be helpful to architects and contractors, as well as to structural engineers. The author's endeavor has been to provide enough materials for successful design procedure.

In conclusion the author takes this opportunity to express his thanks and acknowledgment to all those who have contributed in various ways in preparation and production of this book. Particularly thankful credit is due to the architects Merike and William Phillips for the preparation of the illustrations.

August E. Komendant

Notations

P	concentrated load; load in general
W	resultant of uniform loading
w	total uniformly distributed load
p	pressure; surface loading; ratio
F	force in general; Airy's function
R	vertical reaction
H	horizontal reaction; thrust
M	bending moment
L, l	length of span; characteristic length
c	length of bay; crown; factor
s	unit length of curvature
R, r	radius of curvature
x, y, z	cartesian coordinates
$\xi, \xi' = x/L, x'/L$	
$\eta = z/z_0$	
$\lambda = z_0/L$	
$\zeta, \zeta' = z/h, z'/h, x_m/l, x_m'/l,$	
α, β, ϑ	angles
γ, ω, ψ	angles; factors
A	cross-sectional area
d_0	total depth of section
d	effective depth of section; diameter
d'	distance from extreme fiber to center of reinforcement
b	width of flange or section; span of shell curvature ($B = 2b$)
b_0	width of web or stem
t	thickness
ψ	section factor, $\psi = e_0 z/i^2$
Q	static moment
η	center of gravity of section
I_0	moment of inertia about center of gravity
S	section modulus; stiffness
E	modulus of elasticity
G	shear modulus
i	radius of gyration, $i^2 = I/A$
ν	Poisson's ratio; ratio of triangular to parabolic stress distribution; factor
j	ratio, z_0/d
k	subsoil or spring constant; ratio of distance from extreme fiber to neutral plane
γ	shear distortion angle; capillary constant
$\rho = EI/GI_T$	
m	mass; midspan
$n = E_c/E_s$; number of redundants	

e	eccentricity from center of gravity
μ, κ	factors
$a, b, c, \ldots$	coefficients in general; constants; parameters
$A, B, C, \ldots$	differential coefficients; joint designations
ν	volume of voids in concrete; displacement
v	factor of safety; factor
α	ratio of strains; coefficient of linear expansion
t	time
ΔT	change of temperature
N	normal force
V	shearing force; vertical force
D	diagonal force; stiffness
S	ridge force; shear
C	compressive force
T	tensile force; horizontal shear force
X	statically redundant force or moment
φ, ψ	geometrically redundant rotation and displacement; ratio
σ	normal stress in plane
f	extreme fiber stress; coefficient of friction
f'	ultimate strength
τ, v	shearing stress
$\sigma_{1,2}$	principal stresses
U	bond stress
∂	differential
δ	relative displacement of a definite point or relative angular rotation; vertical deflection
ϵ	strain; relative moisture
w, v, u	elastic displacements in direction of three axes
Δ	referring to change; difference
$F.S.$	factor of safety

Indices and Sign Conventions

xyt	In general, the first index (x) indicates origin, location, or place under consideration; the second index (y), reason or direction; the third index (t), time of this action
$M_K^{(J)}$	moment at K in direction of J
ik	any point i on a system where an action occurs which is caused by a force applied at a position k
ki	vice versa
α, β, ω	location and direction of an action considered in shells
o	indicating loads in statically determinate principal system; starting time of action; center or plane of gravity of entire concrete section (steel area not deducted, i.e., gross area of concrete); perimeter of bars (Σo)
i	ideal or transformed; number of prestressing units; number of statically indeterminate forces internally
n	number of statically indeterminate forces externally; time at which plastic flow is finished
m	moving load or force; any point on a system
r	denoting a member of truss in general; any joined boundary (shells)
d	design (stress, load)
cu	cube
cl	cylinder
$a, b, c, \ldots$	denoting supports
$1, 2, 3, \ldots$	denoting prestressed reinforcement units; denoting span concerned
$J, K \ldots$	denoting joints
c	referring to concrete
s	referring to steel
(n)	denoting action in an n-times statically indeterminate principal system
D	dead load
L	live load
S	surcharge; strain stiffness
F	flexural stiffness
T	total load
pr	denoting prestressing
sr	shrinkage
σ	stress
el	elastic
pl	plastic flow
f	friction
ϵ	strain
τ	shearing stress
c	compression fiber
T	tension fiber; top fiber; torsion; total
B	extreme bottom fiber; state of breaking or rupture
t	time

+ moment acting in clockwise direction; vertical loads or forces acting downward; reaction acting upward; shearing forces related to positive (clockwise) rotation of the end of a member; slope of tangent to the elastic curve related to an increasing moment or in direction of acting force; deflection downward or in direction of acting force; tension in concrete; tension in steel
− vice versa
∓ algebraic sum (signs disregarded)
≃ approximates
→ approaches or becomes equal

GREEK ALPHABET

Alpha (a) ...... A a or α Nu (n) .... N ν
Beta (b) .... B β or ς Xi (x) .... Ξ ξ
Gamma (g) ...... Γ γ Omicron (o) .... O o
Delta (d) ...... Δ δ or ∂ Pi (p) .... Π π
Epsilon (e) ...... E ϵ Rho (r) .... P ρ
Zeta (z) .... Z ζ Sigma (s) ...... Σ σ or ς
Eta (h) .... H η Tau (t) .... T τ
Theta (th) ...... Θ θ or ϑ Upsilon (u) ...... Υ υ
Iota (i) .... I ι Phi (ph) .... Φ φ or ϕ
Kappa (k) ...... K κ or $\varkappa$ Chi (ch) .... X χ
Lambda (l) ...... Λ λ Psi (ps) .... Ψ ψ
Mu (m) .... M μ Omega (o) ...... Ω ω

ABBREVIATIONS

C.G.   center of gravity
k   kips (1,000 lb)
k'   kip-foot
psi   pounds per square inch
ksi   kips per square inch
psf   pounds per square foot
ksf   kips per square foot
N.P.   neutral plane
C.J.   construction joint
i.p.   inflection point
S.S.   stressteel
W.W.F. welded wire fabric
P.S.   pipe sleeve

# PART 1

THEORY OF STRUCTURES

1
DESIGN AND ITS CRITICAL EVALUATION

1. APPROACH TO A DESIGN

Any creative activity within man's experience involves four basic phases:

1. *Image*—the processing of ideas, concepts, and meanings, the weighing of possibilities, and the analysis of meanings. The process is purely abstract, free from limitations. It is an entirely immeasurable phase of a design.
2. *Design*—the realization of the image. This phase is not abstract. It involves practical consideration and is largely controlled by limitations.
3. *Construction*—the building of reality.
4. *Critical evaluation* of the finished product. This, too, is an immeasurable phase.

The last phase—critical evaluation—must be also an integral part of the first three phases, as guidance to the creator himself. Without it, the finished product—building, bridge, etc.—is not even worth evaluating.

There are two basic approaches to developing the first two phases of a design: artistic and philosophic. The artistic approach makes use of forms and patterns as the universal elements in the creation of architecture as well as art. In this approach, architecture is regarded as a research for right forms for making useful spaces. Thus, creative architectural impulse is primarily an impulse to form.

Using form and pattern as the operational bases for architecture and assuming that the design phase (from abstraction to reality) is correct, the following questions immediately arise: Is form really immeasurable and impersonal? What produces emotional response—forms and their relations? If so, what are the right forms and relations?

An objective, direct answer to these questions, so fundamental in architecture, is not possible. Supporters of the artistic approach, not being able to answer these vital questions, claim that such questions are irrelevant and meaningless in architecture. They use intuition for guidance in their aims and decisions and, very often, undefined architectural or structural reasons as justification of their choices and actions. Therefore, the artistic approach in architecture leads to confusion, aimlessness, and experimentation. It is highly subjective, being based primarily upon feeling and thus lacking the objective principles required for any applied art, such as architecture, and as such it is disputable.

The philosophic approach considers form not as a primary element in architecture, but rather as a result of abstract and physical relations whose treatment is not any more elementary. The form does not necessarily determine the end in view because the form is only part of the meaning. The order and position of elements define the form; that is, how the materials or elements are ordered determines what the form means. The vital union of form and matter does not necessarily produce a complete integration or a desired harmony. Also, it may not realize the intended meaning.

Thus a combination of forms and elements can become mere artifice, an added attraction or concealment, empty of content.

The only basis which allows us to answer all vital questions objectively and which provides guidance through all creative phases of a design is the *function*. The function is dynamic, it represents and cannot be represented, and as such it is immeasurable. The upper boundary of a function is a mathematical idea—a pure concept, very abstract and rationally manipulatable, whose meaning does not flow from the accident of its existence. The lower boundary of the function is represented by certified facts and thus independent of our control; it is the last instance against which our competence and validity of abstract concepts can only be checked.

Since a function is dictated by an "inner reasoned certainty," it is not disputable, and this makes all the difference. Thus the philosophic approach, using function as the basis for all phases of design, is conclusive and convincing. It is the only acceptable approach for the contemporary mind because it creates emotional response—"mysterious present—field of functions"—and guides the creator from the beginning to the end of his work. The ever-present field of functions is the only medium by which the creator and the observers of his work are related.

2. THE MEANS OF REALIZATION

As the image, the design, and their realization are closely related, the quality of the finished product—the structure or building—is determined already in the design phase by the choice of structural and mechanical systems, materials, and methods of construction. If these choices are not correctly made, intentions and accomplishments will commonly be out of scale and the work will not carry out the intention.

Since the structure is basically a servant to some purpose, the purpose will most likely limit the choice of structural and mechanical systems, and these, in turn, will control the materials and methods of construction to be used.

There are four basic carrying actions—suspension, arch, beam, and shell—and a variety of combinations of these basic actions, such as beam-arch action and beam-suspension action. Also, there are four basic materials: wood, masonry, steel, and reinforced concrete. The physical characteristics of the materials and the structural peculiarities of the carrying actions determine their relationship. For example: *Suspension action* involves only tensile stresses and can be accomplished at present by steel cables and strands. *True arch action* involves only compressive stresses and can be accomplished most efficiently with masonry, concrete, or steel. *Beam action* involves three types of stresses—compressive, tensile, and shear—and can be accomplished with wood, prestressed concrete, and steel. Beam action accomplished with reinforced concrete involves an additional stress, namely, bond stress. *Shell action* involves mainly two types of stresses—shear and compressive—and can be accomplished with reinforced and prestressed concrete.

The number of stresses involved indicates the span limits and the economy of the carrying action. Thus the most uneconomical carrying action with limited spans is beam action. The longest spans that are economically possible can be obtained by suspension and arch actions or by a combination of both, that is, by a trussed system.

Rarely can an efficient and acceptable structural system be accomplished by one carrying action. Beam action is generally required to obtain the stability and flexibility needed to meet local conditions and satisfy architectural requirements.

The shape of the structural member or members making up the system is controlled mainly by the physical characteristics of the material, the type of stresses involved, and their distribution over the cross-sectional area and along its length. The color and texture of surfaces and the pattern of members, as well as the shapes and connections, are atmosphere-creating qualities and are controlled by the image.

Selection of the method of construction is controlled mainly by considerations of economy, but whatever method is used, the finished product must clearly indicate what that method was, that is, how the structure was built and the required quality obtained.

Engineering science at present is at such a level that the strength of materials produced, the deformations of a structural element or system under any known loading it may be subject to, and its stability and carrying capacity can be computed with sufficient accuracy that any conceivable project can be carried out with reasonable safety and quality. As a result, methods and systems change more rapidly than ever before. In this progress appears the ability and face of contemporary architecture and engineering.

3. CRITICAL EVALUATION

Before critical evaluation of any design can be started, the basis of such an evaluation must be established. The more conscious the designer is about what constitutes a good design, the more successful he can be in pursuing his aim, the sounder his concept will be of the correct way of deciding what is right or wrong, and the more rational his decisions are likely to be. It is quite possible to be a good designer without ever philosophizing about basic principles of engineering, but the advancement of engineering depends upon someone's deep and persistent thinking about methods and meanings, about why and how to apply certain theories, using assumptions and weighing validities, justifications, etc. Thus to be a good and progressive engineer, it is not enough to accumulate a vast amount of facts, data, and experience in one's specialized field. Facts and data that are not ordered by reason and understanding into constructive relations may be called "knowledge," but they are not significant in themselves unless they suggest ideas combined into theories. Engineering as an applied science is simply not a two-dimensional affair, composed of factual knowledge lacking depth and stability. The third dimension, which is required for stability, is provided by theories, which are the only means of advancing engineering and of discovering what the advancement signifies.

Generally the evaluation of any design makes use of the following three criteria:

1. Rational solution
2. Aesthetic quality
3. Economy

One is immediately confronted with the following question: What are the relative degrees of importance of these criteria? For example, is a sound structural solution more important than aesthetic quality or economy? Probably no one could lay down beforehand a complete set of such rated criteria to take care of all eventualities. Even if that were possible, what about the value judgment of the end in view itself—the finished product whose quality is based upon a single order that will somehow take account of all three criteria? It is this that usually creates the conflict among the engineer, architect, public, and owner. In most cases, the owner's part in this dispute is assumed to be canceled out by fixing limits for expenditures and establishing the program to be completed within the financial limits. However, confining the program and funds available does not satisfy the third criterion—economy—it merely establishes the size of the project.

Before the problem can be investigated further, the value-judgment criteria must be clarified, defined, and analyzed explicitly in detail.

Rational Solution

What is understood by a "rational" structural solution? The term "rational" has many meanings and therefore must be fixed in the context of this particular usage. Let us assume, for example, that a preliminary image and schedule are created for a particular design, to be executed in a definite location under local conditions, and that data such as materials, equipment, and labor available and soil, seismic, and climatic conditions are known. There is, in this case, only one rational structural solution: The solution which carries out the purpose of the project and satisfies all conditions with maximum efficiency. Furthermore, a rational structural system (or systems) is one that does not interfere with the functions the structure is designed for. It must correspond to the image or architectural solution, and thus it should be aesthetically acceptable. All these factors considered, any other structural solution will have a lower degree of rationality. A further question that arises in connection with these statements is: What sort of evidence is required to prove that the structural solution is rational under given conditions? This question cannot be answered directly; it requires a broad and thorough study based upon the following fundamental concepts:

1. The structural system and shapes must correspond with the special physical characteristics of the material to be used.
2. Each structural member as a component of the structural system must be designed so that its relative function, type of carrying action, and degree of importance in the structural system are clearly pronounced.
3. The number of members or elements to accomplish the overall carrying action of a structural system should be the minimum possible.
4. A structural member serving functions additional to its structural one has a higher degree of importance, which must be indicated.
5. The use of materials should be efficient and balanced within the structural system.
6. The materials of the members used should be stable and durable under normally existing and possible expected conditions.
7. The design must be simple, honest, and easy to construct under given conditions, and the method of construction should be visible.

The simultaneous satisfaction of all these fundamental concepts is a rather complex problem. It requires both a profound structural and mathematical knowledge and the understanding and experience that result in a trained intuition. By "trained intuition" we mean a faculty for contemplating a set of possibilities and then arriving almost at once at a valid rational solution, which is suggested by the given architectural image and set of local conditions. Without such intuition, the resulting design seldom approaches the set standards or required results.

Aesthetic Quality

The second criterion, *aesthetic quality*, is more difficult to define than the first because the term is subjective and therefore one cannot be fully guided by reason. But even when one cannot formulate the concept exactly, it is possible to distinguish between external and internal evidence of the aesthetic quality of a structure.

External evidence of aesthetic quality can be derived from direct inspection of the structure, whereas the internal evidence of quality requires a knowledge of background, function, and relations and of the architectural image and design itself. The external qualities can be seen by anyone who has had some previous acquaintance with the general idea of the structure or the method of construction. It would not be difficult to select a set of characteristic qualities in a finished structure which the majority of observers would feel similarly satisfied about. However, such qualities must be real, that is, must embody the structure's own characteristics, and not simply the illusory qualities which, by ordinary standards, one may expect or hope to see because the structure has been designed by a well-known architect or engineer having a high degree of prestige or for some other, similar reason. Also, the decorative effect of a structure, which very often is confused with its external quality, must be taken into consideration. A decorative effect usually does not have duration; it is like fashion without depth, creating only temporary response. Real aesthetic quality is self-contained, independent, free of artificial effects, and capable of maintaining its own value; as such, it is timeless and universal.

For the characteristics of a finished structure that may serve as evidence for aesthetic quality, the following postulates are suggested:

2a. *Expressiveness*
2b. *Atmosphere*
2c. *End-means relation*

Before one can make fair sense of these qualitative characteristics, they must be explained, if possible, in terms of their meaning, their interrelationship, and the means by which a designer can control them.

Expressiveness in this context means the relationship between a design and its objective or the relationship between a structural member or members and their real function. If a building or structure is designed for a particular purpose, the design must exhibit the characteristics that indicate its purpose, function, and performance capacity. The appearance of a structure should express all its embodied qualities and, without any mystery, should make the observer directly aware of what the structure is or is expected to be. The shapes and forms should grow freely out of the nature of the purpose itself instead of being forced or borrowed from elsewhere.

Atmosphere is a quality which cannot be analyzed into simpler qualities or defined precisely because the term is not only subjective but also psychological, closely related to the perceptual conditions on which it largely depends. Therefore, it can be made meaningful more by explanations and proper illustrations than by definitions. For example, atmosphere makes the difference between a house and a home, between a warehouse and a library, or between a theater and a nightclub, which have similar functions for different people at different times, namely, relaxation, enjoyment, etc. Thus the real difference is in the atmosphere. Churches, museums, schoolhouses, factories, laboratories, etc., differ by their atmosphere, as do apartment houses and office buildings. Of two structural systems designed for the same purpose and having the same numerical factor of safety, one may create a safe atmosphere and the other may lack it or have it in a lesser degree, even though the second structure may appear more massive than the first (a quality which is often assumed to be synonymous with strength). Nobody can argue the simple truth that the proper atmosphere of a building is an indispensable quality which makes a design significant. When a building, whatever its purpose may be, lacks the expected atmosphere and rationality, it reveals only its struggle between hope and despair in every aspect.

The *end-means* relation is, in the true sense, an alternative means to the same end: to obtain the highest degree of expressiveness and desired atmosphere by natural, nonintellectual, and material means. The first two postulates—expressiveness and atmosphere—are both more or less relative and subjective in nature, involving the observer's response to the structure. This type of evidence for the aesthetic quality of a structure is as often asserted as denied by different people. Therefore, subjective external evidence does not carry positive conviction unless it is supported by more objective internal qualities—qualities which are supported by reason and which are expressed by the third postulate, the end-means relation.

Now, what are these internal qualities? Before this question can be answered, a more profound problem must be clarified. Postulates 2a and 2b are, in a sense, effects, that is, results of a certain cause or a complex of causes. Very often, the significance in a design of the cause-effect or, more explicitly, end-means relation is overlooked or not understood. Thus, confusingly, the structural design is regarded only as satisfying all the structural fundamentals (concepts 1 to 7), and the engineer's task as a whole is considered to be like that of a mathematician—solving a problem described by a set of impersonal equations with given boundary conditions. However, it is true that a proper structural solution is almost by its very nature aesthetically satisfying; even when a structural solution is based only on the principle of efficiency and on the satisfaction of functional purpose, the results, while they may be aesthetically insignificant, are never irritating. But also nobody can argue the simple truth that, however well a structure performs its function, some alternative structural solution, almost equally efficient, can often add to the structure as a whole something that raises its quality. This "something" is the difference between statical computations—dimensioning the carrying elements and analyzing the stresses—and making a structural design. To make this more explicit, the structural design requires, besides the technical data and factual knowledge, a deep understanding of structural systems, related materials, their behavior under conditions the structure will be subjected to, mechanical installation, construction methods, etc., which are inseparable parts of any design. The problems involved in a contemporary design are extremely complicated. It is not a process in which an idea simply forms in the designer's mind and is mechanically worked out in accordance with some accepted method and in available materials.

As in any other creative work, the concept of a design most suitable to accomplish an architectural image develops and ripens gradually in the mind before the actual mechanical designing process starts. The process involves recognizing and weighing the possibilities inherent in the particular structural system, materials, and method of construction to be used. The designer must be aware of what causes a certain effect and what effects a structural system, members, arrangement of members, and shape and form of members are capable of yielding. Also, distinction between the quality of effects is most significant.

The effects caused by the design itself—such as by choice of structural system and materials and by the form and shape of members, which result in effects of completeness, power, order, degree of heterogeneity, etc.—or effects caused by surface characteristics and color of material without any addition to the design besides those elements that contribute to its capacity to serve its purpose have aesthetic value of a higher degree because the external qualities are supported by internal evidence or rational reasoning. All other effects, produced by addition to the structural design for concealment, etc., have a lesser degree of aesthetic value or no value at all. It is perhaps not feasible to try to account for all the casual features which might influence the aesthetic quality of a design, but at least these are factors that are objectively true, that is, connected with knowledge and understanding and not based only on purely subjective judgments, such as liking. The perception of the aesthetic quality of a design, especially of a structural design, must be based upon conceptual content and must be free from irrelevant subjective distortions. Therefore, a progressive and good engineer must always be conscious of what he is doing and must control his actions by asking the philosophical questions "why" and "if": What are the results and what are the rational means to obtain desired results—safety, efficiency, expressiveness, and atmosphere—without losing sight of the end-means relation?

Expressiveness:

The essentials of expressiveness are simplicity, clarity, and above all modesty. These qualities can be obtained by means of the following principles: The function and degree of importance of each element in the structural system must be meaningful and unmistakably pronounced by mass-space relations or even more explicitly by scale, form, and shape of the elements. The arrangement of the elements in the overall system must show the flow of forces and how they are led into the supporting elements. The type of loads the system is subjected to must appear from the shape as well as from the arrangement of elements in the carrying system. The overall carrying system must be homogeneous, have novelty, etc., and individual secondary elements should not dominate but should fuse freely and naturally into the system so that not a single element could be removed without the carrying system losing its wholeness.

The degree of expressiveness can be controlled by the choice of carrying system, by variation of mass-space relation, by the number and combination of elements constituting a carrying system, by the shape of elements, by the choice of materials the elements are made of, and by the surface qualities (color, texture, etc.).

Atmosphere:

The atmospheric qualities are obtained by the choice of qualities of expressiveness and as such can be controlled also by proper choice of the structural system, materials, surface texture, color, and arrangement of lighting.

For example, a quiet, neutral atmosphere is promoted by completeness of the structural system, by simplicity and simple order—all that is needed must be there modestly. The elements should be balanced and in equilibrium. The system should not be analyzable into more simple secondary systems. The number of elements in the system should be minimum. The materials should be concrete, masonry, or wood.

A vivid atmosphere can be most successfully produced by complexity: two-directional space systems, trusses, systems composed with harmonically ordered elements. Also, heterogenic systems with periodically ordered members are suitable.

In general, a similarity of members or elements creates harmony, clarity, and completeness—in other words, safety.

Expressiveness and atmosphere are closely associated, and their relationship can be explained most simply by using an example. Let us analyze a structural system for a church. The desired atmosphere requires that the system should have a relatively low degree of expressiveness, that it should not dominate the visual field and should not attract any direct attention. However, it must create an effect of power—even, in an indirect way, of mystery and greatness. The proper structural solution by which such qualities can be obtained is to use arches and shells, relatively flat folded plates, or a highly unified structural system composed of similar, simple, harmoniously ordered elements fusing naturally into unity.

Whatever the purpose of a building, the relationship between expressiveness and atmosphere is established by the architectural image. The structural system must correspond directly to that image without affecting carrying action and structural clarity. To the degree of conformity of the perceptual and conceptual images the design owes its power and aesthetic value.

End-Means Relation:

The end-means relation postulate is truly a critical accomplishment, connecting quality judgment directly with knowledge and understanding. To satisfy the end-means postulate requires a high degree of imagination and a thoroughly up-to-date knowledge, both theoretical and practical, of parts-whole relationships, space-mass relationships, etc., because this type of knowledge is indispensable to the achievement of desired qualities and also to the balancing and ordering of complex qualities into the single quality of the finished product. The basis for evaluation of the end-means relation involves, besides originality and novelty, the principle of intellectual economy: the most by the least.

Many critics deny that knowledge is essential for evaluating the aesthetic qualities of a structure. They insist that such judgments should be based only upon the appearance or perceptual image, and thus upon external evidence only. If this were true, how could one judge a dishonest structure? For example, in terms of appearance, a structure may seem to be a reinforced-concrete shell when structurally it is no shell at all but only a shell-shaped plastic ceiling hung from a truss. Before one is able to pass any judgment of aesthetic quality, one must distinguish the aesthetic quality itself from its effects. As another example, the color of a surface may be the natural color of the material or it may be a painted imitation. The paint may actually be more attractive to some observers, but for one having knowledge of how the texture and color are obtained, it will have no quality at all.

On the other hand, a building with a high degree of aesthetic quality may not serve the purpose it was designed for or may be too rich for this particular purpose—for example, a factory designed as a library or a laboratory that looks like a monument. When the observer is aware of such a discrepancy, even the most beautiful building will lose all its aesthetic qualities and look only funny.

Economy

The third criterion, economy, depends mainly on the efficient use of materials, on the simplicity and efficiency of the structural system, and on the individual members of the system—factors already included in the first criterion. However, in some cases the efficiency of the structure has to be sacrificed to realize the image, and here lies the difference between a proper structural system and a structural system which satisfies only the structural needs. Besides these main factors in obtaining economy, there are secondary factors, such as assigning other services to a structural member, for example, embedment of ducts, utilities, etc. The installation and maintenance of such services should be easy. Furthermore, extensive economy is obtained by using efficient construction methods and materials which correspond to the structural solution and time schedule.

In many designs the aesthetic criterion, based on the perceptual image only, is overemphasized at the expense of the conceptual image and other criteria, or vice versa. In the first case, the resulting design tends to be decorative, and in the second case, "functional." Commonly both such designs lack significance because the intention and accomplishment contradict each other. Such conflicts seem to be almost common among advanced and complex contemporary designs.

To conclude, the value judgment of a design is based upon the evaluation of all the criteria and their relations in terms of rationality—maximum results at minimum cost within the limits of the funds available. This principle is the only sound basis for a value judgment of any design because it combines all the essential and rational factors: knowledge, understanding, imagination, local conditions, and the technical and cultural standards of the country in which the structure under evaluation is located.

CARRYING ACTIONS

There are four basic carrying actions: beam, arch, suspension, and shell. Seldom is a single carrying action adequate to accomplish a structural purpose or even to create a proper structural system. In most cases at least two carrying actions are involved in contemporary structural solutions: one as the primary carrying action and the other as at least a secondary carrying action. For example, in typical trusses the carrying ability is mainly through arch and suspense actions, both having an equal degree of importance, and the secondary carrying action is beam action. In a Vierendeel truss the beam action, commonly considered as secondary action, actually is a controlling carrying action. In frames the beam action and arch action are both prime carrying actions, but in most cases the beam action controls the design. In arch beams the arch action controls. In a typical reinforced-concrete arch—whether a three-hinge, two-hinge, or fixed arch—the arch action is the basic carrying action, but beam action, as a secondary carrying action, influences the behavior of these arches to such a degree that it cannot be disregarded. Beam action is the action most commonly present in any structural system or systems and therefore will be discussed in detail first.

2
BEAM ACTION

4. BEAM THEORY

Beam action is a carrying action in which a set of external loads acting upon a structural element is resisted and balanced by resultants of internal stresses developed in the beam due to the bending of the element and transmitted into the supporting element by internal shear. Thus, beam action is characterized by three types of stresses: compressive, tensile, and shear.

The relation between external loads $w(x)$, $w(z)$, P_k, W_k, reaction R, shear V, moment M and the internal stresses or their resultants C, T, V is established by the three fundamental equations of equilibrium

$$\sum Z = \sum X = \sum M = 0 \tag{1}$$

and by the assumptions that the strain-stress relationship of concrete and steel obeys Hooke's law $\epsilon = \sigma/E$ and that a plane of a section perpendicular to the axis of a beam remains plane after the bending (Navier).

To demonstrate the beam action, let us consider two different simply supported beams acted upon by uniformly distributed loads $w(x)$, $w(z)$ and concentrated loads P_k and W_k, as illustrated in Fig. 1.

Equilibrium [Eq. (1)] requires that the algebraic sum of all external vertical and horizontal loads and forces to which the beam is subjected must equal zero ($\Sigma Z = \Sigma X = 0$) and that the algebraic sum of all external moments of the external loads and forces, lying in one plane, about any point in the plane must equal zero ($\Sigma M = 0$).

By the application of these three equations, the external forces (R, V) and moments (M) due to the external loads (w, P, W) can be determined for a simply supported beam.

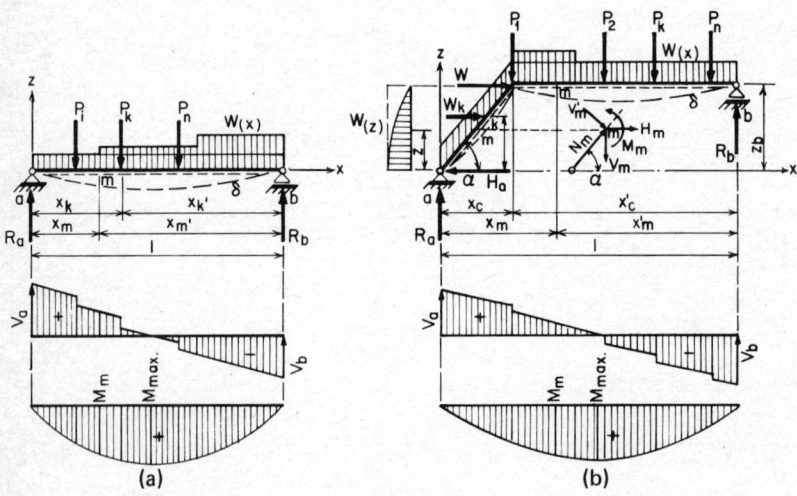

Fig. 1

Vertical loads (Fig. 1a and 1b):

$$R_a + R_b = \sum_0^l w(x)\Delta x + \sum_{k=1}^{k=n} P_k \qquad k = 1, 2, \ldots, n$$

$$R_a = \frac{1}{l}\sum_0^l w(x) x' \Delta x + \frac{1}{l}\sum_{k=1}^{k=n} P_k x'_k$$

(2)

$$R_b = \frac{1}{l} \sum_0^l w(x)\, x \Delta x + \frac{1}{l} \sum_{k=1}^{k=n} P_k x'_k$$

$$V_m = R_a - \sum_0^{x_m} w(x)\, \Delta x - \sum_{k=1}^{k<m} P_k$$

$$H_a = 0$$

$$V'_m = V_m \cos\alpha_m$$

$$N_m = -V_m \sin\alpha_m$$

$$M_m = R_a x_m - \sum_0^{x_m} w(x)(x_m - x)\, \Delta x - \sum_{k=1}^{k<m} P_k(x_m - x_k) \qquad (2)$$
(Cont.)

Horizontal loads (Fig. 1b):

$$-R_a = R_b = \frac{1}{l} \sum_0^{z_b} w(z)\, z\, \Delta z + \frac{1}{l} \sum_{k=1}^{k=n} W_k z_k$$

$$-H_a = \sum_{z=0}^{z=z_b} w(z)\, \Delta z$$

$$H_m = -H_a + \sum_0^{z=m} w(z)\, \Delta z + \sum_{k=1}^{k=n} W_k \qquad (3)$$

$$V'_m = H_m \sin\alpha_m + V_m \cos\alpha_m$$

$$N_m = H_m \cos\alpha_m - V_m \sin\alpha_m$$

$$M_m = H_a z_m - \sum_0^{z=m} w(z)(z_m - z)\, \Delta z - \sum_{k=1}^{k<m} W_k(z_m - z_k)$$

The algebraic sum of the vertical and horizontal external forces V_m, V'_m, N_m and moment M_m at any section m must be in equilibrium with resultants of the internal stresses in the section, as illustrated in Fig. 2.

$N = 0$:

$$V_m - V_{im} = 0$$
$$\Sigma C = C_c + C_s = \Sigma T = T_c + T_s$$
$$M_m = \Sigma C z_0 = \Sigma T z_0$$

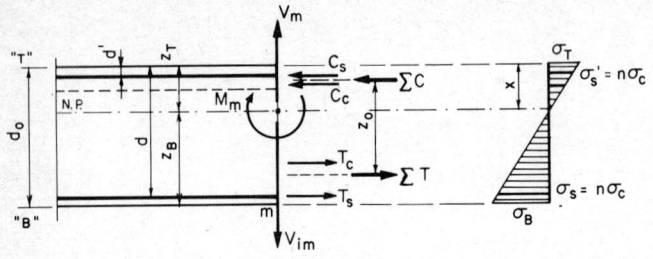
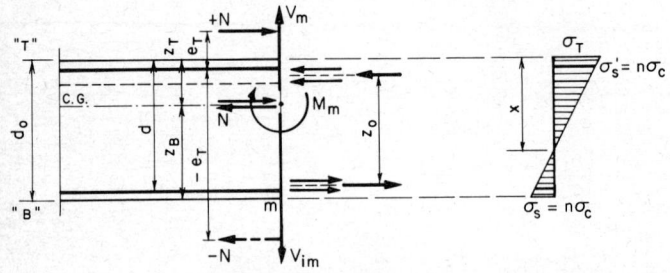

Fig. 2

$N \neq 0$:

$$V_m - V_{im} = 0$$
$$\Sigma T - \Sigma C \pm N = 0$$
$$\Sigma M - \Sigma C z_0 = 0$$
$$\Sigma M = M_m + N(d - z_r)$$

$$e_m = \frac{M_m}{N} = e_r + z_r$$

$$e_m = -\frac{M_m}{N} = -e_T + z_T \tag{4}$$

The participation of concrete and steel in the carrying action of a beam is in accordance with their relative strain and flexural stiffnesses.

These stiffnesses are

$$\begin{array}{ll}
\text{For concrete:} & S_{cs} = E_c A_c \\
& S_{cF} = E_c I_c \\
\text{For steel:} & S_{ss} = E_s \Sigma A_s \\
& S_{sF} = E_s \Sigma I_s \\
& n = E_s / E_c
\end{array}
\qquad \begin{array}{l}
\text{where } c = \text{concrete} \\
s = \text{steel} \\
s = \text{strain stiffness} \\
F = \text{flexural stiffness} \\
E = \text{modulus of elasticity} \\
A = \text{area}
\end{array} \tag{5}$$

Thus the normal force N and moment M carried by steel and concrete are

$$N_s = N \frac{S_{ss}}{S_{cs} + S_{ss}} = N\alpha_s$$

$$N_c = N \frac{S_{cs}}{S_{cs} + S_{ss}} = N(1 - \alpha_s)$$

$$M_s = M \frac{S_{sF}}{S_{cF} + S_{sF}} = M\alpha_F \quad (6)$$

$$M_c = M \frac{S_{cF}}{S_{cF} + S_{sF}} = M(1 - \alpha_F)$$

$$\alpha_s = \frac{nA_s}{A_c + nA_s}$$

$$\alpha_F = \frac{nI_s}{I_c + nI_s}$$

As we have seen, the three equations of equilibrium allowed us to determine the three external quantities V_m, N_m, and M_m required to describe the stress condition at any section of the simply supported straight, sectional, or curved beam. When a beam is supported also at intermediate points besides at its ends, such a beam is called a *continuous beam* and the three equations of equilibrium are not sufficient to determine the moment or reaction at the intermediate support. The continuous beam is externally statically indeterminate. The support moments (X) are most conveniently taken as redundants. The method used to compute these moments is to cut the beam over the intermediate supports, thus reducing, for purposes of analysis, the beam to a statically determinate "principal system" consisting of a number of simple beams and acted upon at the cut faces by the redundants. The redundants $X_{1,2,\ldots,n}$ are computed from the conditions of continuity of the beam. Thus,

$$\delta_{10} - X_1 \delta_{11} - X_2 \delta_{12} - \cdots - X_n \delta_{1n} = 0$$
$$\delta_{20} - X_1 \delta_{21} - X_2 \delta_{22} - \cdots - X_n \delta_{2n} = 0 \quad (7)$$
$$\cdots\cdots\cdots\cdots\cdots\cdots\cdots\cdots\cdots\cdots$$
$$\delta_{n0} - X_1 \delta_{n1} - X_2 \delta_{n2} - \cdots - X_n \delta_{nn} = 0$$

The relative beam-end rotation quantities δ in Eqs. (7) are

$$EI_c \delta_{n0} = \sum \int_0^L M_0 M_n \frac{I_c}{I(x)} dx + \sum 2\kappa \frac{I_c}{A_c} \int_0^L V_0 V_n \frac{A_c}{A(x)} dx - EI_c \sum R_{nn} \Delta_n$$

$$EI_c \delta_{12} = \sum \int_0^L M_1 M_2 \frac{I_c}{I(x)} dx + \sum 2\kappa \frac{I_c}{A_c} \int_0^L V_1 V_2 \frac{A_c}{A(x)} dx \qquad (8)$$

$$EI_c \delta_{nn} = \sum \int_0^L M_n M_n \frac{I_c}{I(x)} dx + \sum 2\kappa \frac{I_c}{A_c} \int_0^L V_n V_n \frac{A_c}{A(x)} dx$$

$$\kappa \sim 1.2, \quad G \simeq E/2$$

The actual moments M_m and reactions R are obtained at any place in the beam, after the redundants (X) have been determined by solving the linear Eqs. (7), by applying the principle of superposition:

$$M_m = M_0 - X_1 M_1 - X_2 M_2 - \ldots - X_n M_n$$
$$R_n = R_{n0} - X_1 R_1 - X_2 R_2 - \ldots - X_n R_n \qquad (9)$$

where $M_{1,2,\ldots,n}$ are the moments and $R_{1,2,\ldots,n}$, $V_{1,2,\ldots,n}$ the reactions and shear forces due to the dummy moments $-X_{1,2,\ldots,n} = 1$ and where M_0 are the moments and R_0, V_0 the reactions and shear forces due to the external loads $w(x)$, $w(z)$, P_k, and W_k acting on the statically determinate principal system. I_c is the arbitrarily chosen constant moment of inertia, and A_c is the cross-sectional area of the beam. $I(x)$ and $A(x)$ are the actual moments of inertia and area of the spans. Thus $I_c/I(x)$ and $A_c/A(x)$ are the relative flexural and shear stiffnesses of the spans. Δ_n is the support settlement.

The influence of shear upon the redundants in common cases is negligible. However, the influence of the support settlement can be considerable. Equations (7) can be written in matrix form for use on computers, but they can also be directly solved very quickly by iteration because the convergence of this type of linear equation is generally very good. In the case of nonyielding supports and constant I in span limits, the redundants can also be easily obtained by the Cross method.

However, for greater convenience, especially for preliminary designs, the redundants $(X_{1,2,\ldots,n})$ and maximum span moments $(M_{1,2,\ldots,n})$ can be computed for various span loadings by use of the coefficients given for uniform span loads in Table 1 and for concentrated loads in Table 2. The given coefficients are computed on the basis of constant flexural stiffness $l'_{1,2,\ldots} = l_n/I_n = \text{const}$ for all spans and free rotation at supports. These assumptions are not satisfied in a monolithically poured reinforced-concrete system. However, compared with the other uncertainties also involved in so-called "exact" solutions, such as the variations in I due to change in reinforcing, extent of floor-slab participation, or crack formation, the deviations of the computed moments M_x from the exact values have no great significance, as will be shown later. The use of these coefficients is fully justified as long as the variation of the spans remains within reasonable limits (a maximum variation of 20 percent).

TABLE 1 Moment coefficients for continuous beam over three to six supports subjected to uniform various loading at spans $l_c = l_2 = l_3 = \cdots l_n =$ constant. Spans are loaded with $w_1 \cdots w_5$ to obtain maximum moments for supports (X_n) as well as for span moments (M_n).

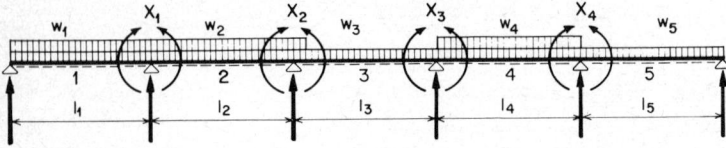

Fig. 3

Number of spans	Moment (max)	Spans participating				
		$w_1 l_1^2$	$w_2 l_2^2$	$w_3 l_3^2$	$w_4 l_4^2$	$w_5 l_5^2$
2	x_1	−0.0625	−0.0625			
	M_1	+0.0950	−0.0250			
	M_2	−0.0250	+0.0950			
3	x_1	−0.0625	−0.0500	+0.0156		
	x_2	+0.0156	−0.0500	−0.0625		
	M_1	+0.0950	−0.0200	+0.0063		
	M_2	−0.0234	+0.0750	−0.0234		
	M_3	+0.0063	−0.0200	+0.0950		
4	x_1	−0.0625	−0.0500	+0.0125	−0.0039	
	x_2	+0.0156	−0.0500	−0.0500	+0.0156	
	x_3	−0.0039	+0.0125	−0.0500	−0.0625	
	M_1	+0.0950	−0.0200	+0.0050	−0.0016	
	M_2	−0.0234	+0.0750	−0.0188	+0.0059	
	M_3	+0.0059	−0.0188	+0.0750	−0.0234	
	M_4	−0.0016	+0.0050	−0.0200	+0.0950	
5	x_1	−0.0625	−0.0500	+0.0125	+0.0031	+0.0010
	x_2	+0.0156	−0.0500	−0.0500	+0.0125	−0.0039
	x_3	−0.0039	+0.0125	−0.0500	−0.0500	+0.0156
	x_4	+0.0010	−0.0031	+0.0125	−0.0500	−0.0625
	M_1	+0.0950	−0.0200	+0.0050	−0.0013	+0.0004
	M_2	−0.0234	+0.0750	−0.0188	+0.0047	−0.0015
	M_3	+0.0059	−0.0188	+0.0750	−0.0188	+0.0059
	M_4	−0.0015	+0.0047	−0.0188	+0.0750	−0.0234
	M_5	+0.0004	−0.0013	+0.0050	−0.0200	+0.0950

TABLE 2 Moments X and M and reaction R due to a moving load $P = 1.0$
$l_1 \cdots l_n = $ constant

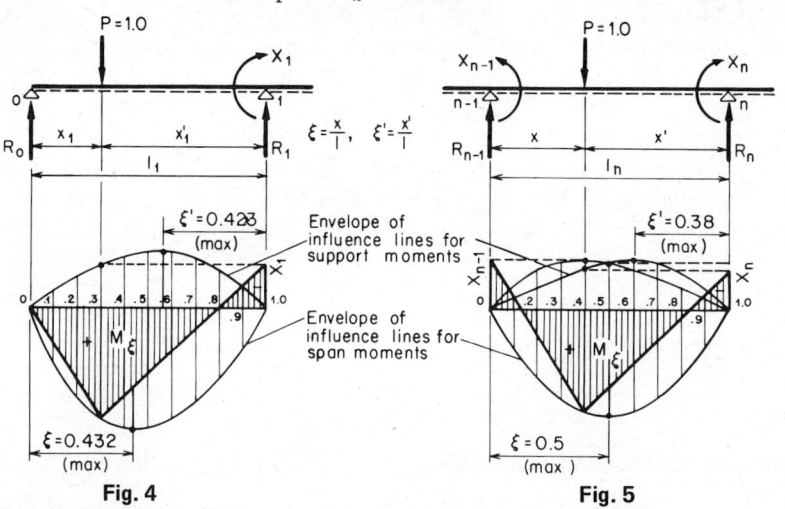

Fig. 4   Fig. 5

End spans ($M = \alpha l$):*

ξ	X_1	$M_{1\xi}$	R_1	ξ	X_1	$M_{1\xi}$	R_1	ξ	X_1	$M_{1\xi}$	R_1
0.00	0.0000	0.0000	1.000	0.35	0.0763	0.2006	0.573	0.70	0.0892	0.1475	0.211
0.05	0.0125	0.0469	0.937	0.40	0.0840	0.2064	0.516	0.75	0.0820	0.1260	0.168
0.10	0.0248	0.0875	0.875	0.45	0.0897	0.2071	0.460	0.80	0.0720	0.1024	0.128
0.15	0.0366	0.1220	0.813	0.50	0.0937	0.2031	0.406	0.85	0.0590	0.0774	0.091
0.20	0.0480	0.1504	0.752	0.55	0.0959	0.1948	0.354	0.90	0.0427	0.0515	0.057
0.25	0.0586	0.1728	0.691	0.60	0.0960	0.1824	0.304	0.95	0.0231	0.0255	0.027
0.30	0.0682	0.1895	0.632	0.65	0.0938	0.1665	0.256	1.00	1.0000	0.0000	0.000

Intermediate spans ($M = \alpha l$):*

ξ	X_{n-1}	X_n	M_n	R_{n-1}	ξ	X_{n-1}	X_n	M_n	R_{n-1}
0.00	0.0000	0.0000	0.0000	1.000	0.55	0.0701	0.0784	0.1728	0.442
0.05	0.0214	0.0071	0.0268	0.964	0.60	0.0640	0.0800	0.1664	0.384
0.10	0.0390	0.0150	0.0534	0.924	0.65	0.0569	0.0796	0.1558	0.328
0.15	0.0531	0.0234	0.0788	0.880	0.70	0.0490	0.0770	0.1414	0.272
0.20	0.0640	0.0320	0.1024	0.832	0.75	0.0406	0.0719	0.1234	0.219
0.25	0.0719	0.0406	0.1234	0.781	0.80	0.0320	0.0640	0.1024	0.168
0.30	0.0770	0.0490	0.1414	0.728	0.85	0.0234	0.0531	0.0788	0.120
0.35	0.0796	0.0569	0.1558	0.673	0.90	0.0150	0.0390	0.0534	0.076
0.40	0.0800	0.0640	0.1664	0.616	0.95	0.0071	0.0214	0.0268	0.036
0.45	0.0784	0.0701	0.1728	0.558	1.00	0.0000	0.0000	0.0000	0.000
0.50	0.0750	0.0750	0.1750	0.500					

*If the location of load application differs from ξ, interpolation can be used.

The use of Table 1 is illustrated in computing the maximum moment at support 2 and the center of the second span of a four-span continuous beam:

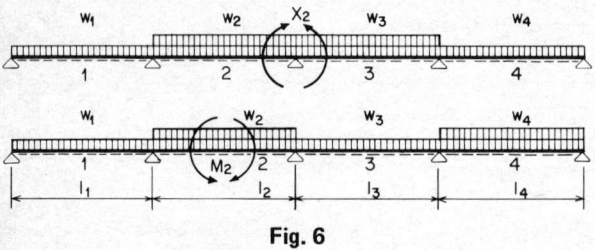

Fig. 6

$$X_{2,\max} = +0.0156\, w_1 l_1^2 - 0.500\, w_2 l_2^2 - 0.500\, w_3 l_3^2 + 0.0156\, w_4 l_4^2$$
$$M_{2,\max} = -0.0234\, w_1 l_1^2 + 0.0750\, w_2 l_2^2 - 0.0188\, w_3 l_3^2 + 0.0059\, w_4 l_4^2$$

The use of Table 2 is illustrated in computing the support and span moments for $P = 1.0$ at $\xi = 0.4$ for the second span of a four-span continuous beam. The support moments follow the law $(-1/4, +1/4^2, -1/4^3, +1/4^4)(-X_n)$:

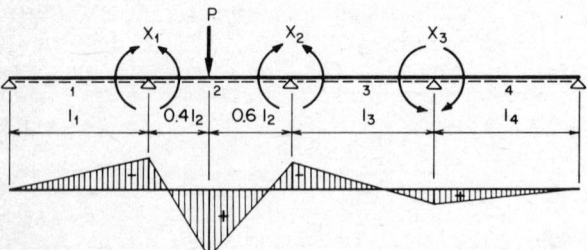

Fig. 7

$$X_1 = -0.080\, Pl_2$$
$$X_2 = -0.064\, Pl_2$$
$$X_3 = -\tfrac{1}{4}(-X_2) = +0.016\, Pl_2$$
$$M_2 = +0.2064\, Pl_2$$

In case of several loads in one or more spans, superposition is used.

In many cases, instead of a uniformly distributed load w, triangular or trapezoidal loads are present (Fig. 8). Using the values from Tables 1 and 2, the moments and reaction can be easily computed by applying superposition. For example, the second span of four-span continuous beam is loaded with a triangular load having a maximum density of w_2. Using w_2 as the uniformly distributed load at the second span, compute the moments X and M_x. Then compute the moments due to the added loading:

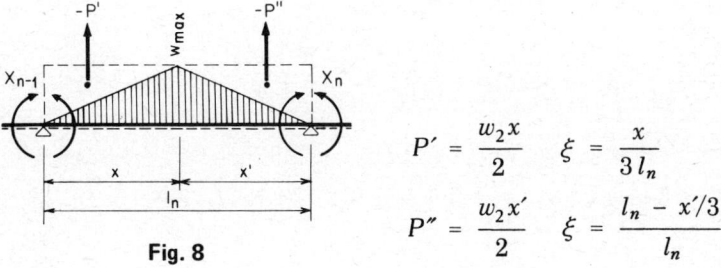

Fig. 8

$$P' = \frac{w_2 x}{2} \quad \xi = \frac{x}{3 l_n}$$

$$P'' = \frac{w_2 x'}{2} \quad \xi = \frac{l_n - x'/3}{l_n}$$

by Table 2. The final moments and reaction are obtained by the use of superposition.

Analysis of a Two-span Continuous Beam

To demonstrate the use, meaning, and significance of Eqs. (7) to (9), a two-span continuous beam will be analyzed.

Geometric data and loading:

$$l_1 = 42.0 \text{ ft} \quad l_2 = 50.0 \text{ ft}$$

$$\xi = \frac{x}{l} \quad \xi' = \frac{x'}{l} = 1 - \xi$$

$$I_1(x) = I_2(x) = \text{const} \quad \frac{I_c}{I(x)} = 1.0$$

$$w = 1.0 \, k/\text{ft} \quad w_D = 0.56 \, k/\text{ft}, \quad w_L = 0.44 \, k/\text{ft}$$

Principal system: two simple beams
Redundant: $X_1 = -M_c$
Continuity requirement:

$$\delta_{10} - X_1 \delta_{11} = 0 \quad X_1 = \frac{\delta_{10}}{\delta_{11}}$$

28 Theory of Structures

Moments and reactions in principal systems:

$$M_{10} = \frac{w_1 l_1^2}{8} = \frac{1.0 \times 42.0^2}{8} = 221\,k'$$

$$R_{10} = \frac{1.0 \times 42.0}{2} = 21.0\,k$$

$$M_{20} = \frac{w_2 l_2^2}{8} = \frac{1.0 \times 50.0^2}{8} = 313\,k'$$

$$R_{20} = \frac{1.0 \times 50.0}{2} = 25.0\,k$$

$$M_{0x} = 4M_{max}\,\xi\xi'$$

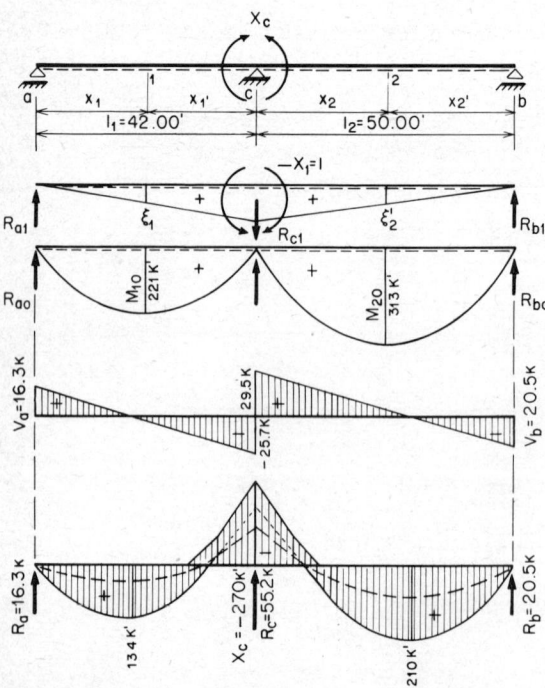

Fig. 9

$X_1 = -1$: $M_{x1} = \xi_1$ $M_{x2} = \xi_2'$ $R_{11} = \pm 1/l_1$ $R_{21} = \pm 1/l_2$

$$EI_c \delta_{11} = \sum_a^b \int M_1^2 dx = \tfrac{1}{3}[1.0\,(42.0 + 50.0)] = 30.67$$

$$\begin{aligned}
EI_c \delta_{10} = \sum_a^b \int M_1 M_0 dx &= \tfrac{1}{3}[1.0^2 \times 0.56\,(221 \times 42.0 + 313 \times 50.0)] \\
&= 4{,}639 \text{ (dead load)} \\
&+ \tfrac{1}{3}(1.0 \times 0.44 \times 221 \times 42.0) \\
&= 1{,}358 \text{ (span 1}: w_L) \\
&+ \tfrac{1}{3}(1.0 \times 0.44 \times 313 \times 50.0) \\
&= 2{,}283 \text{ (span 2}: w_L) = 8{,}280
\end{aligned}$$

$$X = \frac{4{,}639 + 1{,}358 + 2{,}283}{30.67} = 151.0 + 44.5 + 74.5$$

$$= 270\,k'\,(268\,k')$$

$$M = M_0 - X_1 M_1$$
$$R = R_0 - X_1 R_1$$

Moments:
$$\begin{aligned}
M_{1,\min} &= M_{\xi D} - X_1 \xi = 80 - (151.0 + 74.5)0.203 \simeq 34\,k' \\
M_{1,\max} &= M_{\xi 0} - X_1 \xi = 210 - (151.0 + 44.5)0.390 \\
&= 134\,k'\,(132.5\,k') \\
M_{2,\min} &= M_{\xi' D} - X\xi' = 161 - (151.0 + 44.5)0.360 \\
&= 90\,k' \\
M_{2,\max} &= M_{\xi' 0} - X_1 \xi' = 303 - (151.0 + 74.5)0.410 \\
&= 210\,k'\,(212.3\,k')
\end{aligned}$$

Reactions:
$$\begin{aligned}
R_{a,\max} &= R_{a,0} - X_1 R_{11} = 21.0 - (151.0 + 44.5)/42.0 \\
&= 16.3\,k \\
R_c^{(1)} &= \sum_D^L W_1 - R_{a,\max} = 42.0 - 16.3 = 25.7\,k \\
R_{b,\max} &= R_{b,0} - X_1 R_{21} = 25.0 - (151.0 + 74.5)/50.0 \\
&= 20.5\,k \\
R_c^{(2)} &= \sum_D^L W_2 - R_{b,\max} = 50.0 - 20.5 = 29.5\,k \\
R_{c,\max} &= R_{c,\max}^{(1)} + R_{c,\max}^{(2)} = 21.0 + 270/42.0 \\
&\quad + 25.0 + 270/50.0 \\
&= 27.4 + 30.4 = 57.8\,k
\end{aligned}$$

The values computed by using Table 1 are given in parentheses. Although the lengths of the spans differ approximately 20 percent, the difference in moments is about ±1 percent.

5. TWO-WAY BEAM ACTION

Slabs may be supported on two, three, or four sides and on points or small areas. One-way slabs are analyzed in the same manner as that used for simply supported or continuous beams [Eqs. (1) to (9)].

For the two-way slab external moments (M_x, M_y, and $M_{x,y}$), shears V_x, V_y and reactions R_x, R_y are statically indeterminate, and the tree equations of equilibrium [Eq. (1)] are not enough to determine all the unknowns. Therefore, the external moments and forces have to be computed as functions of the deflections (δk). The deflections are related with the loading (w) by the partial fourth-order differential equation

$$\frac{\partial^4 \delta}{\partial x^4} + 2\frac{\partial^4 \delta}{\partial x^2 \partial y^2} + \frac{\partial^4 \delta}{\partial y^4} = \frac{w}{D} \tag{10}$$

where $\quad D = \dfrac{EI}{(1-\nu^2)} = \dfrac{Et^3}{12(1-\nu^2)}$

E = modulus of elasticity

$I = 1.0 t^3/12$ = moment of inertia of slab

ν = Poisson's ratio (approx 0.167 for reinforced concrete)

Since all these quantities are constant for a slab, D is called the *characteristic constant* in slab theory.

Applying Eq. (10) to practical problems leads to mathematical difficulties, especially in case of concentrated loads (P) and nonuniform loading. To overcome the mathematical difficulties, the differentials $\partial/\partial x$ and $\partial/\partial y$ can be expressed by finite geometric functions of the deflections δ_k. This means that the infinitely small quantities ∂x and ∂y are replaced by the finite quantities Δx and Δy, the differentials $\partial/\partial x$ and $\partial/\partial y$ by the difference quotations $\Delta/\Delta x$ and $\Delta/\Delta y$, and the differential equation by a difference equation. This gives

$$\frac{\Delta^4 \delta_k}{\Delta x^4} + 2\frac{\Delta^4 \delta_k}{\Delta x^2 \Delta y^2} + \frac{\Delta^4 \delta_k}{\Delta y^4} = \frac{w_k}{D} \tag{11}$$

The difference quotations required to describe the deformation and stress condition in a two-way slab are* (Fig. 10)

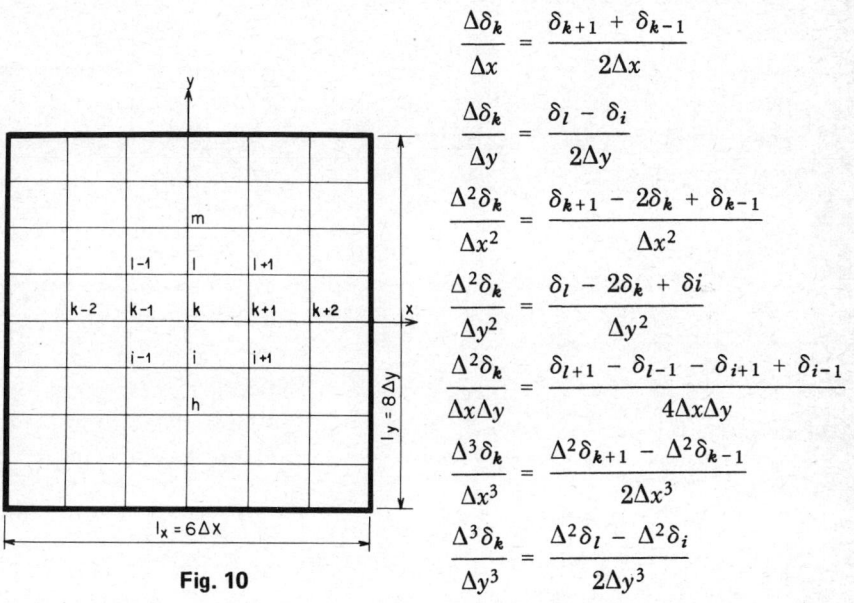

Fig. 10

$$\frac{\Delta \delta_k}{\Delta x} = \frac{\delta_{k+1} + \delta_{k-1}}{2\Delta x}$$

$$\frac{\Delta \delta_k}{\Delta y} = \frac{\delta_l - \delta_i}{2\Delta y}$$

$$\frac{\Delta^2 \delta_k}{\Delta x^2} = \frac{\delta_{k+1} - 2\delta_k + \delta_{k-1}}{\Delta x^2}$$

$$\frac{\Delta^2 \delta_k}{\Delta y^2} = \frac{\delta_l - 2\delta_k + \delta i}{\Delta y^2}$$

$$\frac{\Delta^2 \delta_k}{\Delta x \Delta y} = \frac{\delta_{l+1} - \delta_{l-1} - \delta_{i+1} + \delta_{i-1}}{4 \Delta x \Delta y}$$

$$\frac{\Delta^3 \delta_k}{\Delta x^3} = \frac{\Delta^2 \delta_{k+1} - \Delta^2 \delta_{k-1}}{2 \Delta x^3}$$

$$\frac{\Delta^3 \delta_k}{\Delta y^3} = \frac{\Delta^2 \delta_l - \Delta^2 \delta_i}{2 \Delta y^3}$$

$$\frac{\Delta^4 \delta_k}{\Delta x^4} = \frac{\delta_{k+2} - 4\delta_{k+1} + 6\delta_k - 4\delta_{k-1} + \delta_{k-2}}{\Delta x^4}$$

$$\frac{\Delta^4 \delta_k}{\Delta y^4} = \frac{\delta_m - 4\delta_l + 6\delta_k - 4\delta_i + \delta_h}{\Delta y^4}$$

$$\frac{\Delta^4 \delta_k}{\Delta x^2 \Delta y^2} = \frac{4\delta_k - 2(\delta_{k+1} + \delta_{k-1} + \delta_l + \delta_i) + (\delta_{i-1} + \delta_{i+1} + \delta_{l+1} + \delta_{l-1})}{\Delta x^2 \Delta y^2} \quad (12)$$

Substituting the quotations of Eqs. (12) into Eq. (11) and selecting $\Delta x = \Delta y = \lambda$ and the load at grid points P_k ($P_k = \lambda^2 w$ or $P_k = P$), we obtain

*F. Bleich and E. Melan, "Die gewöhnlichen und partiellen Differenzengleichungen der Baustatik," Springer-Verlag OHG, Berlin, 1927.

$$20\delta_k - 8(\delta_{k-1} + \delta_{k+1} + \delta_l + \delta_i) + 2(\delta_{i-1} + \delta_{i+1} + \delta_{l-1} + \delta_{l+1})$$
$$+ (\delta_{k-2} + \delta_{k+2} + \delta_m + \delta_h) = \frac{\lambda^4}{D} P_k \tag{13}$$

Equation (13) has to be written for every grid point k. In case of symmetry (boundaries and grid), the number of equations is reduced. The moment and forces in the slab are for a grid point k:

$$M_{xk} = -D\left(\frac{\Delta^2 \delta_k}{\Delta x^2} + \nu \frac{\Delta^2 \delta_k}{\Delta y^2}\right)$$

$$M_{yk} = -D\left(\nu \frac{\Delta^2 \delta_k}{\Delta x^2} + \frac{\Delta^2 \delta_k}{\Delta y^2}\right)$$

$$M_{xy,k} = -D \frac{\Delta^2 \delta_k}{\Delta x \Delta y} (1 - \nu)$$

$$V_{xk} = \frac{1}{2\Delta x} (M_{k+1} - M_{k-1})$$

$$V_{yk} = \frac{1}{2\Delta y} (M_l - M_i) \tag{14a}$$

$$R_{x,k} = -\frac{1}{2\Delta x} (M_{k+1} - M_{k-1} + M_{xyl} - M_{xyi})$$

$$R_{y,k} = -\frac{1}{2\Delta y} (M_l - M_i + M_{xyk+1} - M_{xyk-1})$$

The boundary conditions for a simply supported, fixed, or unsupported edge are illustrated in Fig. 11.

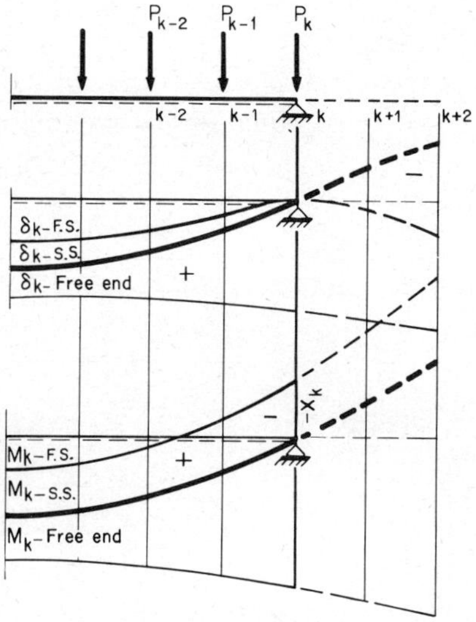

Simple support:

$$\delta_k = \delta_i = \delta_l = 0$$
$$\delta_{k+1} = -\delta_{k-1}$$
$$\delta_{k+2} = -\delta_{k-2}$$
$$M_k = M_i = M_l = 0$$
$$M_{k+1} = -M_{k-1} \quad (14b)$$

Fixed edge:

$$\delta_k = \delta_i = \delta_l = 0$$
$$\delta_{k+1} = \delta_{k-1}$$
$$\delta_{k+2} = \delta_{k-2}$$
$$M_{xyk} = 0$$

Unsupported edge:

$$M_{kx} = M_{xyk} = 0$$
$$R_{xk.} = R_{yk} = 0$$

Fig. 11

Example:

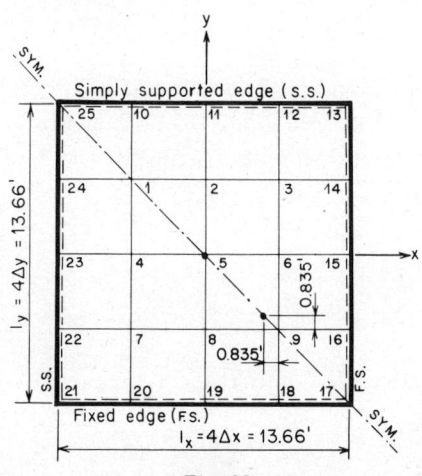

Fig. 12

$k = 1 \ldots 25$ (Fig. 12)

$\Delta x = \Delta y = \lambda = 3.415'$

$P = 12.0 k$

$P_1 = P_2 = P_3 = P_4 = P_5 = 0$

$P_6 = P_8 = \dfrac{P}{2} \cdot \dfrac{0.835}{3.415} = 1.45 k$

$P_9 = P \dfrac{3.415 - 0.835}{3.415} = 9.10 k$

34 Theory of Structures

Boundary condition:

δ_{10} to $\delta_{25} = 0$

M_{10} to M_{13} and M_{21} to $M_{25} = 0$

Because of symmetry:

$\delta_2 = \delta_4 \quad \delta_3 = \delta_7 \quad \delta_6 = \delta_8$

Equation (13) written for $9 - 3 = 6$ grid points is arranged in matrix form. The solution of this matrix gives

	δ_1	δ_2	δ_3	δ_5	δ_6	δ_9	$P\lambda^2/D$	
1	18	−16	2	2	−	−	0	$\delta_1 = 0.132\lambda^2/D$
2	−8	21	−8	−8	3	−	0	$\delta_2 = 0.235\lambda^2/D = \delta_4$
3	1	−8	20	2	−8	1	0	$\delta_3 = 0.210\lambda^2/D = \delta_7$
5	2	−16	4	20	−16	2	0	$\delta_5 = 0.427\lambda^2/D$
6	−	3	−8	−8	23	−8	$1.45\lambda^2/D$	$\delta_6 = 0.519\lambda^2/D = \delta_8$
9	−	−	2	2	−16	22	$9.10\lambda^2/D$	$\delta_9 = 0.726\lambda^2/D$

The moments M_x computed by Eqs. (12) and (14) are illustrated in Fig. 13. Because of symmetry, the moments at grid line 10-11-12 in the y direction are the same as the moments at grid line 24-23-22 in the x direction. For example, the moments for M_{x5} and M_{x15} are

$$M_{x5} = \frac{D}{\lambda^2}[-\delta_{k-1} + 2\delta_k - \delta_{k+1} + \nu(-\delta_i + 2\delta_k - \delta_l)]$$

$$= [-0.235 + 2 \times 0.427 - 0.519 + 0.167(-0.519 + 2 \times 0.427 - 0.235)]$$

$$= 0.222 \, k'$$

$$M_{x15} = -2\delta_6 = -2 \times 0.519 = -1.038 \, k'$$

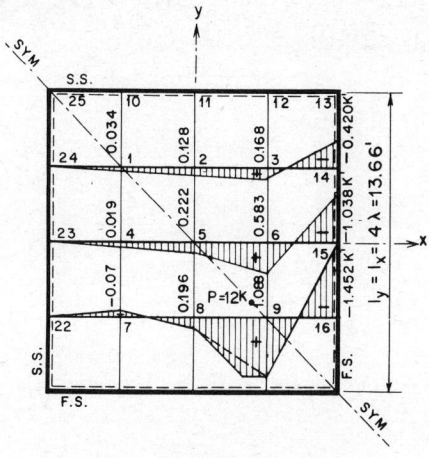

Fig. 13

As demonstrated by this example, the application of the differential method is rather simple. However, it is time-consuming, especially for uniform loading (w). Therefore, for uniform loading the moments and forces can be determined most simply by Marcus' method.*

Marcus' method at first disregards M_{xy} to obtain M'_x and M'_y. Then M''_x and M''_y caused by M_{xy} are determined and expressed as functions of M'_x and M'_y. The final moments are obtained by superposition:

$$M_x = M'_x + M''_x = M'_x(1 - \varphi_x)$$
$$M_y = M'_y + M''_y = M'_y(1 - \varphi_y)$$
(15)

The M'_x and M'_y moments are computed by considering the two-way slab as composed of a series of equal-width strips acting independently in the x as well as the y direction. The intersection of the centerlines of each strip, loaded by w_x and w_y, must satisfy the conditions

$$\delta'_x = \delta'_y$$
$$w_x + w_y = w$$
(16)

*H. Marcus, "Vereinfachte Berechnung biegsamer Platten," Springer-Verlag OHG, Berlin, 1929.

36 Theory of Structures

The δ values depend upon the loading w_x, w_y, the spans l_x, l_y, and the boundary conditions. There are three principal sets of boundary conditions: simply supported, fixed-end, or a combination of one end simply supported and the other end fixed.

The corresponding midspan deflections δ' for a constant EI are

Simply supported: $\quad EI\delta'_x = \dfrac{5}{384} w_x l_x^4 \quad EI\delta'_y = \dfrac{5}{384} w_y l_y^4$

Fixed-end: $\quad EI\delta'_x = \dfrac{1}{384} w_x l_x^4 \quad EI\delta'_y = \dfrac{1}{384} w_y l_y^4 \quad (17)$

Simple and fixed: $\quad EI\delta'_x = \dfrac{1}{192} w_x l_x^4 \quad EI\delta'_y = \dfrac{1}{192} w_y l_y^4$

Applying Eqs. (16) and taking $l_y/l_x = \epsilon$ and $w_x/w = \kappa_0$, the w_x and w_y relationships for two-way symmetrical boundary conditions become

$$w_x = \dfrac{l_y^4}{l_x^4 + l_y^4} w = \kappa_0 w \quad \kappa_0 = \dfrac{\epsilon^4}{1 + \epsilon^4}$$

$$w_y = (1 - \kappa_0) w \quad (18)$$

The corresponding moments for simple, fixed, and simple-fixed supports are

$$M'_x = \dfrac{\kappa_0 w l_x^2}{8} \quad M'_x = \dfrac{\kappa_0 w l_x^2}{24} \quad M'_x = \dfrac{9\kappa_0 w l_x^2}{128}$$

$$M'_y = \dfrac{(1 - \kappa_0) w l_y^2}{8} \quad M'_y = \dfrac{(1 - \kappa_0) w l_y^2}{24} \quad M'_y = \dfrac{9(1 - \kappa_0) w l_y^2}{128}$$

$$(19)$$

For simple symmetry, there are three boundary cases, as given in Fig. 14. The loading and maximum midspan moments are

Case 1: $\quad w_{1x} = \dfrac{5\epsilon^4}{2 + 5\epsilon^4} w = \kappa_1 w \quad M_{x,\max} = \dfrac{9\kappa_1 w l_x^2}{128}$

$$w_{1y} = (1 - \kappa_1) w \quad M_{y,\max} = \dfrac{(1 - \kappa_1) w l_y^2}{8} \quad (20)$$

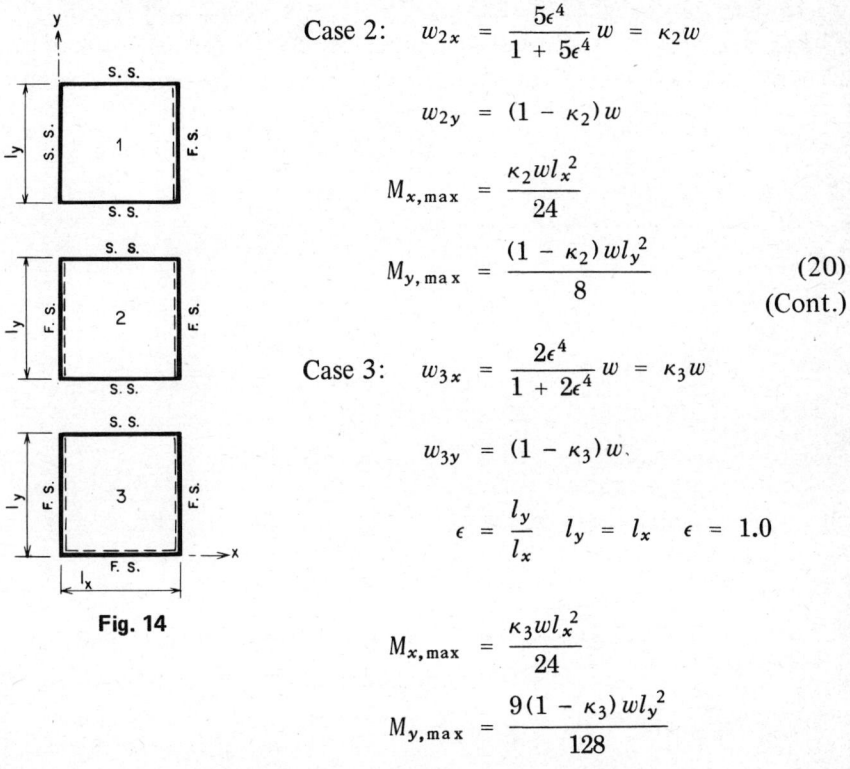

Fig. 14

Case 2: $w_{2x} = \dfrac{5\epsilon^4}{1 + 5\epsilon^4} w = \kappa_2 w$

$w_{2y} = (1 - \kappa_2) w$

$M_{x,\max} = \dfrac{\kappa_2 w l_x^2}{24}$

$M_{y,\max} = \dfrac{(1 - \kappa_2) w l_y^2}{8}$ \hspace{1em} (20)
(Cont.)

Case 3: $w_{3x} = \dfrac{2\epsilon^4}{1 + 2\epsilon^4} w = \kappa_3 w$

$w_{3y} = (1 - \kappa_3) w.$

$\epsilon = \dfrac{l_y}{l_x} \quad l_y = l_x \quad \epsilon = 1.0$

$M_{x,\max} = \dfrac{\kappa_3 w l_x^2}{24}$

$M_{y,\max} = \dfrac{9(1 - \kappa_3) w l_y^2}{128}$

The continuity of the slab is established by the action of $M_{x,y}$ moments [Eqs. (14)]:

$$M_{x,y} = -D \frac{\Delta^2 \delta}{\Delta x \Delta y} (1 - \nu)$$

From strip to strip in the x or y direction, the changes in $M_{x,y}$ are

$$\frac{\Delta M_{x,y}}{\Delta y} = -D \frac{\Delta^3 \delta}{\Delta x \Delta y^2} (1 - \nu)$$

$$\frac{\Delta M_{x,y}}{\Delta x} = -D \frac{\Delta^3 \delta}{\Delta x^2 \Delta y} (1 - \nu)$$

(21)

These changes in M_{xy} counteract as a force couple (Fig. 15) the deflections δ'_x or δ'_y corresponding to reduction in M'_x, M'_y by M''_x and M''_y. In accordance with Marcus, the M'' moments can be expressed in terms of $-M''_x = M'_x \cdot \varphi_x$ and $-M''_y = M'_y \cdot \varphi_y$, where

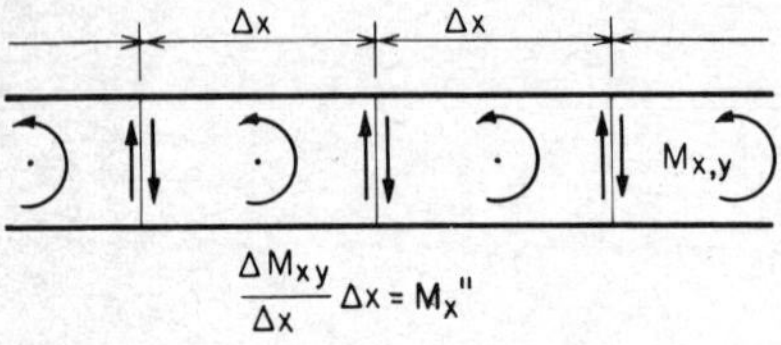

$$\varphi_x = \frac{5}{6}\epsilon^2 \frac{M'_{x,max}}{M_{0x,max}} \qquad (22)$$

$$\varphi_y = \frac{5}{6}\epsilon^2 \frac{M'_{y,max}}{M_{0x,max}}$$

Fig. 15

M_{x0} and M_{y0} are maximum simply supported beam moments ($M_0 = wl^2/8$), subjected to full design loading (w).

For two-way symmetry:

Simply supported slab: $\quad \varphi_x = \varphi_y = \dfrac{5}{6}\dfrac{\epsilon^2}{1+\epsilon^4}$

Fixed-edge slab: $\quad \varphi_x = \varphi_y = \dfrac{5}{18}\dfrac{\epsilon^2}{1+\epsilon^4}$

Simple and fixed slab: $\quad \varphi_x = \varphi_y = \dfrac{15}{32}\dfrac{\epsilon^2}{1+\epsilon^4}$

For simple symmetry:

Case 1: $\varphi_x = \dfrac{75}{32}\dfrac{\epsilon^2}{2+5\epsilon^4}$

$\varphi_y = \dfrac{5}{3}\dfrac{\epsilon^2}{2+5\epsilon^4}$

(23)

Case 2: $\varphi_x = \dfrac{25}{18}\dfrac{\epsilon^2}{1+5\epsilon^4}$

$\varphi_y = \dfrac{5}{6}\dfrac{\epsilon^2}{1+5\epsilon^4}$

Case 3: $\varphi_x = \dfrac{5}{9} \dfrac{\epsilon^2}{1 + 2\epsilon^4}$

$\varphi_y = \dfrac{15}{32} \dfrac{\epsilon^2}{1 + 2\epsilon^4}$ (23)
(Cont.)

Since the M_{xy} at the supports are zero, the φ values are zero. Therefore, the maximum support moments x can be computed as the span moments M'_x, M'_y. They are for fixed and simple-fixed edges, respectively:

$$-X_x = \kappa w l_x^2/8$$
$$-X_y = (1 - \kappa) w l_y^2/8$$
$$-X_x = \kappa w l_x^2/12$$
$$-X_y = (1 - \kappa) w l_y^2/12$$

(24)

The reactions R or shear V at the supports are approximately parabolic and roughly total

$$R_x \simeq \tfrac{1}{2} \kappa w l_x l_y$$
$$R_y \simeq \tfrac{1}{2}(1 - \kappa) w l_x l_y$$

(25)

To demonstrate the use of Marcus' method, let us compute the slab, illustrated in Fig. 12, for a uniform loading $w = 0.180$ ksf:

$l_x = l_y = 13.66$ ft $\epsilon = 1.0$

$\kappa_0 = \dfrac{\epsilon^4}{1 + \epsilon^4} = 0.50$

$M'_{x,\max} = \dfrac{9 \times 0.50 \times 0.180 \times 13.66^2}{128} = 1.182\,k' = M'_{y,\max}$

$\varphi_x = \varphi_y = \dfrac{15}{32} \dfrac{\epsilon^2}{1 + \epsilon^4} = 0.234$

$M_{x,\max} = M'_x(1 - \varphi_x) = 1.182 \times 0.766 = 0.91\,k' = M_{y,\max}$

$X_x = X_y = -\dfrac{0.50 \times 0.180 \times 13.66^2}{8} = -2.10\,k'$

For a continuous slab, the use of symmetry and antisymmetry allows us to obtain the maximum span moments without any modification of the above methods. The method is illustrated in principle in Fig. 16.

With symmetrical parts, all slabs are loaded uniformly by half the live load ($w_L/2$) and the moments M_x, M_y are computed as discussed above. In antisymmetric loading, the adjacent slabs are loaded $\pm w_L/2$ and, as can easily be understood, the slabs behave as independent, simply supported slabs. The final results are obtained by superposition (Fig. 16).

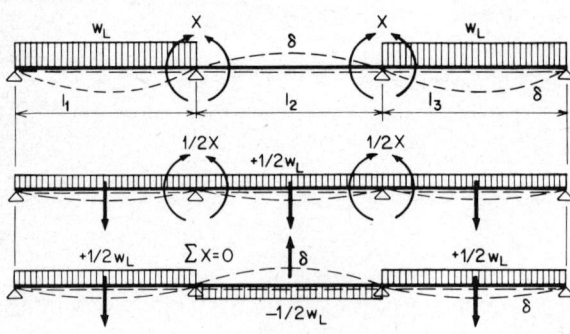

Fig. 16

Flat plates can also be computed by the difference method, but because the number of difference equations becomes relatively large and convergence is poor, they must be solved by algebraic methods or by computer. However, the most convenient method is to estimate the moments and forces of flat plates by frame analysis (see Sec. 15). The accuracy of this method is quantitatively within the acceptable limits.

6. BEAM-GRID SYSTEMS

A beam grid consists of two or three sets of beams lying on the same plane and intersecting each other at an angle. The angle may vary at any intersection. Also, the angle at which the grid beams meet the perimeter may vary. The ends of the grid beams may be simply supported or may be fixed at their supports. The static behavior of the beam grid is similar to that of a two-way slab, described before. However, the significance of the M_{xy} moments in the beam grid is negligible in comparison with the slab.

The number of redundants at each intersection is three: vertical deflection (δ_k) and two rotation angles $\varphi_{ak}, \varphi_{bk}$, which are functions of the torsional stiffness of the grid beams (a, b). The beam grid illustrated in Fig. 17 has 18 intersection points. Thus the system is $n = 3 \times 18 = 54$ times indeterminate. If the beam grid is supported at its corners only, the number of redundants would be $n = 108$.

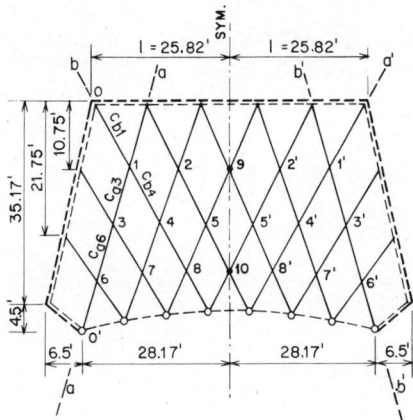

Fig. 17

To reduce the number of redundants to a somewhat more manageable level, use can be made of symmetry for uniform loading (w) and antisymmetry for unsymmetric concentrated loading (P_k). But regardless of this, the number of continuity equations remains high even for a relatively small beam grid. Therefore, the use of an approximate method is unavoidable for analyzing beam grids.

The most rational method is to set the redundants $\varphi_{ak} = \varphi_{bk} = 0$. This means that the interacting beams are considered not as rigidly connected but as transmitting loads from each other so that the continuity requirement $\delta_{ak} = \delta_{bk}$ is satisfied. The assumption $\varphi_{ak} = \varphi_{bk} = 0$ is well justified for almost all grid points except perimeter intersections, because the φ_{ak} and φ_{bk} rotations are resisted by the torsional stiffness of the beams, which in most cases is rather small and therefore does not influence the deflections δ_k appreciably. However, when the perimeter beams are rigidly connected to heavy columns or concrete walls, the φ values must be considered. Commonly, the influence of the φ values is analyzed separately and the final moments $M_k^{(a)}$, $M_k^{(b)}$ and shear $V_k^{(a)}$, $V_k^{(b)}$ are obtained by superposition.

The principal system ($X_k = 0$) consists of series of beams (a, b) supported at their ends—simply or fixed—similar to the individual strips in the two-way slab approximate analysis. The redundants (X_k) are the interacting forces between the beams (a, b) at the intersections (k). They are computed from the continuity equations [Eqs. (7)]:

$$\delta_{k0} - \sum X_i \delta_{ki} = 0 \quad k, i = 1 \cdots n$$

where δ_{kk}, δ_{ki}, and δ_{k0} are the deflections due to dummy load $-X_k = 1$ ($k = 1 \cdots n$) and the external loading (w, P_k) in the principal system. Thus

$$EI_c \delta_{kk} = \sum \int_0^L M_k^2 \frac{I_c}{I_k} ds$$

$$EI_c \delta_{ki} = \sum \int_0^L M_k M_i \frac{I_c}{I_k} ds \quad (26)$$

$$EI_c \delta_{k0} = \sum \int_0^L M_k M_0 \frac{I_c}{I_k} ds$$

The moments $M_k^{(a)}, M_k^{(b)}$ and shear $V_k^{(a)}, V_k^{(b)}$ are obtained by superposition:

$$\begin{aligned} M_k^{(a)} &= M_{k0}^{(a)} - \sum M_{ki}^{(a)} X_k \\ M_k^{(b)} &= M_{k0}^{(b)} - \sum M_{ki}^{(b)} X_k \\ V_k^{(a)} &= V_{k0}^{(a)} - \sum V_{ki}^{(a)} X_k \\ V_k^{(b)} &= V_{k0}^{(b)} - \sum V_{ki}^{(b)} X_k \end{aligned} \quad (27)$$

Analysis of a Beam Grid

To demonstrate this method, the beam grid illustrated in Fig. 17 loaded with a uniform load w will be considered.

Geometric data:

$$L_k = \sum_{1}^{k=n} c_k$$

$L_1^{(a)} = L_1^{(b)'} = 11.76 + 10.46 + 9.75 + 9.68 = 41.65 \text{ ft}$

$L_1^{(b)} = L_1^{(a)'} = 13.09 + 11.40 + 9.68 + 8.18 = 42.35 \text{ ft}$

$L_3^{(b)} = L_3^{(a)'} = 11.77 + 10.40 + 8.88 = 31.05 \text{ ft}$

$L_6^{(b)} = L_6^{(a)'} = 10.68 + 9.37 = 20.05 \text{ ft}$

$$\xi_k = \frac{\sum_0^k c_k}{L_k} \qquad \xi_{\bar{k}} = \frac{L_k - \sum_0^k c_k}{L_k} \qquad \xi_k + \xi_k' = 1$$

$I_a = I_b = \text{const} \qquad I_c/I_k = 1.0$

Redundants:

The beam-grid is simply supported at its perimeter. Thus the redundants—interacting normal forces between the grid beams:

$X_k \quad k = 1, 2, \cdots n$
$\varphi_{ak} = \varphi_{bk} = 0$ $\Big| \quad n = 18$

Principal system:

Two sets (a, b) of simply supported beams $(X_k = 0)$. Because of symmetry, the number of redundants is reduced to $n = 8$:

$X_1 = X_1' \qquad X_4 = X_4' \qquad X_7 = X_7'$
$X_2 = X_2' \qquad X_5 = X_5' \qquad X_8 = X_8'$
$X_3 = X_3' \qquad X_6 = X_6' \qquad X_9 = X_{10} = 0$

$-X_k = 1$: Moments M_k and shear V_k in principal system (Fig. 18, $k = 1, k = 3, k = 6, \ldots$)

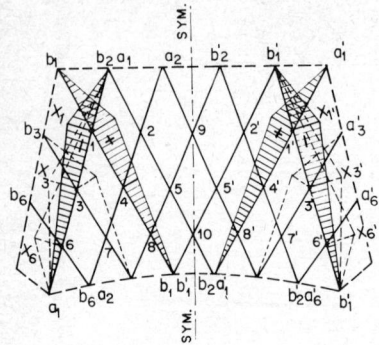

Fig. 18

$k = 1$:

$$\xi_1^{(a)} = \frac{11.76}{41.65} = 0.283 = -R_{01}^{(a)'}$$

$$\xi_1^{(a)'} = (1 - 0.283) = 0.717 = -R_{01}^{(a)}$$

$$\xi_1^{(b)} = \frac{13.09}{42.35} = 0.309 = R_{01}^{(b)'}$$

$$\xi_1^{(b)'} = (1 - 0.309) = 0.691 = R_{01}^{(b)}$$

$$M_{11}^{(a)} = M_{11}^{(b)'} = \xi_1^{(a)} \xi_1^{(a)'} L_1^{(a)} = -8.45\,k'$$
$$M_{31}^{(a)} = M_{31}^{(b)'} = -5.47\,k', \quad M_{61}^{(a)} = M_{61}^{(b)'} = -2.74\,k'$$
$$M_{11}^{(b)} = M_{11}^{(a)'} = 0.309 \times 0.691 \times 42.35 = 9.05\,k'$$
$$M_{41}^{(b)} = M_{41}^{(a)'} = 5.53\,k', \quad M_{81}^{(b)} = M_{81}^{(a)'} = 2.53\,k'$$

$k = 3$:

$$\xi_3^{(a)} = \frac{22.22}{41.65} = 0.534 = -R_{03}^{(a)'}$$

$$\xi_3^{(a)'} = (1 - 0.534) = 0.466 = -R_{03}^{(a)}$$

$$\xi_3^{(b)} = \frac{11.77}{31.05} = 0.379 = R_{03}^{(b)'}$$

$$\xi_3^{(b)'} = (1 - 0.379) = 0.621 = R_{03}^{(b)}$$

$$M_{33}^{(a)} = M_{33}^{(b)'} = \xi_3^{(a)} \xi_3^{(a)'} L_1^{(a)} = -10.38\,k'$$

$$M_{13}^{(a)} = M_{13}^{(b)'} = -5.48\,k', \quad M_{63}^{(a)} = M_{63}^{(b)'} = -5.16\,k'$$

$$M_{33}^{(b)} = M_{33}^{(a)'} = 0.379 \times 0.621 \times 31.05 = 7.36\,k'$$

$$M_{73}^{(b)} = M_{73}^{(a)'} = 3.37\,k'$$

$k = 6$:

$$\xi_6^{(a)} = \frac{31.97}{41.65} = 0.765 = -R_{06}^{(a)}$$

$$\xi_6^{(a)'} = (1 - 0.765) = 0.235 = -R_{06}^{(a)}$$

$$\xi_6^{(b)} = \frac{10.68}{20.05} = 0.532 = R_{06}^{(b)}$$

$$\xi_6^{(b)'} = (1 - 0.532) = 0.468 = R_{06}^{(b)}$$

$$M_{66}^{(a)} = M_{66}^{(b)'} = 0.765 \times 0.235 \times 41.65 = -7.50\,k'$$

$$M_{16}^{(a)} = M_{16}^{(b)'} = 2.76\,k' \quad M_{36}^{(a)} = M_{36}^{(b)'} = -5.23\,k'$$

$$M_{66}^{(b)} = 0.532 \times 0.468 \times 20.05 = 4.98\,k'$$

$k = 4$:

$$M_{44}^{(b)} = 10.35\,k' \quad M_{14}^{(b)} = 5.55\,k' \quad M_{81}^{(b)} = 4.62\,k'$$

$$R_{04}^{(b)} = 0.567\,k \quad R_{04}^{(b)} = 0.433\,k$$

$k = 8$:

$$M_{88}^{(b)} = 6.58\,k' \quad M_{18}^{(b)} = 2.53\,k' \quad M_{48}^{(b)} = 4.72\,k'$$

$$R_{08}^{(b)} = 0.807\,k \quad R_{08}^{(b)} = 0.193\,k$$

Matrix (Eqs. 7):

	X_1	X_2	X_3	X_4	X_5	X_6	X_7	X_8	δ_{k0}
1	δ_{11}	–	δ_{13}	δ_{14}	–	δ_{16}	–	δ_{18}	δ_{10}
2	–	δ_{22}	–	δ_{24}	δ_{25}	–	δ_{27}	–	δ_{20}
3	δ_{31}	–	δ_{33}	–	–	δ_{36}	δ_{37}	–	δ_{30}
4	δ_{41}	δ_{42}	–	δ_{44}	–	–	δ_{47}	δ_{48}	δ_{40}
5	–	δ_{52}	–	–	δ_{55}	–	–	δ_{58}	δ_{50}
6	δ_{61}	–	δ_{63}	–	–	δ_{66}	–	–	δ_{60}
7	–	δ_{72}	δ_{73}	δ_{74}	–	–	δ_{77}	–	δ_{70}
8	δ_{81}	–	–	δ_{84}	δ_{85}	–	–	δ_{88}	δ_{80}

δ_{kk}, δ_{ki} *values:*

$$EI_c\, \delta_{11} = \sum \int_0^L M_1^2 dx = 2\tfrac{1}{3}(8.435^2 \times 41.65 + 9.05^2 \times 42.35) = 4{,}280$$

$$EI_c\, \delta_{33} = \sum \int_0^L M_3^2 dx = 2\tfrac{1}{3}(10.38^2 \times 41.65 + 7.36^2 \times 31.05) = 4{,}100$$

$$EI_c\, \delta_{66} = \sum \int_0^L M_6^2 dx = 2\tfrac{1}{3}(7.50^2 \times 41.65 + 4.98^2 \times 20.05) = 1{,}890$$

$$EI_c\, \delta_{13} = \sum \int_0^L M_1 M_3\, dx = 2{,}230 = EI_c\, \delta_{31}$$

$$EI_c\, \delta_{16} = \sum \int_0^L M_1 M_6\, dx = 1{,}363 = EI_c\, \delta_{61}$$

$$EI_c\, \delta_{36} = \sum \int_0^L M_3 M_6\, dx = 1{,}973 = EI_c\, \delta_{63}$$

δ_{k0} values:

Considering that due to symmetry both sets of beams (a, b) carry the total design load w, the uniform load per beam is $w_{a,b} = 0.5\,bw$. When the grid beams are trusses, the joint loads are $P_k = (c_k + c_{k+1})/2$. For example, the joint loads $(k = 1, 3, 6, 4, 8)$ due to dead load w_D for a Vierendeel truss are

$$P_1^{(a)} = P_1^{(b)'} = 4.72\,k \qquad P_3^{(a)} = P_3^{(b)'} = 4.12\,k \qquad P_6^{(a)} = P_6^{(b)'} = 3.76\,k$$
$$P_1^{(b)} = P_1^{(a)'} = 4.72\,k \qquad P_4^{(b)} = P_4^{(a)'} = 3.96\,k \qquad P_8^{(b)} = P_8^{(a)'} = 3.27\,k$$

Thus the moments M_{k0} and shear V_{k0} in the principal system are (Fig. 19)

$$M_{10}^{(a)} = M_{10}^{(b)'} = 73.00\,k' \quad M_{30}^{(a)} = M_{30}^{(b)'} = 88.60\,k' \quad M_{60}^{(a)} = M_{60}^{(b)'} = 62.60\,k'$$
$$R_{00}^{(a)} = V_{10}^{(a)} = 6.20\,k \quad R_{00'}^{(a)} = -V_{00'}^{(a)} = 6.40\,k \quad V_{30}^{(a)} = 1.48\,k \quad V_{60}^{(a)} = -2.64\,k$$
$$M_{10}^{(b)} = M_{10}^{(a)'} = 73.00\,k' \quad M_{40}^{(a)} = M_{40}^{(b)'} = 82.50\,k' \quad M_{80}^{(a)} = M_{80}^{(b)'} = 51.70\,k'$$
$$R_{00}^{(b)} = V_{10}^{(b)} = 5.56\,k \quad R_{00'}^{(b)} = -V_{00'}^{(b)} = 6.39\,k \quad V_{40}^{(b)} = 0.84\,k \quad V_{80}^{(a)} = -3.10\,k$$

$$EI_c\,\delta_{10} = \sum \int_0^L M_1 M_0\,dx = 896 \qquad EI_c\,\delta_{30} = \sum \int_0^L M_3 M_0\,dx = -24{,}100$$

$$EI_c\,\delta_{60} = \sum \int_0^L M_6 M_0\,dx = -19{,}410$$

Matrix

	X_1	X_2	X_3	X_4	X_5	X_6	X_7	X_8	δ_{k0}
1	4,280	–	2,230	2,404	–	1,363	–	1,289	–896
2	–	3,973	–	2,097	2,167	–	1,247	–	+510
3	2,230	–	4,100	–	–	1,973	869	–	+24,100
4	2,404	2,097	–	5,757	–	–	1,762	1,745	–850
5	–	2,167	–	–	5,345	–	–	1,615	–410
6	1,363	–	1,973	–	–	1,890	–	–	+19,410
7	–	1,247	869	1,762	–	–	2,203	–	+11,246
8	1,289	–	–	1,745	1,615	–	–	2,360	–1,478

$X_k = -8.30^k;\ -2.84^k;\ +3.90^k;\ +3.30^k;\ +0.80^k;\ +12.37^k;\ +2.54^k;\ +0.92^k$

The final moments $M_k^{(a)}$, $M_k^{(b)}$ and shear $V_k^{(a)}$, $V_k^{(b)}$ are computed by Eqs. (26). For beams b_k' and a_k' (k = 1, 3, 6, and 1, 4, 8),

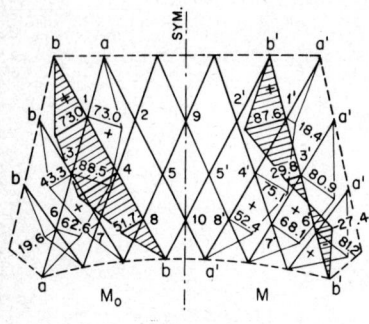

Fig. 19

$$M_k = M_{k0} - \Sigma X_k M_{ki}$$

$M_1^{(b)'} = 73.00 + 8.30 \times 8.44 - 3.90 \times 5.47 - 12.37 \times 2.76 = 87.6\,k'$

$M_3^{(b)'} = 88.60 + 8.30 \times 5.47 - 3.90 \times 10.38 - 12.37 \times 5.16 = 29.8\,k'$

$M_6^{(b)'} = 62.60 + 8.30 \times 2.74 - 3.90 \times 5.16 - 12.37 \times 7.50 = -27.4\,k'$

$M_1^{(a)'} = 73.00 - 8.30 \times 9.05 + 3.30 \times 5.55 + 0.92 \times 2.53 = 18.4\,k'$

$M_4^{(a)'} = 82.50 - 8.30 \times 5.53 + 3.30 \times 10.35 + 0.92 \times 4.72 = 75.1\,k'$

$M_8^{(a)'} = 51.70 - 8.30 \times 2.53 + 3.30 \times 4.72 + 0.92 \times 6.58 = 52.4\,k'$

$M_3^{(a)'} = 43.30 + 3.90 \times 7.39 + 2.54 \times 3.47 = 80.9\,k'$

$M_7^{(a)'} = 38.60 + 3.90 \times 3.36 + 2.54 \times 6.44 = 68.1\,k'$

$M_6^{(a)'} = 19.7 + 12.37 \times 4.96 = 81.2\,k'$

$$V_k = V_{k0} - \Sigma X_k V_{ki}$$

$V_{01}^{(b)'} = 6.20 + 8.30 \times 0.717 - 3.90 \times 0.466 - 12.37 \times 0.235$
$\quad = 7.43\,k = R_{01}^{(b)'}$

$V_{0'1}^{(b)'} = -6.40 - 8.30 \times 0.283 + 3.90 \times 0.534 + 12.37 \times 0.765$
$\quad = 2.82\,k = -R_{0'1}^{(b)'}$

$V_{01}^{(a)'} = 1.40\,k$, $V_{0'1}^{(a)'} = -6.40\,k$

The moment diagrams of M_k and corresponding M_{k0} moments in the principal system are illustrated in Fig. 19.

Due to the restricted rotation and difference in deflection of grid points, the beams are subjected to torsion. The torsional moments of marginal beams can be considerable, especially when one end of the grid beams is simply supported and other fixed (Fig. 20). In this particular

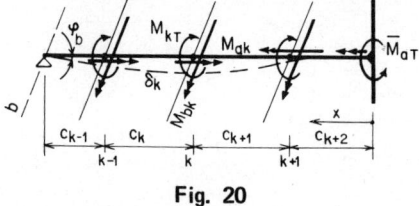

Fig. 20

case, the total torsional moment (ΣM_{kT}), induced by the transversal set of beams, is resisted by fixed end only. Thus,

$$\sum_{0}^{k=n} M_{kT} = \overline{M}_{aT} \tag{28}$$

The rough estimate of the magnitude of M_{kT} values will be discussed under "Torsion" (Sec. 7).

7. TORSION

Under torsion a beam is twisted by a moment M_T, as illustrated in Fig. 21. Due to the action of M_T, a straight line parallel to the x axis at the surface becomes the helix of angle ϑ and a radius from the center of gravity to a point A on the surface rotates through angle α, where α is proportional to ϑx:

$$r\alpha = \vartheta x$$

For convenience, r is taken as unity.

The displacement of any surface point in the z direction and y direction, respectively, is

$$\begin{aligned} \Delta y &= r\alpha \sin \alpha_A = \vartheta xz \\ \Delta z &= r\alpha \cos \alpha_A = \vartheta xy \end{aligned} \tag{29}$$

where z and y are the coordinates of the point considered.

In accordance with St. Vernant, ϑ is proportional to the torsional

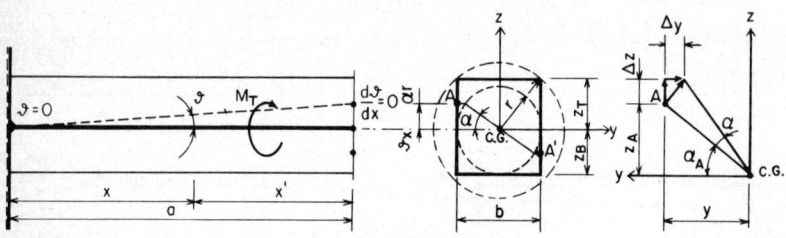

Fig. 21

moment M_T:

$$\vartheta = \frac{M_T}{GI_T} \tag{30}$$

where G is the shear modulus [$G = E/2(1 - \nu^2)$] and I_T is the moment of inertia for torsion.

For rectangular sections $d/b = n > 1$, in accordance with Weber

$$I_T = \frac{1}{3}\left(n - 0.630 + \frac{0.052}{n^4}\right)b \tag{31}$$

and the shear stresses

$$\tau_{max} = \frac{M_T}{I_T}b\left(1 - \frac{0.65}{1 + n^2}\right) \tag{32}$$

The shear stresses are parabolic: maximum at the surface and zero at the center of gravity and at the corners of the section.

For composite sections, where rectangular sections are rigidly connected to each other, use can be made of superposition:

$$I_T \simeq \sum_{}^{n} I_{Tn} \quad \Delta M_{Tn} = \frac{I_{Tn}}{\Sigma I_{Tn}} \quad n = 1, 2, \ldots \tag{33}$$

For sections composed of relatively thin rectangular units, Eq. (30) has to be extended because of the arching of units. Arching increases the resistance to twisting and is required to establish continuity. For example, in a box section illustrated in Fig. 22, the displacements due to V_y and V_z are

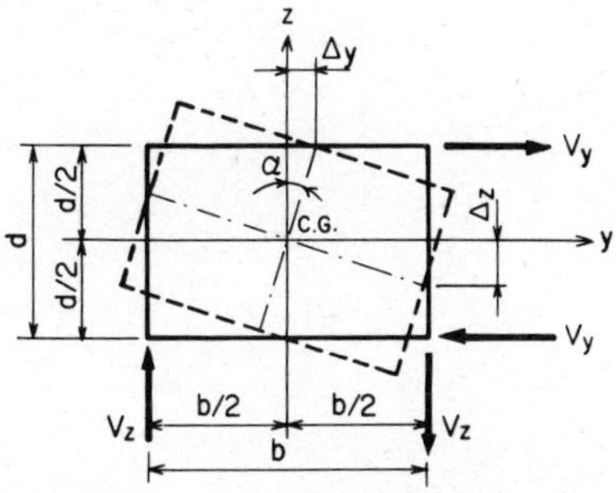

Fig. 22

$$\vartheta = \frac{\Delta y}{xz} = \frac{\Delta z}{xy} \qquad \alpha = f(\vartheta)$$

$$= \frac{2}{d}\frac{d\Delta y}{dx} = \frac{2}{b}\frac{d\Delta z}{dx}$$

$$V = EI\frac{d^3\Delta}{dx^3} \qquad \Delta = \text{deflection}$$

$$V_y = EI_y \frac{d}{4}\frac{d^2\vartheta}{dx^2} \qquad V_z = EI_z \frac{b}{4}\frac{d^2\vartheta}{dx^2}$$

Thus,

$$M_T'' = V_y d + V_z b = \frac{E}{4}(d^2 I_y + b^2 I_z)\frac{d^2\vartheta}{dx^2}$$

In case of an I section, $V_z = 0$. Adding Eq. (30) to this equation, we obtain the total resisting torsional moment:

$$M_T = M_T' + M_T''$$
$$= GI_T \vartheta + \frac{E}{4}(d^2 I_y + b^2 I_z)\frac{d^2\vartheta}{dx^2} \tag{34a}$$

The solution of this differential equation is

$$\vartheta = \frac{M_T}{GI_T} - (C_1 e^{\beta x} + C_2 e^{-\beta x})$$

$$\beta = 2\sqrt{\frac{GI_T}{E(d^2 I_y + b^2 I_z)}}$$

The integration coefficients C_1 and C_2 are determined from the supporting conditions:

$$x = 0: \ \vartheta = 0 \qquad x = a: \ \frac{d\vartheta}{dx} = 0$$

$$\vartheta = \frac{M_T}{GI_T}\left[1 - \frac{e^{\beta x} + e^{-\beta(2a-x)}}{1 + e^{\beta 2a}}\right] \tag{34b}$$

The rotation angle of the section $\alpha_k = \vartheta(x/z)$ must be equal to the deflection angle φ_k of the beam (Fig. 23a). The supporting conditions are

Fixed support: $\quad \alpha_0 = \varphi_0 = 0$
Simple support: $\quad \alpha_0 = \varphi_0$

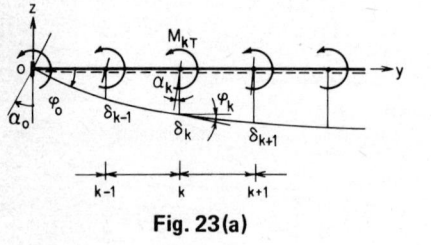

Fig. 23(a)

$$\alpha_k = \varphi_k = \vartheta \frac{x}{z} = \frac{M_{kT}}{GI_T} \frac{x}{z}$$

$$\varphi_k = \frac{\delta_k - \delta_{k-1}}{c_k} - \frac{\delta_{k+1} - \delta_k}{c_{k+1}}$$

$$M_{kT} = GI_T \frac{z}{x} \varphi_k \tag{35}$$

The distribution of the torsional moments to which a beam is subjected can be computed most simply by applying the M_{kT} moments as loads. The reactions of these moments (loads) are the maximum torsional moments at each end of the beam $\left(R_T = \sum_{0}^{k} M_{kT}\right)$. The shear diagrams of M_{kT} moments represent the torsional moments along the beam (Fig. 23b).

The maximum torsional resisting moment can be estimated from the allowable shear stress (Eq. 32):

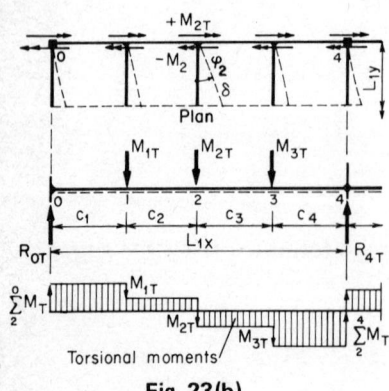

Fig. 23(b)

$$\overline{M}_{kT,\,max} = \frac{\tau_{max}\, I_T}{b[1 - 0.65/(1 + n^2)]}$$

Beyond the shear strength of concrete (Eq. 556a), cracks occur.

8. STRESS ANALYSIS

The application of the three equations of equilibrium between external forces V, N, moments M, and internal stresses leads to the following relations, formulas, and equations (Fig. 24):

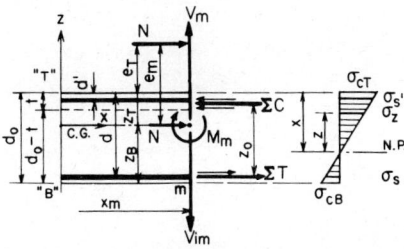

Fig. 24

$$\int \sigma_z \, dA = \frac{\sigma_{cT}}{x} \int z \, dA = \frac{\sigma_{cT}}{x} Q = N$$

$$\int \sigma_z z \, dA = \frac{\sigma_{cT}}{x} \int z^2 \, dA = \frac{\sigma_{cT}}{x} I = M \tag{36}$$

$$\int \tau_z z \, dA = V_m \qquad \tau_z = \frac{V_m Q_z}{b_0 I_0}$$

$$M = (e_m - z_T + x)N = M_m + (x - z_T)N$$
$$= (e_T + x)N$$

$$\Sigma C z_0 \stackrel{\wedge}{=} \Sigma T z_0 = M$$
$$\frac{dC}{dx} = \frac{dT}{dx} = \frac{d}{dx}\left(\frac{M}{z_0}\right) = \frac{1}{z_0}\frac{dM}{dx} = \frac{V}{z_0} = \tau_0 b_0 = \Sigma OU$$
$$dM = V dx \tag{37}$$
$$\frac{dV}{dx} = \frac{d^2M}{dx^2} = w$$

For concrete and steel, the static moments Q_0 and moment of inertia I_0 of the section about the center of gravity (–0) are

$$Q_{c0} = \int z\, dA \tag{38}$$

$$\left.\begin{array}{l} Q'_{s0} = nA'_s(z_T - d') \\ Q_{s0} = -nA_s(d - z_T) \end{array}\right| \Sigma Q_0 = \Sigma Q_{c0} + \Sigma Q_{s0}$$

$$I_{c0} = \int z^2\, dA$$

$$\left.\begin{array}{l} I'_{s0} = nA'_s(z_T - d')^2 \\ I_{s0} = nA_s(d - z_T)^2 \end{array}\right| \Sigma I_0 = \Sigma I_{c0} + \Sigma I_{s0} \tag{39a}$$

For convenience, the static moments Q and moments of inertia I are computed about the top fiber T of the section. The moment of inertia at the center of gravity I_0 is

$$I_0 = I_T - z_T^2 \Sigma A \tag{39b}$$

In case $e_m < z_T$, e_T is negative, which means that normal force is located within the section and $x \to d_0$: When $x = d_0$, the neutral plane coincides with the bottom fiber of the section. Thus, the entire section is in compression. When $x < d_0$, the limit of concrete participation in carrying action depends on the tensile capacity of concrete ($f'_{cT} \simeq \frac{1}{10} f'_c$), which means $\sigma_{cB} < \frac{\sigma_{cT}}{4} \simeq \frac{1}{10} f'_c$. Beyond this limit, a cracked section must be considered.

Noncracked Section

$$\Sigma A = \Sigma A_c + n\Sigma A_s = (b - b_0)t + b_0 d_0 + n(A'_s + A_s)$$

$$\Sigma Q_T = \Sigma Q_{cT} + n\Sigma A_{sT} = A'_c \frac{t}{2} + A_c \frac{d_0}{2} + n(A'_s d' + A_s d)$$

$$\Sigma I_T = \Sigma I_{cT} + n\Sigma I_{sT} = A'_c \frac{t^2}{3} + A_c \frac{d_0^2}{3} + n(A'_s d'^2 + A_s d^2) \tag{40a}$$

$$z_T = \frac{\Sigma Q_T}{\Sigma A} \qquad z_B = d_0 - z_T$$

$$\sigma_{cT} = -\frac{(1-\alpha_s)N}{\Sigma A_c} - \frac{(1-\alpha_F)M_m}{\Sigma I_{cT} - z_T^2 \Sigma A_c} \cdot z_T = -\frac{N}{\Sigma A} - \frac{M_m z_T}{\Sigma I_T - z_T^2 \Sigma A}$$

$$\sigma_{cB} = -\frac{(1-\alpha_s)N}{\Sigma A_c} + \frac{(1-\alpha_F)M_m}{\Sigma I_{cT} - z_T^2 \Sigma A_c} z_B = -\frac{N}{\Sigma A} + \frac{M_m z_B}{\Sigma I_T - z_T^2 \Sigma A}$$

$$\sigma_{sT} = -\frac{\alpha_s N}{\Sigma A_s} - \frac{\alpha_F M_m}{\Sigma I_{sT} - z_T^2 \Sigma A_s} z_{sT}$$

$$\sigma_{sB} = -\frac{\alpha_s N}{\Sigma A_s} + \frac{\alpha_F M_m}{\Sigma I_{sT} - z_T^2 \Sigma A_s} z_{sB} \qquad \sigma_{s0} \pm \sigma_{sF} \qquad (40b)$$

$$v_0 = \frac{V_m \Sigma Q_0}{b_0 \Sigma I_0} = \frac{V_m}{b_0 z_0} \qquad z_0 = \frac{\Sigma I_0}{\Sigma Q_0}$$

$$\sigma_{1,2} = -\frac{\sigma_c}{2} \pm \sqrt{\sigma_c^2 - 4v_0^2}$$

$$x = \frac{\sigma_{cT}}{\sigma_{cT} \hat{+} \sigma_{cB}} d_0 = \frac{\epsilon_{cT}}{\epsilon_{cT} + \epsilon_{cB}} d_0 \qquad \epsilon_c = \frac{\sigma_c}{E_c}$$

$$\sigma_{cB} = 0: \quad x \to d_0 \qquad (40c)$$

The bond stresses u develop between steel and concrete to secure participation of both materials in the carrying action. In accordance with their relative stiffness, they are

$$u_x = \frac{\Delta M}{z_0} \frac{1}{\Delta x \Sigma O} = \frac{\Delta T}{\Delta x \Sigma O} = \frac{V_x}{z_0 \Sigma O} \qquad (41)$$

Cracked Section

The first crack formation may occur in tension fibers of concrete under instant loading when the elongation $\epsilon_c = \sigma_c/E_c$ reaches approximately 0.15 percent. Assuming $E_c = 5.0 \times 10^6$ psi for concrete and $E_s = 30.0 \times 10^6$ psi for steel, the corresponding stresses are 750 and 4,500, respectively. Up to this strain limit, bond stresses U_x are continuous and steady. After the crack formation, the distributions v, σ_{cT} of bond stresses and tensile stresses in concrete between the cracks are illustrated in Fig. 25.

The bond stresses left of a crack increase gradually to a maximum and then decrease to zero. In the range from maximum to zero, the elongations of steel and concrete are not equal; shear in the vertical direction is lost, and thus the elastic-beam theory no longer applies. However, it is valid with sufficient accuracy until the average bond stresses are sufficient to satisfy Eqs. (40), even when the steady character of stresses is lost. The apparent validity of the elastic theory for a cracked section is explained by the phenomena illustrated in Fig. 26. As can be seen, the sections between the cracks act as cantilevers fixed in the compressive zone and acted upon by ΔT. The force couple ΔT_z and ΔC_z is in balance with $\Delta M = \Delta T z_0$, and flexural shear is not required to establish equilibrium.

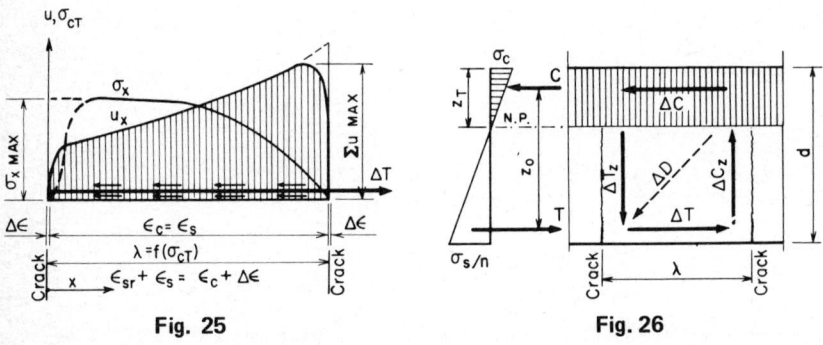

Fig. 25   Fig. 26

The superposition of the two tensile stresses perpendicular to each other is the reason for the inclination of the upper ends of the cracks toward the beam center. To increase the cantilever capacity, stirrups are required.

When the tensile capacity of the cantilever is reached, the cracks occur; but equilibrium is reestablished by the compressive force C_D in the diagonal direction (Fig. 26). In the region of supports the beam action does not apply for any loading conditions. Here the arch action controls.

Under further loading, the bond stress increases and starts traveling to allow for the steel elongation. Finally the bond is overcome entirely. The slippage starts from the midspan, where the elongation of steel is maximum. The bond stresses from the slippage toward the supports increase to balance the loss in bond. Also the ΔT increments from the crack toward the midspan are no longer proportional to the loading. Due to this fact, the deflection increases from now on more rapidly, and even the apparent beam action loses its validity. The beam action converts to arch action where the reinforcing acts as tie rods. In cases in which the depth-span ratio is small, the beam action converts to suspension action, provided that end anchorage and supports exist.

The converted tied arch action is most clearly pronounced in continuous beams at intermediate supports (Fig. 27).

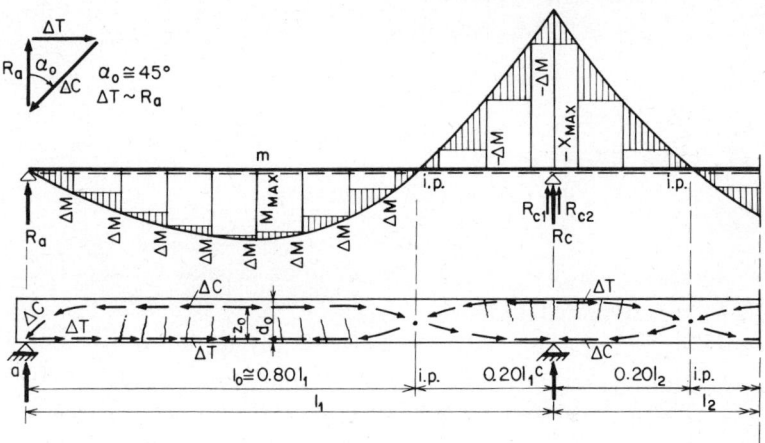

Fig. 27

In accordance with the elastic theory, the negative moments for uniform loading are approximately two times as large as in midspan and increase very rapidly toward inflection points. Due to this, the increments ΔM are rather large in comparison with simple span.

The distance x of neutral plane from the compressive fiber can be computed from the equilibrium requirement (Fig. 24):

$$x^3 + 3e_T x^2 + \frac{6}{b_0} x(e_T \Sigma A_{sc'} + \Sigma Q_{sc'T}) \\ - \frac{6}{b_0}(e_T \Sigma Q_{sc'T} + \Sigma I_{sc'T}) = 0 \tag{42}$$

where

$$\Sigma A_{sc'} = n\Sigma A_s + A_{c'} = n(A'_s + A_s) + (b - b_0)t$$
$$\Sigma Q_{sc'T} = n\Sigma Q_{sT} + Q'_{cT} = n(A'_s d' + A_s d) + A'_c t/2$$

Theory of Structures

$$\Sigma I_{sc'T} = n\Sigma I_{sT} + I_{c'T} = n(A'_s d'^2 + A_s d^2) + A'_c t^2/3$$

$$e_T = \frac{M_m}{N} - z_T$$

Since the z_T value of a cracked section can be determined when x is known, the problem can be solved by successive approximation. First, assuming $e'_T = e_m - d/2$ and computing x,

$$z'_T = \frac{b_0 x^2/2 + \Sigma Q_{sc'}}{b_0 x + \Sigma A_{sc'}}$$

$$e''_T = e_m - z'_T$$

Introducing the e''_T into Eq. (42), the obtained second x value will be already within sufficient accuracy.

The third-degree equation (42) can most simply be solved by trial. If the N lies within the section or is a tensile force, the e_T is negative. For a rectangular section, the slab part in $\Sigma A_{sc'}$, $\Sigma Q_{sc'T}$, and $\Sigma I_{sc'T}$ is zero.

For simple bending ($N = 0$), the location of neutral plane from the compressive fiber for double-reinforced T sections is

$$x = \frac{bt^2/2 + n(A'_s d' + A_s d)}{bt + n(A'_s + A_s)} \tag{43}$$

and for rectangular sections,

$$x = -f_s + \sqrt{f_s^2 + \frac{2n}{b}(A'_s d' + A_s d)}$$

The moment of inertia for concrete and steel about the top fiber is

$$\Sigma I_{cT} = \frac{b_0 x^3}{3} + \frac{(b - b_0)t^3}{3} \tag{44a}$$

$$\Sigma I_{sT} = A'_s d'^2 + A_s d^2$$

and about the center of gravity is

$$\Sigma I_{c0} = \Sigma I_{cT} - z_T^2 \Sigma A_c$$
$$\Sigma I_{s0} = \Sigma I_{sT} - z_T^2 \Sigma A_s$$
$$\Sigma I_0 = \Sigma I_{c0} + \Sigma I_{s0} \qquad (44b)$$

The stresses for concrete and steel are obtained by Eqs. (40).

The vertical shear stresses v_y in the slab at plane a-a are, with sufficient accuracy,

$$v_y = v_{max} \frac{b_0}{t} \frac{\Delta C}{\Sigma C}$$

$$v_{max} = \frac{VQ_0}{\Sigma I b_0} \sim \frac{V}{z_0 b_0} \qquad (45)$$

where ΔC is the resultant of concrete stresses at the area left of the plane and ΣC is the total compressive force.

The flexural shear v is maximum at the center of gravity of a section. It is, as can be seen from Fig. 28, slightly less at the neutral plane $v_0 = V/z_0 b_0$. However, the difference is negligible, and the simplification of substituting v_0 for v_{max} is well justified.

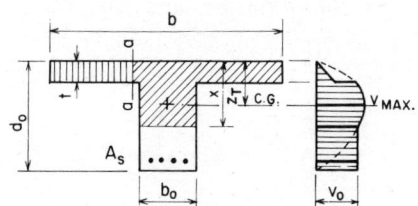

Fig. 28

The shear stress v_0, as given by Eqs. (40b), is valid only for beams with constant depth. For variable-depth beams, the internal forces C and T are not parallel, as illustrated in Fig. 29.

Equilibrium requires

$$\Sigma X = T + C \cos\alpha = 0$$
$$\Sigma Z = V_z - V_i - C \sin\alpha = 0$$
$$\Sigma M_0 = M + V_i x' - V_z x' = 0$$

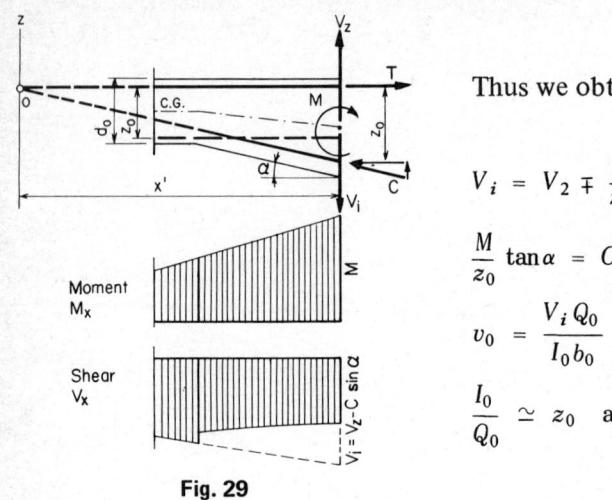

Fig. 29

Thus we obtain

$$V_i = V_2 \mp \frac{M}{z_0} \tan\alpha \qquad x' = \frac{z_0}{\tan\alpha}$$

$$\frac{M}{z_0} \tan\alpha = C \sin\alpha$$

$$v_0 = \frac{V_i Q_0}{I_0 b_0} = \frac{V_i}{z_0 b_0}$$

$$\frac{I_0}{Q_0} \simeq z_0 \quad \text{at neutral plane} \qquad (46)$$

Equation (46) indicates that the shear stresses are decreased in a beam of variable depth. However, this is true only when the increase of moment and depth of beam follow the same direction. In the opposite case, the shear stresses are increased by $(M/z_0) \tan\alpha$.

In the case of an unsymmetric section, as illustrated in Fig. 30, the centroid of the compressive force C does not lie on the same plane with the tensile force T. This results in a torsional moment

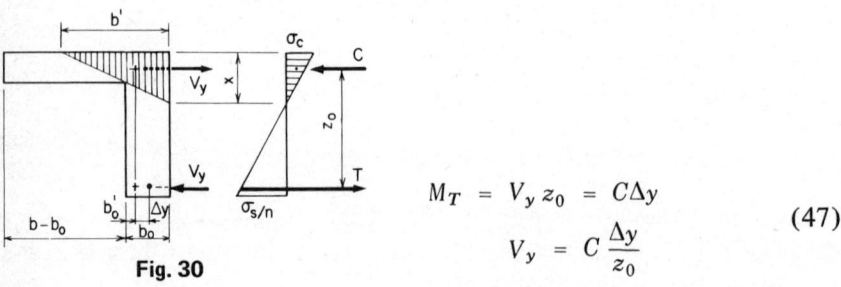

Fig. 30

$$M_T = V_y z_0 = C\Delta y$$
$$V_y = C \frac{\Delta y}{z_0} \qquad (47)$$

For uniform loading w, the torsional moment is affined to the external moment M_m and is resisted by a triangular compressive-stress distribution having its centroid at the plane of the tensile force. Thus,

$$T = \frac{M_m}{z_0} = C = \frac{b'x}{2}\frac{\sigma_c}{3} \quad b' = 3(b_0 - b'_0) \tag{48}$$

$$\sigma_c = \frac{2T}{(b_0 - b'_0)x}\left(x = \frac{\epsilon_c}{\epsilon_c + \epsilon_s} = \frac{\sigma_c}{\sigma_c + \sigma_s/n}\right)$$

This means that the slab escapes the carrying action and the σ_C stresses increase. For long spans and unrestricted rotations of the beam, considerable crack formation may occur if this phenomenon is not considered.

Bond Stresses

In homogeneous sections, bond stresses u_0 are proportional to the increase of the internal force:

$$u_0 = \frac{\Delta T_0}{\Delta x \, \Sigma O} \tag{49}$$

where $\Delta T_0 = \Delta M/z_0$ and ΣO is the contact area between concrete and steel. After the cracks occur, the linear relationship between the bonding stress u and ΔT is lost. The increase in u_0 can be computed from the strain relationship of concrete and steel.

Equilibrium requires that (Fig. 31)

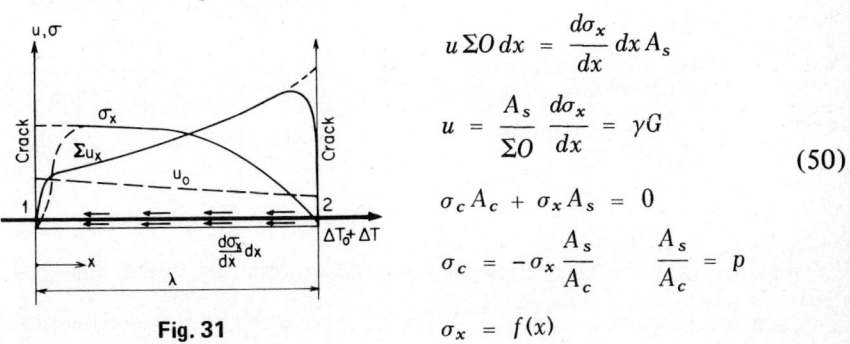

Fig. 31

$$u \Sigma O \, dx = \frac{d\sigma_x}{dx} dx \, A_s$$

$$u = \frac{A_s}{\Sigma O}\frac{d\sigma_x}{dx} = \gamma G \tag{50}$$

$$\sigma_c A_c + \sigma_x A_s = 0$$

$$\sigma_c = -\sigma_x \frac{A_s}{A_c} \quad \frac{A_s}{A_c} = p$$

$$\sigma_x = f(x)$$

where γ is the shear strain and $G = E_0/2(1 + \nu)$ is the modulus of shear.
Continuity requires that

$$\frac{d\gamma}{dx} = \frac{\epsilon_s - \epsilon_c}{\kappa} \tag{51}$$

where κ is a material constant obtained by pullout tests. Its value is dependent on the surface characteristics of the steel and the quality of the concrete; it may vary from 30 to 60. Taking $\epsilon_c = \sigma_c/E_0$ and $\epsilon_s = (\sigma_x - \Delta\sigma_s)/E_s$, $\Delta\sigma_s = \Delta T/A_s$. Substituting these values and the shear strain from Eq. (50) into Eq. (51), we obtain the differential equation

$$\frac{d^2\sigma_x}{dx^2} - \beta^2(\sigma_x - \Delta\sigma_s) = 0 \qquad (52)$$

$$\beta = \sqrt{\frac{\Sigma O}{A_s} \cdot \frac{1}{2(1+\nu)\kappa n}}$$

The solution of this differential equation is

$$\sigma_x = \Delta\sigma_s (C_1 e^{\beta x} + C_2 e^{-\beta x})$$

and thus

$$u = \frac{A_s}{\Sigma O} \Delta\sigma_s \beta (C_1 e^{\beta x} - C_2 e^{-\beta x}) \qquad (53)$$

The differential constants are determined from boundary conditions $x = 0 : u = 0$ and $x = \lambda : \sigma_x = 0$. Substituting into Eq. (53), we obtain

$$\sigma_x = \Delta\sigma_s (1 - e^{-\beta(\lambda - x)})$$

$$u_x = -\frac{A_s}{\Sigma O} \Delta\sigma_s \beta e^{-\beta(\lambda - x)}$$

$$= \frac{\Delta T}{\Sigma O} \beta e^{-\beta(\lambda - x)} \qquad (54)$$

Considering that the hair cracking starts approximately under $0.35 \sum_D^L M$, the final bond stress is obtained for a cracked section by superposition:

$$\Sigma u_x = u_0 + u = \frac{\Delta T}{\Sigma O} \left(\frac{0.35}{\lambda} + \beta e^{-\beta(\lambda - x)} \right) \qquad (55)$$

The σ_x and Σu_x diagrams are represented in Fig. 31. The maximum Σu_x

can be up to 1,500 psi before notable slippage (deformed bars) occurs. The $\sigma_{x,\max}$ is approximately $\frac{1}{10} f'_c$.

Prestressing Stresses*

The stresses in a prestressed section are computed by the equations for a noncracked section [Eqs. (40)]. This is explained by the fact that the tensile zone of the concrete is precompressed by prestressing force H_{pr} and, therefore, is able to carry tensile stresses to the extent of precompression. The point of application of the prestressing force, as an external load, is always within the section. The tensile stresses, after the precompression is exhausted, remain in the limit of concrete-tensile-strength capacity for design load condition. Thus, in Eqs. (40) the normal force $N \rightarrow H_{pr}$ and the external moment $\Sigma M \rightarrow M_m - e_0 H_{pr}$, where e_0 is the center of gravity of the tendons from the center of gravity of the noncracked section. The sign of e_0 is negative because the point of application of H_{pr} is in the tensile zone opposite to e_m, which is on the compression side from the center of gravity of the section.

Commonly the stresses due to loading and prestressing are computed separately and the final stresses are obtained by superposition. This is required because the magnitude of prestressing force H_{pr} and prestressing moment $M_{pr} = -e_0 H_{pr}$ are determined from the stresses due to loading ($w = w_D + w_L$), so that no cracking and undue final stresses occur in the section.

The stresses due to prestressing are

$$\sigma_{pr} = -\frac{H_{pr}}{\Sigma A} \pm \frac{M_{pr}}{\Sigma I_0} z$$

$$= -\sigma_{0\,pr}\left(1 \mp \frac{e_0 z}{i^2}\right) \qquad \sigma_{0\,pr} = \frac{H_{pr}}{\Sigma A} \tag{56}$$

where $-\sigma_{0\,pr}$ is the normal stress and $M_{pr} = H_{pr} e_0$ is due to prestressing. $i^2 = \Sigma I/\Sigma A$, and z is the distance of plane considered from the center of gravity of the section.

---

*A. E. Komendant, "Prestressed Concrete Structures," McGraw-Hill Book Company, New York, 1952.

For convenience, substituting $\psi = 1 \mp e_0 z/i^2$ into the Eq. (56) gives

$$\sigma_{T_{pr}} = \left(1 - \frac{e_0 z_T}{i^2}\right)\sigma_{pr0} = \psi_T \sigma_{0\,pr}$$
$$\sigma_{B_{pr}} = \left(1 + \frac{e_0 z_B}{i^2}\right)\sigma_{pr0} = \psi_B \sigma_{0\,pr} \tag{57}$$

The value of ψ_T can be positive or negative $(1 > e_0 z_T/i^2, 1 < e_0 z_T/i^2)$. The final fiber stresses then will be

$$\sum_{pr}^{D+L} f_T = f_{cT} \pm \sigma_{T_{pr}}$$
$$\sum_{pr}^{D+L} f_B = f_{cB} - \sigma_{B_{pr}} = 0 \quad \text{or} \quad \leq \tfrac{1}{10} f'_c \tag{58}$$

The magnitude of the prestressing force is computed from the condition

$$f_{cB} - 0.10 f'_c = \sigma_{B_{pr}}$$

To satisfy this condition requires

$$H_{pr} = \frac{f_{cB} - 0.10 f'_c}{\psi_B} \Sigma A \tag{59}$$

Generally the tensile strength $(0.10 f'_c)$ is disregarded. However, for prestressing of building elements, it is not recommended because the specified full live load w_L seldom occurs, and in normal loading conditions the elements are usually over-prestressed, which results in an upward deflection of the beams due to plastic flow, as will be discussed under "Deformations" (Sec. 11).

The magnitude of the shear V_{pr} due to prestressing, which counteracts the shear v_0 developed by loading (w, P), is computed for a tendon as illustrated in Fig. 32.

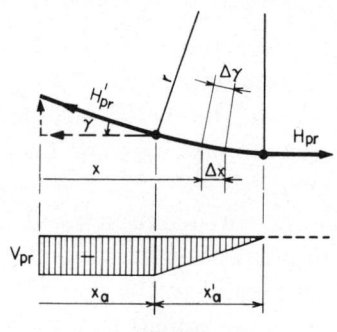

Fig. 32

For curved section x'_a,

$$V_{pr} x'_a = T_{pr} \sin \Delta\gamma = T_{pr} \frac{\Delta x}{r} = \text{const} \tag{60}$$

and for section x_a,

$$V_{pr} x_a = T_{pr} \sin \gamma$$
$$H_{pr} = T_{pr} \cos \gamma \tag{61}$$

The actual shear in the section x_a is obtained by superposition:

$$\Sigma V_x = V_{x0} - V_{prx}$$
$$v_x = \frac{\Sigma V_x Q}{I_x b_0} \simeq \frac{\Sigma V_x}{z_0 b_0} \tag{62}$$

The magnitude and direction of principal stresses are obtained by Eqs. (40), substituting σ_C for $\Sigma\sigma_c = \sigma_c + \sigma_{pr}$ and v_0 for $\Sigma V = V_0 - V_{pr}$:

$$\sigma_{1,2} = -\frac{\Sigma\sigma_C}{2} \pm \frac{1}{2}\sqrt{\Sigma\sigma_c^2 + 4\Sigma v^2}$$
$$\tan 2\varphi_{1,2} = \frac{2\Sigma v}{\Sigma\sigma_c} \tag{63}$$
$$\tau_{1,2} = \frac{1}{2}\sqrt{\Sigma\sigma_c^2 + 4\Sigma v^2}$$
$$\tan 2\varphi_{1,2} = \frac{2\Sigma\sigma_c}{\Sigma v}$$

The principal stresses computed by Eqs. (63) are not applicable at the vicinity of the supports because of the interaction of the reactions. The actual stress condition in these areas can be computed by Airy's stress function, as will be discussed later.

9. PLASTIC FLOW AND SHRINKAGE

Because of the pseudosolid nature of concrete, it is subject to volume change under sustained constant stresses. The magnitude of the change for any given concrete is a function of stress and time. The volume change due to sustained loading is known as *plastic flow* (ϵ_{pl}), and that caused by capillary action is known as *shrinkage* (ϵ_{sr}).

Under instant loading or loading of relatively short duration ($t = 0$), concrete behavior is elastic (ϵ_e). Its modulus of elasticity is a function of the strength, age, and degree of stress or strain existing at the time the increments of stress or strain are assumed to be measured. Extensive research has provided enough comprehensive information to determine the modulus of elasticity E_0 of concrete with sufficient accuracy. The influence of age appears in the strength of concrete. Due to continued hydration and crystallization, the 28-day strength, which is the standard, may increase up to 150 percent. The magnitude of the increase in strength depends on climatic conditions, moisture available, and the type of cement. The influence of stress upon the modulus of elasticity seems negligible up to stresses one-third of the ultimate strength, but above this limit it is significant and should be considered if higher stresses are used for live load. Due to this, for loading of short duration, E_0 can be considered constant in the design load range. For high-quality standard concrete, it is approximately

$$E_0 = 8.15 \times 10^6 \frac{f'_c}{2300 + f'_c} \tag{64}$$

and for lightweight concrete (natural sand),

$$E_0 = 3.65 \times 10^6 \frac{f'_c}{1500 + f'_c} \tag{65}$$

where f'_c is the 28-day cylinder strength and the numerical factors are average values from numerous tests.

As determined by tests, plastic flow is proportional at any given time to the sustained stress, and thus it obeys Hooke's law. Due to this, the influence of plastic flow ϵ_{pl} can be expressed in terms of elastic strain ϵ_e:

$$\begin{aligned} \Sigma \epsilon_c &= \epsilon_e + \epsilon_{pl} \\ &= (1 + \varphi_t)\epsilon_e \quad \varphi_t = \frac{\epsilon_{pl}}{\epsilon_e} \end{aligned} \tag{66}$$

The final value of $\varphi_t \to \varphi_n$ depends upon the quality of concrete, commonly characterized by the 28-day ultimate strength f'_c, and may vary from 1.5 to 3.0 for high-quality standard concrete and from 2.5 to 4.0 for lightweight concrete.

The apparent modulus of elasticity E_t of plain concrete for $t = n$ is

$$E_t = \frac{E_0}{1 + \varphi_t} \tag{67}$$

While plastic flow is a function of time, the apparent magnitude E_t at any given time t can be expressed with sufficient accuracy by the following exponential function (Fig. 33):

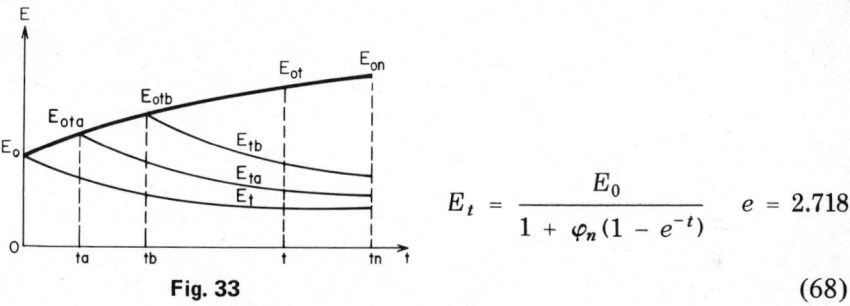

Fig. 33

$$E_t = \frac{E_0}{1 + \varphi_n(1 - e^{-t})} \qquad e = 2.718 \tag{68}$$

For $t = 0$; $\varphi_t = 0$: $E_t \to E_0$

Considering the origin of the plastic behavior of concrete, there is a definite limit for volume change; beyond this limit no plastic deformation takes place. As a conclusion,

$$\Sigma\epsilon \simeq \epsilon_{pl} + \epsilon_{sr} = \text{const} \tag{69}$$

and the two phenomena, plastic flow and shrinkage, are related and interacting for any given type of concrete. Therefore, it is justified to assume shrinkage ϵ_{sr} proportional to plastic flow and to express it as a ratio of plastic flow $(\epsilon_{sr}/\varphi_n)$.

In reinforced concrete, steel is not subject to plastic deformations under design load condition. Therefore, steel resists the free plastic flow (φ_t) and shrinkage $(\epsilon_{sr}\varphi_t/\varphi_n)$ and counteracts the loss of stress in concrete from $t = 0$ to $t = n$.

Denoting the internal forces [Eqs. (6)] transferred from concrete to steel by ΔN_t and the free shrinkage and plastic flow by ϵ_{sr} and φ_t, respectively, the ΔN_t value can be obtained by assuming that the shortening of concrete must be equal to the shortening of steel at any given time t. This condition is expressed by the following differential equation:

$$\frac{\epsilon_{sr}}{\varphi_n}\frac{d\varphi_t}{dt} + (N_{co} - \Delta N_t)\frac{1}{S_{cs}}\frac{d\varphi_t}{dt} - \frac{d\Delta N_t}{dt}\frac{1}{S_{cst}} = \frac{d\Delta N_t}{dt}\frac{1}{S_{ss}} \quad (70)$$

The right side of the equation represents the shortening of steel, and the left side the shortening of concrete for the time interval dt. The first factor on the left is the shortening due to shrinkage, where ϵ_{sr} and φ_n are the maximum values of free shrinkage and plastic flow. The second factor is the elastic shortening of concrete. The third element represents the influence of the retransfer of ΔN_t during the time dt. Assuming $S_{cs} \simeq S_{cst}$, the solution of this linear differential equation gives

$$\Delta N_t = \left(N_{co} + \frac{\epsilon_{sr}}{\varphi_n}S_{cs}\right)(1 - e^{-\alpha_s \varphi_t})$$

$$\alpha_s = \frac{E_s A_s}{E_s A_s + E_0 A_c} = \frac{nA_s}{nA_s + A_c} = \frac{S_{ss}}{S_{ss} + S_{cs}} \quad (71)$$

This equation is strictly valid for plain concrete only because it is based on the assumption that plastic flow is proportional to the stress in concrete. But during the transfer, the stress in concrete decreases steadily, resulting in decreased plastic flow φ'_t [Eq. (70)]. Also, during the period $t = 0$ to $t = n$, there is a steady increase in concrete strength (up to 1.5 f'_c), which results in an increased modulus of elasticity of concrete. As a thorough analysis by Dischinger indicates, the φ_n values are reduced by more than 50 percent when a high percentage of compressive reinforcing is present in a section. However, the influence of the increase of concrete strength is practically negligible because the modulus of elasticity [Eq. (64)] is not directly proportional to the cylinder strength f'_c and, therefore, the increase of E_0 in high-quality concrete is rather small and can be disregarded.

The influence of reinforcing steel on the reduction of plastic flow and shrinkage can be estimated directly from the actual strain relationships $\epsilon_{st}/\epsilon_{so} = \epsilon_{co}/\epsilon_{ct}$:

$$\epsilon_{st} = \epsilon_{so} + \Delta\epsilon_{st}$$

$$= \epsilon_{so} + \epsilon_{so} \frac{1-\alpha_s}{\alpha_s}(1 - e^{-\alpha_s \varphi_t})$$

$$\frac{\epsilon_{st}}{\epsilon_{so}} = \frac{\epsilon_{so}\left[1 + \frac{1-\alpha_s}{\alpha_s}(1 - e^{-\alpha_s \varphi_t})\right]}{\epsilon_{so}} = \frac{\sigma_{st}}{\sigma_{so}} \quad (72)$$

$$= 1 + \frac{1-\alpha_s}{\alpha_s}(1 - e^{-\alpha_s \varphi_t}) = 1 + \varphi'_t$$

$$\varphi'_t = \frac{1-\alpha_s}{\alpha_s}(1 - e^{-\alpha_s \varphi_t})$$

where φ'_t is the reduced value of plastic flow ($\varphi_t \to \varphi'_t$).

Considering that $S_{cs} = N_{co}\dfrac{E_0}{\sigma_{co}} = N_{co}\dfrac{E_s}{n\sigma_{co}}$ and $N_{co} = \dfrac{1-\alpha_s}{\alpha_s}N_{so}$, the normal force carried by concrete and steel will be, at any time t,

$$N_{ct} = (1 - \alpha_s)N_{co} - \Delta N_t$$

$$= N_{co}\left[1 - \left(1 + \frac{\epsilon_{sr}}{\varphi_n}\frac{E_0}{\sigma_{co}}\right)(1 - e^{-\alpha_s \varphi_t})\right] \quad (73)$$

$$= N_{co}\left[1 - \frac{\alpha_s}{1-\alpha_s}\left(1 + \frac{\epsilon_{sr}}{\varphi_n}\frac{E_0}{\sigma_{co}}\right)\varphi'_t\right]$$

$$N_{st} = \alpha_s N_{so} + \Delta N_t$$

$$= N_{so}\left[1 + \frac{1-\alpha_s}{\alpha_s}\left(1 + \frac{\epsilon_{sr}}{\varphi_n}\frac{E_0}{\sigma_{co}}\right)(1 - e^{-\alpha_s \varphi_t})\right] \quad (74)$$

$$= N_{so}\left[1 + \left(1 + \frac{\epsilon_{sr}}{\varphi_n}\frac{E_0}{\sigma_{co}}\right)\varphi'_t\right]$$

The moment transfer from concrete to steel is obtained when substituting $\Delta M_t, M_{co}\alpha_F$ for $\Delta N_t, N_{co}\alpha_s$ in Eqs. (71) and considering that only unequal shrinkage ($\Delta \epsilon_{sr}$) of either fiber in distance $\pm z$ from the neutral plane causes the rotation of the section. Thus

$$\Delta M_t = \left(M_{co} \pm \frac{\Delta \epsilon_{sr}}{\varphi_n z} S_{cF} \right)(1 - e^{-\alpha_F \varphi_t})$$

$$\alpha_F = \frac{E_s I_s}{E_s I_s + E_0 I_c} = \frac{n I_s}{n I_s + I_c} = \frac{S_{sF}}{S_{sF} + S_{cF}} \qquad (75)$$

$$\varphi_t' = \frac{1 - \alpha_F}{\alpha_F}(1 - e^{-\alpha_F \varphi_t})$$

and the moment carried by concrete and steel at any time t is

$$M_{co} = (1 - \alpha_F)M$$

$$M_{ct} = M_{co} - \Delta M_t \qquad \sigma_{co} = \frac{M_{co}}{I_c} z \qquad (76)$$

$$\sigma_{ct} = \sigma_{co}\left[1 - \frac{\alpha_F}{1 - \alpha_F}\left(1 \pm \frac{\Delta \epsilon_{sr}}{\varphi_n z} \frac{E_0}{\sigma_{co}}\right)\varphi_t'\right]$$

$$M_{so} = \alpha_F M$$

$$M_{st} = M_{so} + \Delta M_t \qquad \sigma_{so} = \frac{M_{so}}{I_s} z \qquad (77)$$

$$\sigma_{st} = \sigma_{so}\left[1 + \left(1 \pm \frac{\Delta \epsilon_{sr}}{\varphi_n z} \frac{E_0}{\sigma_{co}}\right)\varphi_t'\right]$$

When a section is subjected to normal force and moment, the combined stresses are obtained by superposition. It must be noted that Eqs. (73) to (77) are strictly valued only for homogeneous materials and, therefore, must be modified for cracked sections.

Analysis of the Span Section and Support Section of a Two-span Continuous Beam

To demonstrate the influence of plastic flow and shrinkage upon stresses of a cracked section, the span section m_2 and support section c of the two-span continuous beam, illustrated in Fig. 9, will be analyzed. Design data, as computed by standard practice, are given in Figs. 34 and 35.

Span section:

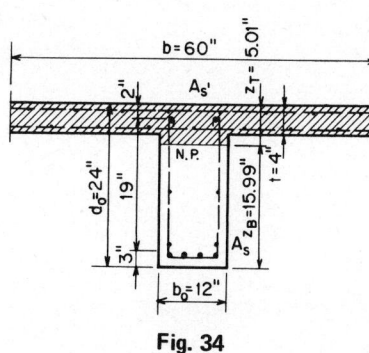

Fig. 34

$$N = 0$$

$$\sum_{D}^{L} M_{m2} = 210\,k'$$

$$A'_s = 1.58 \text{ in.}^2 \quad (2 - {}^\#8)$$

$$A_s = 7.82 \text{ in.}^2 \quad (2 - {}^\#8 + 4 - {}^\#11)$$

$$\Sigma A_s = 9.40 \text{ in.}^2$$

$$E_0 = 5.0 \times 10^6 \text{ psi}$$
$$E_s = 30.0 \times 10^6 \text{ psi} \quad \Big| \quad n = \frac{E_s}{E_0} = 6$$

Sectional coefficients:

The distance z_T from the extreme top fiber to the neutral plane is computed from the condition $\Sigma Q_0 = 0$.

Concrete:

$$A'_c = b't = (60 - 12)4 = 192 \text{ in.}^2$$
$$+ b_0 z_T = 12\, z_T \text{ in.}^2$$
$$Q'_{co} = A'_c (z_T - 2) = 192\, z_T - 384 \text{ in.}^3$$
$$+ \frac{b_0 z_T^2}{2} = 6\, z_T^2 \text{ in.}^3$$

Steel:

$$(n - 1) A'_s = (6 - 1) 1.58 = 7.90 \text{ in.}^2$$
$$n A_s = 6 \times 7.82 = 46.92 \text{ in.}^2$$
$$Q_{so} = 7.90 (z_T - 2) = 7.90\, z_T - 15.80 \text{ in.}^3$$
$$- 46.92 (21 - z_T) = -985.32 + 46.92\, z_T \text{ in.}^3$$
$$6 z_T^2 + 246.82\, z_T - 1{,}385.12 = 0$$
$$z_T = \frac{1}{2 \times 6} \left(-246.82 + \sqrt{246.82^2 + 4 \times 1{,}385.12 \times 6} \right)$$
$$= 5.01 \text{ in.} \quad z_{sT} = 3.01 \text{ in.} \quad z_{sB} = 15.99 \text{ in.}$$

Theory of Structures

$$I_{co} = \frac{192 \times 4^2}{12} + (192 - 1.58)3.01^2 + \frac{12 \times 5.01^3}{3} = 2,486 \text{ in.}^4$$

$$nI'_{so} = 6 \times 1.58 \times 3.01^2 = 86 \text{ in.}^4$$
$$nI_{so} = 46.92 \times 15.99^2 = 11,980 \text{ in.}^4$$

$$\Sigma I_s = 12,066 \text{ in.}^4$$
$$I_0 = \underline{14,552 \text{ in.}^4}$$

Stresses:

$t = 0$:
$$\sum_D^L \sigma_{cT} = -\frac{210 \times 12,000}{14,552} \, 5.01 = -870 \text{ psi}$$

$$\sum_D^L \sigma_{sT} = -6 \times 870 \times 3.01/5.01 = -3,130 \text{ psi}$$

$$\sum_D^L \sigma_{sB} = 6 \times 870 \times 15.99/5.01 = 16,650 \text{ psi}$$

$t = n$:

$\varphi_n = 2.0 \qquad \Delta \epsilon_{sr} = 10 \times 10^{-5}$ — top fiber

$$I_{cB} = \frac{192 \times 4^3}{12} + (192 - 1.58)19^2 + \frac{60.12 \times 5.01^3}{12} + 60.12 \times 18.5^2$$
$$= 89,712 \text{ in.}^4$$

$$nI'_{sB} = 6 \times 1.58 \times 19^2 = 3,420 \text{ in.}^4 \qquad \sum_c^{s'} I_{sB} = \underline{93,132 \text{ in.}^4}$$

$$\alpha_F = \frac{3,420}{93,132} = 0.0367$$

$$\varphi'_n = \frac{1 - 0.0367}{0.0367}(1 - 2.718^{-0.0367 \times 2.0}) = 1.86$$

$$\bar{\sigma}_{cD} = k'_D \sum_D^L \sigma_{cT} = -0.67 \times 870 = -583 \text{ psi}$$

$$k'_D = k_D + 0.25 k_L = 0.67$$

$$\bar{\sigma}_{cL} = k'_L \sum_D^L \sigma_{cT} = -0.33 \times 870 = -287 \text{ psi}$$

$$k'_L = 1 - k'_D = 0.33$$

$$\sigma_{cD} = -583\left[1 - \frac{0.0367}{1-0.0367}\left(1 + \frac{10 \times 10^{-5}}{2.0 \times 5.01} \frac{5.0 \times 10^6}{583}\right)1.86\right]$$

$$= -537 \text{ psi}$$

$$\sum_{D}^{L} \sigma_{cT} = \sigma_{cD} - \bar{\sigma}_{cL} = -537 - 287 = -824 \text{ psi}$$

$$\bar{\sigma}_{sD} = k'_D \sum_{D}^{L} \sigma_{sT} = -0.67 \times 3{,}130 = -2{,}100 \text{ psi} \Bigg| \text{ sustained stresses}$$

$$\bar{\sigma}_{sL} = k'_L \sum_{D}^{L} \sigma_{sT} = -0.33 \times 3{,}130 = -1{,}030 \text{ psi} \Bigg|$$

$$\sum_{D}^{L} \sigma_{sB} = 16{,}650 \text{ psi}$$

$$\sigma_{sT} = -2{,}100\left[1 + \left(1 + \frac{10 \times 10^{-5}}{2.0 \times 5.01} \frac{5.0 \times 10^6}{583}\right)1.86\right]$$

$$= -6{,}320 \text{ psi}$$

$$\sum_{D}^{L} \sigma_{sT} = \sigma_{sT} + \bar{\sigma}_{sL} = -6{,}320 - 1{,}030 = -7{,}350 \text{ psi}$$

Support section:

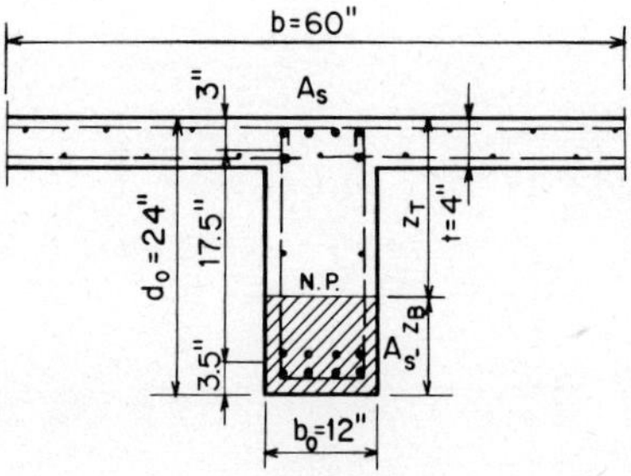

$$N = 0:$$

$$\sum_{D}^{L} M_c = -270 \, k'$$

$A_s = 9.36 \text{ in.}^2 \, (6 - \#11)$

$A'_s = 12.48 \text{ in.}^2 \, (8 - \#11)$

$$\Sigma A_s = 21.84 \text{ in.}^2$$

$nA_s = 6 \times 9.36 = 56.16 \text{ in.}^2$

$(n-1)A'_s = 5 \times 12.48 = 62.40 \text{ in.}^2$

$A_c = b_0 z_B \text{ in.}^2$

Fig. 35

Sectional coefficients:

$$Q_{sB} = -56.16\,(21 - z_B) = -1,180 + 56.16\,z_B \text{ in.}^3$$
$$\phantom{Q_{sB} =} + 62.40\,(z_B - 3.5) = 62.40\,z_B - 218.40 \text{ in.}^3$$

$$Q_{cB} = 12\,\frac{z_B^2}{2} = 6z_B^2$$

$$6z_B^2 + 118.56\,z_B - 1,398.40 = 0$$

$$z_B = \frac{1}{2 \times 6}\left(-118.56 + \sqrt{118.56^2 + 4 \times 1,398.40 \times 6}\right)$$

$$= 8.3 \text{ in.} \quad z_{sB} = 4.8 \text{ in.} \quad z_{sT} = 12.70 \text{ in.} \quad z_B > kd$$

$$I_{co} = \frac{12 \times 8.3^3}{3} - 12.48 \times 4.8^2 \simeq 2,000 \text{ in.}^4$$

$$I'_{so} = 6 \times 12.48 \times 4.8^2 = 1,725 \text{ in.}^4 \quad \Sigma I_{so} = 10,785 \text{ in.}^4$$
$$I_{so} = 56.16 \times 12.7^2 = 9,060 \text{ in.}^4 \quad I_0 = \underline{12,785 \text{ in.}^4}$$

Stresses:

$t = 0$:

$$\sum_D^L \sigma_{cB} = -\frac{270 \times 12,000}{12,785}\,8.3 = -2,100 \text{ psi}$$

$$\sum_D^L \sigma_{sB} = -6 \times 2,100 \times 4.8/8.3 = -7,280 \text{ psi}$$

$$\sum_D^L \sigma_{sT} = 6 \times 2,100 \times 12.7/8.3 = 19,300 \text{ psi}$$

$t = n$:

$$I_{cT} = \frac{12 \times 8.3^3}{12} + 100 \times 16.85^2 - 12.48 \times 17.5^2 = 25,150 \text{ in.}^4$$

$$I'_{sT} = 6 \times 12.48 \times 17.5^2 = 22,900 \text{ in.}^4 \quad \Sigma I_T = 48,050 \text{ in.}^4$$

$$\alpha_F = \frac{22,900}{48,050} = 0.476$$

$$\varphi'_n = \frac{1 - 0.476}{0.476}\,(1 - 2.718^{-0.476 \times 2.0}) = 0.675$$

$\bar{\sigma}_{cD} = -0.67 \times 2{,}100 = -1{,}410$ psi $\quad$ sustained stresses
$\bar{\sigma}_{cL} = -0.33 \times 2{,}100 = -690$ psi

$$\sigma_{cD} = -1{,}410\left[1 - \frac{0.476}{1 - 0.476}\left(1 - \frac{10 \times 10^{-5}}{2.0 \times 15.7}\cdot\frac{5.0 \times 10^{6}}{1{,}410}\right)0.675\right]$$

$\qquad = -550$ psi

$\sum_{D}^{L} \sigma_{cB} = -550 - 690 = -1{,}240$ psi

$\bar{\sigma}_{sD} = -0.67 \times 7{,}280 = -4{,}870$ psi $\quad \sigma_{sL} = -0.33 \times 7{,}280$
$\qquad\qquad\qquad\qquad\qquad\qquad\qquad\qquad = -2{,}400$ psi

$$\sigma_{sD} = -4{,}870\left[1 + \left(1 - \frac{10 \times 10^{-5}}{2.0 \times 15.7}\cdot\frac{5.0 \times 10^{6}}{1{,}410}\right)0.675\right] = -8{,}130 \text{ psi}$$

$\sum_{D}^{L} \sigma_{sB} = -8{,}130 - 2{,}400 = -10{,}530$ psi

$\sum_{D}^{L} \sigma_{sT} = 12{,}940 + 6{,}360 = 19{,}300$ psi

The above computations clearly indicate that considerable transfer of stresses from concrete to steel, due to shrinkage and plastic flow of concrete, takes place in a course of time ($t = 0$ to $t = n$). The extent of stress transfer from concrete to steel in low E_0 value, high percentage of reinforcing in section, high shrinkage and plastic flow can be such that concrete in the compression zone escapes entirely from carrying action and compressive steel stresses approach the yield point. This phenomenon explains the occasionally observed crack formation in sections under compression when rolled sections encased in fine-aggregate concrete are used.

Loss in Prestress

Under sustained loading and prestressing, the concrete stresses due to plastic flow and shrinkage change from $t = 0$ to $t = n$. This change results in loss in the prestressing force. As previously discussed, the loss can be estimated from relative stiffnesses of the prestressing steel and concrete at the plane of action of the prestressing force H_{pr}. Since the strain stiffness S_{cs} is defined as the force S which causes a unit strain 1, it is at the plane of a prestressing tendon:

$$\epsilon_c = \frac{S_{cs}}{E_0 A_c} + \frac{S_{cs} e_0}{E_0 I_{co}} = 1 \tag{78}$$

Thus

$$S_{cs} = \frac{E_0 A_c}{1 + e_0^2 A_c / I_{co}} = \frac{E_0 A_c}{1 + e_0^2 / i^2}$$

$$i^2 = \frac{I_{co}}{A_c}$$

$$S_{prs} = E'_{pr} A_{pr}$$

The resultant of the sustained tensile stress T_D is counteracting and reduces the prestressing force. The T_D value is

$$T_D = \frac{M_D}{z_0}$$

The lever arm of the internal forces z_0 can be estimated by the Busemann method, as illustrated in Fig. 36:

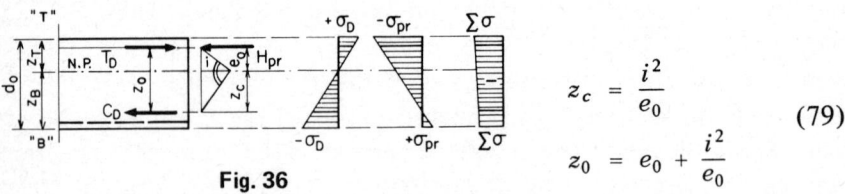

$$z_c = \frac{i^2}{e_0}$$

$$z_0 = e_0 + \frac{i^2}{e_0} \tag{79}$$

Fig. 36

The relationship $z_c e_0 = i^2$ means that if a force is applied at a distance e_0 from the center of gravity (C.G.) of a section, having the radius of gyration i, the neutral plane (N.P.) lies at the distance z_c from the other side of the center of gravity.

Introducing the proper values into Eqs. (71), the loss in prestressing force for any time t is

$$\Delta H_{prt} = \left(H_{pro} - T_D + \frac{\epsilon_{sr}}{\varphi_n} S_{cs} \right) (1 - e^{-\alpha_s \varphi'_t}) \tag{80}$$

and the actual prestressing force for $t = n$ is

$$H_{prn} = H_{pro} - \Delta H_{prn}$$

For example, the support section of the beam in Fig. 9 will be analyzed as a prestressed section (Fig. 37).

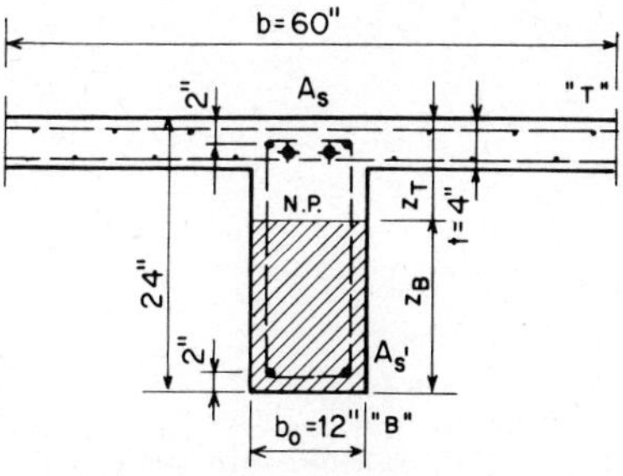

Fig. 37

$$\sum_{D}^{L} M_c = -270\,k' \quad k'_D = 0.67$$

$$k'_L = 0.33$$

$A_s = A'_s = 1.58\text{ in.}^2 \quad \Sigma A_s = 3.16\text{ in.}^2$

$A'_{pr} = 3.60\text{ in.}^2 \quad$ grouted

$$n = \frac{E_s}{E_0} = \frac{30.0 \times 10^6}{5.0 \times 10^6} = 6$$

Sectional coefficients:

$$A_c = 48 \times 4 + 12 \times 24 - (3.16 + 3.60) = 473\text{ in.}^2$$
$$n\Sigma A_s = 6 \times 3.16 \simeq 19\text{ in.}^2$$
$$\sum_{s}^{c} A = 492\text{ in.}^2$$

$$\sum_{s}^{c} Q_T = 192 \times 2.0 + 288 \times 12.0 + 19 \times 12.0$$
$$- (3.16 \times 12.0 + 3.60 \times 3.15) = 4{,}019\text{ in.}^3$$

$$z_T = \frac{4{,}019}{492} = 8.15\text{ in.} \quad z_B = 15.85\text{ in.}$$

$$z_{sT} = 6.15\text{ in.} \quad z_{sB} = 13.85\text{ in.}$$

$$I_{co} = \frac{192 \times 4^2}{12} + 192 \times 6.15^2 + \frac{288 \times 24^2}{12} + 288 \times 3.85^2$$
$$- 1.58(6.15^2 + 13.85^2) = 25{,}220\text{ in.}^4$$

$$I_{so} = 6 \times 1.58(6.15^2 + 13.85^2) = 2{,}180\text{ in.}^4$$

$$\sum_{s}^{c} I_0 = 27{,}400\text{ in.}^4$$

Stresses due to loading:

$$\sum_{D}^{L} \sigma_{cT} = \frac{270 \times 12{,}000}{27{,}400}\, 8.15 = 965\text{ psi}$$

$$\sum_{D}^{L} \sigma_{cB} = -\frac{270 \times 12{,}000}{27{,}400}\, 15.85 = -1{,}875\text{ psi}$$

Prestressing:

$$e_0 = z_T - 3.15 = 5.0 \text{ in.}$$

$$i^2 = \frac{27{,}400}{492} = 56 \text{ in.}^2 \quad z_0 = e_0 + \frac{i^2}{e_0} = 16.2 \text{ in.} = 1.35 \text{ ft}$$

$$\psi_T = 1 + \frac{5.0 \times 8.15}{56} = 1.728$$

$$\psi_B = 1 - \frac{5.0 \times 15.85}{56} = -0.415$$

$$H_{pm} = \frac{\Sigma \sigma_T \Sigma A}{\psi_T} = \frac{0.965 \times 492}{1.728} = 275 \, k$$

Use: $2 \times 6 \text{-} \tfrac{1}{2}''$ - strand cables. $A_{pr} = 1.84 \text{ in.}^2$

$$H_{pro} = 2 \times 173.4 \times 0.94 = 326 \, k \quad 6 \text{ percent friction and relaxation}$$

$$\sigma_{opr} = -\frac{H_{pro}}{\Sigma A} = -\frac{326 \times 1{,}000}{492} = -660 \text{ psi}$$

$$\sigma_{Tpr} = \psi_T \sigma_{opr} = -1.728 \times 660 = -1{,}140 \text{ psi}$$

$$\sigma_{Bpr} = \psi_B \sigma_{opr} = 0.415 \times 660 = 275 \text{ psi}$$

Stresses:

$t = 0:$

$$\sum_{pr}^{D+L} \sigma_{cT} = 965 - 1{,}140 = -175 \text{ psi}$$

$$\sum_{pr}^{D+L} \sigma_{cB} = -1{,}875 + 275 = -1{,}600 \text{ psi}$$

Loss in prestress:

$$\varphi_n = 2.0 \quad \epsilon_{sr} = 25 \times 10^{-5} \quad \Delta \epsilon_{sr} = 0$$

$$M_D = k'_D \sum_D^L M_c = -0.67 \times 270 = -181 \, k' \text{ - sustained moment}$$

$$T_D = \frac{M_D}{z_0} = \frac{181}{1.35} = 134 \, k$$

$$S'_{cs} = E_0 A_c = 5.0 \times 10^6 \times 473 = 2{,}360 \times 10^6 \text{ lb}$$

$$S'_{ss} = E_s A_s = 30.0 \times 10^6 \times 3.16 = 95 \times 10^6 \text{ lb}$$

$$\alpha'_s = \frac{S'_{ss}}{S'_{ss} + S'_{cs}} = \frac{95}{95 + 2{,}360} = 0.0388$$

$$\varphi'_n = \frac{1 - \alpha'_s}{\alpha'_s}[1 - e^{-\alpha'_s \varphi_n}]$$

$$= \frac{1 - 0.0388}{0.0388}(1 - 2.718^{-0.0388 \times 2.0}) = 1.85$$

$$S_{prs} = E_s A_{pr} = 28.0 \times 10^3 \times 1.84 = 0.515 \times 10^5 \text{ k}$$

$$S_{cs} = \frac{E_o \Sigma A}{1 + i^2/e_0^2} = \frac{5.0 \times 10^3 \times 492}{1 + 56/5.0^2} = 7.6 \times 10^5 \text{ k}$$

$$\alpha_{pr} = \frac{0.515}{0.515 + 7.6} = 0.0635$$

$$\Delta H_n = \left(326 - 134 + \frac{25 \times 10^{-5}}{2.0} \cdot 7.6 \times 10^5\right)(1 - 2.718^{-0.0635 \times 1.85})$$

$$= 287 \times 0.107 \simeq 31 \text{ k}$$

$$\Delta\sigma_{opr} = \frac{31 \times 1{,}000}{492} = 63 \text{ psi}$$

$$\Delta\sigma_{Tpr} = 1.728 \times 63 = 110 \text{ psi}$$

$$\Delta\sigma_{Bpr} = -0.415 \times 63 = -26 \text{ psi}$$

Stresses:

$t = n:$

$$\sum_{pr}^{D+L} \sigma_{cT} = -175 + 110 = -65 \text{ psi}$$

$$\sum_{pr}^{D+L} \sigma_{cB} = -1{,}600 - 26 = -1{,}626 \text{ psi}$$

In the case of nonbonded tendons, the loss in prestressing force due to plastic flow and shrinkage has to be computed from the total shortening of the beam because here the elongation of tendons and shortening of concrete are independent of each other. For continuous tendons, from end to end of the beam, the loss in prestressing force is computed at the center of gravity and T_D is taken zero. However, for this condition, Eq. (80) gives rather accurate loss in prestress also for nonbonded tendons. When a different length of tendons is used, the computation of losses in each tendon must be determined separately.*

Change in Redundants

It is standard practice to assume an uncracked homogeneous plain concrete section for computing the moment of inertia I_c and the elastic deformations δ_{kk}, δ_{ik}, δ_{ko} to determine the redundants x_n. This assumption is necessary because the extent of cracking and the required layout and amount of reinforcing can be determined only after the redundants x_n are known.

The inaccuracy caused by this assumption is best demonstrated by the preceding numerical examples. The redundant X_1 (Fig. 9) was computed for $I_c/I = 1.0$ for both spans. When the reinforcing is considered, for uncracked sections the relative stiffnesses are, for span L_1, $I_c/I_1 = 1.22$ and, for span L_2, $I_c/I_2 = 1.130$. In cracked sections the relative stiffnesses are 1.34 and 0.88, respectively. Considering that concrete cracks after the tensional stresses have reached approximately $1/10 f'_c = 500$ psi, the relative stiffnesses change, resulting in considerable increase in redundant X_1. However, in accordance with experience, the design loading seldom occurs, and under working load conditions ($w_D + 0.25 w_L$) the reduction of stiffness due to hair cracking in a properly reinforced concrete beam results in only a slight increase in redundants.

Due to these facts, it is fully justified to first compute the redundants approximately (Tables 1 and 2) and design the beam. Then the actual quantities of I and δ_{kk}, δ_{ik}, δ_{ko} must be computed and the redundants X_n obtained from continuity equations (7). Commonly the relatively small changes in reinforcing, if required, do not change the redundants appreciably.

Using the data from previous examples (Figs. 9, 34, and 35), the change in redundant X_1 due to various stiffnesses of beam subjected to sustained uniform loading ($k'_D = 0.67$) are, for $t = 0$,

$$M_{1,0} = 0.67 \times 221 = 148 \, k'$$
$$M_{2,0} = 0.67 \times 313 = 210 \, k'$$

*Ibid.

Case 1. Uncracked section, reinforcing not considered (Fig. 9), $I_c/I = 1$:

$$X_1 = 0.67 \times 270 = 181 \, k' = -M_c$$

Case 2. Uncracked section, reinforcing considered:

$$I_c = 1.0 \quad \frac{I_c}{I_1} = 1.22 \quad \frac{I_c}{I_2} = 1.13$$

$$E_0 I_c \delta_{11} = \sum_a^b \int M_1^2 \, dx = 32.76$$

$$E_0 I_c \delta_{10} = \sum_a^b \int M_1 M_0 \, dx = 6{,}421$$

$$X_1 = \frac{\delta_{10}}{\delta_{11}} = \frac{6{,}421}{32.76} = 196 \, k' = -M_c$$

Case 3. Cracked section ($f'_{cT} = 1/10 f'_c$):

$$I_c = 12{,}785 \text{ in.}^4 \quad I_1 = 9{,}555 \text{ in.}^4 \quad I_2 = 14{,}552 \text{ in.}^4$$

$$E_0 I_c \delta_{11} = \sum_a^b \int M_1^2 \, dx = 25.9$$

$$E_0 I_c \delta_{10} = \sum_a^b \int M_1 M_0 \, dx = 5{,}175$$

$$X_1 = \frac{\delta_{10}}{\delta_{11}} = \frac{5{,}175}{25.9} = 200 \, k' = -M_c$$

The moment diagrams for cases 1 to 3 are plotted in Fig. 38.

82 Theory of Structures

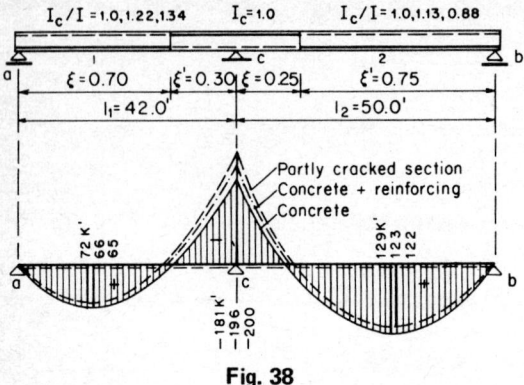

Fig. 38

The influence of plastic flow and shrinkage upon the redundants will be considered in the following (Fig. 39).

Rotation of the plane of a section is, for $t = 0$,

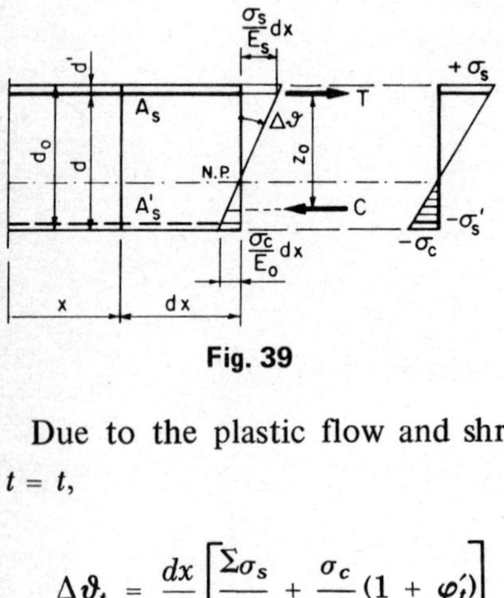

Fig. 39

$$\Delta\vartheta_0 = \frac{dx}{d}\left(\frac{\sigma_s}{E_s} + \frac{\sigma_c}{E_0}\right)$$

$$n = \frac{E_s}{E_0} \tag{81}$$

$$\frac{dx}{d}\frac{\sigma_c}{E_0} = \Delta\vartheta_0 \frac{1}{1 + \sigma_s/\sigma_c n}$$

Due to the plastic flow and shrinkage, the rotation angle will be, for $t = t$,

$$\Delta\vartheta_t = \frac{dx}{d}\left[\frac{\Sigma\sigma_s}{E_s} + \frac{\sigma_c}{E_0}(1 + \varphi'_t)\right]$$

$$= \Delta\vartheta_0 + \frac{dx}{d}\frac{\sigma_c}{E_0}\varphi'_t \tag{82}$$

$$= \Delta\vartheta_0\left(1 + \frac{\varphi'_t}{1 + \Sigma\sigma_s/\sigma_c n}\right)$$

Since the continuity equations (7) were satisfied for $\Delta\vartheta_0$, this means that there will be no change in the redundants x_n provided that the value

$\varphi'_t(1 + \Sigma\sigma_s/\sigma_c n)$, is constant over the entire length of the beam. Commonly such is not the case, because reinforcing and the φ'_t value vary. Thus, for $t = n$,

$$X_n \to X_0 + \Delta X$$

For time $t = 0$, $\varphi'_t = 0$, $\Delta x = 0$; and for $t = n$, Δx becomes maximum. For moderate change in $\varphi'_t(1 + \Sigma\sigma_s/\sigma_c n)$, the influence of plastic flow and nonuniform shrinkage upon the moments is small. The magnitude of ΔX_t can easily be estimated from Eq. (82):

$$\Delta X_t = EI_0 \Delta \vartheta_0 \frac{\varphi'_t}{1 + \Sigma\sigma_s/\sigma_c n} \tag{83}$$

In accordance with Eq. (82), the moments caused by yielding or rotation of supports from $t = 0$ to $t = n$ are greatly reduced by plastic flow.

By differentiation of Eq. (82) with respect to dt, we obtain

$$\frac{d\Delta\vartheta_t}{dt} = \frac{dx}{d}\frac{\sigma_c}{E_0}\frac{d\varphi'_t}{dt}$$
$$= \Delta\vartheta_0 \frac{1}{1 + \Sigma\sigma_s/\sigma_c n}\frac{d\varphi'_t}{dt} \tag{84}$$

Thus, by analogy, the moment ΔM caused by differential support settlement, rotation, and nonuniform shrinkage, as computed by elastic theory, will be for $t = t$

$$\Delta M_t = \Delta M e^{-\varphi'_t/(1+\Sigma\sigma_s/\sigma_c n)} \tag{85}$$

For example, the ratio $\Sigma\sigma_s/\sigma_c n$ for high-quality concrete is approximately 2.5, and assuming

$$\varphi'_n = \frac{1 - \alpha_F}{\alpha_F}(1 - e^{-\alpha_F \varphi_n}) \simeq 1.8$$

the value of ΔM is reduced by plastic flow about 40 percent.

10. ULTIMATE CARRYING CAPACITY AND FACTOR OF SAFETY

Generally, a reinforced-concrete beam or girder may lose its carrying capacity for any of the following reasons:

1. The yield point of the reinforcing steel is reached; at this point, the steel first exhibits an appreciable increase in strain without notable increase in stress. The large strain of the steel causes rapid decrease of the compression area of the concrete; subsequent crushing of the concrete results in rupture. In a failure of this type, the limit of carrying capacity is distinguished by the formation of large cracks in tension areas and notable deflection of the beam.
2. The ultimate strength of the concrete is reached although the steel is not stressed to the yield point. Rupture follows promptly and is not preceded by noticeable crack formation.
3. Excessive diagonal stresses develop, resulting in the formation of diagonal cracks in the web or stem of the beam. The limit of carrying capacity of the beam is indicated by the opening of inclined cracks in the web and the partial separation of the main reinforcing steel from the concrete.
4. Excessive local stresses are caused by concentrated loads.

In bonded prestressed beams, rupture is usually caused by a sudden break of the reinforcement. This occurs because the strainability of the steel is partly used in obtaining the prestressing and the portion remaining is not sufficient to absorb the excessive strain which develops when the concrete ruptures. The limit of the carrying capacity is indicated by small cracks.

In nonbonded prestressed beams or trussed girders, the rupture of the beams occurs slowly and is accompanied by the formation of large cracks. In such beams, the yield strength of the prestressed steel is seldom reached at the rupture. The failure of the beam is caused, therefore, by the crushing of the concrete or the opening up of diagonal cracks. In this type of beam, the diagonal cracks are caused by the independent deformation of the tension steel, in which the elongation does not increase with the increase in compressive stresses in the concrete; thus a rapid increase of the principal stresses results in the tension portion of the beam.

In prestressed trusses, the loss of carrying capacity is usually caused by the great increase in secondary stresses as a result of the increase in deflection of the truss under excessive loads. The failure first appears at the joints of the truss, and rupture is finally caused by breaking of the compression chord.

Since the safety of the structure depends to a great extent on the character of the rupture, the factor of safety should be considered differently for each of the types of failure. Following from the above, the amount and properties of steel at critical sections are governing in the determination of the factor of safety in bonded prestressed beams, whereas in nonbonded prestressed beams, the concrete controls.

Since the uniformity and quality of concrete vary more than those of steel, the uncertainties in the concrete must be considered when computing the ultimate carrying capacity of a structure. As indicated by numerous tests, the quality of concrete may vary approximately in the range of 15 percent. Therefore, the beam should be so designed that the breaking moment, as governed by the concrete, is at least 15 percent higher than that computed on the basis of the steel.

In the following, each type of failure will be discussed in the context of the ultimate carrying capacity and the related factor of safety.

Flexural Failures (Types 1 and 2)

Failure types 1 and 2 are generally denoted as *bending* or *flexural* failures. The location of the neutral plane at the time of rupture determines the type of failure (1 or 2) which may occur. Assuming that the linear relationship between stresses and strains is valid at rupture, the location of neutral plane x_{UL} as a function of strain k_ϵ is (Fig. 40)

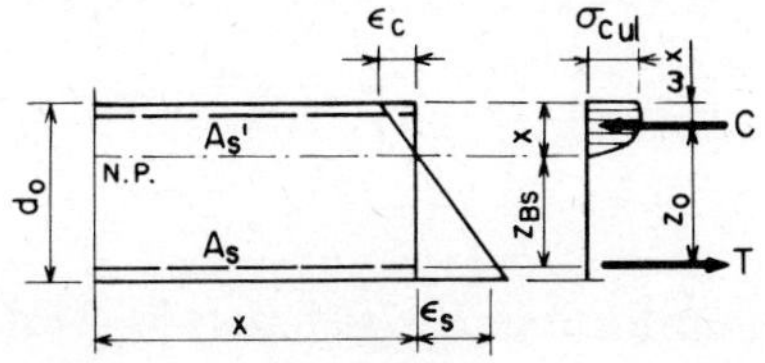

Fig. 40

$$x = k_\epsilon d$$

$$k_\epsilon = \frac{\epsilon_{cb}}{\epsilon_{cb} + \epsilon_{sy}} \tag{86}$$

The maximum strain for concrete (ϵ_{cb}) is approximately 2.0 percent and for steel at yield point is $\epsilon_{sy} = \sigma_{sy}/E_s$. The yield stress σ_{sy} varies considerably for various steels and, therefore, has to be obtained from the steel manufacturer. However, for intermediate-grade reinforcing steel, it is approximately 48,000 psi. Up to the yield point, the moment and stresses are proportional and the neutral plane has a definite location, but beyond the yield point it moves toward the compression fiber, whereby ϵ_s increases under almost constant stress σ_{sy}. Rupture occurs when the strain of the concrete (ϵ_c) reaches its maximum ($\epsilon_{c,\max} \sim 2\permil$). Since the ϵ_s is unknown in Eq. (86), the k value as a function of stress (k_σ) must first be computed from the equilibrium requirement $T_b = C_{UL}$ by ignoring Eq. (86) and assuming that the lever arm z_0 of internal forces for the rectangular section and the T section, respectively, is

$$z_0 = d(1 - \omega k) \simeq 0.85d \text{ to } 0.90d \tag{87}$$

Then the required tensile force at rupture is

$$T_b = \frac{v \Sigma M}{z_0} = \Sigma A_s \sigma_s \quad v\text{-factor of safety} \tag{88}$$

where ω is a factor locating the center of gravity of the compressive force C_{UL}. With plain concrete, ω is approximately 0.33 for a T section and 0.45 for a rectangular section. In the case of compressive reinforcing, its value is slightly less. The shape of the compressive stress diagram is approximately a parabola, with maximum ordinate $\sigma_{cUL} = 0.85 f'_c$. The steel breaking stress $\sigma_{sb} = \sigma_{sy}$ is known.

Thus $\Sigma X = 0$:

$$T_b = C_{UL} = \sigma_{cUL}\left[(b - b_0)t + \left(\tfrac{2}{3} b_0 d k_\sigma - A'_s\right)\right] + A'_s \sigma_{sy}$$

$$k_\sigma = \frac{T_b - [A'_s \sigma_{sy} + \sigma_{cUL}(b't - A'_s)]}{\tfrac{2}{3} db_0 \sigma_{cUL}} \tag{89}$$

$$T_b = A_s \sigma_{sy} \quad b - b_0 = b'$$

For simplification, σ_{cUL} has been taken uniform in the slab part (t). Test results prove this to be fully justified, especially considering all other uncertainties involved.

Taking $k_\sigma = k_\epsilon$, the strain of steel is

$$\epsilon_{sb} = \frac{\epsilon_{c,\max}(1 - k_\sigma)}{k_\sigma} \tag{90}$$

The steel strain ϵ_{sb} indicates the extent of cracking and determines also the type (1 or 2) of failure.

In the case of prestressing, the k_σ value can be computed in a manner similar to that for reinforced-concrete sections. The k_ϵ value, i.e., the location of the neutral plane, must be computed from the "residual strainability" of both materials. However, for convenience, one starts here from strains at design load (d) condition.

Thus,

$$k_\sigma = \frac{\Sigma T_b - [A_c' \sigma_{sy} + \sigma_{cUL}(b't - A_s')]}{\frac{2}{3} d b_0 \sigma_{cUL}} \qquad \Sigma T_b = \sigma_{prUL} A_{pr} + \sigma_{sy} A_s$$

$$k_\epsilon = \frac{\epsilon_{cUL} - \epsilon_{cd}}{\epsilon_{cUL} - \epsilon_{cd} + \epsilon_{prUL} - \epsilon_{pr}d} \tag{91}$$

If $k_\epsilon > k_\sigma$, steel controls the factor of safety; and if $k_\epsilon < k_\sigma$, concrete controls. In a balanced design, $k_\epsilon = k_\sigma$. The strain of concrete (ϵ_{cd}) can be computed by Hooke's law or by Eq. (558), and the strain of prestressing steel ($\epsilon_{prd}, \epsilon_{prUL}$) can be computed from the strain-stress diagram of the manufacturer. The influence of conventional steel upon the factor of safety beyond the yield point is zero because the strain occurs under constant stress. Thus k_ϵ value in this region depends entirely on the strain of prestressing steel.

The ultimate strength of nonbonded prestressing depends on the magnitude of precompression of the tension zone and on the amount of conventional steel in the tension zone. This is explained by the fact that a nonbonded tendon practically does not participate in the carrying action. The $\epsilon_{pr\sigma}$ in Eqs. (91) may increase, where curved tendons are used, up to approximately 10 percent from the design stress to rupture. Therefore, Eqs. (91) do not apply for nonbonded prestressing and must be modified. They are

$$k_\sigma = \frac{\Sigma T' - [A'_s \sigma_{sy} + \sigma_{cUL}(b't - A'_s)]}{\frac{2}{3} db_0 \sigma_{cUL}} \tag{92}$$

$$k_\epsilon = \frac{\epsilon_{cUL} - \epsilon_{cd}}{\epsilon_{cUL} - \epsilon_{cd} + \epsilon_{sb} + 1.10\,\epsilon_{prd}}$$

where $\Sigma T' = 1.10 H_{prn} + A_s \sigma_{sy}$ and the value of ϵ_{sb} is given by Eq. (90). It is obvious that $\Sigma T'$ can be increased at the expense of conventional steel to obtain a reasonable factor of safety.* When analyzing Eqs. (91), it is evident that cracks appear where the nonbonded prestressed beam is subjected to a temporary overloading. The limit of crack formation under a fixed condition of conceivable overload serves not only as the factor of safety but also as the basis for determining the amount of conventional reinforcing required.

The greater k value determines the neutral plane and the lever arm of internal forces at rupture. After z_0 is computed for concrete, the ultimate moment is

$$M_{UL} = A'_s \sigma_{sy} z_s + C_{cUL} z_0 \tag{93}$$

and the factor of safety is

$$\text{F.S.} = \frac{M_{UL}}{\sum_D^L M} \geq 1.8$$

Under the ultimate strength of continuous beams plastic hinges are formed. In the section in which they occur, plastic deformation takes place first, whether they occur at midspans or at support points is not significant, resulting in moment redistribution. This balancing of moments proceeds until at least two sections have reached ultimate strength. Thus the factor of safety is approximately the average of the ultimate strength of these sections.

Shear Failure (Type 3)

Type 3 failure is commonly called *shear* failure. As the shear—or, more properly, diagonal-tension—failure is closely related to bond failure, they will be analyzed together.

Ibid.

In a conventional reinforced-concrete beam under ultimate load condition $\left(v \sum_{D}^{L} w, P\right)$, depending upon span length L and shear V, cracks may or may not develop in the support regions where the moments are small. If the diagonal tension σ_1 is less than $\frac{1}{10} f'_c$, there will be no cracks and Eqs. (40) will apply. In shorter spans with relatively large shear V, the sections most certainly will be cracked and Eqs. (40) will not, in the strict sense, apply. The amount of reinforcing and the factor of safety can be estimated most conveniently by the cantilever method, previously described in the discussion of cracked sections (Sec. 8). In prestressed beams, crack formation in the support region depends on tendons' location and anchorage force, as concentrated loads, which will be discussed under failure type 4.

The change in tensile force ΔT is proportional to the change in moment ΔM and must be equal to the bond capacity for any section under consideration. The force diagram is illustrated in Fig. 41.

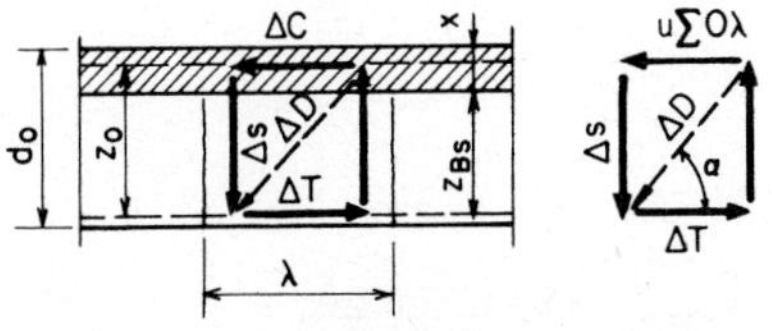

Fig. 41

$$\Delta T = \frac{\Delta M}{z_0}$$

$$\frac{\Delta T}{\Delta x} = u\Sigma O \tag{94}$$

The angle α depends mainly on the depth of the beam (d_0) and the tensile strength of the concrete. It may vary from 20 to 60°. The lever arm z_0 of internal forces can be estimated by Eqs. (79), (89), (91), or (92), and $u\Sigma O\lambda$ by Eq. (55).

The bond condition under ultimate load in conventional reinforced-concrete beams and in prestressed beams with bonded and nonbonded tendons will be considered in the following.

$L = 40.0 \text{ ft} \quad d_0 = 24 \text{ in.} \quad v = 1.8 \text{ (F.S.)}$

$\Sigma W = 1.0 \, k/\text{ft} \quad M_{max} = 200 \, k' \quad R = 20 \, k$

$M_{ck} = v M_{max} = 1.8 \times 200 = 360 \, k'$

$M_k = 4 M_{max} \xi_k \xi'_k \quad \xi = \dfrac{x}{L} \quad \xi' = 1 - \xi$

$T_k = \dfrac{M_k}{z_0} \quad \Delta T_k = T_k - T_{k-1} \quad z_0 = 1.65 \text{ ft}$

Reinforced-concrete beam:

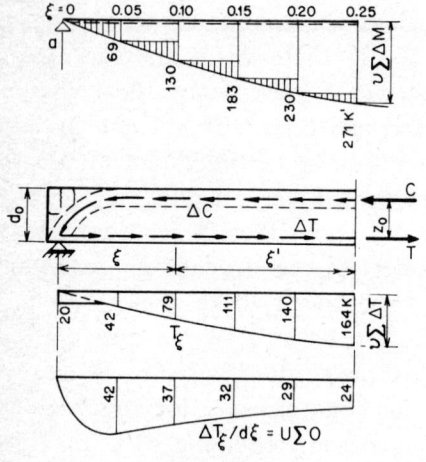

Fig. 42(a)

$A_s = 6.25$ in.²   4 - #11, $\Sigma O = 17.7$ in.
2 - #11, $\Sigma O = 8.9$ in. $c_{min} = 6$ in. (code)

$$\sum_0^{0.5} \Delta T = \frac{M_{UL}}{z_0} = \frac{360}{1.65} = 218\, k \text{ - to be balanced by bond}$$

$$T_{0.25} = \frac{271}{1.65} = 164\, k \quad T_{0.2} = \frac{230}{1.65} = 140\, k$$

$$\Delta T_{0.25} = 164 - 140 = 24\, k$$

$$u\Sigma O = \frac{\Delta T_{0.25}}{\xi L} = \frac{24}{2.0} = 12\, k/ft \quad u \simeq 56 \text{ psi}$$

$$u\Sigma O_{max} = \frac{42}{2.0} = 21\, k/ft \quad u = \frac{21{,}000}{12 \times 8.9} = 197 \text{ psi}$$

$$T_0 = vR \tan \alpha = 1.8 \times 20 = 36\, k \quad \alpha = 45°$$

$$u_0 = \frac{T_0}{c\Sigma O} = \frac{36{,}000}{6 \times 8.9} = 673 \text{ psi}$$

Prestressed beam—bonded strands:

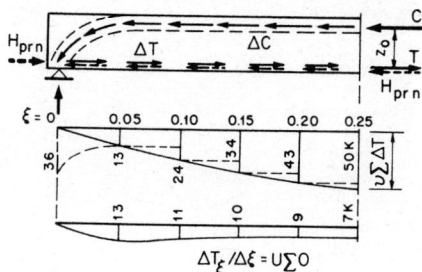

Fig. 42(b)

$H_{prn} = 151 k \quad H_{pr\,UL} = 218 k = T_{0.5}$

$6 - \frac{1}{2}" - 7$ wire strands $\Sigma O \sim 12.0$ in.

$$\sum_0^{0.5} \Delta T_{pr} = H_{pr\,UL} - H_{prn} = 218 - 151 = 67 k$$
$$\text{to be balanced by bond}$$

$\Delta T_{0.2} = 50 - 43 = 7 k, \quad u = \dfrac{7{,}000}{2.0 \times 12 \times 12.0} = 24$ psi

$u\Sigma O_{max} = \dfrac{13}{2.0} = 6.5 k/\text{ft}$

$u_{max} = \dfrac{6{,}500}{12 \times 12.0} = 45$ psi

$u_0 = \dfrac{36{,}000}{6 \times 12.0} = 500$ psi $\quad -u_{pr}$ not considered

Prestressed beam—nonbonded cables:

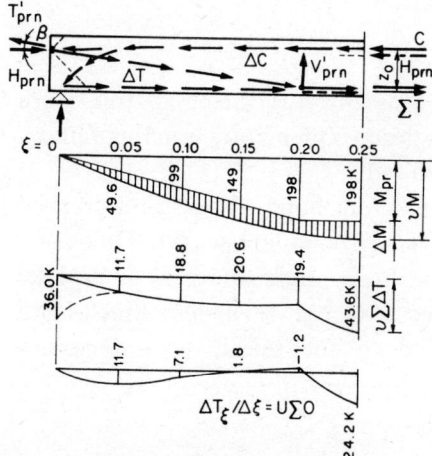

Fig. 42(c)

[See next page for equations for Fig. 42(c)]

$T'_{prn} = 151\,k$, $-H_{prn} = T_{prn} = T'_{prn}\cos\beta = 149\,k$

$V_{prn} = T'_{prn}\tan\beta = 24.8\,k$, $\beta = 9.3°$

$\Delta M_{UL} = M_\xi - V_{prn}\xi L$, $V_{prn} = 0.0$ from $\xi_{0.2}$ to $\xi = 0.5$

$\Delta M_{0.5} = M_{UL} - M_{pr,\max} = 360 - 198 = 162\,k'$

$\Delta M_{0.25} = 270 - 198 = 72\,k'$, $T_{0.25} = \dfrac{72}{1.65} = 43.6\,k$

$\Delta M_{0.20} = 230 - 198 = 32\,k'$, $T_{0.20} = \dfrac{32}{1.65} = 19.4\,k$

$\displaystyle\sum_0^{0.5}\Delta T = \dfrac{162}{1.65} \simeq 98\,k$ to be balanced by bond

$A_{s,\min} = \dfrac{T_{0.5}}{\sigma_{sy}} = \dfrac{98}{48} = 2.04$ in.^{2}

$3 - {}^\#8$, $\Sigma O = 9.4$ in. $2 - {}^\#8$, $\Sigma O = 6.3$ in.

$\Delta T_{0.25} = 43.6 - 19.4 = 24.2\,k$

$u\Sigma O = \dfrac{24.2}{2.0} = 12.1\,k/\text{ft}$, $u = \dfrac{12{,}100}{12\times 9.4} = 107$ psi

$\Delta T_{0.05} = 11.7 - 0.0 = 11.7\,k$

$u\Sigma O = \dfrac{11.7}{2.0} = 5.85\,k/\text{ft}$, $u = \dfrac{5{,}850}{12\times 6.3} \simeq 78$ psi

$T_0 = vR - V_{pr} = 36 - 24.8 = 11.2\,k$

$u_0 = \dfrac{11{,}200}{6\times 6.3} = 297$ psi, $(-u_{pr} \sim 0)$

As a conclusion from this rough analysis (Fig. 42), the bond stresses are critical at the end suppports and may lead to premature bonding failure, especially if the bonding length c is limited.

In accordance with Eq. (55) the maximum bond stresses in a cracked section can be twice as high as the average bond stresses (u). Therefore, to avoid slippage failures, the average bond stress under ultimate load condition should not be more than 500 psi. If higher stresses are unavoidable, stirrups must be welded to the tensile bars to secure equilibrium (Fig. 41).

Failure under Concentrated Loads (Type 4)

Failure type 4 is the most complicated to analyze because the stresses must be computed by Airy's stress function. There are four basic types of concentrated load acting upon the beam (Fig. 43). For mathematical analysis, to estimate the σ_x, σ_z, and τ_{xz} stresses, the concentrated load acting under an angle on the beam will be divided into its components P_x and P_z. The final stresses are obtained by superposition.

The equilibrium requires that

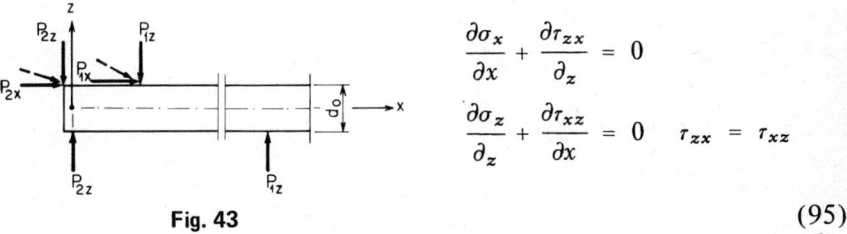

Fig. 43

$$\frac{\partial \sigma_x}{\partial x} + \frac{\partial \tau_{zx}}{\partial z} = 0$$

$$\frac{\partial \sigma_z}{\partial z} + \frac{\partial \tau_{xz}}{\partial x} = 0 \qquad \tau_{zx} = \tau_{xz} \tag{95}$$

Expressing the stresses by Airy's function F,

$$\sigma_x = \frac{\partial^2 F}{\partial z^2} \qquad \sigma_z = \frac{\partial^2 F}{\partial x^2} \qquad \tau_{xz} = \frac{\partial^2 F}{\partial x \partial z} \tag{96}$$

and introducing these into Eqs. (95), we obtain

$$\frac{\partial^4 F}{\partial x^4} + 2\frac{\partial^4 F}{\partial x^2 \partial z^2} + \frac{\partial^4 F}{\partial z^4} = 0 \tag{97}$$

In polar coordinates, the equation is

$$\left(\frac{\partial^2}{\partial r^2} + \frac{1}{r}\frac{\partial}{\partial r} + \frac{1}{r^2}\frac{\partial^2}{\partial \alpha^2}\right)^2 F = 0 \tag{98}$$

and the stresses are

$$\sigma_r = \frac{1}{r}\frac{\partial F}{\partial r} + \frac{1}{r^2}\frac{\partial^2 F}{\partial \alpha^2} \qquad \sigma_t = \frac{\partial^2 F}{\partial r^2}$$

$$\tau_{rt} = -\frac{\partial}{\partial r}\left(\frac{1}{r}\frac{\partial F}{\partial \alpha}\right) \tag{99}$$

The function F (Airy's function) must satisfy the partial differential equation (97) or (98) and the boundary stresses σ_x, σ_z, τ_{xz} or σ_r, σ_t, τ_{rt} caused by the boundary forces x, z. To determine the function F for this condition, the equilibrium of the forces and stresses on the boundary r will be considered (Fig. 44).

The equilibrium requires that

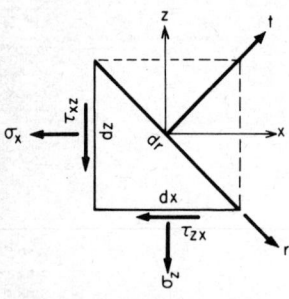

Fig. 44

$$\sigma_x\,dz + \tau_{zx}\,dx = X\,dr$$
$$\sigma_z\,dx + \tau_{xz}\,dz = Z\,dr$$

Expressing the stresses in F [Eqs. (96)], we obtain

$$\frac{\partial^2 F}{\partial z^2}\,dz + \frac{\partial^2 F}{\partial x\,\partial z}\,dx = X\,dr$$
$$\frac{\partial^2 F}{\partial x^2}\,dx + \frac{\partial^2 F}{\partial x\,\partial z}\,dz = Z\,dr$$
(100)

and by integration

$$\frac{\partial F}{\partial z} = \int X\,dr = P_x$$

$$\frac{\partial F}{\partial x} = \int Z\,dr = P_z \quad \text{Absolute values}$$
(101)

Thus:
$$F = \int (P_x\,dz + P_z\,dx)$$

The integration constants in these equations have been left out because by choice of integration in the direction of r they are of higher order and do not influence the stresses appreciably.

Case 1. $P_z = 1.0$, $P_x = 0$ (Fig. 45):
Boundary condition, $z = 0$: $\sigma_z = 0$, $\tau_{xz} = 0$

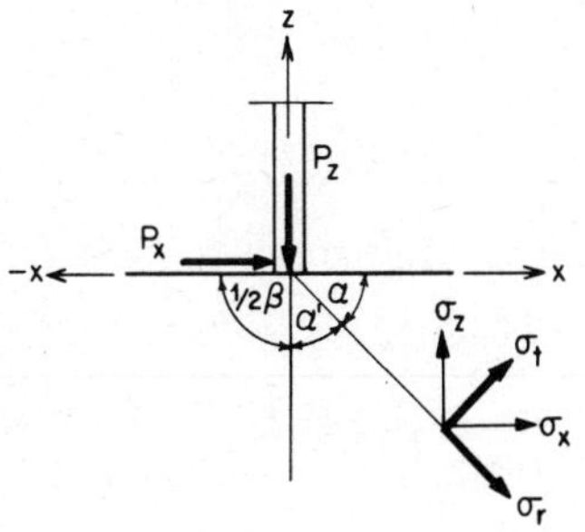

$$F = \frac{P_z}{\pi} r\alpha \cos\alpha$$

$$\sigma_r = \frac{1}{b}\left(\frac{1}{r}\frac{\partial F}{\partial r} + \frac{1}{r^2}\frac{\partial^2 F}{\partial \alpha^2}\right)$$

$$= -\frac{2P_z}{\pi b}\frac{\sin\alpha}{r}$$

$$\sigma_t = \tau_{rt} = 0$$

Fig. 45

$$\cos\alpha = \sin\alpha' \quad \alpha + \alpha' = 90°$$

$$\sigma_x = \sigma_r \cos^2\alpha = -\frac{2P_z}{\pi b}\frac{\sin\alpha \cos^2\alpha}{r} = -\frac{2P_z}{\pi b}\frac{zx^2}{(x^2+z^2)^2} \quad (102a)$$

$$\sigma_z = \sigma_r \sin^2\alpha = -\frac{2P_z}{\pi b}\frac{\sin^3\alpha}{r} = -\frac{2P_z}{\pi b}\frac{z^3}{(x^2+z^2)^2}$$

$$\tau_{xz} = -\sigma_r \sin\alpha \cos\alpha = \frac{2P_z}{\pi b}\frac{\sin^2\alpha \cos\alpha}{r} = \frac{2P_z}{\pi b}\frac{z^2 x}{(x^2+z^2)^2}$$

For appreciably large contact area A_L, the P_z value can be sub-divided into $dP = p_0\, dx$. For this case (Fig. 46),

$$\sigma_x = \frac{p_0}{2\pi}[2(\alpha'_2 - \alpha'_1) - \sin 2\alpha'_2 + \sin 2\alpha'_1]$$

$$\sigma_z = \frac{p_0}{2\pi}[2(\alpha'_2 - \alpha'_1) + \sin 2\alpha'_2 - \sin 2\alpha'_1] \quad (102b)$$

$$\tau_{xz} = \frac{p_0}{2\pi}(\cos 2\alpha'_2 - \cos 2\alpha'_1)$$

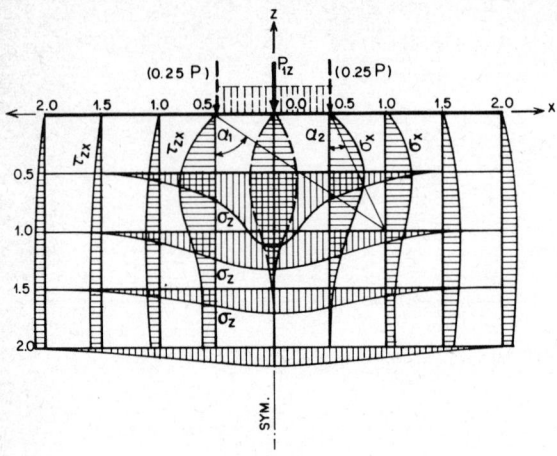

Fig. 46

Case 2. $P_z = 0$, $P_x = 1.0$:

$$F = -\frac{P_x}{\pi} r\alpha \sin\alpha \qquad \sigma_t = \tau_{rt} = 0$$

$$\sigma_r = -\frac{2P_x}{\pi b} \frac{\cos\alpha}{r}$$

$$\sigma_x = -\frac{2P_x}{\pi b} \frac{\cos\alpha \cos^2\alpha}{r} = -\frac{2P_x}{\pi b} \frac{x^3}{(x^2 + z^2)^2}$$

$$\sigma_z = -\frac{2P_x}{\pi b} \frac{\cos\alpha \sin^2\alpha}{r} = -\frac{2P_x}{\pi b} \frac{xz^2}{(x^2 + z^2)^2} \qquad (103)$$

$$\tau_{xz} = \frac{2P_x}{\pi b} \frac{\cos^2\alpha \sin\alpha}{r} = \frac{2P_x}{\pi b} \frac{x^2 z}{(x^2 + z^2)^2}$$

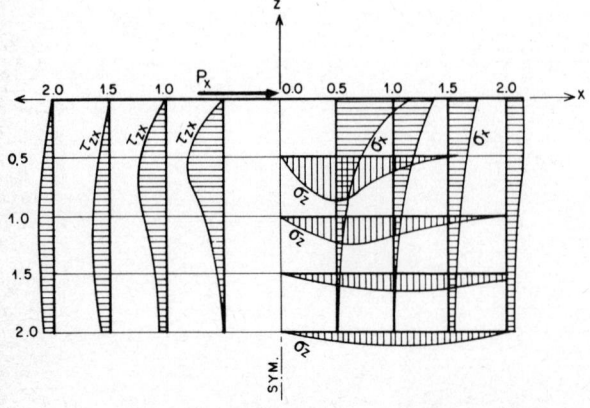

Fig. 47

If the boundary slopes symmetrically from the point of load application and has a central angle β, the Eqs. (102a) and (103) remain the same, except that $\pi \to \beta \pm \sin\beta$.

The stresses computed for point loads P_z and P_x are plotted in Figs. 46 and 47.

Case 3. $P_x = 1.0$, $P_z = 0$:

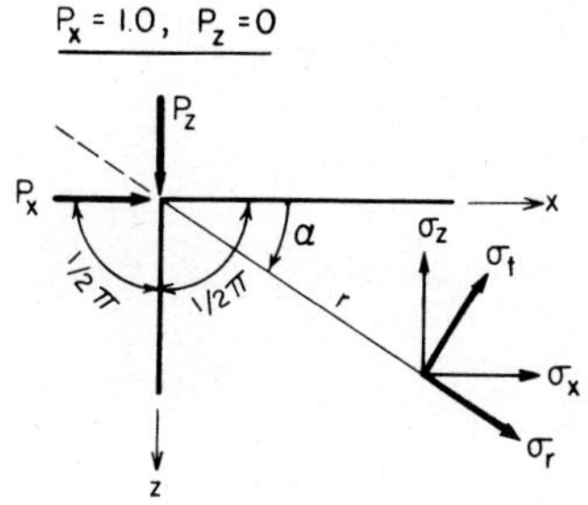

Fig. 48

$$F = -\frac{P_x}{\pi^2/4 - 1} r\alpha\left(\frac{\pi}{2}\sin\alpha + \cos\alpha\right)$$

$$\sigma_t = \tau_{rt} = 0$$

$$\sigma_r = -\frac{2P_x}{(\pi^2/4 - 1)b} \frac{(\pi/2)\cos\alpha - \sin\alpha}{r}$$

$$\sigma_x = -\frac{2P_x}{(\pi^2/4 - 1)b} \frac{(\pi/2)\cos\alpha - \sin\alpha}{r} \cos^2\alpha$$

$$= -\frac{2P_x}{(\pi^2/4 - 1)b} \frac{[(\pi/2)x - z]x^2}{(x^2 + z^2)^2}$$

$$\sigma_z = -\frac{2P_x}{(\pi^2/4 - 1)b} \frac{(\pi/2)\cos\alpha - \sin\alpha}{r} \sin^2\alpha \qquad (104)$$

$$= -\frac{2P_x}{(\pi^2/4 - 1)b} \frac{[(\pi/2)x - z]z^2}{(x^2 + z^2)^2}$$

$$\tau_{xz} = \frac{2P_x}{(\pi^2/4 - 1)b} \frac{(\pi/2)\cos\alpha - \sin\alpha}{r} \cdot \sin\alpha \cos\alpha$$

$$= \frac{2P_x}{(\pi^2/4 - 1)b} \frac{[(\pi/2)x - z]zx}{(x^2 + z^2)^2}$$

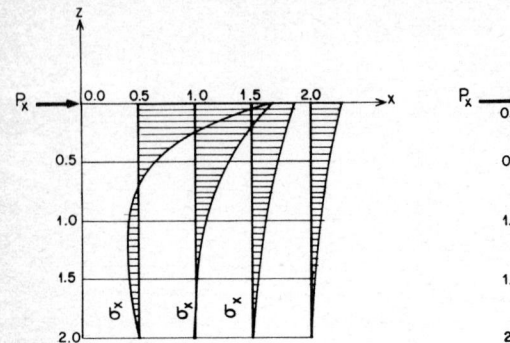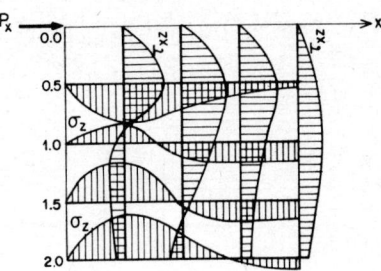

Fig. 49

Case 4. $P_x = 0$, $P_z = 1.0$:

$$F = -\frac{P_z}{\pi^2/4 - 1} r\alpha \left(\sin\alpha + \frac{\pi}{2}\cos\alpha\right)$$

$$\sigma_t = \tau_{rt} = 0$$

$$\sigma_r = -\frac{2P_z}{(\pi^2/4 - 1)b} \frac{\cos\alpha - (\pi/2)\sin\alpha}{r}$$

$$\sigma_x = -\frac{2P_z}{(\pi^2/4 - 1)b} \frac{\cos\alpha - (\pi/2)\sin\alpha}{r} \cos^2\alpha$$

$$= -\frac{2P_z}{(\pi^2/4 - 1)b} \frac{[x - (\pi/2)z]x^2}{(x^2 + z^2)^2} \qquad (105)$$

$$\sigma_z = -\frac{2P_z}{(\pi^2/4 - 1)b} \frac{\cos\alpha - (\pi/2)\sin\alpha}{r} \sin^2\alpha$$

$$= -\frac{2P_z}{(\pi^2/4 - 1)b} \frac{[x - (\pi/2)z]z^2}{(x^2 + z^2)^2}$$

$$\tau_{xz} = \frac{2P_z}{(\pi^2/4 - 1)b} \frac{\cos\alpha - (\pi/2)\sin\alpha}{r} \sin\alpha\cos\alpha$$

$$= \frac{2P_z}{(\pi^2/4 - 1)b} \frac{[x - (\pi/2)z]zx}{(x^2 + z^2)^2}$$

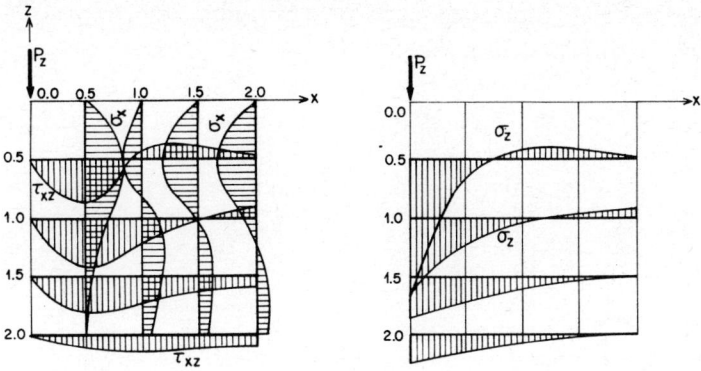

Fig. 50

The point-load equations give rather accurate stresses also for relatively small contact areas. However, within a short distance from the load application they are slightly higher than as computed by Eq. (102b).

The maximum allowable local stress $\sigma_L{}^*$ beneath the load can be assumed to be

$$\sigma_L = 0.85 f'_c \sqrt{\frac{A_L}{A_c}} \quad -\sigma_L = p_0 \tag{106a}$$

where A_L is the contact area of the load and A_c is the symmetrically surrounding concrete area. It is recommended that, for practical design purposes, the ratio A_L/A_c should not be taken to be more than 3.0, assuming that adequate reinforcing is provided. It must be kept in mind that the stresses in transversal direction are

$$\sigma_y = \nu(\sigma_x + \sigma_z) \tag{106b}$$

where ν is the Poisson's ratio ($\nu = 0.167$).

11. DEFORMATIONS

Assuming that the cross-sectional planes remain plane during bending due to external loads and temperature changes, then the planes can only

---

*K. H. Middendorf, Practical Aspects of End Zone Bearing of Post-tensioning Tendons, *PCI J.*, vol. 8, 1963.

displace parallel to each other and undergo a rotation with respect to their neutral axes, as illustrated in Fig. 51.

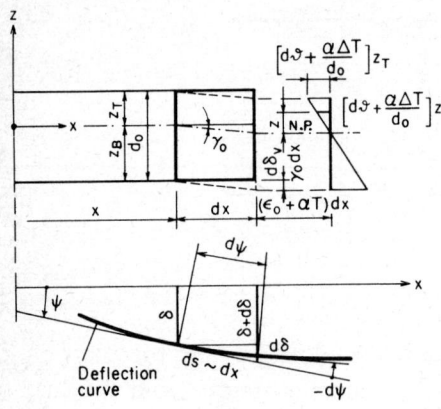

Fig. 51

$$\epsilon \, dx = \left(\epsilon_0 + \frac{d\vartheta}{dx} z\right) dx + \left(\alpha T + \frac{\alpha \Delta T}{d_0}\right) dx$$

$$\epsilon = \frac{\sigma}{E} + \alpha T = \frac{1}{E}\left(\frac{N}{A} \mp \frac{M}{I} z\right) + \alpha T \tag{107}$$

$$\gamma_0 = \frac{\kappa V}{GA} \qquad G = \frac{E}{2(1-\nu^2)} \qquad -d\psi = \frac{ds}{r} \sim \frac{dx}{r}$$

Introducing the ϵ value, expressed in terms of external forces and temperature change, into Eqs. (107) and considering that the uniform temperature change αT and the normal force N acting at neutral plane do not influence the rotation of the cross-sectional plane, we obtain

$$\frac{d\vartheta}{dx} = -\frac{d\psi}{dx} = \frac{M}{EI} + \frac{\alpha \Delta T}{d_0} - \frac{\kappa \, dV}{GA \, dx} \tag{108}$$

Expressing the moment of inertia I in relative terms (I_c/I), the differential equation of the deflection curve is

$$-EI_c \frac{d^2\delta}{dx} = \frac{I_c}{I} M + EI_c \left(\frac{\alpha \Delta T}{d_0} - \frac{\kappa dV}{GA\, dx} \right) \quad (109)$$

The ordinates of the deflection curve δ are obtained by double integration. Commonly, the influence of shear V is relatively small in comparison with deflection due to the external moment M and therefore is disregarded. Thus

$$-EI_c \delta = \iint M \frac{I_c}{I} dx\, dx + EI_c \iint \frac{\alpha \Delta T}{d_0} dx\, dx \quad (110)$$

The interpretation of Eq. (110) leads to the following. Applying $M(I_c/I)$ (moment) as external loading and computing the reactions and moments, the reactions will give the end rotation $\psi_{a,b}$ of the beam and $\bar{M}$ will give the deflections δ.

However, the $\psi_{a,b}$ and δ values can be obtained most conveniently by the method of virtual work. In accordance with this method, a so-called *unit dummy load* is applied, acting at the point where the deflection is desired [Eq. (8)]. The dummy load $P_k = I_k$ causes a moment $\bar{M}_k$ and shear $\bar{V}_k$. Denoting the moment due to the external loading by M_0 and the shear by V_0, we obtain

$$\begin{aligned} EI_c \delta_0 &= \int M_0 \bar{M}_k \frac{I_c}{I(x)} dx + EI_c \int \frac{\alpha \Delta T}{d_0} \bar{M}_k\, dx \\ &\quad + \frac{EI_c}{G} \kappa \int V_0 \bar{V}_k \frac{1}{A} dx \end{aligned} \quad (111)$$

If $M_0, \bar{M}_k, V_0, \bar{V}_k$, and $I(x)$ are steady functions of x, direct integration over the entire span l is possible. If this is not the case, numerical integration (Simpson's rule) has to be used. Denoting the quantities $M_0 \cdot \bar{M}_k \cdot I_c / I(x) = \eta$ thus,

$$EI_c \delta = \frac{\Delta x}{3} (\eta_0 + 4\eta_1 + 2\eta_2 + 4\eta_3 + \cdots + 2\eta_{n-2} + 4\eta_{n-1} + \eta_n) \quad (112)$$

For continuous beams, the deflection can be computed by similar means. However, direct integration is time-requiring and rather complicated because of the shape of the moment areas $M_0^{(n)}$. This difficulty can be overcome if one considers that in an n-times indeterminate system, in accordance with Maxwell,

$$1_k^{(n)} \delta_{k0}^{(n)} = 1_k^{(n)} \delta_{k0}^{(0)} \tag{113}$$

We can write

$$EI_c \delta_k^{(n)} = \sum \int M_0^{(0)} \overline{M}_k^{(n)} \frac{I_c}{I(x)} dx + EI_c \sum \int \frac{\alpha \Delta T}{d_0} \overline{M}_k^{(n)} dx$$
$$+ EI_c \sum \frac{\kappa}{G} \int V_0^{(0)} \overline{V}_k^{(n)} dx \tag{114}$$

Equation (114) means that the moment $M_0^{(0)}$ and shear $V_0^{(0)}$ due to loading as well as to unequal temperature change $\alpha \Delta T / d_0$ can be computed in a principal system and $\overline{M}_k^{(n)}$ and $\overline{V}_k^{(n)}$ in an n-times indeterminate system. Using Table 2 (influence-line envelopes) to obtain dummy support and span moment due to $1_k^{(n)}$ at the span, the $\delta_k^{(n)}$ is required, and applying the law of the support-moment X_n distribution (Fig. 7), direct integration becomes rather simple.

The only case in which the M_0 moments have to be computed in an indeterminate system are the moments due to support settlements:

$$EI_c \delta_{ks}^{(n)} = -EI_c \sum \int M_{0s}^{(n)} \overline{M}_k^{(n)} dx \tag{115}$$

The $M_{0s}^{(n)}$ moments and the dummy-load moments $\overline{M}_k^{(n)}$ are always simple functions of x, and direct integration is convenient. The redundants X_{ns} due to support settlement are (Eqs. 8)

$$X_{ns} = -\frac{\sum R_{nn}^{(0)} \Delta n}{\delta_{nn}} \tag{116}$$

where $R_{nn}^{(0)}$ are the reactions due to the dummy load in the principal system at the support (n) where the settlement Δn occurs.

The integrals for most common loading conditions are given in Table 3. Using superposition, almost all commonly necessary integrals can be

Table 3

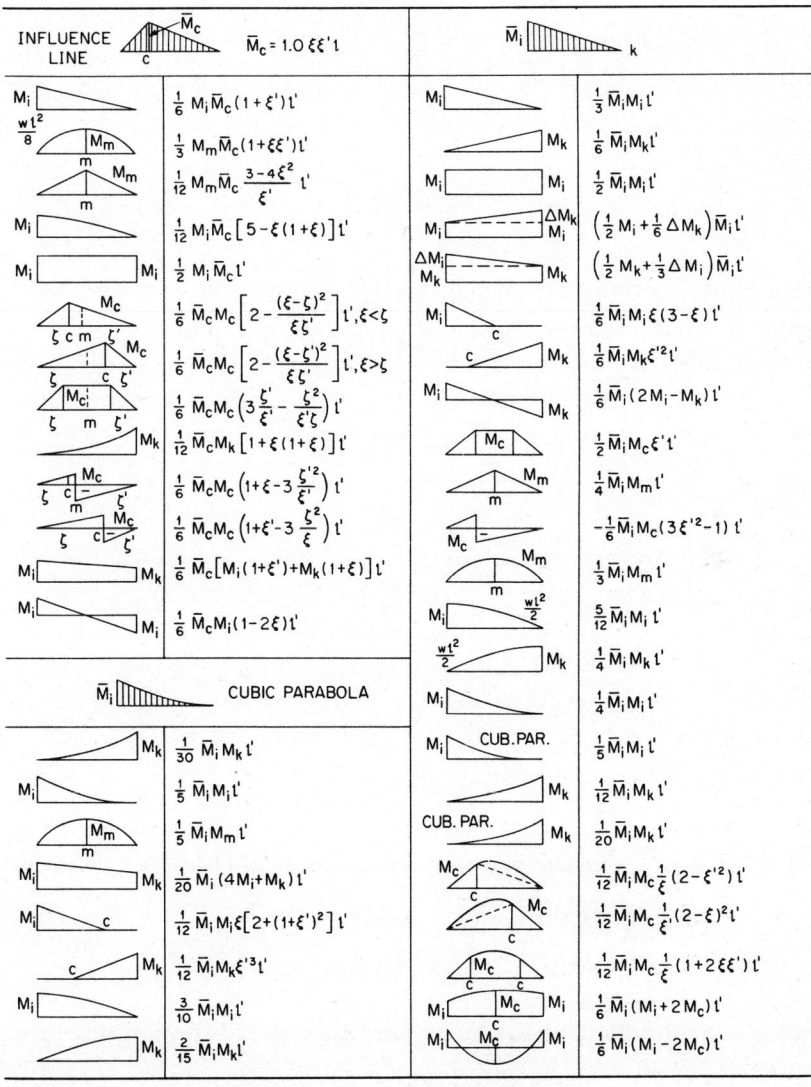

solved by the given integrals. If $I(x)$ is not a constant but a function of x, the area of $I_c \bar{M}/I(x)$ will be similar to one or a combination of two moment areas given in Table 3. The solution of the integral consisting of three variables can be obtained rather simply in this manner. For more complex integrals, numerical integration must be used [Eq. (112)].

Beams on Elastic Foundation

The differential equation for deflection of a beam with constant depth is

$$\frac{EI}{b}\frac{d^4\delta}{dx^4} + k\delta = w(x) \qquad (117)$$

where k is the modulus of subgrade reaction, $w(x)$ is the load the beam is subjected to, and $\bar{p} = k\delta$ is the soil pressure.

The solution of this differential equation consists of a particular solution (δ_0), representing the deflection of a simply supported beam resting on elastic foundation, and a homogeneous solution

$$\frac{EI}{bk}\frac{d^4\delta_1}{dx^4} + \delta_1 = 0$$

or denoting

$$L^4 = \frac{4EI}{bk} \qquad L = \sqrt[4]{\frac{4EI}{bk}} \qquad \xi = \frac{x}{L}$$

$$\frac{d^4\delta_1}{d\xi^4} + 4\delta_1 = 0$$

This homogeneous equation has four roots, and its solution is

$$\delta_1 = [C_1 \cos\xi \cosh\xi + C_2 \cos\xi \sinh\xi \\ + C_3 \sin\xi \cosh\xi + C_4 \sin\xi \sinh\xi] \qquad (118)$$

Thus the final deflection is obtained by the superposition of these two solutions:

$$\delta = \delta_0 + [C_1 \cos\xi \cosh\xi + C_2 \cos\xi \sinh\xi \\ + C_3 \sin\xi \cosh\xi + C_4 \sin\xi \sinh\xi]$$

and the moment and shear are obtained by differentiation:

$$M = -EI\left(\frac{d^2\delta_0}{d\xi^2} - \frac{d^2\delta_1}{d\xi^2}\right)$$

$$V = -EI\left(\frac{d^3\delta_0}{d\xi^3} - \frac{d^3\delta_1}{d\xi^3}\right) \tag{119}$$

For an infinitely long beam subjected to a concentrated load P_0, the integration constants can be determined from the boundary condition $\xi = \infty: \delta = 0, d\delta/d\xi = 0$ and $\xi = 0: d\delta/d\xi = 0, -V = \frac{1}{2}P_0$. Expressing the hyperbolic functions by exponential ones,

$$\cosh\xi = \frac{1}{2}(e^\xi + e^{-\xi})$$
$$\sinh\xi = \frac{1}{2}(e^\xi - e^{-\xi})$$

the deflection, moment, and shear are, for an infinitely long beam, P_0 at $\xi = 0$ (Fig. 52):

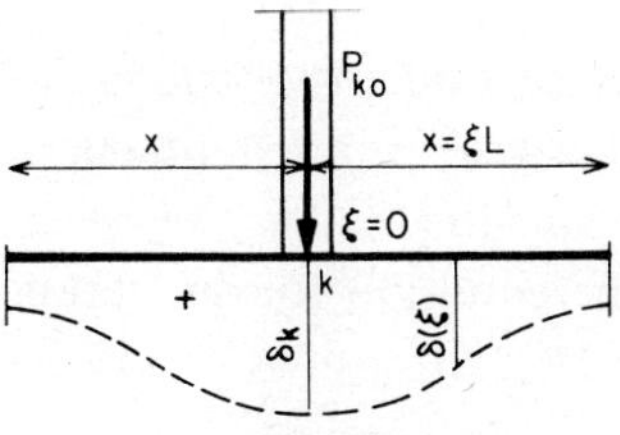

Fig. 52

$$\delta_\xi = \frac{P_0}{2Lbk}e^{-\xi}(\cos\xi + \sin\xi)$$

$$M_\xi = \frac{P_0 L}{4}e^{-\xi}(\cos\xi - \sin\xi) \tag{120}$$

$$V_\xi = -\frac{P_0}{2}e^{-\xi}\cos\xi$$

and for semi-infinitely long beam, P_0 at $\xi = 0$ (Fig. 53):

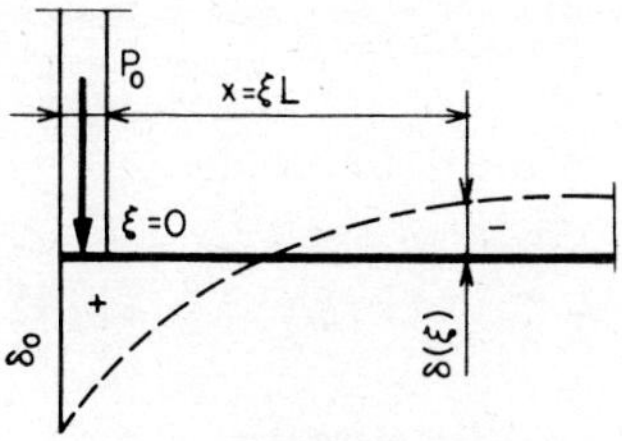

Fig. 53

$$\delta_\xi = \frac{2P_0}{Lbk}e^{-\xi}\cos\xi$$

$$M_\xi = -P_0 L e^{-\xi}\sin\xi \tag{121}$$

$$V_\xi = -P_0 e^{-\xi}(\cos\xi - \sin\xi)$$

M_0 at $\xi = 0$ (Fig. 54):

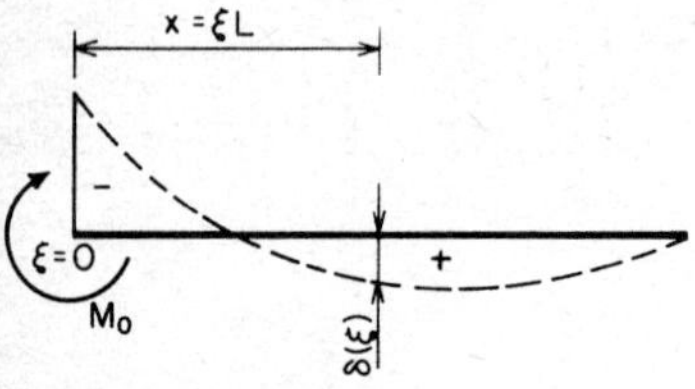

Fig. 54

$$\delta_\xi = -\frac{2M_0}{L^2 bk} e^{-\xi}(\cos\xi - \sin\xi)$$

$$M_\xi = M_0 e^{-\xi}(\cos\xi + \sin\xi) \qquad (122)$$

$$V_\xi = -\frac{2M_0}{L} e^{-\xi} \sin\xi$$

In practice, there are no infinitely long beams; therefore, in a finitely long beam we obtain moments M_E and shear V_E at the beam ends. To make them zero in order to satisfy the boundary conditions, the end moments and shear computed by Eqs. (120) will be applied as external forces with countersigns in Eqs. (121) and (122). The final values of $\Sigma\delta$, ΣM, and ΣV will be obtained by superposition. The boundary condition commonly will be satisfied by one operation only.

For an infinitely large slab acted upon by a concentrated load, the deflections, soil pressure, moments, and shear can be roughly estimated using the same analogy.

Assuming that the deflection curve δ is affined and proportional to that of the infinitely long beam and that the deflections are damping in accordance with harmonic vibrations, for this case the distance between zero point of the deflection curve is πL and the equation for deflection curve becomes

$$\delta = \frac{4P_0}{\pi^3 L^2 k} e^{-\xi}(\cos\xi + \sin\xi) \qquad (123)$$

$$\xi = \frac{r}{L} \qquad L = \sqrt[4]{\frac{D}{k}} \qquad D = \frac{E d_0^3}{12(1-\nu^2)}$$

The soil pressure at any point is proportional to the deflection

$$\bar{p}_\xi = k\delta_\xi \qquad (124)$$

The corresponding moments in the radial and tangential directions are

$$M_r = -D\left(\frac{d^2\delta}{d\xi^2} + \nu\frac{1}{r}\frac{d\delta}{d\xi}\right)$$

$$M_a = -D\left(\nu\frac{d^2\delta}{d\xi^2} + \frac{1}{r}\frac{d\delta}{d\xi}\right) \tag{125}$$

$$V_{rz} = \frac{dM_r}{dr} + \frac{M_r - M_a}{r}$$

The deflections and moments decrease from $\xi = 0$ rather rapidly for relatively small D and high k values.

To illustrate this method, the deflections δ, soil pressure $\bar{p}$, and moments M in a beam on elastic soil will be computed (Fig. 55).

Assuming:

$E_0 = 4.5 \times 10^6$ psi
$I_0 = 1.52 \times 10^5$ in.^{4} $b = 132$ in. $d_0 = 24$ in.
$k_1 = 180$ p/in.^{3} $k_2 = 350$ p/in.^{3} $k_3 = 590$ p/in.^{3}

Characteristic length:

$$L = \sqrt[4]{\frac{4EI}{bk}} \quad L_1 = 9.0 \text{ ft}, \quad L_2 = 7.5 \text{ ft} \quad L_3 = 6.4 \text{ ft}$$

Using Eqs. (120), the deflections, moments, and shear due to $P_{1...n}$ are computed. Then, making use of superposition, the results for an infinitely long beam are obtained. At the end, however, the boundary conditions are violated because shear V_0 and moment M_0 have definite values and are not zero. By introducing the V_0 and M_0 with opposite signs into Eqs. (121) and (122) and adding these results to the values of the infinitely long beam, the final results are obtained. These results are plotted in Fig. 55.

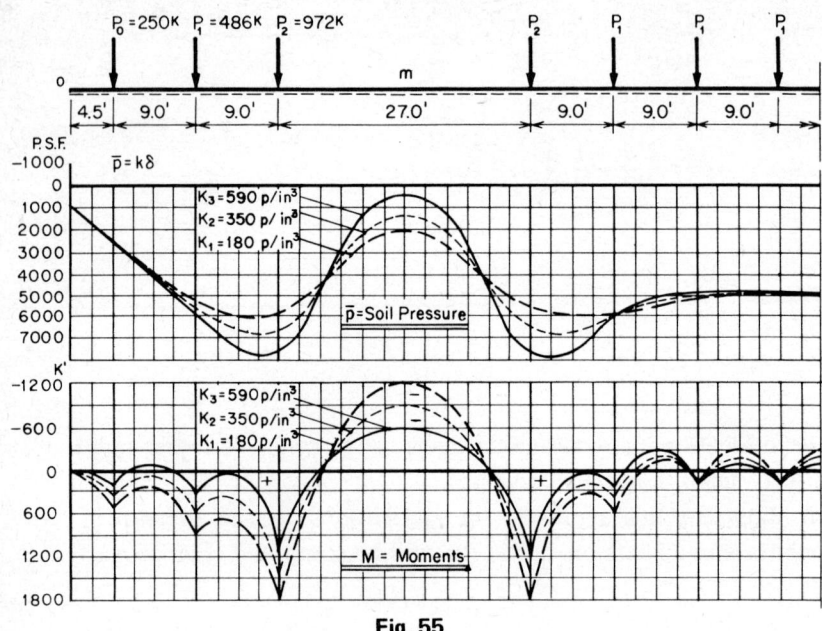

Fig. 55

Time Deflection Due to Plastic Flow and Shrinkage

Plastic flow and shrinkage cause a change in deflection and a shortening of the beam. Defects in a structural system can result if both phenomena are not considered in the design. In deriving equations necessary for estimating the magnitude of change in deformation, let us consider a prestressed cross section acted upon by prestressing force H_{pr} and dead-load moment M_D (Fig. 56).

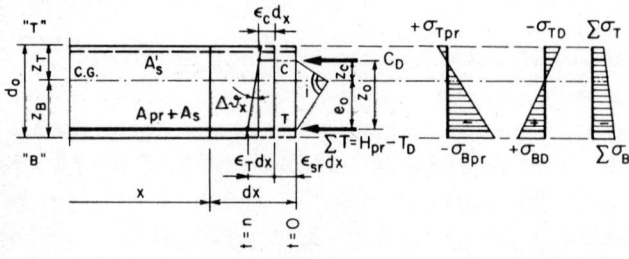

Fig. 56

$$z_0 = e_0 + z_c, \quad z_c = \frac{i^2}{e_0} = \frac{\Sigma I_0}{e_0 \Sigma A} \quad i = \sqrt{\frac{I}{A}}$$

$$= e_0 + \frac{i^2}{e_0}$$

$$\Delta \vartheta_x = \frac{(\epsilon_r - \epsilon_c) dx}{z_0} \tag{126}$$

$$\epsilon_T = \frac{\Delta H_{pr\,x}}{E_{pr} A_{pr}} = \frac{\Delta M_{pr}}{n E_0 \, e_0 \, A_{pr}} \qquad n = \frac{E_{pr}}{E_0}$$

$$\epsilon_c = \frac{M_{Dx}}{\Sigma I_0 E_0} z_c \, \varphi'_n \qquad \varphi'_n = \frac{1 - \alpha_s}{\alpha_s} (1 - e^{-\alpha_s \varphi_n})$$

$$\epsilon_{pr\,c} = 0$$

Substituting the ϵ values into Eqs. (126), we obtain

$$\frac{\Delta \vartheta_x}{dx} = -\frac{\Delta M_x}{E_0 \Sigma I_0} = \frac{1}{z_0 E_0} \left(\frac{\Delta M_{pr\,x}}{n e_0 A_{pr}} - \frac{M_{Dx}}{\Sigma I_0} z_c \, \varphi'_n \right)$$

$$\Delta M_x = -\frac{\Sigma I_0}{z_0} \left(\frac{\Delta M_{pr\,x}}{n e_0 A_{pr}} \frac{\Sigma I_0}{z_c} - M_{Dx} \varphi'_n \right) \tag{127}$$

$$= -\frac{i^2}{i^2 + e_0^2} \left(\frac{\Delta M_{pr}}{n A_{pr}/\Sigma A} - M_{Dx} \varphi'_n \right)$$

The loss $\Delta M_{pr\,x} = \Delta H_{pr\,x} e_0$ in prestressing force due to plastic flow and shrinkage is given by Eq. (80). For curved tendons, in accordance with Leonhardt, ΔM_x follows a parabola very closely. Thus

$$\overline{\Delta M}_{\max} = -\frac{i^2}{i^2 + e_0^2} \left(\frac{\Delta M_{pr\,n}}{n A_{pr}/\Sigma A} - M_{D,\,\max} \varphi'_n \right) \tag{128}$$

The maximum deflection, in accordance with Eq. (111) or (114), is

$$\Delta \delta_{\max} = \frac{1}{EI_c} \int \overline{\Delta M}_{\max} \overline{M}_1 dx \tag{129}$$

In the case of straight tendons, the shape of ΔM_{pr} is no longer a parabola. The loss in prestress is maximum at the beam ends and minimum at midspan, where the prestressing stresses are reduced by dead load. The deflection $\Delta \delta_x$ can most conveniently be obtained by numerical integration (Eq. 112).

The shortening of the beam centerline is, for $t = 0$,

$$\Delta l_0 = \frac{\sigma_{0\,pr}}{E_0} l \tag{130}$$

and for $t = n$,

$$\Delta l_1 = \Delta l_0 (1 + \varphi'_n) + \left(\epsilon_{sr} - \frac{\Delta \sigma_{pr}}{E_0} \right) l \tag{131}$$

Equations (128) and (129) are also valid for reinforced concrete when ΔM_{pr} is set zero and $e_0 \to z_{sB}$ is the distance of the tensional reinforcement from the neutral plane.

3
ARCH ACTION

Arch action is the equilibrium between the external loads the arch is subjected to and the internal compressive stresses in the arch, which are uniformly distributed over the cross-sectional area from the crown to the springing of the arch. Such an equilibrium condition is possible when the centerline of the arch coincides with the external-load "force polygon" (line of pressure), as shown in Fig. 57a.

True arch action is accomplished by one type of stress—compressive stress. Due to this fact, and because the stresses are uniformly distributed over the entire cross-sectional area, the arch action is one of the most economical of the carrying actions.

Arch action is best demonstrated graphically, as in Fig. 57. The load to be carried by a symmetrical arch is divided into 10 divisions: $\Delta x =$ one-tenth of the span L. These loads $w_1, w_2, \ldots, w_{10}$, acting at the center of gravity of the division, are plotted as a force diagram (Fig. 57b). Other forces acting on the arch are the horizontal thrust H at the crown and normal forces N at the springings. The shear V_c at the crown is zero

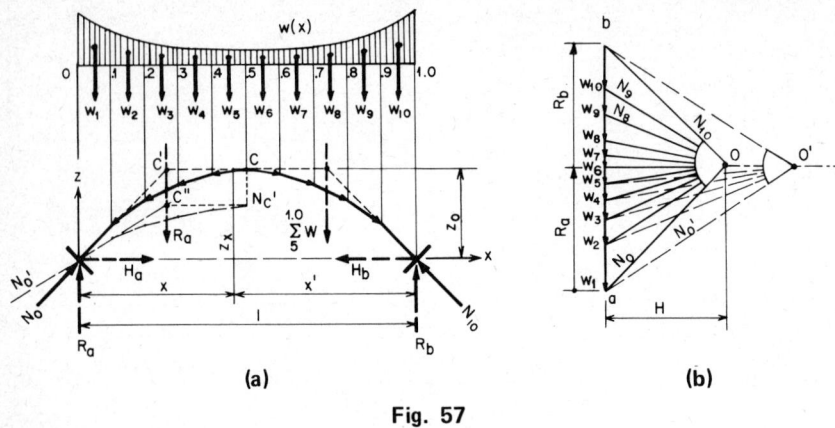

Fig. 57

because of symmetry. The centerline of the arch is determined graphically by choosing an arbitrary pole o' on the horizontal line drawn from the end of $R_a = \sum_{1}^{5} w$ connecting it with a and b (Fig. 57b). Starting at the centerline of the springing, a parallel is drawn to N'_o until it intersects the load w_1. From this intersection, a parallel is drawn to N'_1 until it intersects the load w_2, and so on until the final ray intersects the crown horizontally. The point of intersection c'' of the end lines N'_o and N'_6 of the polygon locates the resultant R_A of the loads $w_1 \cdots w_5$. The intersection of R_A and the horizontal line through center of crown c locates the point c'. The line parallel to $c'A$ through a locates the correct pole 0 and determines the magnitude of the thrust H. The second polygon, with the correct H and carried out in the same manner as the first polygon, which determines the location of the resultant R_A, is the centerline of arch. The rays N_o to N_{10} are the normal forces in the arch and act at center of the section.

For this centerline of the arch, the equilibrium condition requires that

$$Rx - \sum_{1}^{k} w_k(x - a_k) - Hz_x = 0 \tag{132}$$

for any section of the arch. If this condition is not satisfied, it means that the centerline and pressure line of the arch deviate from each other. This may be caused by differences in actual and assumed loading, by elastic and plastic shortening of the arch, by settlement of supports, and finally by temperature changes. In such a case, the true arch action is lost and equilibrium is established by beam action. The beam action in an arch is represented by the moment

$$M_x = Rx - \sum_1^k w_k(x - a_k) - Hz_k \qquad (133)$$

The deviation of the line of pressure from the arch centerline is

$$e_x = \frac{M_x}{H} \qquad (134)$$

As long as e_x remains within the limits of the kern $\pm d_0/6$ from the centerline, no tensile stresses will occur in the section.

The compressive and tensile stresses in the arch are

$$f_x = -\frac{N_x}{A_x} \mp \frac{M_x}{S_x} \qquad (135)$$

where A_x is the cross-sectional area, N_x is the normal force, and $S_x = I_x/z$, the section modulus of the arch (at location x).

This graphical representation was performed to demonstrate the nature of arch action. The design of contemporary reinforced-concrete arches must be done analytically because the curvature of the true arch is very often aesthetically as well as functionally not acceptable; the change of thrust (ΔH) due to elastic and plastic deformation, temperature variations, and support settlement is a function of time and difficult to analyze graphically.

There are four types of arches: three-hinged, two-hinged, one-hinged, and hingeless or fixed-end. Each type of arch has its characteristic shape. Selection of the proper type of arch for a particular structural system depends mainly on the perceptual image, the foundation, and the construction method intended. For contemporary reinforced-concrete design, especially in precasting, the most important are three-hinged, two-hinged, and fixed arches. Therefore, the special characteristics of these three arch types will be discussed in detail.

Theoretically, any arched curvature is possible for any of these arch types. However, for economical reasons the beam action should be kept to a minimum in an arch. Thus, the curvature selected should not deviate much from the line of pressure.

When $-N_x/A_x \leq (+\Sigma M_x/I_x)z$, tensile stresses are developed in the section. As the factor of safety of an arch is controlled largely by the extent of crack formation, tensile stresses must be avoided in arches. In view of the fact that the efficiency of beam action M_x is only one-sixth that of arch action N_x, the depth of the arch must be increased in accordance with the M_x present to avoid cracking and to keep the compressive stresses within safe limits. On the other hand, any increase in cross section increases the moment due to temperature change, foundation displacement, etc., as the third power of the depth (d_0^3). The alternative is to use relatively high normal stresses, as illustrated in Fig. 58. But, as a result, the compressive stresses will increase also, the arch

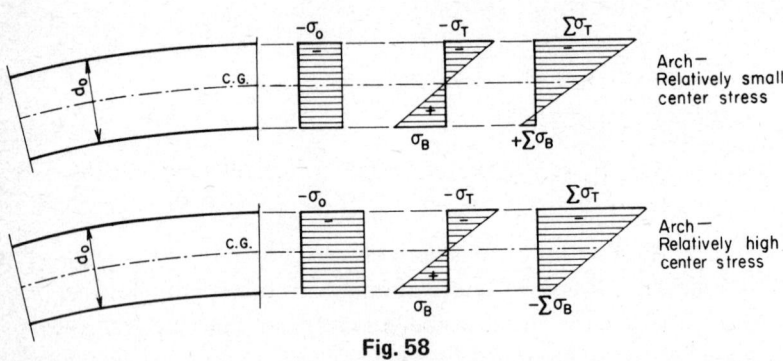

Fig. 58

will become more flexible, and the danger of buckling of the arch may occur. Also, in relatively slender arches the redistribution of moment will be increased, as will be discussed in a later section.

The economical span-rise ratio (L/z_0) for three-hinged arches can be as high as 15, for two-hinged arches up to 12, and for fixed arches up to 8. For halls a circular or parabolic arch is usually chosen. Both arch types are illustrated in Figs. 59 and 60.

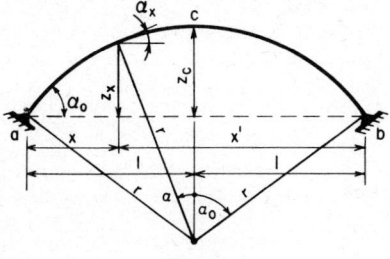

Fig. 59

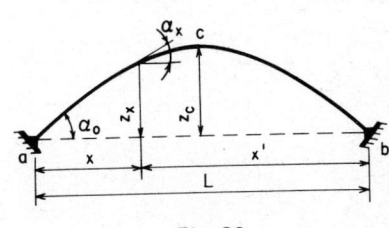

Fig. 60

Circular arch:

$$x = \frac{L}{2} - r \sin \alpha$$
$$z = r(\cos \alpha - \cos \alpha_0)$$
(136)

Parabolic arch:

$$z_x = \frac{4z_c}{L^2} x(L - x) = \frac{4z_c}{L^2} xx'$$

$$= 4z_c \xi \xi', \quad \xi = \frac{x}{L}$$

$$\xi' = (1 - \xi) \qquad (137)$$

12. THREE-HINGED ARCHES

Three-hinged arches are statically determinate, which means that the three equations of equilibrium are sufficient to describe the stresses and deformations in this arch type.

In general, the moments, forces, and stresses in the arch are (Fig. 61):

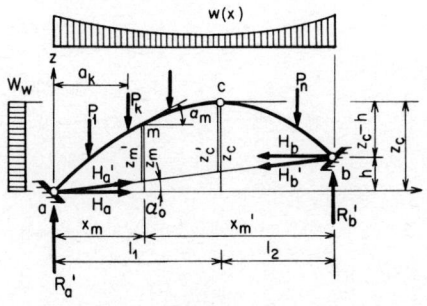

Fig. 61

Vertical Loading

$$R'_a = \frac{1}{L}\left[\sum_0^L w_x(L-x)\Delta x + \sum_1^n P_k a'_k\right]$$

$$R'_b = \left[\sum_0^L w_x \Delta x + \sum_1^n P_k\right] - R'_a$$

$$H_a = H_b = \frac{1}{z'_c}\left[R'_a l_1 - \sum_0^{l_1} w_x(l_1 - x)\Delta x - \sum_0^{l_1} P_k(l_1 - a_k)\right]$$

$$R_a = R'_a + H\tan\alpha_0 \quad \bigg| \quad L = l_1 + l_2$$

$$R_b = R'_b - H\tan\alpha_0 \quad \bigg| \quad H = H'\cos\alpha_0 = \frac{M_{oc}}{z'_c}$$

$$V_{om} = R'_a - \sum_0^{x_m} w_x \Delta x - \sum_1^{x_m} P_k$$

$$M_{om} = R'_a x_m - \sum_0^{x_m} w_x(x_m - x)\Delta x - \sum_1^{x_m} P_k(x_m - a_k)$$

$$M_m = M_{om} - H z'_m$$
$$V_m = V_{om}\cos\alpha_m - H\sin\alpha_m \tag{138}$$
$$-N_m = V_{om}\sin\alpha_m + H\frac{\cos(\alpha_m - \alpha_0)}{\cos\alpha_0}$$

Horizontal Loading

$$R'_b = \frac{1}{L}\frac{w_w z_c^2}{2} \qquad w_w\text{ - wind loading}$$

$$R'_a = w_w z_c \tan\alpha_0 - R'_b$$

$$H_b = \frac{R'_b l_2}{z'_c} \tag{139}$$

$$-H_a = w_w z_c - H_b$$

$$M_m = R'_a x_m - w_w \frac{z_m^2}{2} - H_a z'_m = M_{om} - H_a z'_m$$

$$H_m = -(H_a + w_w z_m)$$

Stresses

$$f_m = -\frac{N_m}{A_m} \mp \frac{M_m}{S_m} \qquad f_{om} = -\frac{N_m}{A_m}$$

$$v_m = \frac{V_m}{bd_0} \cos\alpha_m \qquad\qquad\qquad\qquad (140)$$

$$\sigma_{1,2} = -\frac{f_{om}}{2} \mp \frac{1}{2}\sqrt{f_{om}^2 + 4v_m^2}$$

For a symmetric arch,

$$\alpha_0 = 0$$

$$l_1 = l_2 = \frac{L}{2} \qquad z'_m = z_m \qquad z'_c = z_c$$

The moments and forces due to dead load are

$$R_{a,b} = R_{w_c} + R_{\Delta w} = \frac{w_c L}{2} + \frac{\Delta w_0 L}{6}$$

$$H = H_{w_c} + H_{\Delta w_0} = \frac{w_c L^2}{8z_c} + \frac{\Delta w_0 L^2}{48 z_c}$$

$$M_x = M_{w_c x} + M_{\Delta w_0 x}$$

$$= \frac{w_c L^2}{8}\left(4\frac{xx'}{L^2} - \frac{z_m}{z_c}\right) + \frac{\Delta w_0 L^2}{48}\left[8\frac{xx'}{L^2}\left(1 - 2\frac{xx'}{L^2}\right) - \frac{z_m}{z_c}\right] \quad (141)$$

$$V_m = V_{om}\cos\alpha_m - H\sin\alpha_m$$

$$-N_m = V_{om}\sin\alpha_m + H\cos\alpha_m$$

To demonstrate the difference between circular and parabolic arches, moments due to dead load ($w_c + \Delta w_0$) have been computed for both arch types and are represented in Fig. 62. For uniform load $w = w_c$, the moments in the parabolic arch are zero; the Δw_0 loading causes positive moments in the arch. In the circular arch, the w_c loading causes negative moments and the Δw_0 loading causes positive moments.

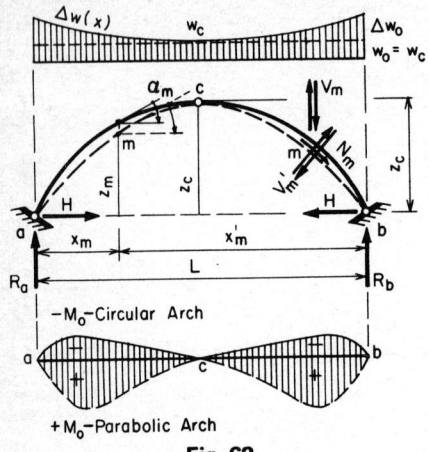

Fig. 62

Temperature Change (ΔT)

If the temperature changes, the centerline length of the arch changes, which results in the deflection of the arch. The sagging of crown is

$$\Delta z_{c\Delta T} = \pm \alpha_t \Delta T \frac{l_1^2 + z_c^2}{z_c} \qquad (142)$$

Due to the deflection, additional moments are introduced in the arch. The maximum temperature moments are

$$\Delta M_{\Delta T} = \pm H_D \Delta z_{\Delta T} \left(\frac{z'}{z_c} - \frac{x}{l_1} \right) \qquad (143)$$

in which

$$z' = z - \Delta z$$

Elastic and Plastic Deformations

The sagging of the crown of a relatively flat and long-span three-hinged arch is considerable in the course of time. The sagging is caused by elastic shortening of the arch length, shrinkage (ϵ_{sr}), plastic flow (φ), and lateral displacement of the supports (ΔL).

For the time when shrinkage and plastic flow are finished (φ_n), the sagging is

$$-\Sigma\Delta z_c = \frac{H_D}{E_0 A}\frac{l_1^2 + z_c^2}{z_c}(1 + \varphi_n) + \epsilon_{sr}\frac{l_1^2 + z_c^2}{z_c} + \frac{\Delta L}{2}\frac{l_1}{z_c} \qquad (144)$$
$$= \Delta z_{ce}(1 + \varphi_n) + \Delta z_{csr} + \Delta z_{cs}$$

and because

$$M_{oc} - H_n(z_c - \Sigma\Delta z_c) = 0$$

the thrust for $t = n$ is

$$H_n = \frac{M_{oc}}{z_c - \Sigma\Delta z_c} \qquad (145)$$

To counteract the increase of moments due to the sagging of the crown ($\Sigma\Delta z_c$), the curvature of the arch must be modified accordingly when the arch is poured. Considering the rotation of supports, the change required from support to crown is

$$\Delta z_x = [\Delta z_c(1 + \varphi_n) + \Delta z_{csr} + \Delta z_{cs}]\left(\frac{z}{z_c} - \frac{x}{l_1}\right) \qquad (146)$$

As a result of the change Δz, positive moments ΔM are introduced in the arch at $t = 0$, which will disappear at time $t = n$:

$$\Delta M_{t=0} = -H_{on}(\Delta z_c \varphi_n + \Delta z_{cs}) \qquad (147)$$

The best results are obtained when the change is of such a magnitude that $\Delta M_{t=0} \cong -\Delta M_{t=n}$. For this case, the change required is

$$\Delta z_x = \frac{1}{2}[\Delta z_c(1 + \varphi_n) + \Delta z_{csr} + \Delta z_{cs}]\left(\frac{z}{z_c} - \frac{x}{l_1}\right) \qquad (148)$$

and

$$\Delta M_t = \mp \frac{H_{Dn}}{2}(\Delta z_c \varphi_n + \Delta z_{cs}) \qquad (149)$$

As can be seen from the above equations, the influence of the deviation from the design curvature of the arch increases when the ratio z_c/L decreases. Therefore, it is recommended that this ratio be limited to 1:10. Further, it is advisable to control the sagging of the crown by providing adequate compressive reinforcing at the bottom of the arch to reduce the plastic flow. This is especially important for slender arches, as are commonly used in precast design, which have relatively small I and A values.

As a conclusion from the above analysis, as can be seen in the moment diagrams (Fig. 62), the three-hinged arch has, in almost any typical loading condition, maximum moments at the quarter points, which decrease toward the crown and spring. The presence of a moment in a section (Fig. 58) reduces the efficient use of material considerably. At the spring and crown, where the moment is zero and only the normal force N is acting at the center of gravity, the material is used 100 percent, but at the quarter point only up to 50 percent is used. To keep the stresses within safe limits, the cross section must be increased to balance the inefficient use of the material.

Due to this fact, the proper shape of a three-hinged arch is as illustrated in Fig. 63.

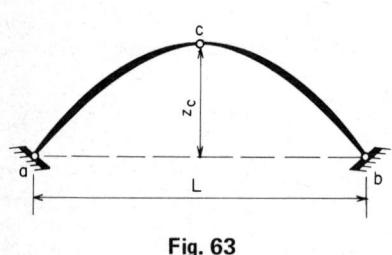

Fig. 63

13. TWO-HINGED ARCHES

A two-hinged arch is one-time statically indeterminate. The redundant is the thrust $H = -X_1$, which can be computed from the deformation of the arch. Thus a simple beam serves as the principal system. The condition for computing X_1 is that the horizontal displacement of the hinge be equal to zero. Thus,

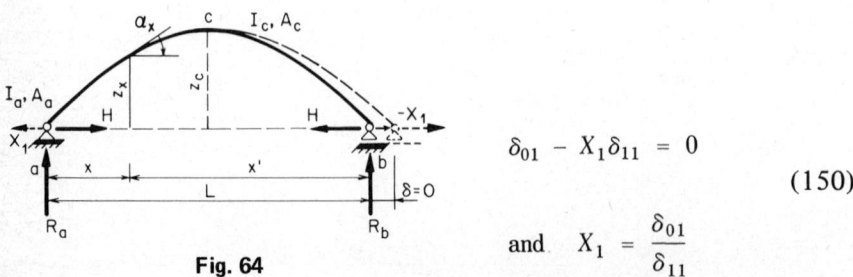

Fig. 64

$$\delta_{01} - X_1 \delta_{11} = 0 \tag{150}$$

$$\text{and} \quad X_1 = \frac{\delta_{01}}{\delta_{11}}$$

in which δ_{11} is the displacement due to $-X_1 = 1$ and δ_{01} is the displacement due to loading. While the horizontal displacements δ depend on the shape and form of the arch, they must be known or estimated before these deformations can be computed. In the case of a parabolic arch, the centerline is determined by the equation

$$z = \frac{4z_c}{L^2} xx' = 4z_c \xi\xi' \quad \xi = \frac{x}{L} \quad \xi' = 1 - \xi$$

The most suitable shape of the arch is assumed to be

$$\frac{I_c}{I \cos\alpha} = 1 - (1 - n)(1 - 2\xi)^2 \tag{151}$$

in which

$$n = \frac{I_c}{I_a \cos\alpha} \quad n = 1: I = \text{constant}$$

The moments M_o and forces N_o in principal system can be computed as for a simple beam [Eqs. (2) and (3)]. $X_1 = -1: \overline{M}_1 = z, \overline{N}_1 = \cos\alpha$ and thus the EI_c times δ values are

$$\delta_{11} = \int z^2 \frac{I_c}{I \cos\alpha} dx + \frac{I_c}{A_c} \int \cos^2\alpha \frac{A_c}{A} ds$$

$$= \frac{8}{15} \frac{6+n}{7} z_c^2 L(1+v) \quad v = \frac{15}{8} \frac{7}{6+n} \frac{1}{z_c^2} \frac{I_c}{A_c} \tag{152}$$

$$\delta_{10} = \int M_o z \frac{I_c}{I \cos\alpha} dx + \frac{I_c}{A_c} \int N_o \cos\alpha \frac{A_c}{A} ds$$

The final moments and normal forces are obtained by superposition

$$\begin{aligned} M_x &= M_{ox} - x_1 z_x \\ N_x &= N_{ox} - x_1 \cos\alpha \end{aligned} \tag{153}$$

The reaction, thrust, and moments for a parabolic symmetrical two-hinged arch for common loadings are given below.

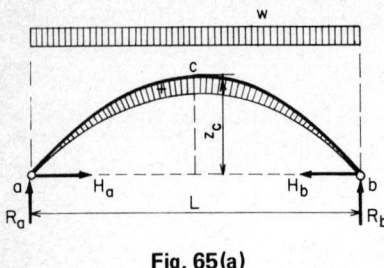

Fig. 65(a)

$$R_a = R_b = \frac{wL}{2}$$
$$H_a = H_b = \frac{wL^2}{8z_c(1+\nu)} \quad (154)$$
$$M_c = \frac{wL^2}{8} \frac{\nu}{1+\nu}$$

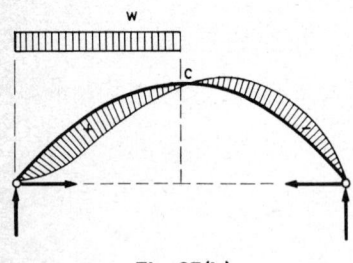

Fig. 65(b)

$$R_a = \frac{3}{8}wL \quad R_b = \frac{1}{8}wL$$
$$H_a = H_b = \frac{wL^2}{16z_c(1+\nu)} \quad (155)$$
$$M_c = \frac{wL^2}{16} \frac{\nu}{1+\nu}$$

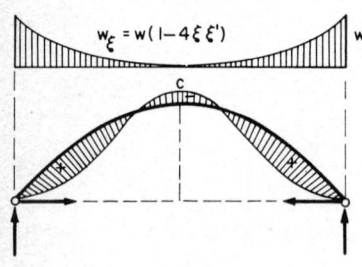

Fig. 65(c)

$$R_a = R_b = \frac{wL}{6}$$
$$H_a = H_b = \frac{wL^2}{36z_c(1+\nu)} \frac{5+n}{6+n}$$
$$M_c = M_{oc} - H_a z_c$$

$$(156)$$

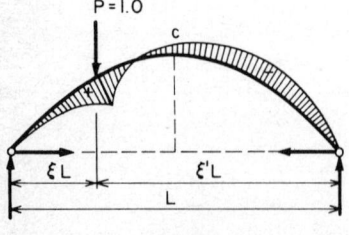

Fig. 65(d)

$$R_a = \xi'P \quad R_b = \xi P$$
$$H_a = H_b = \frac{LP}{8z_c(1+\nu)} \frac{7\xi\xi'}{6+n}$$
$$[5(1+\xi\xi') + (n-1)$$
$$(1 + \xi\xi' - 8\xi^2\xi'^2)] \quad (157)$$

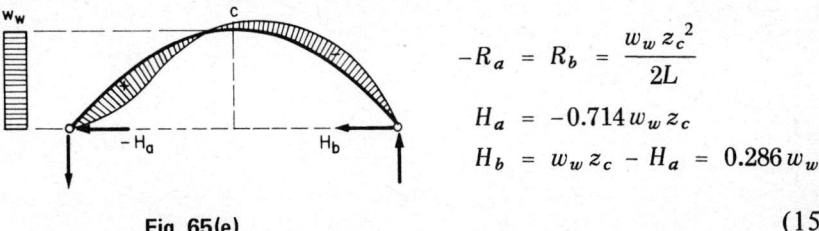

Fig. 65(e)

$$-R_a = R_b = \frac{w_w z_c^2}{2L}$$
$$H_a = -0.714 w_w z_c$$
$$H_b = w_w z_c - H_a = 0.286 w_w z_c \quad (158)$$

When the centerline of the arch is a circle, I and A are constants, and $L = 2l_1$, relatively simple formulas for reactions, thrust, and moments can be derived, as follows:

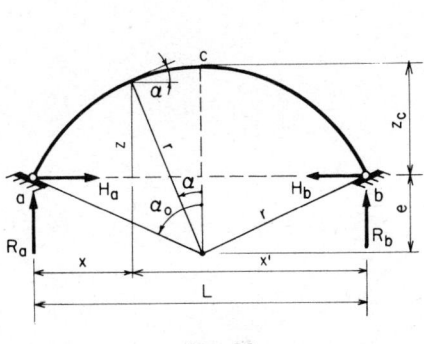

Fig. 66

$$z = r\cos\alpha - (r - z_c)$$
$$x = l_1 - r\sin\alpha$$
$$r = \frac{z_c}{2}\left[1 + \left(\frac{l_1}{z_c}\right)^2\right]$$
$$e = r - z_c \quad s = 2r\alpha_0 \quad (159)$$
$$\sin\alpha_0 = \frac{l_1}{r} \quad \cos\alpha_0 = \frac{e}{r}$$

$$H\delta_{11} - \delta_{10} = 0$$
$$H = \frac{\delta_{10}}{\delta_{11}}$$

$$\delta_{11} = r^3\left[\alpha_0 - 3\frac{l_1 e}{r^2} + 2\alpha_0\left(\frac{e}{r}\right)^2\right] + r\frac{I}{A}\left(\alpha_0 + \frac{l_1 e}{r^2}\right)$$
$$= r^3(\alpha_0 - 3\sin\alpha_0 \cos\alpha_0 + 2\alpha_0 \cos^2\alpha_0) + r\frac{I}{A}(\alpha_0 + \sin\alpha_0 \cos\alpha_0)$$

$$R_a = R_b = \frac{wL}{2}$$

$$\delta_{10} = \frac{wr^4}{2}\left[\sin\alpha_0\left(\frac{4}{3}\sin^2\alpha_0 - \cos^2\alpha_0\right)\right.$$
$$\left. + \alpha_0 \cos\alpha_0(1 - 2\sin^2\alpha_0)\right]$$
$$- \frac{wl_1^3}{3r}\frac{I}{A} \quad (160)$$

$$H_a = H_b = \frac{\delta_{10}}{\delta_{11}}$$
$$M_x = M_{ox} - Hz_x$$

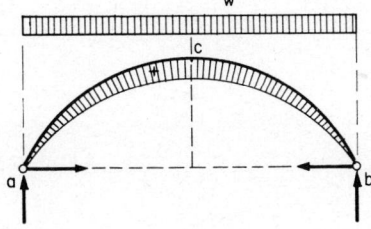

Fig. 67(a)

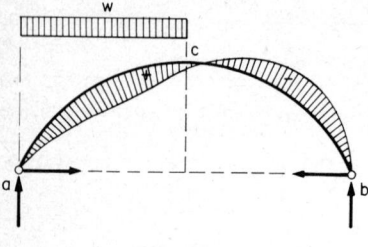

Fig. 67(b)

$$R_a = \tfrac{3}{8} wL \qquad R_b = \tfrac{1}{8} wL$$

δ_{10} = as in Eq. (160)

$$H_a = H_b = \frac{\delta_{10}}{2\delta_{11}} \qquad (161)$$

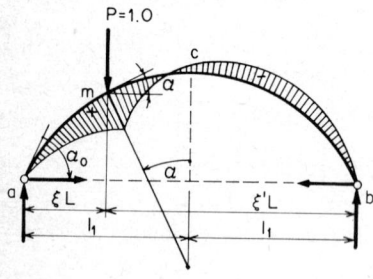

Fig. 67(c)

$$R_a = \xi' P \qquad R_b = \xi P$$

$$H = \frac{\delta_{m1}}{\delta_{11}}$$

$$\delta_{m1} = \frac{rL^2 P}{2} \xi\xi' + er^2[(\cos\alpha + \alpha \sin\alpha) - (\cos\alpha_0 + \alpha_0 \sin\alpha_0)]$$

$$M_x = M_{ox} - H z_x \qquad (162)$$

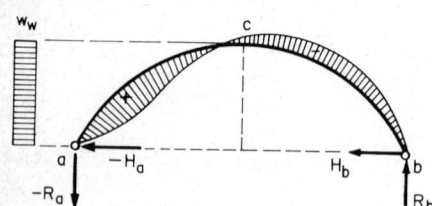

Fig. 67(d)

$$-R_0 = R_b = \frac{w z_c^2}{2L}$$

$$\delta_{10} = \frac{w_w r^4}{2}\left[\alpha_0\left(1 + \tfrac{1}{2}\cos\alpha_0 + 2\cos^2\alpha_0 - \cos^3\alpha_0\right) \right.$$
$$\left. - \sin\alpha_0\left(\tfrac{2}{3} + 3\cos\alpha_0 - \tfrac{7}{6}\cos^2\alpha_0\right)\right] \qquad (163)$$

$$H_a \hat{+} H_b = w_w z_c \qquad H_b \simeq 0.3 w_w z_c$$

An ellipse (Fig. 68) is seldom used as the centerline of the arch because it deviates too much from the pressure line, resulting in high beam action. However, if it is used for halls, the radius and the displacement are

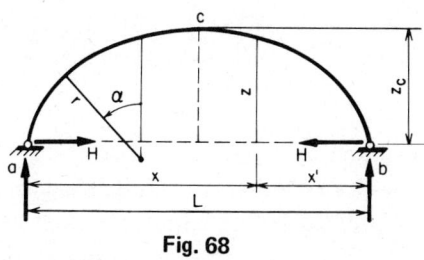

Fig. 68

$$r = \frac{z_c^2 l_1^2}{(l_1^2 \sin^2 \alpha + z_c^2 \cos^2 \alpha)^{3/2}}$$

$$\delta_{11} = \int \frac{z^2 \, dx}{E_0 I(x)} + \int \frac{\cos^2 \alpha}{E_0 A(x)} \, ds$$

(164)

$$\delta_{10} = \int \frac{M_0 z}{E_0 I(x)} \, dx + \int \frac{N_0 \cos \alpha}{E_0 A(x)} \, ds$$

$$H = \frac{\delta_{10}}{\delta_{11}}$$

$$M_x = M_{xo} - H z_x$$

These integrals become too complicated for analytical treatment. Therefore, they must be solved by numerical integration (Simpson's rule). This applies also to other complicated arch curvatures.

Temperature Change (ΔT)

$$\mp \alpha_t \Delta TL + H_{\Delta T} \delta_{11} = 0$$

$$H_{\Delta T} = \pm \frac{\alpha_t \Delta TL}{\delta_{11}}$$

$$= \pm \frac{105 \alpha_t \Delta T}{8z_c^2(6+n)(1+\nu)} \tag{165}$$

$$\Delta M_{\Delta T} = \pm H_{\Delta T} z$$

The analysis of moments for common loadings and temperature change, as given by Eqs. (154) to (165), clearly indicates the significance of the beam-action part in two-hinged arches. The maximum moments are always at the midspan of the arch, and due to this fact the shape of the two-hinged arch is of the utmost importance in controlling the stresses and deformations. The rise-span ratio is largely dependent on the foundation condition unless prestressed tie rods are used. The recommended maximum span-rise ratio is approximately 8.

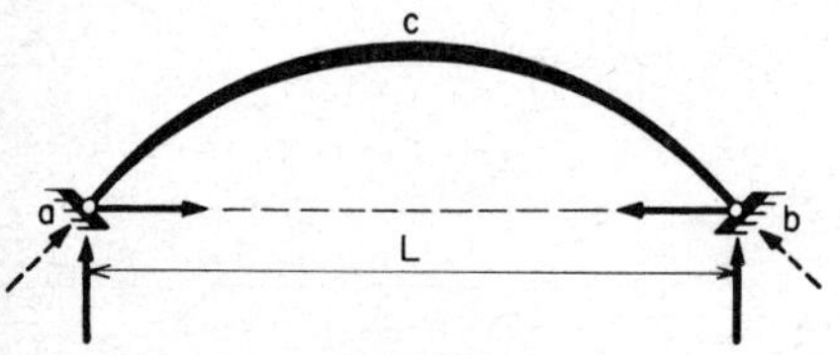

The shape of the arch, as given by Eq. (151) and illustrated in Fig. 69, clearly characterizes the true carrying action of the two-hinged arch.

Fig. 69

14. FIXED ARCHES

Fixed arches are three times statically indeterminate. Span-rise ratio is about 6. Most conveniently, the simple beam serves as the principal system. The redundants X_1, X_2, and X_3, the thrust H, and fixed end moments M_a and M_b (Fig. 70) will be determined from continuity equations for the following conditions:

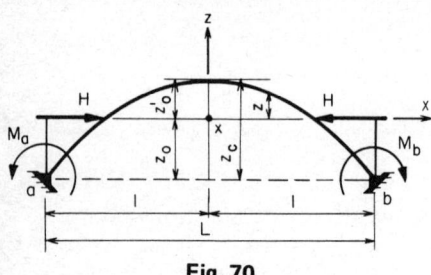

Fig. 70

Lateral displacement: $\Delta L = 0$
End rotations: $\delta_a = \delta_b = 0$

The solution of the problem is somewhat simplified if the redundants can be determined independent of each other, which means $\delta_{ik} = 0$ and thus

$$X_1 = \frac{\delta_{10}}{\delta_{11}} \quad X_2 = \frac{\delta_{20}}{\delta_{22}} \quad \text{and} \quad X_3 = \frac{\delta_{30}}{\delta_{33}} \tag{166}$$

This is possible when $-X_1 = H$ acts at the center of gravity of the arch and the dummy moments $\bar{M}_a$ and $\bar{M}_b$ are combined into a group loading:

$$\begin{aligned} X_1 &= H \\ X_2 &= \tfrac{1}{2}(\bar{M}_a - \bar{M}_b) \\ X_3 &= \tfrac{1}{2}(\bar{M}_a + \bar{M}_b) \end{aligned} \tag{167}$$

For symmetric loading, $M_a = M_b$; and for antisymmetric loading, $M_a = -M_b$. The location of the center of gravity of an arch is

$$z_0 = \frac{\int_a^c z[I_c/I(x)]\,dx}{\int_a^c [I_c/I(x)]\,dx} \tag{168}$$

The actual reactions and moments are obtained by superposition:

$$\begin{aligned} R_a &= R_{a0} + \frac{X_2}{l} \\ R_b &= R_{b0} - \frac{X_2}{l} \\ M_a &= X_1 z_0 - X_2 - X_3 \\ M_b &= X_1 z_0 + X_2 - X_3 \end{aligned} \tag{169}$$

For a symmetric parabolic arch having $I_c/(I \cos\alpha) = 1.0$, $z_0 = \tfrac{2}{3} z_c$ and the EI_c times δ_{kk} are

$$\begin{aligned} \delta_{11} &= \tfrac{4}{45} L z_c^2 (1 + \nu) \quad \nu = \frac{45}{4} \frac{I_c}{A_c z_c^2} \\ \delta_{22} &= \tfrac{1}{3} L \\ \delta_{33} &= L \end{aligned} \tag{170}$$

$$\delta_{10} = \int \overline{M}_1 M_0 \frac{I_c}{I \cos \alpha} dx \quad \overline{M}_1 = 1.0 \, z \quad \int N_o \cos \alpha \frac{A_c}{A} ds \approx o$$

$$\delta_{20} = \int \overline{M}_2 M_0 \frac{I_c}{I \cos \alpha} dx \quad \overline{M}_2 = \pm \left(\frac{x}{l}\right) \int N_o \frac{\sin \alpha}{l} \frac{A_c}{A} ds \approx o$$

$$\delta_{30} = \int \overline{M}_3 M_0 \frac{I_c}{I \cos \alpha} dx \quad \overline{M}_3 = 1.0 \quad \overline{N}_3 = o \qquad (171)$$

where $\overline{M}_1$, $\overline{M}_2$, and $\overline{M}_3$ are the moments due to the $X_1 = 1.0$: $X_2 = X_3 = 0$, $X_2 = 1.0$: $X_1 = X_3 = 0$, and $X_3 = 1.0$: $X_1 = X_2 = 0$.

For typical loadings, the reaction, thrust, and moments are given in the following.

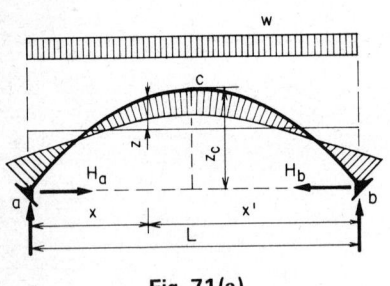

$$R_a = R_b = \frac{wL}{2}$$

$$H_a = H_b = \frac{wL^2}{8z_c} \frac{1}{1+\nu}$$

$$M_a = M_b = -\frac{wL^2}{12} \frac{\nu}{1+\nu} \qquad (172)$$

$$M_c = +\frac{wL^2}{24} \frac{\nu}{1+\nu}$$

Fig. 71(a)

$$R_a = R_b = \frac{w_0 L}{6}$$

$$H = \frac{w_0 L^2}{56 z_c} \frac{1}{1+\nu}$$

$$M_a = M_b = -\frac{w_0 L^2}{420} \frac{7\nu + 2}{1+\nu} \qquad (173)$$

$$M_c = -\frac{w_0 L^2}{1680} \frac{3 - 7\nu}{1+\nu}$$

Fig. 71(b)

Arch Action   129

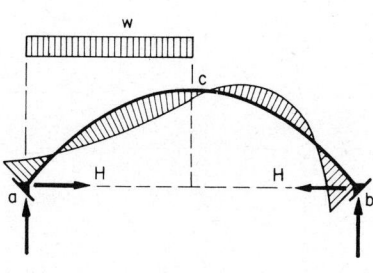

Fig. 71(c)

$$R_a = \frac{13}{32}wL \quad R_b = \frac{3}{32}wL$$

$$H = \frac{wL^2}{16z_c}\frac{1}{1+\nu}$$

$$M_a = -\frac{wL^2}{192}\frac{3+11\nu}{1+\nu} \quad (174)$$

$$M_b = \frac{wL^2}{192}\frac{3-5\nu}{1+\nu}$$

$$M_c = \frac{wL^2}{48}\frac{\nu}{1+\nu}$$

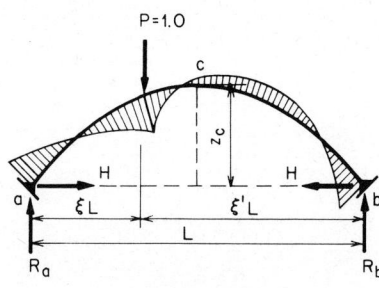

Fig. 71(d)

$$R_a = \xi'^2(1+2\xi)$$
$$R_b = \xi^2(1+2\xi')$$

$$H = \frac{15L}{4z_c}\frac{\xi^2\xi'^2}{(1+\nu)}$$

$$M_a = \xi\xi'^2 LP\left[\frac{5}{2(1+\nu)}\xi - 1\right]$$

$$M_b = \xi^2\xi' LP\left[\frac{5}{2(1+\nu)}\xi' - 1\right]$$

$$M_c = \xi^2\frac{LP}{2}\left[1 - \frac{5}{2(1+\nu)}\xi'^2\right]$$

$$\xi \le \tfrac{1}{2} \quad (175)$$

Circular Fixed Arch (Fig. 72)

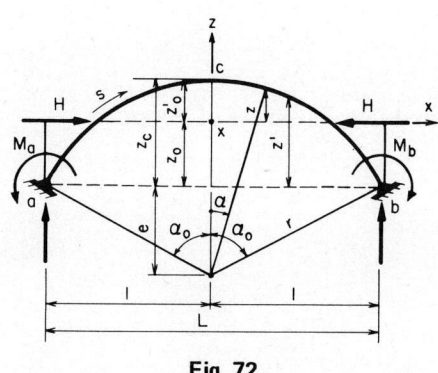

Fig. 72

$$r = \frac{z_c}{2}\left[1 + \left(\frac{l}{z_c}\right)^2\right]$$

$$e = r - z_c \quad \sin\alpha_0 = l/r$$
$$x = r\sin\alpha \quad \cos\alpha_0 = e/r$$
$$z' = r(\cos\alpha - \cos\alpha_0)$$

$$z_0 = r\left(\frac{\sin\alpha_0}{\alpha_0} - \cos\alpha_0\right)$$

$$z = r\left(\cos\alpha - \frac{\sin\alpha_0}{\alpha_0}\right)$$

(176)

$$\frac{I_c}{I \cos\alpha} = 1.0$$

$$EI_c \delta_{11} = 2 \int_0^{\alpha_0} z^2 \, ds + \frac{I_c}{A_c} \int \cos^2 \alpha \, ds$$

$$= r^3 \alpha_0 \left[1 + \frac{\sin\alpha_0}{\alpha_0} \cos\alpha_0 - 2 \left(\frac{\sin\alpha_0}{\alpha_0} \right)^2 \right] + \frac{I_c}{A_c} r\alpha_0 \left(1 + \frac{\sin\alpha_0}{\alpha_0} \cos\alpha_0 \right)$$

$$EI_c \delta_{22} = 2 \int_0^{\alpha_0} \left(\frac{x}{l} \right)^2 ds = \frac{r\alpha_0}{\sin^2\alpha_0} \left(1 - \frac{\sin\alpha_0}{\alpha_0} \cos\alpha_0 \right) \tag{177}$$

$$EI_c \delta_{33} = 2 \int_0^{\alpha_0} ds = 2r\alpha_0$$

$$EI_c \delta_{10} = \int M_0 z \, dx + \frac{I_c}{A_c} \int N_o \cos\alpha \, ds$$

$$EI_c \delta_{20} = \int M_0 \frac{x}{l} dx + \frac{I_c}{A_c} \int N_o \frac{\sin\alpha}{l} ds$$

$$EI_c \delta_{30} = \int M_0 \, dx \qquad \bar{N}_3 = 0$$

Temperature Change (ΔT), Rotation, and Displacement of Supports

$$\delta_{1\Delta T} = EI_c \alpha_t \Delta T L \quad \delta_{2\Delta T} = \delta_{3\Delta T} = 0$$
$$\delta_{1s} = EI_c (z_0 \Delta\alpha_{ab} - \Delta L)$$
$$\delta_{2s} = -EI_c \left[(\Delta\alpha_0 + \Delta\alpha_b) + \frac{2}{L} (\Delta_a - \Delta_b) \right] \tag{178}$$
$$\delta_{3s} = -EI_c \Delta\alpha_{ab}$$

Elastic and Plastic Deformations of Two-hinged and Fixed Arches

In accordance with elastic theory, the changes in thrust (H_D) due to the elastic shortening of the arch (ΔH_e), shrinkage (ΔH_{sr}), and lateral displacement of the supports (ΔH_s) are

$$\Delta H_e = -H_D \frac{\int ds/E_0 A(x)}{\delta_{11}}$$

$$\Delta H_{sr} = -\frac{\epsilon_{sr} L}{\delta_{11}} \tag{179}$$

$$\Delta H_s = -\frac{\Delta L}{\delta_{11}}$$

$$\delta_{11} = \int \frac{z^2 \, ds}{E_0 I(x)} + \int \frac{\cos^2 \alpha}{E_0 A(x)} ds \tag{180}$$

The related moments are obtained by multiplying the ΔH values with the arch-axis ordinates z. Thus for $t = n$,

$$\Sigma \Delta M = +\Sigma \Delta H z \tag{181}$$

Because of plastic flow and shrinkage, the axis of the arch shortens from $t = 0$ to $t = n$. Due to this fact, the crown of the arch has to sag to balance this shortening of axis. The change in thrust ΔH_t, in accordance with Dischinger, can be determined from the following differential equation:

$$\frac{\epsilon_{sr} L}{\varphi_n} \frac{d\varphi_t}{dt} + H_D \frac{d\varphi_t}{dt} \int \frac{ds}{E_0 A(x)} + (\Delta H_e + \Delta H_s) \delta_{11} \frac{d\varphi_t}{dt}$$
$$+ \Delta H_t \delta_{11} \frac{d\varphi_t}{dt} + \frac{d\Delta H_t}{dt} \delta_{11} \frac{E_0}{E_{0n}} = 0 \tag{182}$$

The first four elements of this equation represent, respectively, the shortening of the arch axis by shrinkage, centric plastic flow, change in thrust in accordance with Eqs. (179), and shortening due to the change in centric plastic flow. The last element, $d\Delta H_t \cdot \delta_{11}$, represents the elastic shortening of the arch due to ΔH, and the increased modulus of elasticity from $t = 0$ to $t = n$. The plastic deformations are obtained by multiplying the elastic shortenings by $d\varphi_t/dt$. Considering that

$$\epsilon_{sr} L = -\Delta H_{sr} \delta_{11}$$

$$\Delta H_e \delta_{11} = -H_D \int \frac{ds}{E_0 A(x)}$$

and taking $E_0 \sim E_{0n}$, the differential equation is simplified to

$$\frac{d\Delta H_t}{dt} + \Delta H_t \frac{d\varphi_t}{dt} + \left(\Delta H_s + \frac{\Delta H_{sr}}{\varphi_n}\right) \frac{d\varphi_t}{dt} = 0 \qquad (183)$$

The thrust H_D due to dead load and ΔH_e are cancelled out from the differential equation, which means that the shortening of the arch axis due to centric plastic flow is balanced by the positive moment $\Delta H_e z$.

The solution of this differential equation gives

$$\Delta H_t = \left(\frac{\Delta H_{sr}}{\varphi_n} - \Delta H_s\right)(1 - e^{-\varphi_t}) \qquad (184)$$

and thus the change in moments for time $t = n$ is

$$\Delta M_t = \Delta H_t z \qquad (185)$$

As a conclusion from the above, plastic flow considerably reduces the change in thrust caused by shrinkage and lateral displacement of supports.

4

SUSPENSION ACTION

Suspension action is the simplest of the carrying actions. It involves only one stress type—tension. When a cable is erected and anchored at both ends, a load of any kind can be carried by the cable. Due to the load, the cable deflects because of its elastic properties. The tension in the cable, as well as the deflection, is a function of the load applied. The suspension action for uniform load is best demonstrated and described by the following illustration and formulas.

When the final maximum allowable center deflection and the tensile force in the cable are known, the uniform load the cable carries can be computed as

$$w = \frac{8T(z_0 - z_b/2)}{l^2} \tag{186}$$

Or when the uniform load is known, the tension in the cable at the

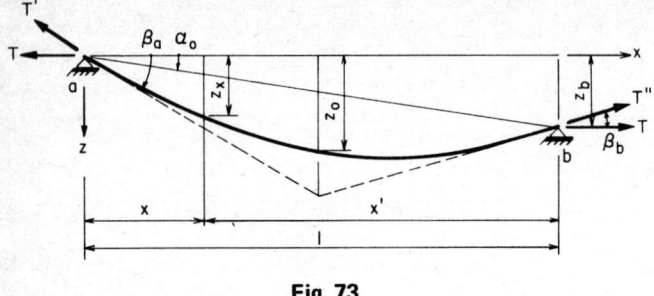

Fig. 73

center is given by

$$T = \frac{wl^2}{8z_0 - 4z_b} \quad \text{or} \quad T' = \frac{T}{\cos\beta_a}$$
$$T'' = \frac{T}{\cos\beta_b} \tag{187}$$

and the stress in the cable is

$$f = \frac{T}{A} \tag{188}$$

in which A is the cross-sectional area of the cable.

The deflection curve of the cable under uniform load is catenary. The ordinates at any point of the curve are

$$z_x = \frac{wxx'}{2T} + x\tan\alpha_0 \quad \tan\alpha_0 = \frac{z_b}{l} \tag{189}$$

in which

$$x' = l - x$$

and the cable slope at either support is

$$\tan\beta_a = \frac{wl}{2T} + \tan\alpha_0$$
$$\tan\beta_b = \frac{wl}{2T} - \tan\alpha_0 \tag{190}$$

The length of cable s with a given center deflection is given with sufficient accuracy by the formula

$$s = \sqrt{l^2 + z_b^2}\left(1 + \frac{8}{3}k^2 - \frac{32}{5}k^4 + \frac{256}{7}k^6\right)$$

$$k = \frac{z_0}{l} = \frac{w_D l \cos^2 \alpha_0}{8T}$$
(191)

The values for level cable are obtained by setting $\alpha_0 = 0$.

In determining the erection tension T_e for a uniformly loaded cable, the values s_e and T_e must satisfy the equation

$$s - s_e = \frac{s}{AE}(T - T_e) \tag{192}$$

where the length of the cable (s_e) under its own load is approximately

$$s_e \simeq \sqrt{l^2 + z_b^2} + \frac{w_e^2(\sqrt{l^2 + z_b^2})^3}{24T_e^2} \tag{193}$$

Substituting this value for s_e in Eq. (192), we obtain

$$s\left(1 + \frac{T_e}{AE}\right) - \left(s\frac{T}{AE} + \sqrt{l^2 + z_b^2}\right) = \frac{w_e^2(\sqrt{l^2 + z_b^2})^3}{24T_e^2} \tag{194}$$

This equation can best be solved by trial for assumed values of T_e. When T_e is obtained, the erection deflection is given by Eq. (189):

$$z_{0e} = \frac{w_e l^2}{8T_e - 4z_b} \tag{195}$$

When, in addition to its uniform load, the cable is loaded with a concentrated load or loads, the deflection of the cable will deviate from the catenary. Let us consider a cable loaded with uniform load w and a concentrated load P at any point, as shown in Fig. 74. The tension T in the cable varies with different positions of the load P and is maximum

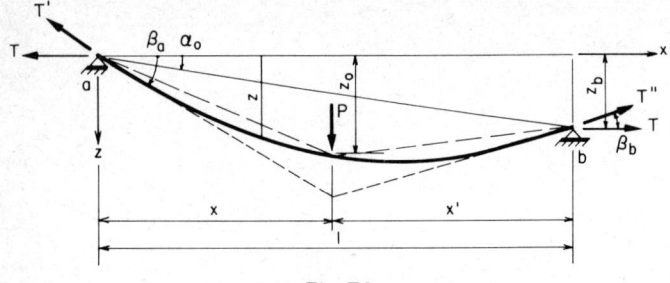

Fig. 74

when P is at the center of the span L:

$$T_{max} = \frac{l(2P + wl)}{8z_0 - 4z_b}$$

$$T'_{max} = \frac{T_{max}}{\cos \beta_a} \quad \tan \beta_a = \frac{P + wl}{2T} + \frac{z_b}{l} \quad \quad (196)$$

$$T''_{max} = \frac{T_{max}}{\cos \beta_b} \quad \tan \beta_b = \frac{P + wl}{2T} - \frac{z_b}{l}$$

Deflection at any point of P is

$$z_x = \frac{xx'(2P + wl)^2}{2T(wl^2 + 4P\sqrt{xx'})} + x \tan \alpha_0 \quad \quad (197)$$

and at the center $x' = x = l/2$,

$$z_0 = \frac{l(2P + wl)}{8T_{max}} + \frac{z_b}{2} \quad \quad (198)$$

The stress in the cable is

$$f = \frac{T}{A_c}$$

where A is the cross-sectional metallic area of the cable.

As can be seen from the above example, the curvature of the cable depends on the loading. Due to this, the suspension action alone is not suitable for creating a stable structural system. As we shall see in Part Three, suspension action is used for long spans in combination with beam action as secondary action to stabilize the cables.

5

COMBINED CARRYING ACTIONS

15. SIMPLE FRAMES

Frames carry and transmit the external loads into the supports by combined arch and beam action. The pressure-line deviation from the center of gravity of the elements of the frame is far more extensive than was the case with arches. Due to this, beam action is more pronounced in frames than it is in arches, but basically frame analysis is the same as for arches. However, most commonly used frames are composed of straight elements with constant or linearly changing depth, and therefore their analysis is rather simple, especially when using the integration results given in Table 3.

The extent of participation of any element of the frame depends upon its relative stiffness, which is expressed by I_c/I_n, where I_c can be the moment of inertia arbitrarily chosen from any one of the elements of the frame and I_n the moment of inertia of the element under consideration. When I_n is not constant, numerical integration must be used or the variable $I_{(z)}$ or $I_{(x)}$ must be replaced by an equivalent constant $I_m = kI_i{}^*$ for variable $I_{(x,\,z)}$. Let us consider an element with linearly changing depth having a moment-of-inertia ratio $n = I_i/I_k$ (Fig. 75):

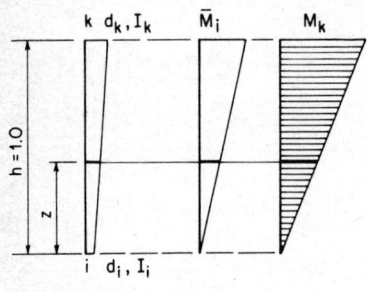

Fig. 75

$$\zeta = \frac{z}{h}$$

$$I_{(z)} = I_i \left[1 + \left(\sqrt[3]{\frac{1}{n}} - 1\right)\zeta\right]^3$$

$$\int \overline{M}_i M_k \frac{I_i}{I_m}\, dz = \frac{1}{3} \overline{M}_i M_k \frac{I_i}{kI_i} h \qquad (199)$$

$$= \overline{M}_i M_k h \int_0^{1.0} \frac{\zeta^2}{[1 + (\sqrt[3]{1/n} - 1)\zeta]^3}\, d\zeta$$

$$k = \frac{(\sqrt[3]{1/n} - 1)^3}{\ln(1/n) + 6\sqrt[3]{n} - 1.5\sqrt[3]{n^2} - 4.5}$$

$$I_m = kI_i = \text{const} \qquad (200)$$

Two-hinged Frames

To demonstrate a frame analysis, the thrust, reactions, and moments will be computed for a simple two-hinged frame having tapered verticals (Fig. 76):

*Kurt Beyer, "Die Statik im Eisenbetonbau," Springer-Verlag OHG, Berlin, 1933.

Combined Carrying Actions   139

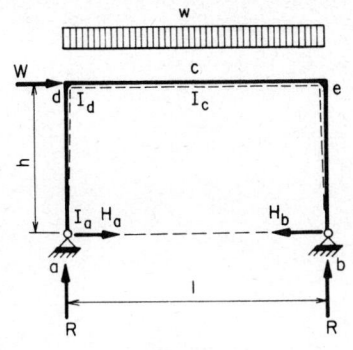

$w = 1.0 \ k/\text{ft}$
$W = 15 \ k$
$l = 50.0 \ \text{ft} \quad d_a = 1.25 \ \text{ft} \quad d_c = 2.50 \ \text{ft}$
$h = 15.0 \ \text{ft} \quad d_d = 2.00 \ \text{ft} \quad b = 1.00 \ \text{ft}$

Fig. 76

$$I_a = \frac{1.0 \times 1.25^3}{12} = 0.163 \ \text{ft}^4$$

$$I_d = \frac{1.0 \times 2.00^3}{12} = 0.666 \ \text{ft}^4$$

$$n = \frac{0.163}{0.666} = 0.245$$

$$I_c = \frac{1.0 \times 2.50^3}{12} = 1.300 \ \text{ft}^4 = l_c$$

$$k = \frac{(\sqrt[3]{1/0.245} - 1)^3}{\ln(1/0.245) + 6\sqrt[3]{0.245} - 1.5\sqrt[3]{0.245^2} - 4.5} \cong 2.90$$

$$I_m = kI_a = 2.90 \times 0.163 = 0.472 \ \text{ft}^4$$

Redundant:

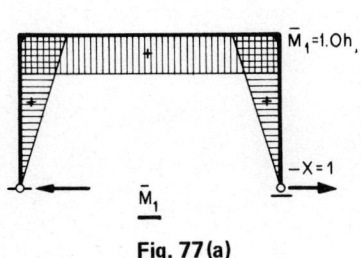

Fig. 77(a)

$$\delta_{10} - X_1 \delta_{11} = 0$$

$$EI_c \delta_{11} = 2 \int_a^d \overline{M}_1^2 \frac{I_c}{I_m} \ dz + \int_d^e \overline{M}_1^2 \frac{I_c}{I_c} \ dx$$

$$= 2 \frac{1}{3} h^2 \frac{1.300}{0.472} h + h^2 1.0 \ l$$

$$= +6,200 + 11,200 = 17,400$$

Uniform load $w = 1.0 \ \text{k/ft}$:

$$M_0 = \frac{1.0 \times 50.0^2}{8} = 313 \ k'$$

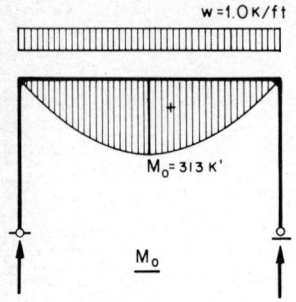

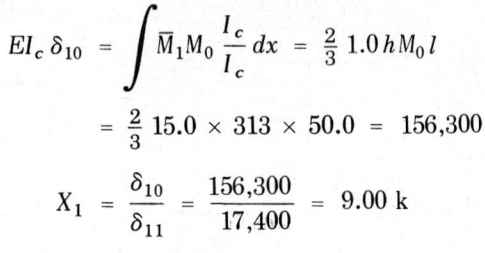

Fig. 77(b)

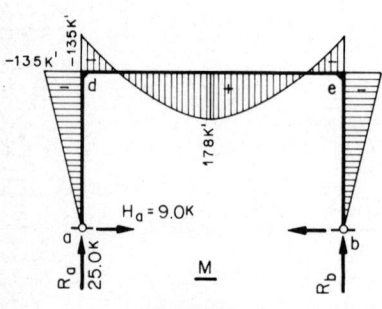

Fig. 77(c)

$$EI_c \delta_{10} = \int \overline{M}_1 M_0 \frac{I_c}{I_c} dx = \tfrac{2}{3} 1.0 h M_0 l$$

$$= \tfrac{2}{3} 15.0 \times 313 \times 50.0 = 156{,}300$$

$$X_1 = \frac{\delta_{10}}{\delta_{11}} = \frac{156{,}300}{17{,}400} = 9.00 \text{ k}$$

$$M = M_0 - X_1 \overline{M}_1$$
$$M_c = 313 - 9.0 \times 15.0 = 178\,k'$$
$$M_d = 0.0 - 9.0 \times 15.0 = -135\,k'$$
$$R_a = R_b = 1.0 \times 50.0/2 = 25.0 \text{ k}$$

Horizontal load $W = 15$ k:

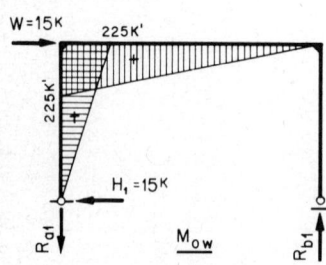

Fig. 77(d)

$$-R_{a1} = R_{b1} = \frac{Wh}{l} = \frac{15.0 \times 15.0}{50.0} = 4.5\,k$$

$$M_d = +15.0 \times 15.0 = 225\,k'$$
$$= 4.5 \times 50.0$$

$$EI_c \delta_{10} = \int \overline{M}_1 M_0 \frac{I_c}{I} ds$$

$$= \tfrac{1}{3} 15.0 \times 225 \, \frac{1.300}{0.472} \, 15.0 + \tfrac{1}{2} 15.0 \times 225 \times 50.0$$

$$= 46{,}400 + 84{,}400 = 130{,}800$$

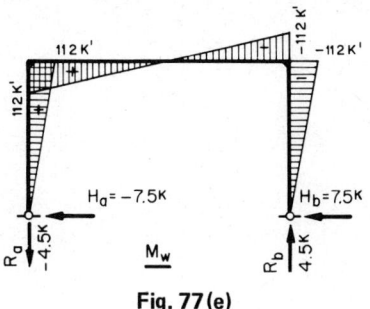

$$X_1 = \frac{\delta_{10}}{\delta_{11}} = \frac{130{,}800}{17{,}400} = 7.50\,k$$
$$H_a = -15.0 + 7.5 \times 1.0 = -7.5\,k$$
$$M_d = -M_e = 7.5 \times 15.0 = 112\,k'$$

Fig. 77(e)

Temperature change:

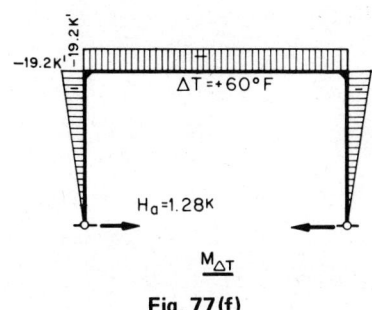

$$\Delta T = \pm 60°\,F$$
$$\alpha_t = 0.0000079$$
$$I_c = 1.300\ \text{ft}^4$$

Fig. 77(f)

$$E_0 = 5.0 \times 10^6\ \text{psi} = 7.2 \times 10^5\ \text{ksf}$$

$$EI_c\,\delta_{\Delta T} = EI_c\,\alpha_t\,\Delta T \times l = \pm 22{,}200$$
$$X_1 = \pm \frac{22{,}200}{17{,}400} = \pm 1.28\,k$$
$$M_d = \mp 1.28 \times 15.0 = \mp 19.2\,k'$$

For common loadings, the reactions and moments for two-hinged frames, illustrated in Figs. 78 and 79, are given in the following:

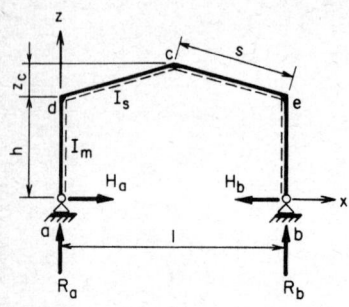
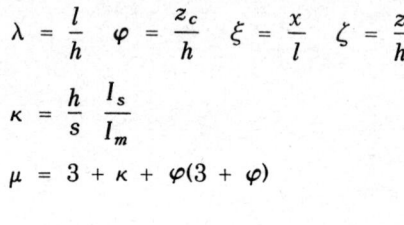

$$\lambda = \frac{l}{h} \quad \varphi = \frac{z_c}{h} \quad \xi = \frac{x}{l} \quad \zeta = \frac{z}{h}$$

$$\kappa = \frac{h}{s} \frac{I_s}{I_m}$$

$$\mu = 3 + \kappa + \varphi(3 + \varphi)$$

Fig. 78

$$R_a = R_b = \frac{wl}{2}$$

$$H_{a,b} = \frac{wl}{8} \lambda \frac{8 + 5\varphi}{4\mu}$$

$$M_{d,e} = -H_{a,b} h \tag{201}$$

$$M_c = \frac{wl^2}{8}\left[1 - (1 + \varphi)\frac{8 + 5\varphi}{4\mu}\right]$$

Fig. 79(a)

$$R_a = \tfrac{3}{8} wl \quad R_b = \tfrac{1}{8} wl$$

$$H_{a,b} = \frac{wl}{16} \lambda \frac{8 + 5\varphi}{4\mu} \tag{202}$$

Fig. 79(b)

$$-R_a = R_b$$

$$H_{a,b} = -\frac{w_w h^2}{2l}\left[1 \pm 1 - \frac{6(2 + \varphi) + 5\kappa}{8\mu}\right]$$

Fig. 79(c)

$$\tag{203}$$

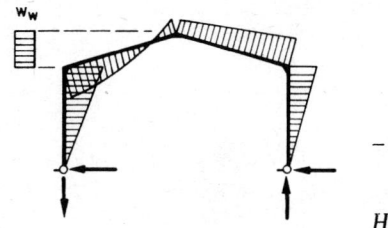

Fig. 79(d)

$$-R_a = R_b = w_w z_c \frac{2h + z_c}{2l}$$
$$H_{a,b} = -\frac{w_w z_c}{2}\left[\frac{\varphi}{8\mu}(4 + 3\varphi) \pm 1\right] \quad (204)$$

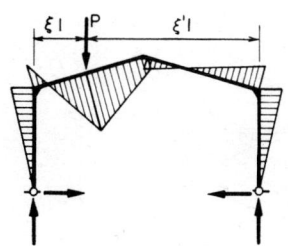

Fig. 79(e)

$$R_a = \xi' P \quad R_b = \xi P$$

$$H_{a,b} = \frac{P}{2}\lambda\frac{\xi}{\mu}[1.5(2 + \varphi) - \xi(3 + 2\varphi\xi)]$$
$$M_{d,e} = -H_{a,b} \times h \quad (205)$$
$$M_c = \frac{Pl}{2}\left\{\xi - (1 + \varphi)\frac{\xi}{\mu}[1.5(2 + \varphi) - \xi(3 + 2\varphi\xi)]\right\}$$

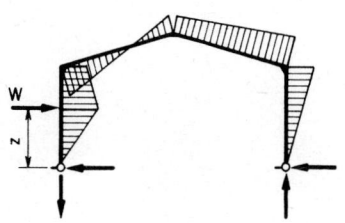

Fig. 79(f)

$$-R_a = R_b = \frac{Wz}{l}$$

$$H_{a,b} = -\frac{W}{2}\left[1 \pm 1 - \frac{\zeta}{2\mu}(6 + 3\varphi + 3\kappa - \kappa\zeta^2)\right] \quad (206)$$

$$M_{d,e} = \frac{Wh}{2}\zeta\left[1 \pm 1 - \frac{6 + 3(\varphi + \kappa) - \kappa\zeta^2}{2\mu}\right]$$

$$M_c = \frac{Wh}{2}\zeta\left\{1 - \frac{1 + \varphi}{2\mu}[6 + 3(\varphi + \kappa) - \kappa\zeta^2]\right\}$$

(206)
(Cont.)

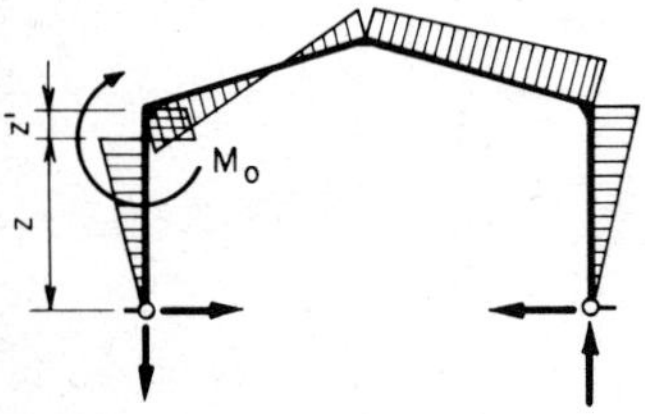

Fig. 79(g)

$$-R_0 = R_b = \frac{M_0}{l}$$

$$H_{a,b} = \frac{3M_0}{4\mu h}[2 + \varphi + \kappa(1 - \zeta^2)]$$

$$M_{d,e} = \frac{M_0}{2}\left[1 \pm 1 - \frac{3}{2\mu}(2 + \varphi + \kappa - \kappa\zeta^2)\right]$$

$$M_c = \frac{M_0}{2}\left[1 - \frac{3(1 + \varphi)}{2\mu}(2 + \varphi + \kappa - \kappa\zeta^2)\right]$$

(207)

Temperature change:

$$R_a = R_b = 0$$

$$H_{a,b} = \frac{3l}{2\mu s}\frac{EI_s}{h^2}\alpha_t \Delta T$$

(208)

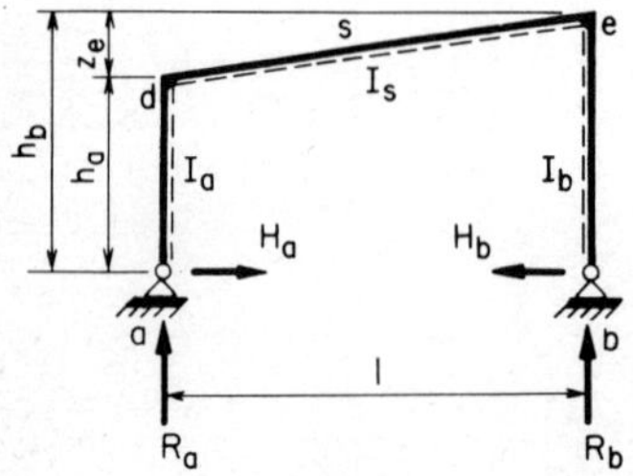

Fig. 80

$$\lambda_a = \frac{h_a}{h_b} \quad \lambda_b = \frac{h_b}{h_a} \quad \varphi_a = \frac{z_e}{h_a} \quad \varphi_b = \frac{z_e}{h_b}$$

$$\nu_a = \frac{l}{h_a} \quad \nu_b = \frac{l}{h_b}$$

$$\kappa_a = \frac{h_a}{s}\frac{I_s}{I_a} \quad \kappa_b = \frac{h_b}{s}\frac{I_s}{I_b}$$

$$\mu = \lambda_a(1 + \kappa_a) + 1 + \lambda_b(1 + \kappa_b)$$

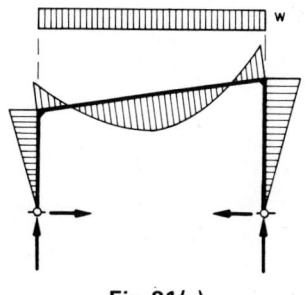

Fig. 81(a)

$$R_a = R_b = \frac{wl}{2}$$
$$H_{a,b} = \frac{wl}{8\mu}(\nu_a + \nu_b) \tag{209}$$

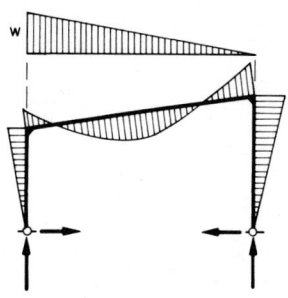

Fig. 81(b)

$$R_a = \frac{wl}{3}, \quad R_b = \frac{wl}{6}$$
$$H_{a,b} = \frac{wl}{120\mu}(7\nu_a + 8\nu_b)$$
$$M_d = H_{a,b} \times h_a$$
$$M_e = H_{a,b} \times h_b \tag{210}$$

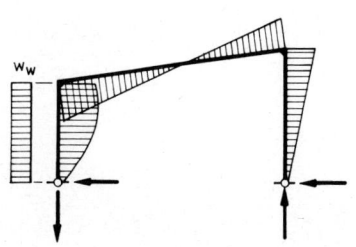

Fig. 81(c)

$$-R_a = R_b = \frac{w_w h_a^2}{2l}$$

$$H_{a,b} = -\frac{w_w h_a}{2}\left\{1 \pm 1 - \frac{1}{4\mu}[2 + \lambda_a(4 + 5\kappa_a)]\right\} \tag{211}$$

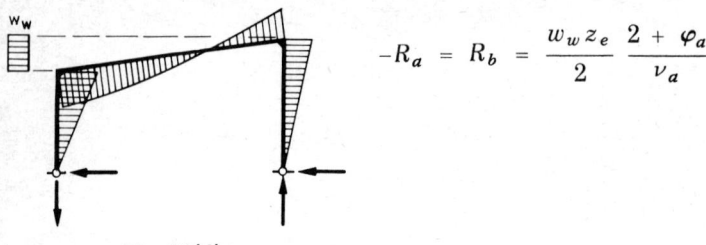

Fig. 81(d)

$$-R_a = R_b = \frac{w_w z_e}{2} \frac{2 + \varphi_a}{\nu_a}$$

$$H_{a,b} = -\frac{w_w z_e}{2} \left[1 \pm 1 - \frac{\lambda_a}{4\mu} (12 + 6\varphi_a + \varphi_a^2 + 8\kappa_a) \right] \tag{212}$$

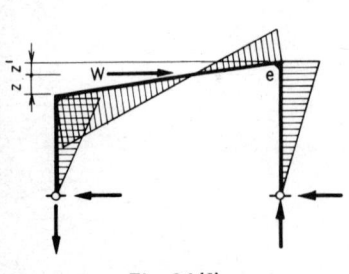

Fig. 81(e)

$$R_a = \xi' P, \quad R_b = \xi P$$

$$H_{a,b} = \frac{P}{2\mu} [\nu_a (\xi - \xi^3) + \nu_b (\xi' - \xi'^3)] \tag{213}$$

Fig. 81(f)

$$z = 0: \quad -R_a = R_b = \frac{W h_a}{l}$$

$$H_{a,b} = -\frac{W}{2} \left[1 \pm 1 - \frac{1 + 2\lambda_a (1 + \kappa_a)}{\mu} \right]$$

$$z = z_e: \quad -R_a = R_b = \frac{W h_b}{l}$$

$$H_{a,b} = -\frac{W}{2} \left[1 \pm 1 - \frac{1 + 2\lambda_a (1 + \kappa_a)}{\mu} \right]$$

$$\tag{214}$$

Temperature change:

$$H_{a,b} = \frac{3}{\mu} \frac{l}{s} \frac{EI_s}{h_a^2} \alpha_t \Delta T \tag{215}$$

Fixed Frames

The hingeless simple frame illustrated in Fig. 82 is three times statically indeterminate. However, making use of symmetry and antisymmetry, the frame can be computed as twice statically indeterminate for symmetric loading and as once statically indeterminate for antisymmetric loading. The redundants are

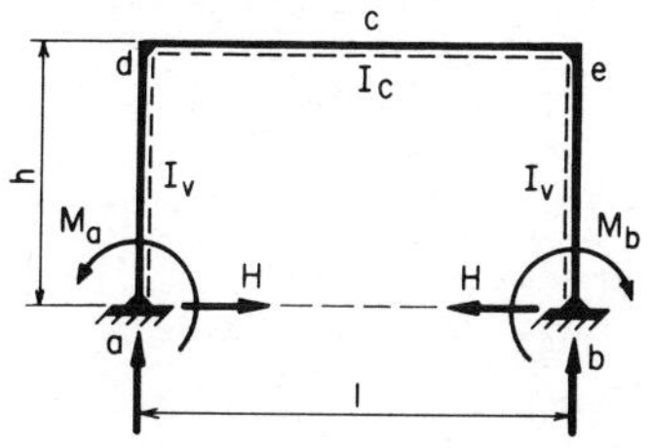

Fig. 82

$$X_1 = H \qquad X_2 = \frac{M_a - M_b}{2} \qquad X_3 = \frac{M_a + M_b}{2} \tag{216}$$

For symmetric loading (X_1 and X_2) and for antisymmetric loading (X_3), the redundants are computed from continuity equations:

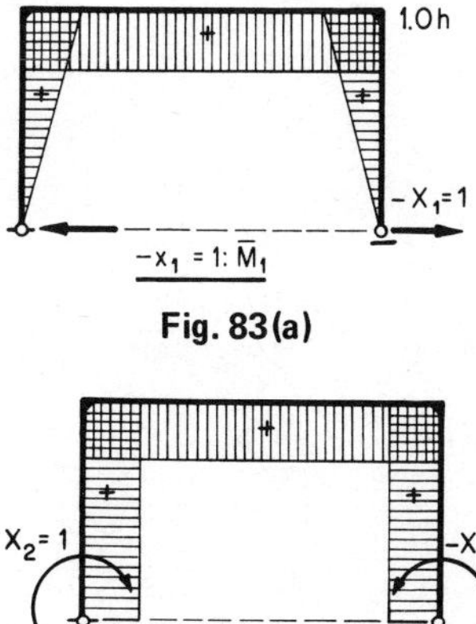

Fig. 83(a)

$$\begin{aligned} \delta_{10} - X_1\delta_{11} - X_2\delta_{12} &= 0 \\ \delta_{20} - X_1\delta_{21} - X_2\delta_{22} &= 0 \\ \delta_{30} - X_3\delta_{33} &= 0 \end{aligned} \tag{217}$$

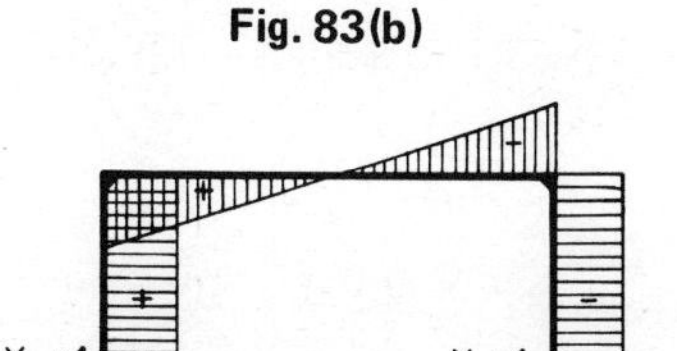

Fig. 83(b)

$$\begin{aligned} I_c &= 1.300 \text{ ft}^4 \qquad I_v = 0.666 \text{ ft}^4 \\ l &= 50.00 \text{ ft} \qquad h = 15.00 \text{ ft} \\ l' &= \frac{I_c}{I_c} l = 50.00 \text{ ft} \qquad h' = \frac{I_c}{I_v} h = 29.30 \text{ ft} \end{aligned} \tag{218}$$

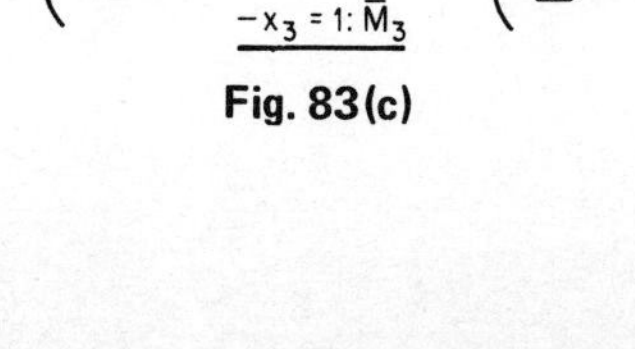

Fig. 83(c)

$$EI_c \delta_{11} = 2\int_a^d \overline{M}_1^2 \frac{I_c}{I_v} dz + \int_d^e \overline{M}_1^2 \frac{I_c}{I_c} dx = \frac{2}{3} h^2 h' + h^2 l' = 15{,}650$$

$$EI_c \delta_{22} = 2\int_a^d \overline{M}_2^2 \frac{I_c}{I_v} dz + \int_d^e \overline{M}_2^2 \frac{I_c}{I_c} dx = 2h' + l' = 108.60$$

$$EI_c \delta_{33} = 2\int_a^d \overline{M}_3^2 \frac{I_c}{I_v} dz + \frac{1}{2}\int_d^e \overline{M}_3^2 \frac{I_c}{I_c} dx = 2h' + \frac{1}{3} l' = 75.25$$

$$EI_c \delta_{12} = 2\int_a^d \overline{M}_1 \overline{M}_2 \frac{I_c}{I_v} dz + \int_d^e \overline{M}_1 \overline{M}_2 \frac{I_c}{I_c} dx = hh' + hl' = 1{,}190$$

$$EI_c \delta_{13} = 0$$
$$EI_c \delta_{23} = 0$$

(218)
(Cont.)

Symmetric loading:

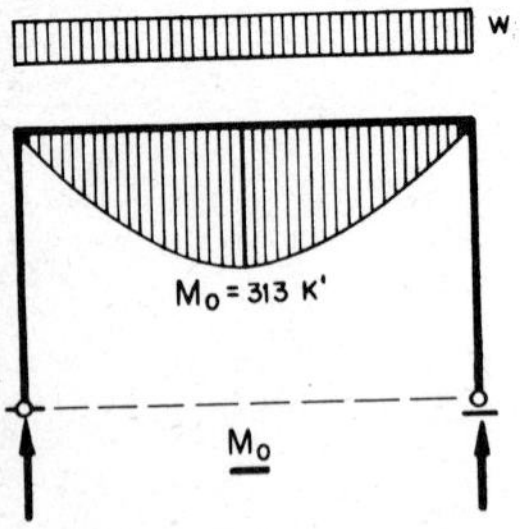

Fig. 83(d)

$w = 1.0 \ k/ft \quad l = 50.0 \ ft$

$R_{a0} = R_{b0} = \frac{1}{2} wl = 25.00 \ k$

$M_0 = \frac{wl^2}{8} = \frac{1.0 \times 50.00^2}{8} \simeq 313 \ k'$

$M_{0x} = 4 M_0 \xi \xi'$

$$EI_c \delta_{10} = \int_d^e \overline{M}_1 M_0 \frac{I_c}{I_c} dx = \frac{2}{3} \overline{M}_1 M_0 l' = \frac{2}{3} \times 15.00 \times 313 \times 50.00 = 156{,}250$$

$$EI_c \delta_{20} = \int_d^e \overline{M}_2 M_0 \frac{I_c}{I_c} dx = \frac{2}{3} \overline{M}_2 M_0 l' = \frac{2}{3} \times 1.0 \times 313 \times 50.00 = 10{,}426$$

$$EI_c \delta_{30} = 0$$

(219)

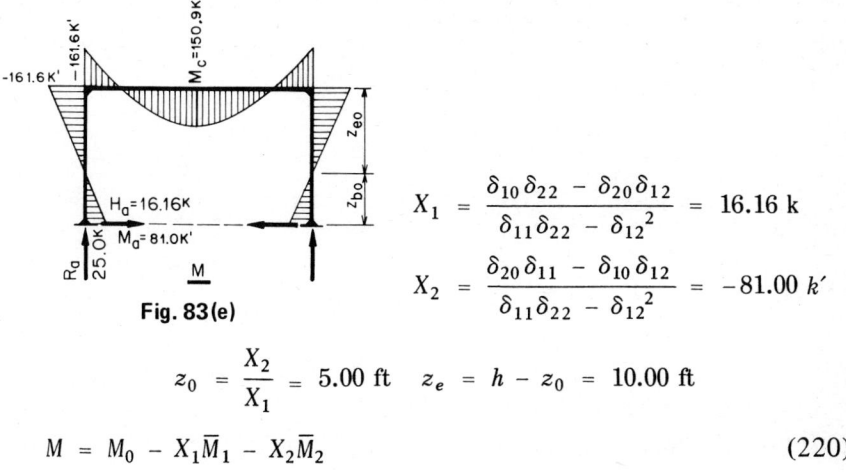

Fig. 83(e)

$$X_1 = \frac{\delta_{10}\delta_{22} - \delta_{20}\delta_{12}}{\delta_{11}\delta_{22} - \delta_{12}^2} = 16.16 \ k$$

$$X_2 = \frac{\delta_{20}\delta_{11} - \delta_{10}\delta_{12}}{\delta_{11}\delta_{22} - \delta_{12}^2} = -81.00 \ k'$$

$$z_0 = \frac{X_2}{X_1} = 5.00 \text{ ft} \quad z_e = h - z_0 = 10.00 \text{ ft}$$

$$M = M_0 - X_1\overline{M}_1 - X_2\overline{M}_2 \tag{220}$$

Antisymmetric loading:

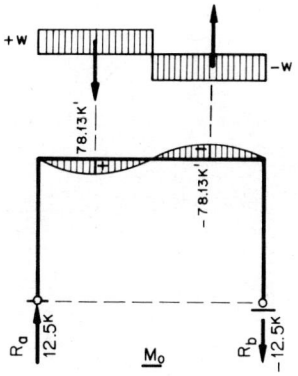

Fig. 84(a)

$$w = \pm 1.0 \ k/\text{ft}$$

$$R_a = -R_b = \tfrac{1}{4} wl = 12.50 \ k$$

$$M_{0,\max} = R_a \frac{l}{4} - \frac{wl^2}{2 \times 16.0} = 78.13 \ k'$$

$$EI_c \, \delta_{30} = 2 \int_d^c \overline{M}_3 M_0 \frac{I_c}{I_c} \, dx$$

$$= \tfrac{2}{3} \overline{M}_3 M_0 \frac{l'}{2} = 1,300$$

Fig. 84(b)

$$X_3 = \frac{\delta_{30}}{\delta_{33}} = \frac{1,300}{75.25} = 17.25 \ k'$$

$$X_1 = X_2 = 0$$

$$M = M_0 - X_3\overline{M}_3 \tag{221}$$

$$H_{a,b} = \frac{1}{2}\Sigma X_1 = +\frac{1}{2}\,16.16 = 8.08\ \text{k}$$

$$M_a = \frac{1}{2}\Sigma M_a = \frac{1}{2}(81.0 - 17.25) \simeq 31.87\ k'$$

$$M_b = \frac{1}{2}\Sigma M_b = \frac{1}{2}(81.0 + 17.25) \simeq 49.13\ k'$$

$$z_{a0} = \frac{31.87}{8.08} = 3.94\ \text{ft} \qquad z_{d0} = h - z_{a0} = 11.06\ \text{ft}$$

$$z_{b0} = \frac{49.13}{8.08} = 6.06\ \text{ft} \qquad z_{e0} = h - z_{b0} = 8.94\ \text{ft}$$

Half-span loading:

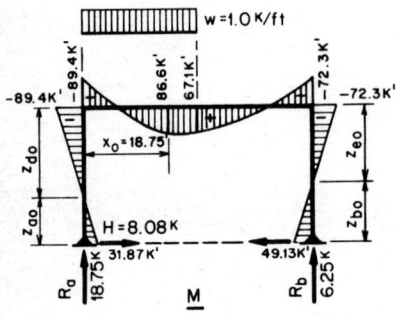

Fig. 85

$$x_0 = \frac{R_a}{w} = 18.75\ \text{ft}$$

$$M_{max} = R_a x_0 - \left(M_d + \frac{w x_0^2}{2}\right) = 86.60\ k'$$

$$M_c = R_0 \frac{l}{2} - \left(M_d + \frac{w l^2}{4}\right) = 67.10\ k'$$

$$M_d = -11.06 \times 8.08 = -89.40\ k'$$

$$M_e = -8.94 \times 8.08 = -72.30\ k'$$

For common loadings the reactions, thrust, and moments are

Combined Carrying Actions 151

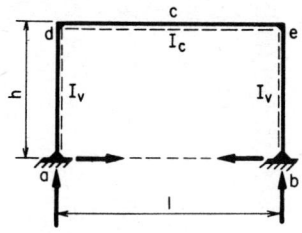

Fig. 86

$$\kappa = \frac{h}{l}\frac{I_c}{I_v} \quad \xi = \frac{x}{l} \; ; \; \xi' = \frac{x'}{l}$$

$$\mu = 2 + \kappa \quad \zeta = \frac{z}{h} \; ; \; \zeta' = \frac{z'}{h}$$

$$\nu = 1 + 6\kappa$$

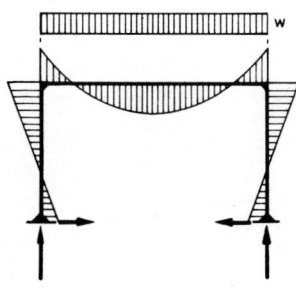

Fig. 87(a)

$$R_a = R_b = \frac{wl}{2}$$

$$H_{a,b} = \frac{1}{4\mu}\frac{wl^2}{h}$$

$$M_{a,b} = \frac{wl^2}{12\mu} \quad\quad (222)$$

$$M_{d,e} = -\frac{wl^2}{6\mu}$$

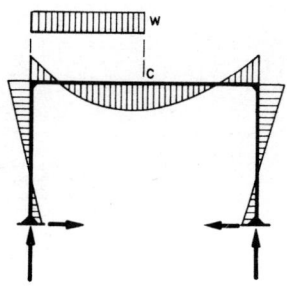

Fig. 87(b)

$$R_a = \frac{wl}{8}\left(3 + \frac{1}{4\nu}\right) \quad R_b = \frac{wl}{8}\left(1 - \frac{1}{4\nu}\right)$$

$$H_{a,b} = \frac{1}{8\mu}\frac{wl^2}{h}$$

$$M_{a,b} = \frac{wl^2}{24}\left(\frac{1}{\mu} \mp \frac{3}{8\nu}\right) \quad\quad (223)$$

$$M_{d,e} = -\frac{wl^2}{24}\left(\frac{2}{\mu} \pm \frac{3}{8\nu}\right)$$

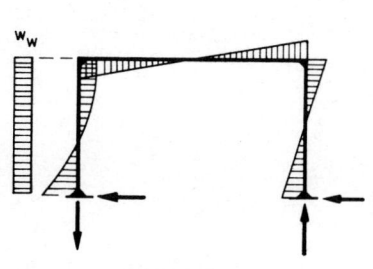

Fig. 87(c)

$$-R_a = R_b = \frac{w_w h^2}{l}\frac{\kappa}{\nu}$$

$$H_{a,b} = -\frac{w_w h}{4}\left(1 \pm 2 + \frac{1}{2\mu}\right)$$

$$M_{a,b} = -\frac{w_w h^2}{4}\left[\frac{3 + \kappa}{6\mu} \pm \left(1 - \frac{2\kappa}{\nu}\right)\right]$$

$$M_{d,e} = -\frac{w_w h^2}{4}\kappa\left(\frac{1}{6\mu} \mp \frac{2}{\nu}\right) \quad\quad (224)$$

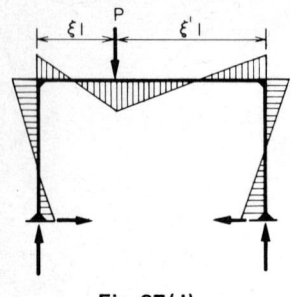

Fig. 87(d)

$$R_a = P\xi'\left(1 + \frac{\xi\xi'}{\nu}\right) \quad R_b = P - R_a$$

$$H_{a,b} = \frac{3}{2}\frac{Pl}{h}\frac{\xi\xi'}{\mu}$$

$$M_{a,b} = \frac{Pl}{2}\xi\xi'\left[\frac{1}{\mu} \mp \frac{1}{\nu}(1 - 2\xi)\right]$$

$$M_{d,e} = -\frac{Pl}{2}\xi\xi'\left[\frac{2}{\mu} \pm \frac{1}{\nu}(1 - 2\xi)\right] \quad (225)$$

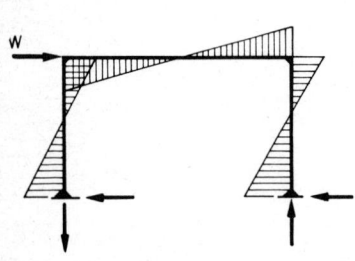

Fig. 87(e)

$$-R_a = R_b = 3\frac{Wh}{l}\frac{\kappa}{\nu}$$

$$H_{a,b} = \mp\frac{W}{2}$$

$$M_{a,b} = \mp\frac{3}{2}Wh\left(\frac{1}{3} - \frac{\kappa}{\nu}\right) \quad (226)$$

$$M_{d,e} = \pm\frac{3}{2}Wh\frac{\kappa}{\nu}$$

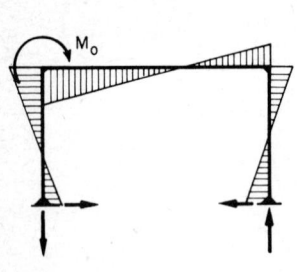

Fig. 87(f)

$$-R_a = R_b = \frac{M_0}{l}\left(1 - \frac{1}{\nu}\right)$$

$$H_{a,b} = \frac{3}{2\mu}\frac{M_0}{h}$$

$$M_{a,b} = \frac{M_0}{2}\left(\frac{1}{\mu} \mp \frac{1}{\nu}\right) \quad (227)$$

$$M_{d,e} = \frac{M_0}{2}\kappa\left(\frac{1}{\mu} \pm \frac{6}{\nu}\right)$$

Temperature change:

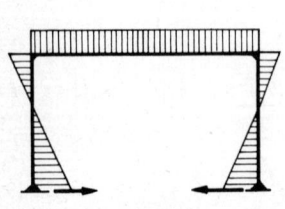

Fig. 87(g)

$$H_{a,b} = \frac{2\kappa + 1}{\kappa}\frac{3}{\mu h^2}EI_c\alpha_t\Delta T$$

$$z_0 = h\left(1 - \frac{\kappa}{2\kappa + 1}\right) \quad z_d = h - z_0$$

$$M_{a,b} = \frac{\kappa + 1}{\kappa}\frac{3}{\mu h}EI_c\alpha_t\Delta T$$

$$M_{d,e} = -\frac{3}{\mu h}EI_c\alpha_t\Delta T \quad (228)$$

16. MULTIBAY AND MULTISTORY FRAMES

These frame types commonly are to a high degree statically indeterminate, and the solution of the linear continuity equation is time-requiring. Due to this, it is more convenient to use a geometrically determinate principal system. For example, the system illustrated in Fig. 88 is six times statically indeterminate but only three times geometrically indeterminate. The redundants are the joint rotations φ_J and φ_K and the lateral displacement ψ_c. In the principal system, $\varphi_J = \varphi_K = \psi_c = 0$.

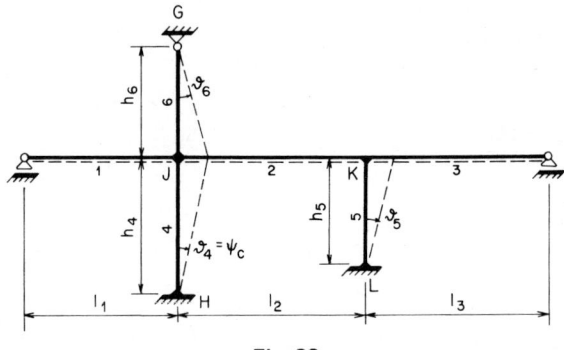

Fig. 88

The relative stiffnesses and angular changes of elements are

$$l'_n = \frac{l_c}{l_n} l_n \quad h'_n = \frac{l_c}{l_n} h_n \quad n = 1, 2, \ldots, 6$$

$$\psi_c = \vartheta_4 = 1.0 \quad \vartheta_5 = \frac{h_4}{h_5} \vartheta_4 \quad \vartheta_6 = -\frac{h_4}{h_6} \vartheta_4 \tag{229}$$

The redundants are determined from the virtual-work equations:

$$\delta A_J = a_{JJ} \varphi_J + \sum a_{JK} \varphi_K + \sum a_{Jc} \psi_c + a_{J0} = 0$$

$$\delta A_c = a_{cc} \psi_c + \sum a_{cJ} \varphi_J + a_{c0} = 0 \tag{230}$$

$$J = A, B, \ldots, K \quad c = 1, 2, \ldots, n$$

The a_{JJ}, a_{JK}, and a_{Jc} values in the first equation ($\delta A_J = 0$) are the virtual work of the joint moments M_{JJ}, M_{JK}, M_{Jc} in the principal system due to $\varphi_J = 1$, $\varphi_K = 1$, and $\psi_c = 1$. The a_{cc} and a_{cJ} values in the second equation ($\delta A_c = 0$) are the virtual work due to $\psi_c = 1$ and $\varphi_J = \varphi_K = 0$.

$\varphi_J = 1: \varphi_K = \psi_c = 0:$

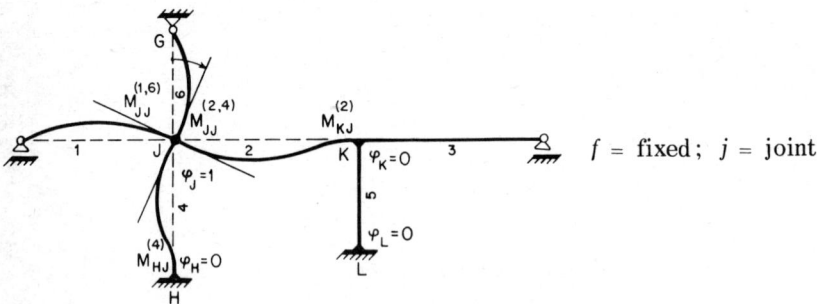

Fig. 89

$$a_{JJ} = -i_J \sum_J \left(M_{JJ}^{(f)} + M_{JJ}^{(j)} \right)$$

$$M_{JJ}^{(f)} = \frac{4}{l'} \quad M_{JJ}^{(j)} = \frac{3}{l'}$$

$$a_{JJ} = -\left(\frac{3}{l'_1} + \frac{4}{l'_2} + \frac{4}{h'_4} + \frac{3}{h'_6} \right) \tag{231}$$

$$a_{JK} = -i_J M_{JK}^{(f)} = -\frac{2}{l'_2}$$

$$M_{JK}^{(f)} = \frac{2}{l'_2} = M_{KJ}^{(f)}$$

$a_{JH} = 0 \quad (\varphi_H = 0)$

$\varphi_K = 1 : \varphi_J = \psi_c = 0:$

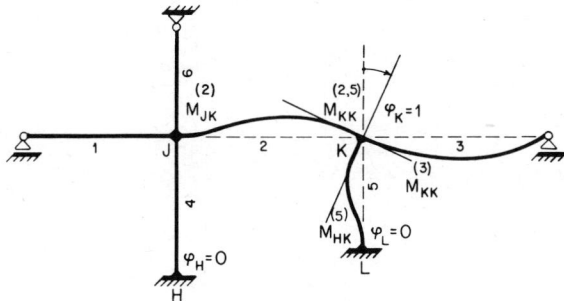

Fig. 90

$$a_{KK} = -\dot{\mathrm{i}}_K \sum_K \left(M_{KK}^{(f)} + M_{KK}^{(j)} \right)$$

$$M_{KK}^{(f)} = \frac{4}{l'} \quad M_{KK}^{(j)} = \frac{3}{l'}$$

$$a_{KK} = -\left(\frac{4}{l'_2} + \frac{3}{l'_3} + \frac{4}{h'_5} \right)$$

$$a_{KJ} = -\dot{\mathrm{i}}_K M_{KJ}^{(f)}$$

$$= -\frac{2}{l'_2} = a_{JK}$$

(232)

$\psi_c = 1 : \varphi_J = \varphi_K = 0$

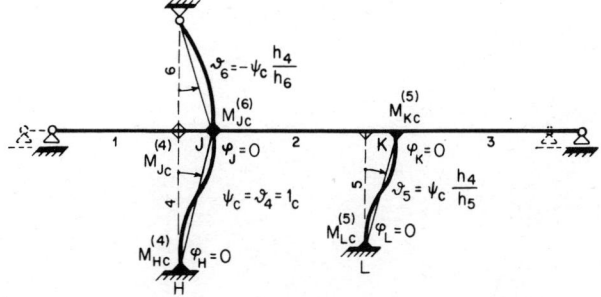

Fig. 91

$$a_{cc} = \dot{\mathrm{i}}_c \sum_c \left[\left(M_{Jc}^{(f)} + M_{Kc}^{(f)} \right) \nu_f + M_{Jc}^{(j)} \nu_j \right]$$

$$M_{Jc}^{(f)} = M_{Kc}^{(f)} = -\frac{6\vartheta_c}{h'} \quad M_{Jc}^{(j)} = -\frac{3\vartheta_c}{h'}$$

(233)

$$a_{cc} = -\dot{i}_c \sum_c \left(\frac{12\vartheta_c}{h'} v_f + \frac{3\vartheta_c}{h'} v_j\right)$$

$$= \left[-\frac{12}{h'_4}\frac{h_4}{h_4} - \frac{12}{h'_5}\frac{h_4^2}{h_5^2} + \frac{3}{h'_6}\frac{h_4}{h_6}\left(-\frac{h_4}{h_6}\right)\right]$$

$$a_{Jc} = a_{cJ} = -\dot{i}_c \sum_c \left(M_{Jc}^{(f)} + M_{Jc}^{(j)}\right) = \dot{i}_c \sum_c \left(\frac{6\vartheta_c}{h'} + \frac{3\vartheta_c}{h'}\right) \quad \text{(233) (Cont.)}$$

$$= +\left[\frac{6}{h'_4} + \frac{3}{h'_6}\left(-\frac{h_4}{h_6}\right)\right]$$

$$a_{Kc} = +\frac{6}{h'_5}\frac{h_4}{h_5}$$

The a_{J0} and a_{c0} values are the virtual work of the joint moments $M_{J0}^{(f)}$, $M_{J0}^{(j)}$ in the principal system due to $\varphi_J = \varphi_K = \psi_c = 0$.

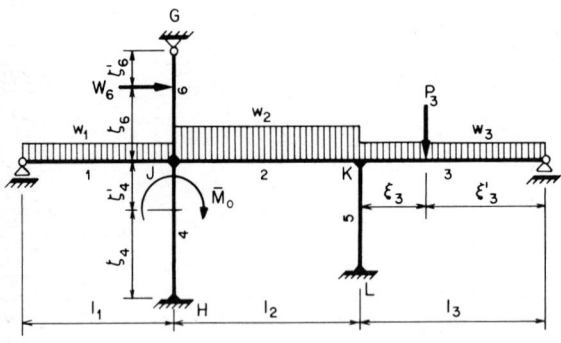

Fig. 92

$$a_{J0} = -\dot{i}_J \sum_J \left(M_{J0}^{(f)} + M_{J0}^{(j)}\right)$$

$$a_{K0} = \dot{i}_K \sum_K \left(M_{K0}^{(f)} + M_{K0}^{(j)}\right)$$

$$M_{J0}^{(f)} = M_{K0}^{(f)} = \pm\frac{wl^2}{12}$$

$$M_{J0}^{(f)} = M_{K0}^{(f)} = \pm Pl\xi\xi'^2 \quad \text{or} \quad Pl\xi^2\xi'$$

$$= \pm\overline{M}_0\,\xi'(2 - 3\xi') \quad \text{or} \quad M\xi(2 - 3\xi)$$

$$M_{J0}^{(j)} = M_{K0}^{(j)} = \pm \frac{wl^2}{8}$$

$$M_{J0}^{(j)} = M_{K0}^{(j)} = \frac{Pl}{2}(\xi' - \xi'^3)$$

$$= \frac{\overline{M}_0}{2}(3\xi' - 1)$$

$$a_{c0} = i_c \sum_c \left[\left(M_0^{(f)} + \overline{M}_0^{(f)} \right) v_f + \left(M_0^{(j)} + \overline{M}_0^{(j)} \right) v_j \right]$$

$$= i_c \left[\left(M_{J0}^{(4)} + M_{H0}^{(4)} + \overline{M}_4 \right) \frac{h_4}{h_4} - \left(M_{J0}^{(6)} \cdot \frac{h_4}{h_6} \right) \right] \tag{234}$$

The joint moments are

$$M_J^{(f)} = M_{J0}^{(f)} + \varphi_J \frac{4}{l'} + \varphi_K \frac{2}{l'} - \vartheta_c \frac{6}{l'} \qquad (l' \to h')$$

$$M_K^{(f)} = M_{K0}^{(f)} + \varphi_J \frac{2}{l'} + \varphi_K \frac{4}{l'} - \vartheta_c \frac{6}{l'} \tag{235}$$

$$M_J^{(j)} = M_{J0}^{(j)} + \varphi_J \frac{3}{l'} - \vartheta_c \frac{3}{l'}$$

To demonstrate this method, the system illustrated in Fig. 92 will be analyzed:

$l_1 = 30.0$ ft $\quad l_2 = 50.0$ ft $\quad l_3 = 40.0$ ft
$h_4 = 20.0$ ft $\quad h_5 = 15.0$ ft $\quad h_6 = 12.0$ ft

$I_1 = I_2 = I_3 = 1.0 = I_c \quad I_4 = I_5 = 0.5 I_c \quad I_6 = 0.33 I_c$

$l' = \dfrac{I_c}{I} l : \; l'_1 = 30.0$ ft ; $\; l'_2 = 50.0$ ft ; $\; l'_3 = 40.0$ ft

$h' = \dfrac{I_c}{I} h : \; h'_4 = 40.0$ ft , $\; h'_5 = 30.0$ ft , $\; h'_6 = 36.0$ ft

MATRIX of the virtual-work equations (230):

	φ_J	φ_K	ψ_c	a_{n0}
J	a_{JJ}	a_{JK}	a_{Jc}	a_{J0}
K	a_{KJ}	a_{KK}	a_{Kc}	a_{K0}
c	a_{cJ}	a_{cK}	a_{cc}	a_{c0}

158 Theory of Structures

$a_{JJ}, a_{cc}, a_{JK}, a_{Jc}$ values:

$$a_{JJ} = -\left(\frac{3}{l'_1} + \frac{4}{l'_2} + \frac{4}{h'_4} + \frac{3}{h'_6}\right) = -\left(\frac{3}{30} + \frac{4}{50} + \frac{4}{40} + \frac{3}{36}\right) = -0.363$$

$$a_{KK} = -\left(\frac{4}{l'_2} + \frac{3}{l'_3} + \frac{4}{h'_5}\right) = -\left(\frac{4}{50} + \frac{3}{40} + \frac{4}{30}\right) = -0.288$$

$$a_{cc} = -\left(\frac{12}{h'_4} + \frac{12}{h'_5} \cdot \frac{h_4^2}{h_5^2} + \frac{3}{h'_6} \cdot \frac{h_4^2}{h_6^2}\right) = -\left(\frac{12}{40} + \frac{12}{30} \cdot \frac{20^2}{15^2} + \frac{3}{36} \cdot \frac{20^2}{12^2}\right)$$
$$= -1.242$$

$$a_{JK} = a_{KJ} = -\frac{2}{l'_2} = -\frac{2}{50} = -0.040$$

$$a_{Jc} = a_{cJ} = +\left(\frac{6}{h'_4} - \frac{3}{h'_6} \cdot \frac{h_4}{h_6}\right) = +\left(\frac{6}{40} - \frac{3}{36} \cdot \frac{20}{12}\right) = +0.011$$

$$a_{Kc} = a_{cK} = +\frac{6}{h'_5} \cdot \frac{h_4}{h_5} = \frac{6}{30} \cdot \frac{20}{15} = +0.267$$

a_{J0}, a_{c0} values:

Uniformly distributed load $w = 1.0$ k/ft on all spans:

$$M_{J0}^{(1)} = +\tfrac{1}{8} 1.0 \times 30.0^2 = +112\,k'$$

$$M_{J0}^{(2)} = -\tfrac{1}{12} 1.0 \times 50.0^2 = -208\,k'$$

$$M_{K0}^{(2)} = -M_{J0}^{(2)} = +208\,k'$$

$$M_{K0}^{(3)} = -\tfrac{1}{8} 1.0 \times 40.0^2 = -200\,k'$$

$$a_{J0} = -\sum M_{J0} = -(+112 - 208) = +96.0$$

$$a_{K0} = -\sum M_{K0} = -(+208 - 200) = -8.0$$

$$a_{c0} = 0$$

MATRIX:

	φ_J	φ_K	ψ_c	a_{n0}	Iteration results:
J	−0.363	−0.040	0.011	−96.0	$\varphi_J = +260.3$
K	−0.040	−0.288	0.267	+8.0	$\varphi_K = -77.1$
c	0.011	0.267	−1.242	0.0	$\psi_c = -14.2$

Moments:

$$M_J^{(1)} = M_{J0}^{(1)} + \frac{3}{l_1'}\varphi_J = +138.0\,k'$$

$$M_J^{(2)} = M_{J0}^{(2)} + \frac{1}{l_2'}(4\varphi_J + 2\varphi_K) = -190.2\,k'$$

$$M_J^{(4)} = \text{———} + \frac{1}{h_4'}(4\varphi_J - 6\psi_c) = +28.2\,k'$$

$$M_J^{(6)} = \text{———} + \frac{3}{h_6'}\left(\varphi_J - \psi_c\frac{h_4}{h_6}\right) = +23.9\,k'$$

$$M_H^{(4)} = \text{———} + \frac{1}{h_4'}(2\varphi_J - 6\psi_c) = +15.2\,k'$$

$$\sum_J M = 0$$

$$M_K^{(2)} = M_{K0}^{(2)} + \frac{1}{l_2'}(4\varphi_K + 2\varphi_J) = +212.2\,k'$$

$$M_K^{(3)} = M_{K0}^{(3)} + \frac{1}{l_3'}3\varphi_K = -205.8\,k'$$

$$M_K^{(5)} = \text{———} + \frac{1}{h_5'}\left(4\varphi_K - 6\psi_c\frac{h_4}{h_5}\right) = -6.5\,k'$$

$$\sum_K M = 0$$

$$M_L^{(4)} = \text{———} + \frac{1}{h_5'}\left(2\varphi_K - 6\psi_c\frac{h_4}{h_5}\right) = -2.3\,k'$$

Moment diagrams

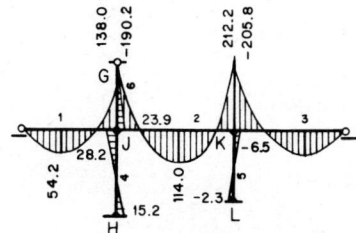

Fig. 93

Temperature change of beams 1, 2, and 3, $\Delta T = +60°\,F$:

$$EI_c\,\alpha_t\,\Delta T = 7.20 \times 10^5 \times 1.0 \times 0.79 \times 10^{-5} \times 60 = 342.0$$

$$\vartheta_{40} = 0 \quad \vartheta_{60} = 0 \quad a_{J0} = 0$$

$$\vartheta_{50} = EI_c\,\alpha_t\,\Delta T\,\frac{l_2}{h_5} = \frac{342.0 \times 50.0}{15} = 1140.0 \qquad (236)$$

160 Theory of Structures

$$a_{K0} = \frac{6\vartheta_{50}}{h_5'} = \frac{6 \times 1140.0}{30} = 228.0$$

$$a_{c0} = -\frac{12\vartheta_{50}}{h_5'} \cdot \frac{h_4}{h_5} = -\frac{12 \times 1140.0}{30} \cdot \frac{20}{15} = -607.5$$

(236)
(Cont.)

MATRIX:

	φ_J	φ_K	ψ_c	a_{n0}
J	−0.363	−0.040	0.011	0.0
K	−0.040	−0.288	0.267	−228.0
c	0.011	0.267	−1.242	+607.5

Iteration results:

$\varphi_J = -59.8$
$\varphi_K = +433.0$
$\psi_c = -396.0$

$$\vartheta_1 = \vartheta_2 = \vartheta = 0$$

$$\vartheta_4 = \vartheta_{40} + \psi_c = 0.0 - 396.0 = -396$$

$$\vartheta_5 = \vartheta_{50} + \psi_c \frac{h_4}{h_5} = 1{,}140 - 396 \frac{20}{15} = 613$$ (237)

$$\vartheta_6 = \vartheta_{60} + \psi_c \frac{h_4}{h_6} = 0.0 - 396\left(-\frac{20}{12}\right) = 660$$

Moments:

$$M_J^{(1)} = \frac{3}{l_1'} \varphi_J = -\frac{3}{30} 59.8 = -5.98\,k'$$

$$M_J^{(2)} = \frac{2}{l_2'}(2\varphi_J + \varphi_K) = \frac{2}{50}(-2 \times 59.8 + 433) = +12.5\,k'$$

$$M_J^{(4)} = \frac{2}{h_4'}(2\varphi_J - 3\vartheta_4) = \frac{2}{40}(-2 \times 59.8 + 3 \times 396) = +53.4\,k'$$

$$M_J^{(6)} = \frac{3}{h_6'}(\varphi_J - \vartheta_6) = \frac{3}{36}(-59.8 - 660) = -60.0\,k'$$

$\sum_J M \simeq 0$

$$M_H^{(4)} = \frac{2}{h_6'}(\varphi_J - 3\vartheta_4) = \frac{2}{40}(-59.8 + 3 \times 396) = +56.3\,k'$$

$$M_K^{(2)} = \frac{2}{l_2'}(2\varphi_K + \varphi_J) = \frac{2}{50}(2 \times 433 - 59.8) = +32.3\,k'$$

$$M_K^{(3)} = \frac{3}{l_3'} \varphi_K = \frac{3}{40} 433 = +32.5\,k'$$

$\sum_K M = 0$

$$M_K^{(5)} = \frac{2}{h_5'}(2\varphi_K - 3\vartheta_5) = \frac{2}{30}(2 \times 433 - 3 \times 613) = -64.8\,k'$$

$$M_L^{(5)} = \frac{2}{h_5'}(\varphi_K - 3\vartheta_5) = \frac{2}{30}(433 - 3 \times 613) = -93.8\,k'$$

Temperature change $\Delta T = +60°F$.

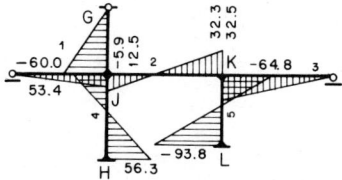

Fig. 94

Analysis of a Multibay, Multistory Structure

To analyze a multistory symmetric frame, use can be made of symmetry and antisymmetry. For illustration, a six-story three-bay garage structure (Fig. 95) will be analyzed.

Relative stiffnesses:

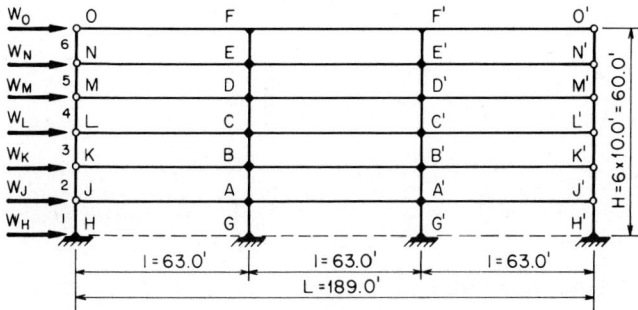

Fig. 95

$I_l = 54{,}000 \text{ in.}^4 = I_c$

$I_v = 13{,}840 \text{ in.}^4$

$l' = \dfrac{I_c}{I_l} l = 63.0 \text{ ft}$

$h' = \dfrac{I_c}{I_v} h = \dfrac{54{,}000}{13{,}840} \cdot 10.0 = 39.0 \text{ ft}$

Principal system (geometric):

Joint rotations: $\varphi_{A, B, \ldots, F} = \varphi_{A', B', \ldots, F'} = 0$
Joint displacements: $\psi_{1, 2, \ldots, 6} = 0$

Symmetric loading—redundants:

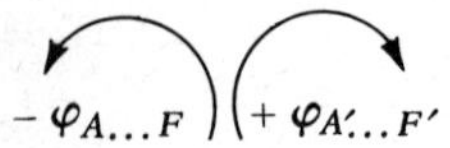

$$\bar{\varphi}_A = \frac{\varphi_A - \varphi'_A}{2} \qquad \bar{\varphi}_B = \frac{\varphi_B - \varphi'_B}{2}$$

$$\bar{\varphi}_C = \frac{\varphi_C - \varphi'_C}{2} \qquad \bar{\varphi}_D = \frac{\varphi_D - \varphi'_D}{2} \qquad (238)$$

$$\bar{\varphi}_E = \frac{\varphi_E - \varphi'_E}{2} \qquad \bar{\varphi}_F = \frac{\varphi_F - \varphi'_F}{2}$$

$$\psi_c = 0 \quad (c = 1, 2, \ldots, 6) \qquad \varphi_A = -\varphi'_A = \bar{\varphi}_A$$

$-\varphi_A \ldots F \quad +\varphi_{A'} \ldots F'$

MATRIX of the virtual-work equations (230):

	$\bar{\varphi}_A$	$\bar{\varphi}_B$	$\bar{\varphi}_C$	$\bar{\varphi}_D$	$\bar{\varphi}_E$	$\bar{\varphi}_F$	
A	a_{AA}	a_{AB}	—	—	—	—	a_{A0}
B	a_{BA}	a_{BB}	a_{BC}	—	—	—	a_{B0}
C	—	a_{CB}	a_{CC}	a_{CD}	—	—	a_{C0}
D	—	—	a_{DC}	a_{DD}	a_{DE}	—	a_{D0}
E	—	—	—	a_{ED}	a_{EE}	a_{EF}	a_{E0}
F	—	—	—	—	a_{FE}	a_{FF}	a_{F0}

$a_{JJ}\, a_{JK}$ values:

$$a_{AA} = -2\left(\frac{2}{l'} + \frac{3}{l'} + \frac{4}{h'} + \frac{4}{h'}\right) = -2\left(\frac{5}{63.0} + \frac{8}{39.0}\right) = -0.5685$$

$$= a_{BB} = a_{CC} = a_{DD} = a_{EE}$$

$$a_{FF} = -2\left(\frac{2}{l'} + \frac{3}{l'} + \frac{4}{h'}\right) = -2\left(\frac{5}{63.0} + \frac{4}{39.0}\right) = -0.3635 \qquad (239)$$

$$a_{AB} = a_{BA} \ldots a_{EF} = a_{FE} = -2\left(\frac{2}{h'}\right) = -\frac{4}{39.0} = -0.1025$$

a_{J0} values:

$$M_{A0}^{(A')} = -\frac{1}{12}wl^2 = -330\,k'$$

$$M_{A'0}^{(A)} = +330\,k' \qquad (240)$$

$$a_{A0} = a_{B0} \ldots a_{F0} = -2M_{A0}^{(A')} = +660$$

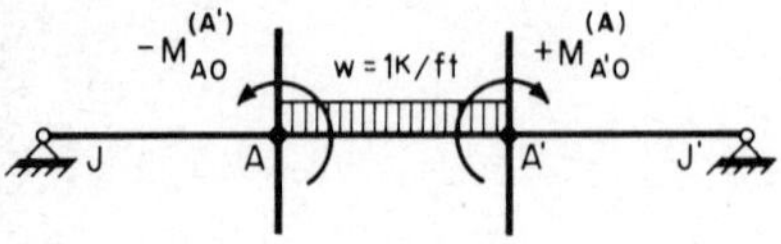

$-M_{A0}^{(A')} \quad w = 1K/ft \quad +M_{A'0}^{(A)}$

J A A' J'

Fig. 96 (a)

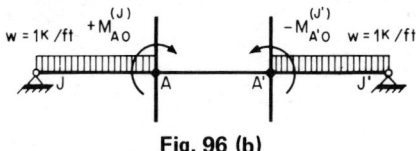

Fig. 96 (b)

$$M_{A0}^{(J)} = +\frac{1}{8}wl^2 = +496\,k'$$
$$M_{A'0}^{(J')} = -496\,k' \tag{241}$$
$$a_{A0} = a_{B0} \ldots a_{F0} = -2M_{A0}^{(J)} = -992$$

MATRIX:

	$\bar{\varphi}_A$	$\bar{\varphi}_B$	$\bar{\varphi}_C$	$\bar{\varphi}_D$	$\bar{\varphi}_E$	$\bar{\varphi}_F$	a	b
A	−0.5685	−0.1025	−	−	−	−	−660	+992
B	−0.1025	−0.5685	−0.1025	−	−	−	−660	+992
C	−	−0.1025	−0.5685	−0.1025	−	−	−660	+992
D	−	−	−0.1025	−0.5685	−0.1025	−	−660	+992
E	−	−	−	−0.1025	−0.5685	−0.1025	−660	+992
F	−	−	−	−	−0.1025	−0.3635	−660	+992

Iteration results:
Loading a:
$$\varphi_A = -\varphi_{A'} = +1{,}012 \quad \varphi_B = -\varphi_{B'} = +821$$
$$\varphi_C = -\varphi_{C'} = +855 \quad \varphi_D = -\varphi_{D'} = +878$$
$$\varphi_E = -\varphi_{E'} = +712 \quad \varphi_F = -\varphi_{F'} = +1{,}612$$

Loading b:
$$\varphi_A = -\varphi_{A'} = -1{,}521 \quad \varphi_B = -\varphi_{B'} = -1{,}240$$
$$\varphi_C = -\varphi_{C'} = -1{,}281 \quad \varphi_D = -\varphi_{D'} = -1{,}320$$
$$\varphi_E = -\varphi_{E'} = -1{,}072 \quad \varphi_F = -\varphi_{F'} = -2{,}422$$

Joint moments:
$$M_J^{(f)} = M_{J0} + \frac{1}{l'}(4\varphi_J + 2\varphi_K)$$
$$M_J^{(i)} = M_{J0} + \frac{1}{l'}3\varphi_J$$

$$\sum_J M = 0$$

164 Theory of Structures

Loading a:

$$M_G = \frac{1}{39.0} 2 \times 1{,}012 = 52.0\,k'$$

$$M_A{}^{(A')} = -330 + \frac{1}{63.0}(4 \times 1{,}012 - 2 \times 1{,}012) = -297.6\,k'$$

$$M_A{}^{(J)} = \underline{} + \frac{1}{63.0} 3 \times 1{,}012 = 48.2\,k'$$

$$M_A{}^{(B)} = \underline{} + \frac{1}{39.0}(4 \times 1{,}012 + 2 \times 821) = 145.4\,k' \qquad \sum_A M = 0$$

$$M_A{}^{(G)} = \underline{} + \frac{1}{39.0} 4 \times 1{,}012 = 104.0\,k'$$

$$M_B{}^{(B')} = -330 + \frac{1}{63.0}(4 \times 821 - 2 \times 821) = -304.0\,k'$$

$$M_B{}^{(K)} = \underline{} + \frac{1}{63.0} 3 \times 821 = 39.0\,k'$$

$$M_B{}^{(C)} = \underline{} + \frac{1}{39.0}(4 \times 821 + 2 \times 855) = 128.0\,k' \qquad \sum_B M = 0$$

$$M_B{}^{(A)} = \underline{} + \frac{1}{39.0}(4 \times 821 + 2 \times 1{,}012) = 137.0\,k'$$

$$M_C{}^{(C')} = -302.7\,k' \quad M_C{}^{(L)} = 40.5\,k' \quad M_C{}^{(D)} = 132.2\,k'$$
$$M_C{}^{(B)} = 130.0\,k'$$

$$M_D{}^{(D')} = -302.2\,k' \quad M_D{}^{(M)} = 41.8\,k' \quad M_D{}^{(E)} = 126.5\,k'$$
$$M_D{}^{(C)} = 133.9\,k'$$

$$M_E{}^{(E')} = -307.4\,k' \quad M_E{}^{(N)} = 34.2\,k' \quad M_E{}^{(F)} = 155.2\,k'$$
$$M_E{}^{(D)} = 118.0\,k'$$

$$M_F{}^{(F')} = -278.8\,k' \quad M_F{}^{(O)} = 77.0\,k' \quad M_F{}^{(E)} = 201.8\,k'$$

Loading b:

$$M_G = -\frac{1}{39.0} 2 \times 1{,}521 = -78.0\,k'$$

$$M_A{}^{(A')} = \underline{} - \frac{1}{63.0}(4 \times 1{,}521 - 2 \times 1{,}521) = -48.0\,k'$$

$$M_A{}^{(J)} = +496 - \frac{1}{63.0} 3 \times 1{,}521 = 423.5\,k'$$

$$M_A{}^{(B)} = \underline{} - \frac{1}{39.0}(4 \times 1{,}521 + 2 \times 1{,}240) = -219.5\,k' \qquad \sum_A M = 0$$

$$M_A{}^{(G)} = \underline{} - \frac{1}{39.0} 4 \times 1{,}521 = -156.0\,k'$$

$M_B^{(B')} = -40.0\,k'$ $M_B^{(K)} = +437.4\,k'$ $M_B^{(C)} = -192.4\,k'$
$M_B^{(A)} = -205.0\,k'$
$M_C^{(C')} = -40.7\,k'$ $M_C^{(L)} = +434.7\,k'$ $M_C^{(D)} = -199.0\,k'$
$M_C^{(B)} = -195.0\,k'$
$M_D^{(D')} = -41.9\,k'$ $M_D^{(M)} = +433.0\,k'$ $M_D^{(E)} = -190.1\,k'$
$M_D^{(C)} = -201.0\,k'$
$M_E^{(E')} = -34.1\,k'$ $M_E^{(N)} = +445.1\,k'$ $M_E^{(F)} = -233.5\,k'$
$M_E^{(D)} = -177.5\,k'$
$M_F^{(F')} = -77.0\,k'$ $M_F^{(O)} = +380.0\,k'$ $M_F^{(E)} = -303.0\,k'$

Moment diagrams for loading a and b are plotted in Fig. 97.

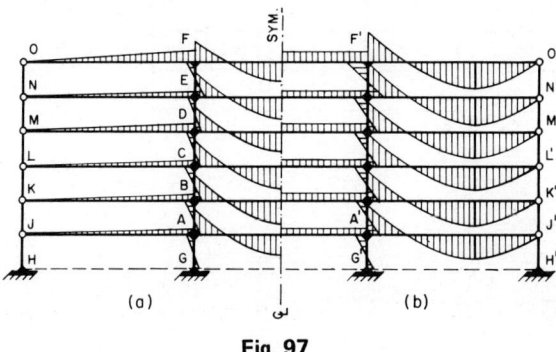

Fig. 97

Critical loading conditions are obtained by superposition of loadings a and b.

Antisymmetric loading—redundants:

$$\bar{\varphi}_A = \frac{\varphi_A + \varphi_{A'}}{2} \qquad \bar{\varphi}_B = \frac{\varphi_B + \varphi_{B'}}{2}$$

$$\bar{\varphi}_C = \frac{\varphi_C + \varphi_{C'}}{2} \qquad \bar{\varphi}_D = \frac{\varphi_D + \varphi_{D'}}{2}$$

$$\bar{\varphi}_E = \frac{\varphi_E + \varphi_{E'}}{2} \qquad \bar{\varphi}_F = \frac{\varphi_F + \varphi_{F'}}{2} \qquad (242)$$

$$\psi_c = \frac{\varphi_c + \varphi_{c'}}{2} \quad (c = 1, 2, \ldots, 6)$$

$$\varphi_A = \varphi_{A'} = \bar{\varphi}_A$$

Theory of Structures

MATRIX of the virtual-work equations (230):

	$\bar\varphi_A$	$\bar\varphi_B$	$\bar\varphi_C$	$\bar\varphi_D$	$\bar\varphi_E$	$\bar\varphi_F$	ψ_1	ψ_2	ψ_3	ψ_4	ψ_5	ψ_6	
A	a_{AA}	a_{AB}					a_{A1}	a_{A2}					a_{A0}
B	a_{BA}	a_{BB}	a_{BC}					a_{B2}	a_{B3}				a_{B0}
C		a_{CB}	a_{CC}	a_{CD}					a_{C3}	a_{C4}			a_{C0}
D			a_{DC}	a_{DD}	a_{DE}					a_{D4}	a_{D5}		a_{D0}
E				a_{ED}	a_{EE}	a_{EF}					a_{E5}	a_{E6}	a_{E0}
F					a_{FE}	a_{FF}						a_{F6}	a_{F0}
1	a_{1A}						a_{11}						a_{10}
2	a_{2A}	a_{2B}						a_{22}					a_{20}
3		a_{3B}	a_{3C}						a_{33}				a_{30}
4			a_{4C}	a_{4D}						a_{44}			a_{40}
5				a_{5D}	a_{5E}						a_{55}		a_{50}
6					a_{6E}	a_{6F}						a_{66}	a_{60}

a_{JJ}, a_{JK} and a_{CC}, a_{JC} values:

$$a_{AA} = -2\left(\frac{3}{l'} + \frac{6}{l'} + \frac{4}{h'} + \frac{4}{h'}\right) = -0.696$$

$$= a_{BB}, \ldots, a_{EE}$$

$$a_{FF} = -2\left(\frac{3}{l'} + \frac{6}{l'} + \frac{4}{h'}\right) = -0.491 \qquad (243)$$

$$a_{AB} = a_{BA} = -2\frac{2}{h'} = -0.1025$$

$$= a_{BC}, \ldots, a_{EF}$$

$$a_{11} = -2\frac{12}{h'} = -0.6154$$

$$= a_{22}, \ldots, a_{66} \qquad (244)$$

$$a_{A1} = a_{1A} = +2\frac{2}{h'} = +0.3077$$

$$= a_{A2}, a_{B2}, \ldots, a_{F6}$$

a_{J0} and a_{c0} values:

Wind (130 mph)

$$a_{A0} = a_{B0}, \ldots, a_{F0} = 0$$

$$W_w = \zeta \frac{\rho}{2} v^2 A_w = 5.30 \, k/\text{ft}$$

$$\zeta = 1.6 = \text{shape factor} \tag{245}$$

$$\rho = \frac{w_A}{G} \simeq 0.125 \text{ kg sec}^2/\text{m}^4$$

$$v = \frac{130 \times 1.66 \times 1{,}000}{60^2} \simeq 60 \text{ m/sec} = \text{wind velocity}$$

$$A_w = 72.5 \text{ ft}^2/\text{fl} = \text{wind area per floor}$$

$$W_H = \tfrac{1}{2} \times 5.30 = 2.65 \text{ kips} = W_0 \quad \Big| \quad \sum_H^0 W_w = 31.80 \, k$$

$$W_{A,\ldots,N} = 5.30 \, k$$

$$a_{10} = 1 h_1 \sum_H^0 W = 10.0 \,(31.80 - 2.65) = 291.5$$

$$a_{20} = 1 h_2 \sum_I^0 W = 10.0 \,(29.15 - 5.30) = 238.5 \tag{246}$$

$$a_{30} = 1 h_3 \sum_K^0 W = 10.0 \,(23.85 - 5.30) = 185.5$$

$$a_{40} = 1 h_4 \sum_L^0 W = 10.0 \,(18.55 - 5.30) = 132.5$$

$$a_{50} = 1 h_5 \sum_M^0 W = 10.0 \,(13.25 - 5.30) = 79.5$$

$$a_{60} = 1 h_6 \sum_N^0 W = 10.0 \,(7.95 - 5.30) = 26.5$$

REDUCED MATRIX:

Because the number of redundants is large and the a_{Jc} values relative to the a_{cc} value are high, the conversion of the matrix is poor. To make it

168 Theory of Structures

possible to solve the equations by iteration, the ψ_c values will be eliminated from the matrix. In general,

$$\psi_c = -\frac{1}{a_{cc}}(\overline{\varphi}_{J-1}a_{J-1,c} + \overline{\varphi}_J a_{Jc} + a_{c0}) \tag{247}$$

$$\bar{a}_{J(J+1)} = a_{J(J-1)} - \frac{a_{Jc}}{a_{cc}} a_{c(J-1)} \quad J = A, \ldots, F, \quad c = 1, \ldots, 6$$

$$\bar{a}_{JJ} = a_{JJ} - \left(\frac{a_{Jc}}{a_{cc}} a_{cJ} + \frac{a_{J(c+1)}}{a_{(c+1)(c+1)}} a_{(c+1)J}\right) \tag{248}$$

$$\bar{a}_{J0} = -\left(\frac{a_{Jc}}{a_{cc}} a_{c0} + \frac{a_{J(c+1)}}{a_{(c+1)(c+1)}} a_{(c+1)0}\right)$$

$\bar{a}_{JJ}$, $\bar{a}_{JK}$, and $\bar{a}_{J0}$ values:

$$\bar{a}_{BA} = -0.1025 + \frac{0.3077}{0.6154} \cdot 0.3077 = +0.051$$

$$= \bar{a}_{AB} \ldots \bar{a}_{EF}$$

$$\bar{a}_{AA} = -0.696 + 2 \cdot \frac{0.3077^2}{0.6154} = -0.389$$

$$= \bar{a}_{BB} \ldots \bar{a}_{EE}$$

$$\bar{a}_{FF} = -0.491 + \frac{0.3077^2}{0.6154} = -0.3375$$

$$\bar{a}_{A0} = -\frac{0.3077}{0.6154}(291.5 + 238.5) = -265.0$$

$$\bar{a}_{B0} = -\frac{0.3077}{0.6154}(238.5 + 185.5) = -212.0$$

$$\bar{a}_{C0} = -\frac{0.3077}{0.6154}(185.5 + 132.5) = -159.0$$

$$\bar{a}_{D0} = -\frac{0.3077}{0.6154}(132.5 + 79.5) = -106.0$$

$$\bar{a}_{E0} = -\frac{0.3077}{0.6154}(79.5 + 26.5) = -53.0$$

$$\bar{a}_{F0} = -\frac{0.3077}{0.6154} \cdot 26.5 = -13.25$$

REDUCED MATRIX:

	$\bar{\varphi}_A$	$\bar{\varphi}_B$	$\bar{\varphi}_C$	$\bar{\varphi}_D$	$\bar{\varphi}_E$	$\bar{\varphi}_F$	$\bar{a}_{J0}$
A	−0.389	+0.051					−265.0
B	+0.051	−0.389	+0.051				−212.0
C		+0.051	−0.389	+0.051			−159.0
D			+0.051	−0.389	+0.051		−106.0
E				+0.051	−0.389	+0.051	−53.0
F					+0.051	−0.3375	−13.25

Iteration results:

$\bar{\varphi}_A = 774 \quad \bar{\varphi}_B = 720 \quad \bar{\varphi}_C = 552 \quad \bar{\varphi}_D = 370 \quad \bar{\varphi}_E = 194 \quad \bar{\varphi}_F = 69$

ψ_c values:

$$\psi_1 = \frac{1}{0.6154}[(774 + 0.0)0.3077 + 291.5] = 860$$

$$\psi_2 = \frac{1}{0.6154}[(774 + 720)0.3077 + 238.5] = 1{,}134$$

$$\psi_3 = \frac{1}{0.6154}[(720 + 552)0.3077 + 185.5] = 930$$

$$\psi_4 = \frac{1}{0.6154}[(552 + 370)0.3077 + 132.5] = 678$$

$$\psi_5 = \frac{1}{0.6154}[(370 + 194)0.3077 + 79.5] = 410$$

$$\psi_6 = \frac{1}{0.6154}[(194 + 69)0.3077 + 26.5] = 175$$

Joint moments:

$$M_G = \frac{1}{h'}(2\bar{\varphi}_A - 6\psi_1) = -92.6\,k'$$

$$M_A^{(G)} = \frac{1}{h'}(4\bar{\varphi}_A - 6\psi_1) = -53.0\,k'$$

$$M_A^{(J)} = \frac{1}{l'}3\bar{\varphi}_A = +36.9\,k'$$

$$M_A^{(A')} = \frac{1}{l'}(4\bar{\varphi}_A + 2\bar{\varphi}_A) = +73.8\,k'$$

$$M_A^{(B)} = \frac{1}{h'}(4\bar{\varphi}_A + 2\varphi_B - 6\psi_2) = -58.0\,k'$$

$\sum_A M \simeq 0$

$M_B^{(A)} = -61.0\,k'$ $\quad M_B^{(K)} = +34.2\,k'$ $\quad M_B^{(B')} = +68.4\,k'$

$M_B^{(C)} = -41.0\,k'$ $\quad M_C^{(B)} = -50.0\,k'$ $\quad M_C^{(L)} = +26.2\,k'$

$M_C^{(C')} = +52.3\,k'$ $\quad M_C^{(D)} = -28.5\,k'$ $\quad M_D^{(C)} = -38.2\,k'$

$M_D^{(M)} = +17.5\,k'$ $\quad M_D^{(D')} = +35.0\,k'$ $\quad M_D^{(E)} = -15.1\,k'$

$M_E^{(D)} = -24.2\,k'$ $\quad M_E^{(N)} = +9.2\,k'$ $\quad M_E^{(E')} = +18.5\,k'$

$M_E^{(F)} = -3.5\,k'$ $\quad M_F^{(E)} = -10.0\,k'$ $\quad M_F^{(O)} = +3.3\,k'$

$M_F^{(F')} = +6.6\,k'$

Moment diagram for wind loading (Fig. 98):

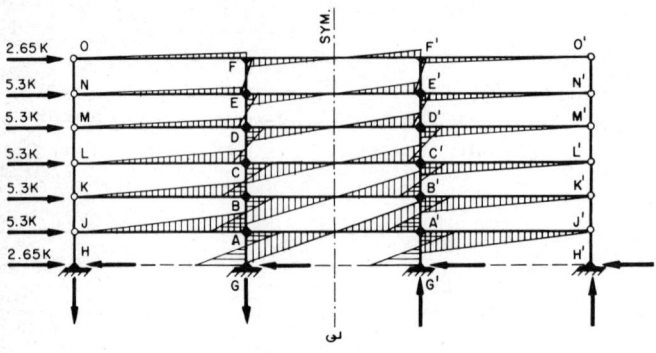

Fig. 98

17. FLAT SLABS

Flat slabs are two-way reinforced slabs supported directly by columns. The connections of the slabs and columns are considered rigid. For heavily loaded flat slabs with variable live loads, the capitals of the columns in most cases are required to keep the shear stresses within safe limits. The column types commonly used are illustrated in Fig. 99.

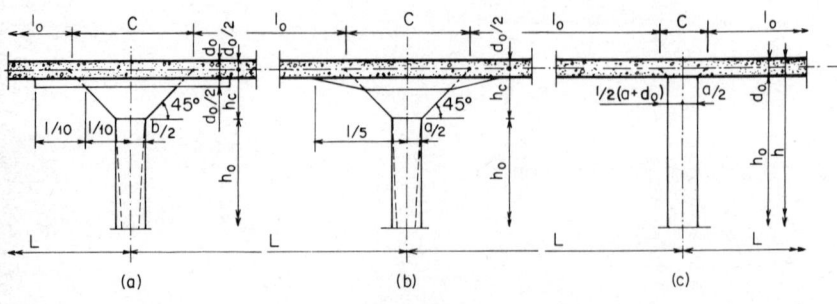

Fig. 99

The portion of the capital which lies outside the 45° angle is considered in structural analysis as nonexistent. The columns can be rectangular, square, or round and straight or tapered. The thickness of the slab must not be less than 1/36 of the largest free span ($l_{0,\,max}$) and not less than 5 in.

Elastic analysis in accordance with slab theory, numerous tests, and experience prove that the structural behavior of flat-slab structures is rather close to that of multistory and multibay frames. Thus, for structural analysis of flat slabs, the structure is considered to be divided into frames, each consisting of a row of columns and the strips of slabs bounded laterally by the centerline of the panel on either side of the centerline of columns. The division and analysis are carried out longitudinally and transversally (Fig. 100).

The lengths of free spans of slabs (l_0) and columns (h_0), for computation of critical moments and relative stiffnesses (l', h') for infinitely rigid capitals, are (Fig. 99):

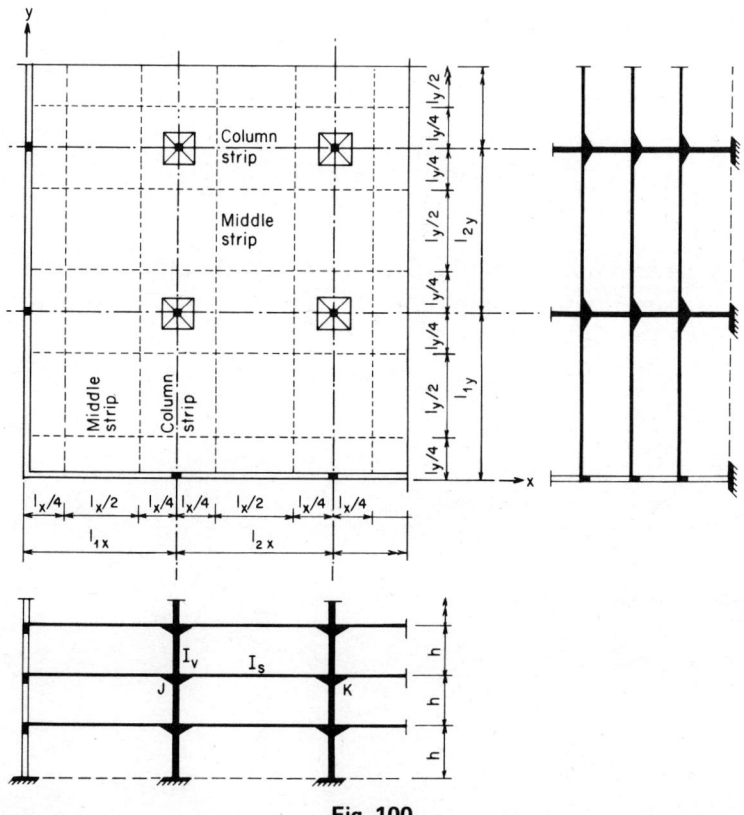

Fig. 100

172   Theory of Structures

$$l_{10} = l_1 - c/2$$
$$l_{20} = l_2 - c$$
$$h_0 = h - (h_c + d_0/2)$$

$$I_{sx} = \frac{l_y d_0^3}{12}$$
$$I_{sy} = \frac{l_x d_0^3}{12} \quad \bigg| \quad I_s = I_c$$

$$I_{vx} = \frac{ba^3}{12}$$
$$I_{vy} = \frac{ab^3}{12}$$

Thus the relative stiffnesses are

$$l'_{x0} = \frac{I_c}{I_{sx}} l_{x0} \quad l'_{y0} = \frac{I_c}{I_{sy}} l_{y0}$$
$$h'_{x0} = \frac{I_c}{I_{vx}} h_{x0} \quad h'_{y0} = \frac{I_c}{I_{vy}} h_{y0}$$
(249)

The joint relations φ_J and φ_K, the angular displacement ψ_c and the end moments at joints J, K are computed as for multistory frames. However, the relatively large influence of the capitals, which are infinitely rigid, cannot be disregarded, as it is in the case of frames and flat slabs with no capitals (Fig. 99c).

The capital's influence can best be demonstrated by a rather simple example (Fig. 101):

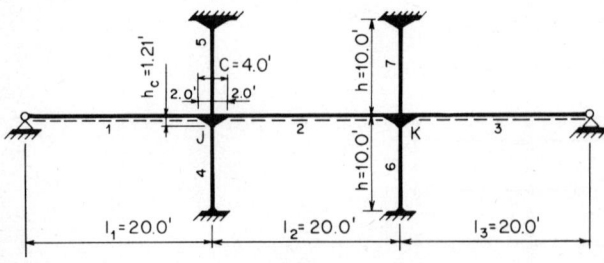

Fig. 101

$$d_0 = 5 \text{ in.}$$
$$a_{5,7} = b_{5,7} = 14 \text{ in.}$$
$$a_{4,6} = b_{4,6} = 16 \text{ in.}$$
$$c = 2.0 + 2.0 = 4.0 \text{ ft}$$
$$h_c = 1.21 \text{ ft} \quad d_0/2 = 0.21 \text{ ft}$$
$$l_{10} = l_{30} = 20.0 - 2.0 = 18.0 \text{ ft}$$
$$l_{20} = 20.0 - 4.0 = 16.0 \text{ ft}$$
$$h_0 = 10.0 - (1.21 + 0.21) = 8.58 \text{ ft}$$
$$= \text{const}$$
$$l_x = l_y = l$$

$$I_s = \frac{20 \times 12 \times 5^3}{12} = 2{,}500 \text{ in.}^4 = I_c$$

$$I_{4,6} = \frac{16 \times 16^3}{12} = 5{,}460 \text{ in.}^4$$

$$I_{5,7} = \frac{14 \times 14^3}{12} = 3{,}200 \text{ in.}^4$$

$$l'_1 = l'_3 = \frac{I_c}{I_s} l_{1,3} = \frac{2{,}500}{2{,}500} \times 18.0 = 18.0 \text{ ft}$$

$$l'_2 = \frac{I_c}{I_s} l_{20} = 16.0 \text{ ft}$$

$$h'_{4,6} = \frac{I_c}{I_{v,4,6}} h_0 = \frac{2{,}500}{5{,}460} \times 8.58 = 3.9 \text{ ft}$$

$$h'_{5,6} = \frac{I_c}{I_{v,5,7}} h_0 = 6.7 \text{ ft}$$

Redundants and a_{JJ}, a_{JK} Values

Making use of symmetry (about the centerline), the system is one-time geometrically indeterminate for symmetric vertical loading. The redundant is

$$\bar{\varphi}_J = \frac{\varphi_J - \varphi_K}{2} \quad \psi_c = 0$$

$$\varphi = 1:$$

$$\vartheta_{11} = \frac{1_J c/2}{l_{10}} = -\frac{2.0}{18.0} = -0.1110 \quad \vartheta_{21} = 0 \quad (250)$$

$$\vartheta_{41} = \frac{1_J h_c}{h_0} = -\frac{1.21}{8.58} = -0.1410$$

$$\vartheta_{51} = \frac{1_J d_0/2}{h_0} = -\frac{0.21}{8.58} = -0.0245$$

(250) (Cont.)

$$a_{JJ} = -2\left[\frac{3}{l'_1}(1+\vartheta_{11}) + \frac{2}{l'_2} + \frac{2}{h'_4}(2+3\vartheta_{41}) + \frac{2}{h'_5}(2+3\vartheta_{51})\right]$$

(251)

$$= -4.343$$

$$a_{JK} = a_{KJ} = 0$$

a_{J0} Value

Loading a

$$w_1 = w_3 = 1\,k/\text{ft}$$

$$M_{J0} = +\frac{1}{8}wl_{10}^2 = 40.5\,k'$$

$$R_{J0} = \frac{5}{8}wl_{10} = 11.25\,k \quad \Delta R_J = w\,c/2 = 2.0\,k$$

$$\overline{M}_J = R_{J0}c/2 + \Delta R_J c/4 = 24.5\,k'$$

$$\sum_J M = 40.5 + 24.5 = 65.0\,k'$$

$$a_{J0} = +2\sum_J M = +130.0$$

$$\overline{\varphi}_J = \frac{a_{J0}}{a_{JJ}} = -29.00 \quad \varphi_J = \overline{\varphi}_J \quad \varphi_K = -\overline{\varphi}_J$$

Moments:

$$M_J^{(1)} = M_{J0}^{(1)} + \frac{3}{l'_1}(1-\vartheta_{11})\varphi_J = +34.96\,k'$$

$$M_J^{(2)} = \frac{2}{l'_2}(2\varphi_J - \varphi_K) = -3.74\,k'$$

$$M_J^{(4)} = \frac{2}{h'_4}(2-3\vartheta_{41})\varphi_J = -37.15\,k'$$

$$M_J^{(5)} = \frac{2}{h'_5}(2-3\vartheta_{51})\varphi_J = -18.57\,k'$$

$$\sum_J M = 0 \quad (252)$$

Loading b

$w_2 = 1.0 \, K/ft$

$M_{J0} = -\frac{1}{12} w l_{20}^2 = -21.33 \, k'$

$R_J = 0.5 w l_{20} = 8.0 \, k \quad \Delta R_J = w c/2 = 2.0 \, k$

$\overline{M}_J = -(R_J \cdot c/2 + \Delta R_j c/4) = -18.0 \, k'$

$\sum_J M = -21.33 - 18.0 = -39.33 \, k'$

$a_{J0} = 2 \sum_J M = -78.66$

$\overline{\varphi}_J = \dfrac{a_{J0}}{a_{JJ}} = +18.11 \quad \varphi_J = \overline{\varphi}_J \quad \varphi_K = -\overline{\varphi}_J$

Moments:

$M_J^{(1)} = \dfrac{3}{18.0}(1 + 0.1110)\,18.11 = +3.35 \, k'$

$M_J^{(2)} = -21.33 + \dfrac{2}{16.0}(2 \times 18.11 - 18.11)$

$\phantom{M_J^{(2)}} = -19.07 \, k'$

$M_J^{(4)} = \dfrac{2}{3.9}(2 + 3 \times 0.1410)\,18.11$

$\phantom{M_J^{(4)}} = +22.51 \, k'$

$M_J^{(5)} = \dfrac{2}{6.7}(2 + 3 \times 0.0245)\,18.11$

$\phantom{M_J^{(5)}} = +11.21 \, k'$

$\sum_J M = -19.07 - 18.00$
$ + 3.35 + 22.51$
$ + 11.21$
$ = 0$

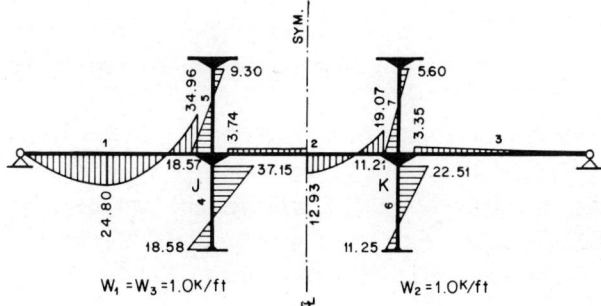

Fig. 102

The moments for uniformly distributed loading (Fig. 103) or critical moments are obtained by superposition of loadings *a* and *b*.

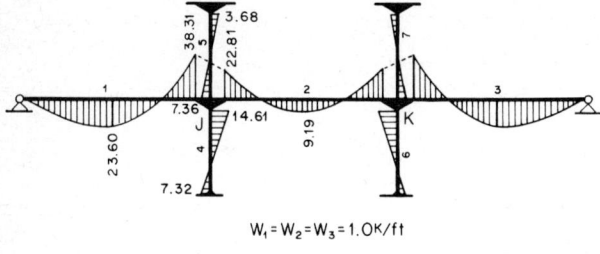

$W_1 = W_2 = W_3 = 1.0 K/ft$

Fig. 103

The critical moments computed in accordance with frame theory must be divided between column strip *c* and middle strip *m* as follows:

Column strip $(b_c = 0.5\,l)$: $M_m^{(c)} = 0.55 M_m$ $M_J^{(c)} = 0.75 M_J$ (253)

Middle strip $(b_m = 0.5\,l)$: $M_m^{(m)} = 0.45 M_m$ $M_J^{(m)} = 0.25 M_J$ (254)

For flat slabs having no capitals, in accordance with slab theory, the width of the column strips should be $0.4 l_x$ or $0.4 l_y$ and the width of the middle strip should be $0.6 l_x$ or $0.6 l_y$ whereas half of the total reinforcing of the column strip should be placed on a width $b = a + 2 d_0$.

When the slabs are restrained at the margin in some degree, the $\sum_J M^{(1)}$ values should be modified accordingly. Where tapered columns are used, the moment of inertia can be estimated by Eq. (200).

Since the slabs are relatively flexible, two adjacent floors above and below and three adjacent bays are considered adequate to compute critical moments. In such an analysis, the columns are assumed fixed or joined at their remote ends. However, for wind loads (nonsymmetric loads), temperature changes, and shrinkage, the entire structure has to be considered.

The analysis, as described above, must be carried out in the transversal (l_y) as well as the longitudinal (l_x) direction of the building.

To reduce dead load, a narrowly spaced grid system can be used instead of a solid slab. Such a system can be structurally analyzed as a flat slab provided that the flexural rigidity of column and middle strips is equal. If this is not the case, the load distribution, as given by Eqs. (253) and (254), is not valid and should be modified in accordance with the rigidity of the strips. Also, for longer spans it is often more economical to use more rigid, prestressed column strips. By prestressing, the deflection can be reduced to zero and the waffle or two-skin grid slab between the column strips can be transferred to a true two-way system. The bays then need not be analyzed in both directions for full load $(w_D + w_L)$ but can be analyzed instead in accordance with two-way beam action [Eqs. (18) to (20)].

18. ARCH BEAMS

A true arch beam action is illustrated in Fig. 104. The beam in this carrying action is comparatively rigid in comparison with the verticals and the arch segments of the system. Since the relatively thin arch is not stable, especially for unsymmetric loading, the beam acts as a stabilizer besides participating directly in the overall carrying action.

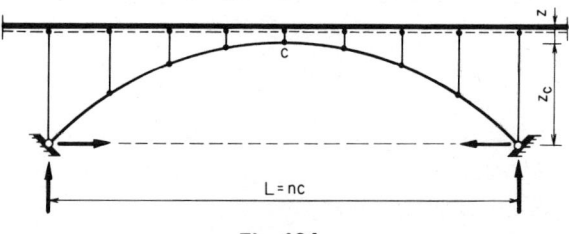

Fig. 104

The exact static analysis of structural systems is rather complex because, due to the flexibility of the arch, the continuous beam is supported by yielding supports. On the other hand, the magnitude of the yielding supports depends on the rigidity of the beam. Due to this fact, an approximate method of analysis of the system will be discussed.

First, the beam is analyzed as a continuous beam on rigid supports $(k = 1, 2, \ldots, n)$ subjected to dead load and live load. Second, the reaction R'_k extended with the dead load of the verticals and arch segments (R''_k) is applied on the arch and resolved into normal forces N_k in the arch.

For a relatively flexible arch ($M \simeq 0$), the normal forces are (Fig. 105)

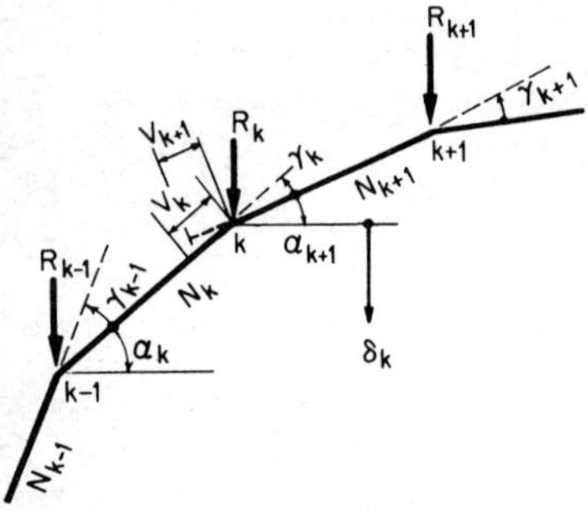

$$N_k = R_k \frac{\cos \alpha_{k+1}}{\sin \gamma_k} - R_{k-1} \frac{\cos \alpha_{k-1}}{\sin \gamma_{k-1}}$$

$$R_k = R'_k + R''_k$$

(255)

Fig. 105

For arches having a relatively large z_c/L ratio, the yielding of supports very often is negligible. If this is not the case, the beam design must be revised by analyzing the support settlement using successive approximations.

The deflections are

$$\delta_k = v_k \frac{\cos \alpha_{n+1}}{\sin \gamma_k} - v_{k+1} \frac{\cos \alpha_k}{\sin \gamma_k} \tag{256}$$

where

$$v_k = \frac{N_k}{EA_k} \frac{c_k}{\cos \alpha_k}$$

The changes in beam moments can be estimated using Eq. (12). Thus they are

$$\Delta M_k = EI_c \frac{\Delta^2 \delta_k}{\Delta x^2} = \frac{EI_c}{c^2} (\delta_{k-1} - 2\delta_k + \delta_{k+1}) \tag{257}$$

In the case of a rigid arch with relatively flexible verticals and beam, moments developed in the arch may be considerable. The carrying action of such a system is not an arch-beam action but a simple arch action, because the superstructure (beam and verticals) does not participate notably in the carrying action.

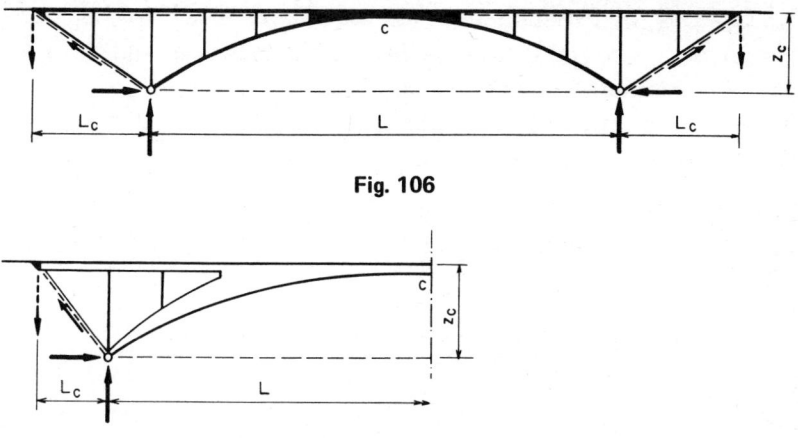

Fig. 106

Fig. 107

For arch-beam systems with relatively rigid beam, verticals, and arch, as illustrated in Figs. 106 and 107, the carrying action is accomplished by all elements: beam, verticals, and arch. By the use of posttensioning and back anchorage, relatively long spans or cantilevers are economically possible. The system can be analyzed most conveniently by the Vierendeel method (see "Trusses," Sec. 20).

19. BEAM-SUSPENSION ACTION

For long spans, beam-suspension systems are the most economical because the loads are carried mostly by cables directly to the supports and thus the moments are no longer proportional to the square of the span. The system illustrated in Fig. 108 and described below will demonstrate this carrying action.

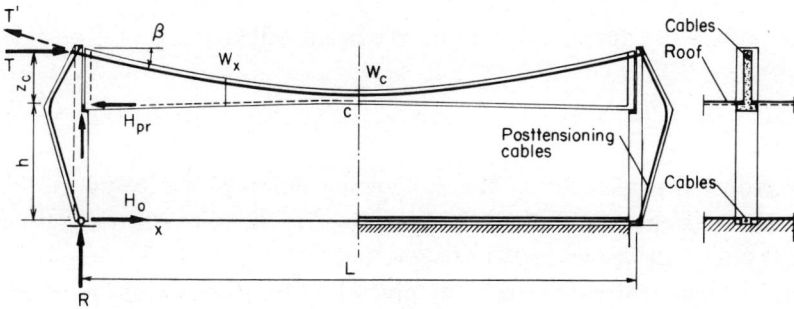

Fig. 108

The cable is first nonbonded to the beam. Then, by applying posttensioning at ground level (H_0), the cable is tensioned and the beam raised and posttensioned simultaneously (H_{pr}).

Denoting the prestressing force applied for columns at the support by H_0, the horizontal component of the cable tension is

$$T = \frac{H_0 h}{z_c} \quad T' = \frac{T}{\cos \beta} \tag{258}$$

and the uniform load, carried directly by cable, by Eq. (186) is

$$w_c = \frac{8 T z_c}{L^2} \tag{259}$$

The residue of load

$$\Delta w_x = \sum_{D}^{L} (w_x - w_c) \tag{260}$$

must be carried by the beam. Thus the maximum beam moment for parabolic cable is

$$R = \frac{1}{L} \sum_{D}^{L} \Delta w_x x' \Delta x$$
$$M_m = R x_m - \sum_{0}^{x_m} \Delta w_x (x_m - x) \Delta x \tag{261}$$

The prestressing force induced into the beam will be

$$H_{pr} = H_0 + T \tag{262}$$

By proper arrangement of the supporting points at the columns, the beam can be kept practically free of bending. To stabilize the deflections of the system, the cables have to be grouted.

Such systems are convenient for precasting because the beam can be made in sections. The prestressing can also be applied by dead load.

However, adjustment of the crown elevation must be made by direct jacking of the cables. For long spans with relatively slender beams, adequate lateral bracing of the beam is required to avoid buckling.

20. TRUSSES (ARCH-SUSPENSION ACTION)

Trusses have found more and more application in contemporary architecture. The truss types most easily built in reinforced and prestressed concrete are illustrated in Fig. 109.

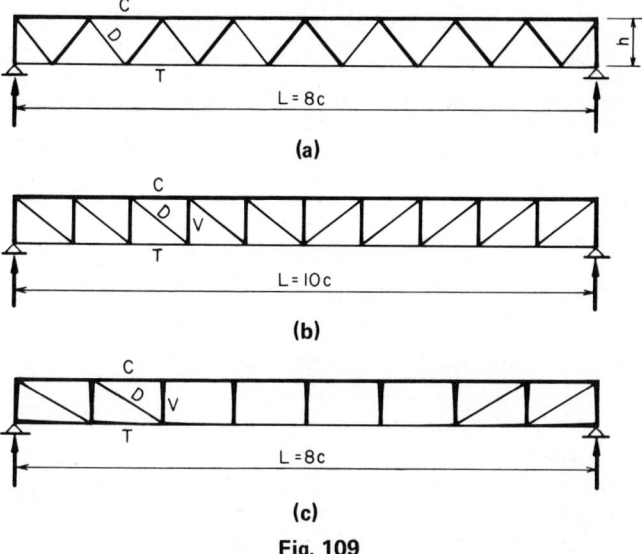

Fig. 109

The web member (verticals and diagonals) can be arranged in many different ways. The arrangement most suitable for long-span prestressed trusses is that in which the diagonal members are in tension and the verticals in compression (Fig. 109b) under both dead and live loads. Such an arrangement clearly emphasizes the suspension part of the joint carrying action. It is especially true when the tension members are designed to be relatively thin in comparison with the compressive members of the truss.

The truss type shown in Fig. 109c is a combination of a simple and a Vierendeel truss. It is most suitable for long spans, for which Vierendeel trusses are uneconomical and become clumsy because of the high shear in the end bays.

The economical ratio of depth to span varies for simple trusses from 1:10 to 1:15 and for continuous trusses from 1:15 to 1:20. The length of the panels (c) should not be more than twice the depth ($h = 0.85\,d_0$) of the truss. For the purpose of dimensioning and preliminary stress analysis, the members are considered pin-connected. However, this arbitrary assumption does not hold for reinforced-concrete trusses. Secondary stresses must be considered in the final analysis. For analysis of the secondary stresses due to the rigid connections of the members at the joints and direct member loading (w in addition to the joint loads P_k), the slope-deflection method is the most suitable (see Sec. 16). The joint rotations $\varphi_J \ldots \varphi_K$ are the unknowns. They can be computed from the condition

$$\sum_J M_r + M_{J0} = 0 \tag{263}$$

where $M_r\,(r = 1, 2, \ldots)$ are the moments at the ends of the member r joining the joint J and M_{J0} is the unbalanced moment due to uniform loading w.

The linear equations for computing the unknowns are

$$\begin{aligned}\sum_J M &= a_{JJ}\,\varphi_J + \sum_J a_{JK}\,\varphi_K - 6\sum_J \frac{\vartheta_r}{s_r'} + M_{J0} = 0 \\ \sum_K M &= a_{KK}\,\varphi_K + \sum_K a_{KJ}\,\varphi_J - 6\sum_K \frac{\vartheta_r}{s_r'} + M_{K0} = 0\end{aligned} \tag{264}$$

where ϑ_r denotes the change of slope of the members due to the external loads. These changes can be computed from the change of length of the member (Δs) caused by normal stresses in the member concerned and from the displacement of joints J, K, assuming pin connections of the members at the joints (Williot displacement diagram).

After the $\varphi_J \ldots \varphi_K$, values are known, the moments in the ends of the members at joints $J, K \ldots$ are obtained by the following equations:

$$\begin{aligned}M_{J_r} &= \frac{2}{s_r'}(2\varphi_J + \varphi_K - 3\vartheta_r) \\ M_{K_r} &= \frac{2}{s_r'}(2\varphi_K + \varphi_J - 3\vartheta_r)\end{aligned} \tag{265}$$

Stresses:

$$f_r = \mp \frac{N_r}{A_r} \mp \frac{M_r}{I_r} z \tag{266}$$

Simple Trusses

The forces in the members of trusses with parallel horizontal chords (Figs. 110 and 111) can be computed by the following equations:

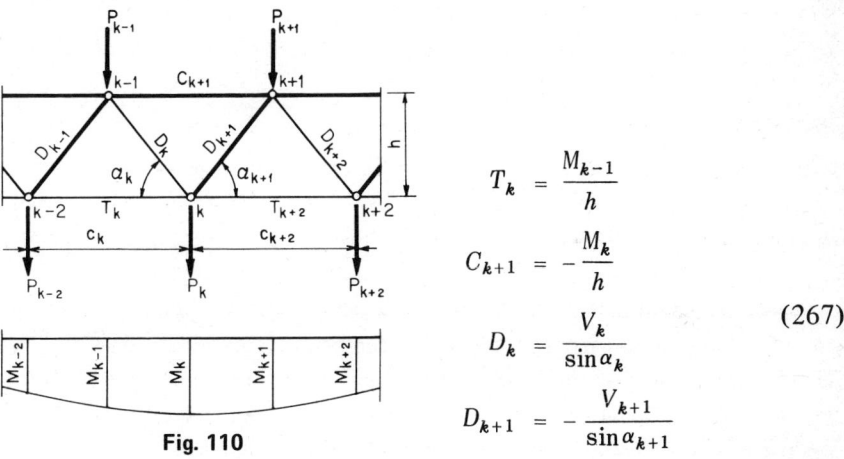

Fig. 110

$$T_k = \frac{M_{k-1}}{h}$$

$$C_{k+1} = -\frac{M_k}{h}$$

$$D_k = \frac{V_k}{\sin \alpha_k} \tag{267}$$

$$D_{k+1} = -\frac{V_{k+1}}{\sin \alpha_{k+1}}$$

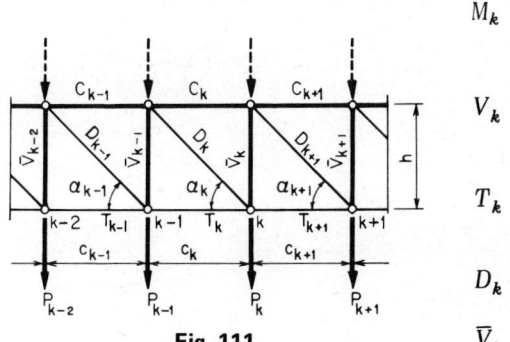

Fig. 111

$$M_k = Rx_k - \sum_0^{k-1} Pc_k$$

$$V_k = R - \sum_0^{k-1} P$$

$$T_k = \frac{M_k}{h} = -C_k \tag{268}$$

$$D_k = \frac{V_k}{\sin \alpha_k}$$

$$\overline{V}_k = -V_k + P_{kB}$$

The forces in the members of a truss with a horizontal top chord and a curved bottom chord are (Fig. 112)

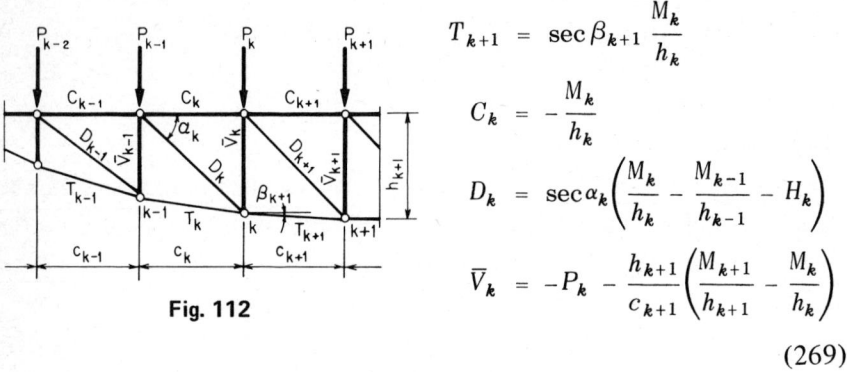

Fig. 112

$$T_{k+1} = \sec\beta_{k+1} \frac{M_k}{h_k}$$

$$C_k = -\frac{M_k}{h_k}$$

$$D_k = \sec\alpha_k \left(\frac{M_k}{h_k} - \frac{M_{k-1}}{h_{k-1}} - H_k \right)$$

$$\bar{V}_k = -P_k - \frac{h_{k+1}}{c_{k+1}} \left(\frac{M_{k+1}}{h_{k+1}} - \frac{M_k}{h_k} \right)$$

(269)

Continuous Trusses

For reasons of accuracy, a statically indeterminate truss, as illustrated in Fig. 113, can be analyzed most easily by cutting the top-chord members at the intermediate supports and thus establishing the principal system as a number of simple trusses, rather than by removing the redundant reactions to establish a single simple truss as the principal system. The cut members introduce the statically redundant forces $X_a, X_b, \ldots, X_n$. Their values are computed from the condition that the relative displacement of the cut faces is equal to zero. Thus as many linear equations can be obtained as there are redundant forces:

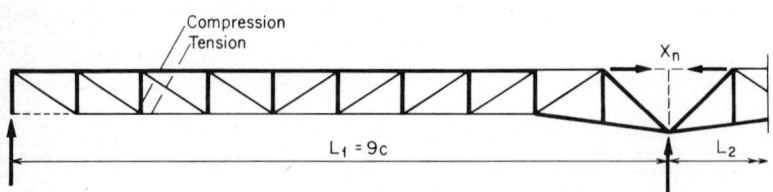

Fig. 113

$$\begin{aligned}
\delta_{a0} - X_a \delta_{aa} - X_b \delta_{ab} - \cdots - X_n \delta_{an} &= 0 \\
\delta_{b0} - X_a \delta_{ba} - X_b \delta_{bb} - \cdots - X_n \delta_{bn} &= 0 \\
\cdots\cdots\cdots\cdots\cdots\cdots\cdots\cdots\cdots\cdots\cdots\cdots\cdots\cdots & \\
\delta_{n0} - X_a \delta_{na} - X_b \delta_{nb} - \cdots - X_n \delta_{nn} &= 0
\end{aligned}$$

(270)

In these equations, the relative displacements are

$$\delta_{n0} = \sum_0^r \frac{N_{rn} N_{r0}}{A_r E} s_r + \sum_0^r N_{rn} \alpha_t \Delta T s_r + \sum_0^r N_{rn} \epsilon_{sr} s_r$$

$$\delta_{nn} = \sum_0^r \frac{N_{rn}^2}{A_r E} s_r \qquad (271)$$

$$\delta_{na} = \sum_0^r \frac{N_{rn} N_{ra}}{A_r E} s_r = \delta_{an}$$

$$\delta_{nb} = \sum_0^r \frac{N_{rn} N_{rb}}{A_r E} s_r = \delta_{bn}$$

where $N_{ra, rb}, \ldots, _{rn}$ are the forces in a member r of the principal system due to the unit dummy loads $(-\bar{X}_{a,b}, \ldots, _n = 1)$ applied at the cut faces of the redundant members and N_{r0} designates the forces in a member r of the principal system due to the external loads, temperature, shrinkage, plastic flow, etc. Σr denotes the number of members in the principal system (the total number of members is $\Sigma r + \sum_1^n X$). The modulus of elasticity of the material of the member considered must be used; this means that, for all the tension members, the modulus of elasticity of steel must be used and, for the compression members, the modulus of elasticity of concrete must be used. The sign of the force in each member should be considered when computing the relative displacements.

The actual forces in members can be obtained by applying the principle of superposition. Thus,

$$N_r = N_{r0} - \sum_1^n X_n N_{rn} \qquad (272)$$

Plastic Flow

Subjecting a compressive member in a truss to bending may result in large deflection of the members. The normal force in the member will then no longer coincide with the center of gravity of the deflected member, resulting in a further increase of the deflection.

If we denote the initial moment due to the dead load of a member r by M_{rD}, the compressive force by N_{rD}, and the deflection by δ_r, then the moment is, for time $t = 0$:

$$M_{r0} = M_{rD} + \delta_r N_{rD} \qquad (273)$$

Since the deflection δ_r follows very closely the sine function and is affined with a buckling wave, the actual moment M_{r0}, as expressed by Dischinger, is

$$M_{r0} = M_{rD} + \delta_r N_{rD} = M_{rD} \frac{v_r}{v_r - 1} \qquad (274)$$

where v_r is the factor of safety against buckling of the member under consideration.

The initial moment M_{r0} increases due to plastic flow in the course of time and reaches its final value when the plastic flow is finished, at time $t = n$:

$$M_{rn} = M_{rD} \frac{v_r}{v_r - 1} e^{\varphi_n/(v_r - 1)} \qquad (275)$$

For illustration, assuming $v_r = 4$ and $\varphi_n = 2$, $M_{rn} = 2.60 M_{rD}$ and for $v_r = 3$ and $\varphi_n = 2$, $M_{rn} = 4.06 M_{rD}$.

As can be seen from this simple illustration, the influence of plastic flow cannot be overlooked in a relatively slender member subjected simultaneously to bending and normal compressive forces.

21. VIERENDEEL GIRDERS (BEAM-ARCH-SUSPENSION ACTION)

The Vierendeel girder is not a truss in the true sense, because the characteristic triangular geometry of a true truss is missing. The carrying capacity of a Vierendeel is accomplished by chords and verticals only. Thus it is more a frame than a truss. A typical Vierendeel girder is illustrated in Fig. 114.

The controlling carrying action here is the beam action, and arch and suspension actions are only secondary actions. To explain the

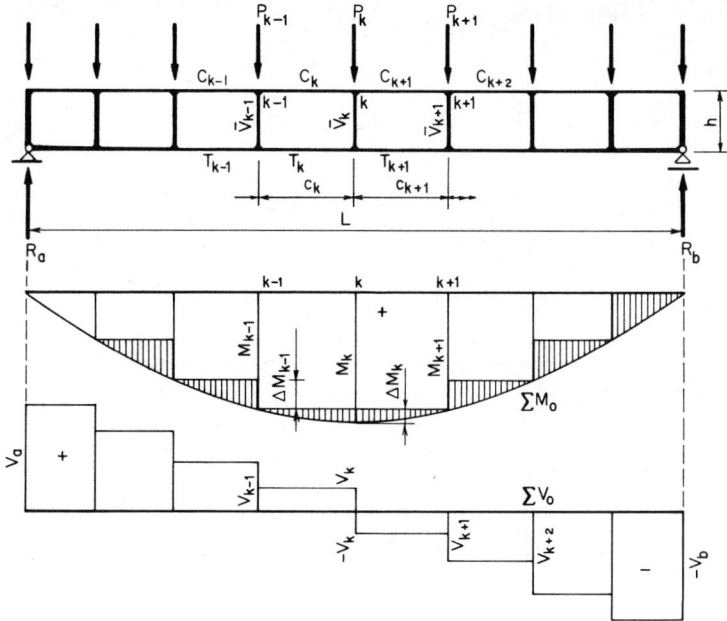

Fig. 114

Vierendeel-girder carrying capacity more explicitly, let us consider a Vierendeel having top and bottom chords equal and with relatively rigid verticals. As the normal forces between joints $k-1$ and k are constant, so also are the internal moments due to the normal forces $N_{kT}h = N_{kB}h =$ constant. But the external moment M_{X0} due to loading is not constant: $M_{(k-1),0} < M_{k0}$. Thus, to obtain equilibrium, the $\Delta M_{k0} = M_{k0} - M_{(k-1),0}$ must be balanced by beam action in the chords and verticals.

In this particular case, the approximate moments and shear in bay $c_{k-1}, c_k, \ldots$ due to joint loads P_k are (Fig. 115)

Top chord: $\quad M^{(T)}_{k-1,k} = M^{(T)}_{k,k-1} = \frac{1}{4}\Delta M_{k0} \quad \bigg| \quad \sum_B^T M_{k,k-1} = \Delta M_{k0} \quad (276)$

Bottom chord: $\quad M^{(B)}_{k-1,k} = M^{(B)}_{k,k-1} = \frac{1}{4}\Delta M_{k0}$

Shear: $\quad V_{kT} = V_{kB} = \dfrac{M_{k,k-1}}{c_k/2} = \dfrac{1}{2}\dfrac{\Delta M_{k0}}{c_k}$

188 Theory of Structures

and in the verticals are

$$-M_{kv}^{(T)} = M_{kv}^{(B)} = \tfrac{1}{4}(\Delta M_{k0} + \Delta M_{k+1,0}) \bigg| \sum_{B}^{T}(M_{k,k-1} + M_{k,k+1} - M_{kv}) = 0$$

$$V_{kv} = \frac{M_{kv}}{h/2} = \frac{1}{2}\frac{\Delta M_{k0} + \Delta M_{k+1,0}}{h}$$

The normal force N_k in the chords is

$$N_{kB} = -N_{kT} = \frac{M_{k0} - \left(M_{k,k-1}^{(T)} + M_{k,k-1}^{(B)}\right)}{h} \tag{277}$$

The moments, shears, and normal forces computed in accordance with Eqs. (276) and (277) are presented in Fig. 115.

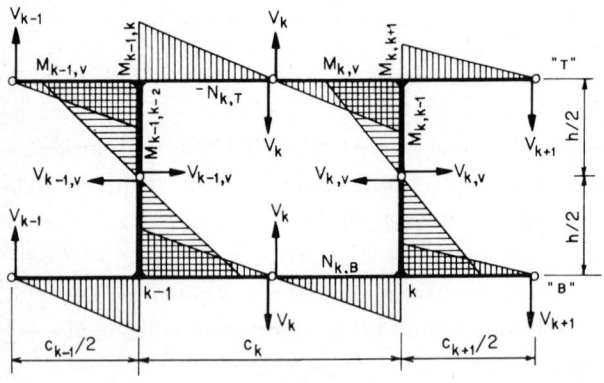

Fig. 115

The zero points of the moments are at the midspan of the members. Also, the four moments in the chords are equal and their algebraic sum is equal to the change of external moment (ΔM_{k0}) from joint $k-1$ to k.

These results mean that the moments in the chords of a bay are independent and are not influenced by the chord moments of adjacent bays. This can be true only if the verticals are infinitely rigid ($I_v \simeq \infty$). Normally this is not the case. Also, the rigidity of the chords may vary considerably from bay to bay. Due to this fact, the chord moments M_k of a bay are related and are dependent on the moments of other bays; as such, they are redundants of a statically indeterminate structure. To obtain qualitatively and quantitatively acceptable results, more accurate methods for analysis of a Vierendeel girder must be used. One of these analyses will be discussed in the following.

Any Vierendeel girder is $3n$ times statically indeterminate, where n is the number of bays in the girder. In the case of symmetry about vertical axes (z), the number of redundants is reduced to half.

The most suitable frame for a principal system is the three-hinged frame, as illustrated for bay $k - 1$, k in Fig. 116. The redundants for bay $k - 1$, k are

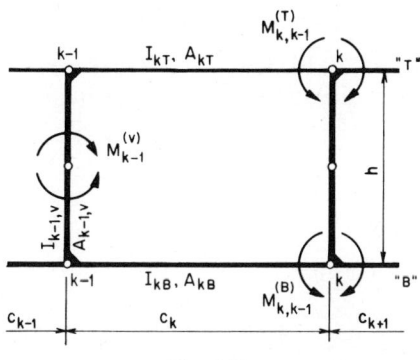

Fig. 116

$M^{(T)}_{k,k-1}$ $M^{(B)}_{k,k-1}$

$M^{(V)}_{k-1}$

The sectional coefficients are indicated for each individual member of the three-hinged frame.

For joint loading ($P_k \ldots _n$) and symmetry about the horizontal axis x, which means that the sectional coefficients of the chords are equal ($I_{kT} = I_{kB}$ and $A_{kT} = A_{kB}$), the location of the zero-moment point in the verticals is at midheight (h). The redundants are $M^{(V)}_{k-1} = 0$ and $M^{(T)}_{k,k-1} = M^{(B)}_{k,k-1}$. Thus the number of redundants for this particular condition is reduced further, down to the number of bays (n). Provided that even such favorable conditions exist, the number of redundants of a long-span Vierendeel girder is relatively large. For example, the Vierendeel girder illustrated in Fig. 114 is eight times statically indeterminate.

To minimize the number n of redundants for cases in which $I_{kT} \neq I_{kB}$ and $A_{kT} \neq A_{kB}$, group moments as redundants must be introduced:

$$X_k = \frac{M^{(T)}_{k,k-1} + M^{(B)}_{k,k-1}}{2} \qquad (278)$$

190 Theory of Structures

The difference between I_{kT} and I_{kB} can be computed by using the average moment of inertia for both chords:

$$I_{kc} = \frac{I_{kT} + I_{kB}}{2} \tag{279}$$

The difference between A_{kT} and A_{kB} does not notably influence the final results and thus can be disregarded. In cases in which I varies within a bay, use can be made of Eqs. (199) and (200). The final moments in chords will be distributed in accordance with their flexural rigidity. The moments in the top and bottom of verticals are the numerical sum of the chord moments at the joints k. The shear V and the zero point of the moments in the chords and verticals are easily computed from the known moments.

The redundants X are computed from the linear continuity equation [Eq. (7)]:

$$X_{k-1}\delta_{k,k-1} + X_k\delta_{kk} + X_{k+1}\delta_{k,k+1} + \delta_{k0} = 0$$

written for each bay (k) of the Vierendeel girder.

$\delta_{k,k-1}, \delta_{kk}, \delta_{k,k+1}$ values:

$-X_k = 1k'$:

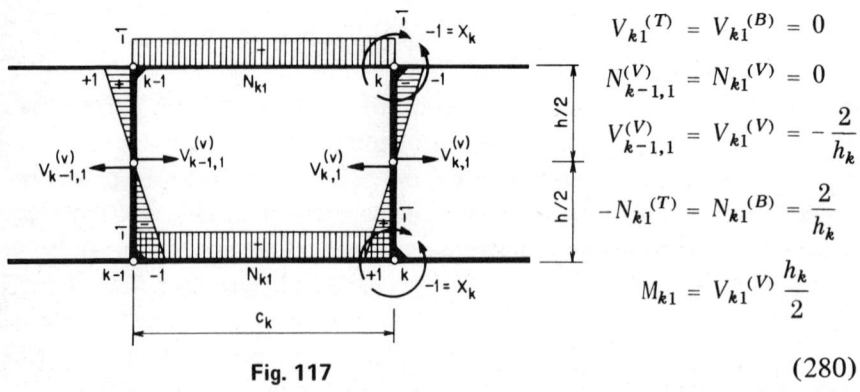

$$\begin{aligned} V_{k1}^{(T)} &= V_{k1}^{(B)} = 0 \\ N_{k-1,1}^{(V)} &= N_{k1}^{(V)} = 0 \\ V_{k-1,1}^{(V)} &= V_{k1}^{(V)} = -\frac{2}{h_k} \\ -N_{k1}^{(T)} &= N_{k1}^{(B)} = \frac{2}{h_k} \\ M_{k1} &= V_{k1}^{(V)} \frac{h_k}{2} \end{aligned} \tag{280}$$

Fig. 117

The moments M_k, normal forces N_k, and shear V_k for the redundants $-X_k = 1 (k = 1, 2, \ldots, n)$ are given by Eq. (280) and are illustrated in Fig. 117. Denoting, in general, the length of a member by s, $v_k = \kappa EI_c/GA_c \sim 2.4 I_c/A_c, (\kappa = 1.2, E/G \sim 2.0); \; h'_k = (I_c/I_v)h_k, c'_k = c_k$.

Making use of Eqs. (8) and the integration table (Table 3), the displacements δ times EI_c are

$$\delta_{k,k-1} = \sum \int_0^s M_k M_{k-1} \frac{I_c}{I_s} ds + \sum \frac{I_c}{A_c} \int_0^s N_k N_{k-1} \frac{A_c}{A_s} ds$$

$$+ \sum \kappa \frac{EI_c}{GA_c} \int_0^s V_k V_{k-1} \frac{A_c}{A_s} ds$$

$$= -\frac{1}{3}\left(h'_{k-1} + 12 \frac{\nu_{k-1}}{h_k}\right) = \delta_{k-1,k}$$

$$\delta_{kk} = \sum \int_0^s M_k^2 \frac{I_c}{I_s} ds + \sum \frac{I_c}{A_c} \int_0^s N_k^2 \frac{A_c}{A_s} ds$$

$$+ \sum \kappa \frac{EI_c}{GA_c} \int_0^s V_k^2 \frac{A_c}{A_s} ds \qquad (281)$$

$$= \frac{1}{3}\left[h'_{k-1} + h'_k + 6c'_k + \frac{12}{h_k}(\nu_{k-1} + \nu_k) + 2.4 \frac{I_c}{A_k} \frac{c_k}{h_k^2}\right]$$

$$\delta_{k,k+1} = \sum \int_0^s M_k M_{k+1} \frac{I_c}{I_s} + \sum \frac{I_c}{A_c} \int_0^s N_k N_{k+1} \frac{A_c}{A_s} ds$$

$$+ \sum \kappa \frac{EI_c}{GA_c} \int_0^s V_k V_{k+1} \frac{A_c}{A_s} ds$$

$$= -\frac{1}{3}\left(h'_k + 12 \frac{\nu_k}{h_{k+1}}\right) = \delta_{k+1,k}$$

MATRIX: The matrix of the linear continuity equations [Eqs. (7)] for the Vierendeel girder illustrated in Fig. 114, $n = 8$, is

192 Theory of Structures

	X_1	X_2	X_3	X_4	X_5	X_6	X_7	X_8	δ_{k0}
1	δ_{11}	δ_{12}	—	—	—	—	—	—	δ_{10}
2	δ_{21}	δ_{22}	δ_{23}	—	—	—	—	—	δ_{20}
3	—	δ_{32}	δ_{33}	δ_{34}	—	—	—	—	δ_{30}
4	—	—	δ_{43}	δ_{44}	δ_{45}	—	—	—	δ_{40}
5	—	—	—	δ_{54}	δ_{55}	δ_{56}	—	—	δ_{50}
6	—	—	—	—	δ_{65}	δ_{66}	δ_{67}	—	δ_{60}
7	—	—	—	—	—	δ_{76}	δ_{77}	δ_{78}	δ_{70}
8	—	—	—	—	—	—	δ_{87}	δ_{88}	δ_{80}

δ_{k0} values:

The moments, forces, and shear in the principal system due to joint loads P_k are given in Eq. (282) and are illustrated in Fig. 118.

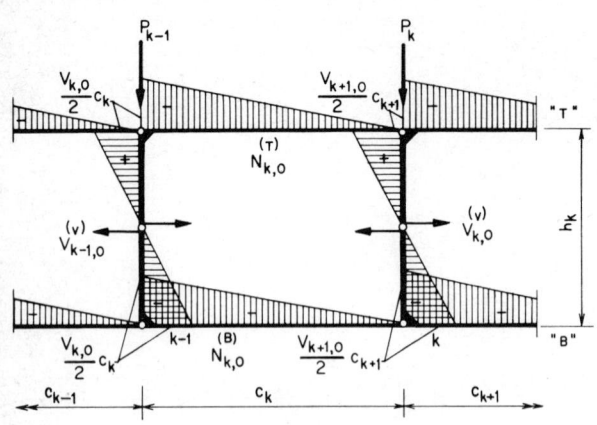

Fig. 118

$$V_{k0} = R_0 - \sum_{0}^{k-1} P$$

$$V_{k0}^{(T)} = V_{k0}^{(B)} = \frac{V_{k0}}{2}$$

$$V_{k-1,0}^{(V)} = -\frac{V_{k0}}{h_{k-1}} c_k \qquad (282)$$

$$V_{k0}^{(V)} = -\frac{V_{k+1,0}}{h_k} c_{k+1}$$

$$-N_{k0}{}^{(T)} = N_{k0}{}^{(B)} = \frac{M_{k0}}{h_k}$$

$$M_{k0}{}^{(V)} = V_{k0}{}^{(V)} \frac{h_k}{2} = \frac{V_{k+1,0}}{2} c_{k+1} \qquad (282)$$
(Cont.)

$$\delta_{k0} = \sum \int_0^s M_{k0} M_k \frac{I_c}{I_s} ds + \sum \frac{I_c}{A_s} \int_0^s N_{k0} N_k \frac{A_c}{A_s} ds$$

$$+ \sum \kappa \frac{EI_c}{GA_c} \int_0^s V_{k0} V_k \frac{A_c}{A_s} ds$$

$$= V_{k0} \frac{c_k}{6h_k} (3 h_k c'_k + h'_{k-1} h_{k-1} + 12 \nu_{k-1})$$

$$- V_{k+1,0} \frac{c_{k+1}}{6} \left(h'_k + 12 \frac{\nu_k}{h_k} \right) + 2 \frac{c_k}{h_k^2} \left(\frac{I_c}{A_k^{(B)}} + \frac{I_c}{A_k^{(T)}} \right) M_{k0} \qquad (283)$$

If the chords of a Vierendeel girder are not parallel, the δ_{k-1}, δ_{kk}, δ_{k+1}, and δ_{k0} values must be modified because, in the principal system (Figs. 117 and 118), the shear V_k, normal force N_k, and moment M_k change. The shear V_{k1} in the chords [Eq. (280)] is no longer zero but is

$$V_{k1}{}^{(T)} = V_{k1}{}^{(B)} \simeq \frac{\Delta h_k}{h_k}$$

Thus:

$$\delta_{k,k-1} = -\frac{1}{3} \left(\frac{h_{k-1}}{h_k} h'_{k-1} + 12 \frac{\nu_{k-1}}{h_k} \right)$$

$$\delta_{k,k+1} = -\frac{1}{3} \left(\frac{h_k}{h_{k+1}} h'_k + 12 \frac{\nu_k}{h_{k+1}} \right)$$

$$\delta_{kk} = \frac{1}{3} \left[2 c'_k \left(\frac{h_{k-1}^2}{h_k^2} + \frac{h_{k-1}}{h_k} + 1 \right) + \frac{h_{k-1}^2}{h_k^2} h'_{k-1} + h'_k \right.$$

$$\left. + 12 \frac{\nu_{k-1} h_{k-1} + \nu_k h_k}{h_k^2} + 12 \frac{c_k}{h_k^2} \left(\frac{I_c}{A_k^{(T)}} + \frac{I_c}{A_k^{(B)}} \right) \right] \qquad (284)$$

The change in shear V_{k0} in the chords and verticals can be considered by multiplying the values in Eqs. (282) with a factor $(c_k/h)\zeta_k$, where

$$\zeta_k = \left(1 - \frac{M_{k0}}{V_{k0}} \frac{\Delta h_k}{h_k c_k}\right)$$

Thus:

$$V_{k0}^{(T)} = V_{k0}^{(B)} = \frac{V_{k0}}{2} \frac{c_k}{h_k} \zeta_k$$

$$V_{k-1,0}^{(V)} = -V_{k0} \frac{c_k}{h_{k-1}} \zeta_k$$

$$V_{k0}^{(V)} = -V_{k+1,0} \frac{c_{k+1}}{h_k} \zeta_{k+1} \qquad (285)$$

$$-N_{k0}^{(T)} \simeq N_{k0}^{(B)} = \frac{M_{k0}}{h_k}$$

The change in normal forces N_{k0} in Eqs. (282) is negligible. However, if considered, the normal force in a sloping chord must be multiplied by the factor:

$$\zeta_{k0} = \frac{c_k}{s_k}\left[1 + \frac{1}{2}\left(\frac{V_{k0}}{M_{k0}} \frac{h_k}{c_k} \Delta h_k + \frac{\Delta h_k^2}{c_k^2}\right)\right]$$

Thus:

$$\delta_{k0} = V_{k0} \frac{\zeta_k c_k}{6 h_k}[c_k'(2h_{k-1} + h_k) + h_{k-1} h_{k-1}' + 12 v_{k-1}]$$
$$- V_{k+1,0} \frac{\zeta_{k+1} c_{k+1}}{6}\left(h_k' + 12 \frac{v_k}{h_k}\right)$$
$$+ 2 \frac{c_k}{h_k^2}\left(\frac{I_c}{A_k^{(T)}} + \frac{I_c}{A_k^{(B)}}\right) M_{k0} \qquad (286)$$

The linear continuity equations (matrix) can best be solved by iteration because of relatively good conversion.

The final moments, normal forces, and shear in members of a parallel-chord Vierendeel are obtained by Eqs. (287):

$$M^{(T)}_{k,k-1} = 2X_k \frac{I_{kT}}{I_{kT} + I_{kB}} = X_k \frac{I_{kT}}{I_{kc}}$$

$$M^{(B)}_{k,k-1} = 2X_k \frac{I_{kB}}{I_{kT} + I_{kB}} = X_k \frac{I_{kB}}{I_{kc}}$$

$$M_{kv}{}^{(T)} = M^{(T)}_{k,k-1} \mp M^{(T)}_{k,k+1}$$

$$M_{kv}{}^{(B)} = M^{(B)}_{k,k-1} \mp M^{(B)}_{k,k+1}$$

$$-N_{kT} = N_{kB} = \frac{M_{k0} - 2X_k}{h_k}$$

$$V_{kT} = V_{k0} \frac{I_{kT}}{I_{kT} + I_{kB}} = V_{k0} \frac{2I_{kT}}{I_{kc}} = \frac{M^{(T)}_{k,k-1}}{x'_T} = \frac{M^{(T)}_{k-1,k}}{x_T}$$

$$V_{kB} = V_{k0} \frac{I_{kB}}{I_{kT} + I_{kB}} = V_{k0} \frac{2I_{kB}}{I_{kc}} = \frac{M^{(B)}_{k,k-1}}{x'_B} = \frac{M^{(B)}_{k-1,k}}{x_B}$$

$$V_{kv} = \frac{M_{kv}{}^{(T)}}{z'} = \frac{M_{kv}{}^{(B)}}{z} = N_{k+1} - N_k \quad c = x + x', \ h = z + z'$$

$$= -V_{k+1,0} \frac{c_{k+1}}{h} + \frac{2}{h}(X_{k+1} - X_k)$$

(287)

For a fixed-end or continuous Vierendeel girder, when the external moments M_{k0} and shear V_{k0} are computed or obtained by the use of Table 1 or 2, the analysis of moments, normal force, and shear is the same as for a simply supported Vierendeel.

Analysis of a Vierendeel Girder

To demonstrate the application of the Vierendeel theory, a girder subjected to external moments M_{k0} and shear V_{k0}, as illustrated in Fig. 119, will be analyzed.

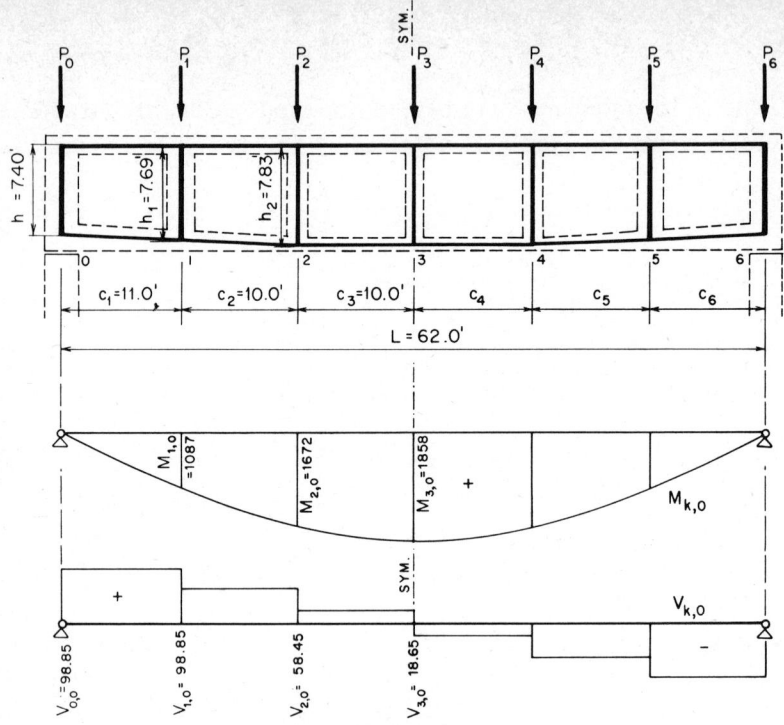

Fig. 119

Geometric data and relative stiffnesses:

$A_{kT} = 544$ in.^{2} = const. $I_{kT} = 35,075$ in.^{4} = const. = I_c
$\quad = A_c$ $I_{1B} = 8,034$ in.^{4} = I_{5B}, $I_{2B} = 3,482$ in.^{4}
$\qquad\qquad = I_{3B} = I_{4B}$, $I_{6B} = 27,000$ in.^{4}

$$I_{1C} = \frac{35,075 + 8,034}{2} = 21,553 \text{ in.}^4 = I_{5C}$$

$A_{0B} = 544$ in.^{2} = A_{6B} $I_{2C} = 19,277$ in.^{4} = $I_{3C} = I_{4C}$
$\qquad\qquad\qquad\qquad\quad I_{6C} = 31,036$ in.^{4} = I_{0C}
$A_{1B} = 400$ in.^{2} = A_{5B} $I_{0v} = 70,000$ in.^{4} = I_{6v}
$\qquad\qquad\qquad\qquad\quad I_{1v} = 40,500$ in.^{4} = I_{5v}
$A_{2B} = 328$ in.^{2} = A_{3B} $I_{2v} = 20,700$ in.^{4} = I_{4v}, $I_{3v} = 8,750$ in.^{4}
$\qquad\qquad\quad = A_{4B}$

$A_{0v} = 648$ in.^{2} = A_{6v} $\nu_{0,6} = 2.4 \dfrac{I_c}{A_{0v}\,12^2} = 0.904$

$A_{1v} = 540$ in.^{2} = A_{5v} $\nu_{1,5} = 1.082$
$A_{2v} = 432$ in.^{2} = A_{4v} $\nu_{2,4} = 1.352$
$A_{3v} = 324$ in.^{2} $\nu_3 = 1.803$

$h'_0 = h'_6 = \dfrac{I_c}{I_{0v}} h_0 = 3.70$ ft, $h'_1 = h'_5 = 6.65$ ft, $h'_2 = h'_4 = 13.26$ ft,

$h'_3 = 31.40$ ft, $c'_{1c} = \dfrac{I_c}{I_{1c}} c_{1c} = 17.90$ ft, $c'_2 = 18.20$ ft, $c'_3 = 18.20$ ft $= c'_4$,

$c'_5 = 16.30$ ft, $c'_6 = 12.45$ ft.

$\delta_{k,k+1}$, δ_{kk} values:

$$\delta_{12} = -\frac{1}{3}\left(\frac{h_1}{h_2} h'_1 + \frac{12\nu_1}{h_2}\right) = -\frac{1}{3}\left(\frac{7.69}{7.83} 6.65 + \frac{12 \times 1.082}{7.83}\right) = -2.720$$

$$\delta_{23} = -\frac{1}{3}\left(\frac{h_2}{h_3} h'_2 + \frac{12\nu_2}{h_3}\right) = -\frac{1}{3}\left(\frac{7.83}{7.83} 13.26 + \frac{12 \times 1.352}{7.83}\right) = -5.110$$

$\delta_{34} = -11.390$, $\delta_{45} = -5.120$, $\delta_{56} = -2.890$

$$\delta_{k,k-1} = \delta_{k-1,k}$$
$$\delta_{k,k+1} = \delta_{k+1,k}$$

$$\delta_{11} = \frac{1}{3}\left[2c'_1\left(\frac{h_0^2}{h_1^2} + \frac{h_0}{h_1} + 1.0\right) + \frac{h_0^2}{h_1^2} h'_0 + h'_1 + \frac{12}{h_1^2}(\nu_0 h_0 + \nu_1 h_1) \right.$$
$$\left. + 12\frac{c_1}{h_1^2}\left(\frac{I_c}{A_{1T}} + \frac{I_c}{A_{1B}}\right)\right]$$

$$= \frac{1}{3}\left[2 \times 17.90\left(\frac{7.40^2}{7.69^2} + \frac{7.40}{7.69} + 1.0\right) + \frac{7.40^2}{7.69^2} 3.70 + 6.65 \right.$$

$$+ \frac{12}{7.69^2}(0.904 \times 7.40 + 1.082 \times 7.69)$$

$$\left. + 12\frac{11.00}{7.83^2}\left(\frac{35{,}075}{628} + \frac{35{,}075}{400}\right)\frac{1}{12^2}\right] = 39.12$$

$$\delta_{22} = \frac{1}{3}\left[2 \times 18.20\left(\frac{7.69^2}{7.83^2} + \frac{7.69}{7.83} + 1.0\right) + \frac{7.69^2}{7.83^2} 6.65 + 13.26 \right.$$

$$+ \frac{12}{7.83^2}(1.082 \times 7.69 + 1.352 \times 7.83)$$

$$\left. + 12\frac{10.00}{7.83^2}\left(\frac{35{,}075}{628} + \frac{35{,}075}{328}\right)\frac{1}{12^2}\right] = 44.25$$

$$\delta_{33} = \frac{1}{3}\left[6 \times 18.20 + 1.0 \times 13.26 + 31.40 + \frac{12}{7.83^2}(1.352 + 1.803)\right.$$

$$\left. + 12\frac{10.00}{7.83^2}\left(\frac{35{,}075}{628} + \frac{35{,}075}{328}\right)\frac{1}{12^2}\right] = 53.64 = \delta_{44}$$

$$\delta_{55} = \frac{1}{3}\left[2 \times 16.30\left(\frac{7.83^2}{7.69^2} + \frac{7.83}{7.69} + 1.0\right) + \frac{7.83^2}{7.69^2}13.26 + 6.65\right.$$

$$+ \frac{12}{7.69^2}(1.352 \times 7.83 + 1.082 \times 7.69)$$

$$\left. + 12\frac{10.00}{7.69^2}\left(\frac{35{,}075}{628} + \frac{35{,}075}{400}\right)\frac{1}{12^2}\right] = 41.92$$

$$\delta_{66} = \frac{1}{3}\left[2 \times 12.45\left(\frac{7.69^2}{7.40^2} + \frac{7.69}{7.40} + 1.0\right) + \frac{7.69^2}{7.40^2}6.65 + 3.70\right.$$

$$+ \frac{12}{7.40^2}(1.352 \times 7.69 + 0.904 \times 7.40)$$

$$\left. + 12\frac{11.00}{7.40^2}\left(\frac{35{,}075}{628} + \frac{35{,}075}{544}\right)\frac{1}{12^2}\right] = 31.29$$

δ_{k0} *values:*

$\zeta_1 = 0.962 \quad \zeta_2 = 0.949 \quad \zeta_{3,4,6} = 1.0 \quad \zeta_5 = 0.966$

$\zeta_{10} = 1.010 \quad \zeta_{20} = 1.002 \quad \zeta_{30,40,60} = 1.0 \quad \zeta_{50} = 1.003$

$$\delta_{10} = V_{10}\frac{\zeta_1}{6}\frac{c_1}{h_1}[c_1'(2h_0 + h_1) + h_0'h_0 + 12\nu_0]$$

$$- V_{20}\frac{\zeta_2}{6}c_2\left(h_1' + 12\frac{\nu_1}{h_2}\right) + 2\frac{c_1}{h_1^2}\left(\frac{I_c}{A_{1T}} + \frac{I_c}{A_{1B}}\right)\frac{M_{10}}{12^2}$$

$$= 98.85\frac{0.962}{6}\frac{11.00}{7.69}[17.9(2 \times 7.40 + 7.69) + 3.70 \times 7.40$$

$$+ 12 \times 0.904] - 58.45\frac{0.949}{6}10.00\left(6.65 + 12\frac{1.082}{7.69}\right)$$

$$+ 2\frac{11.00}{7.69^2}\left(\frac{35{,}075}{628} + \frac{35{,}075}{400}\right)\frac{1{,}087}{12^2} = +9{,}620$$

$$\delta_{20} = 58.45 \frac{0.949}{6} \frac{10.00}{7.83} [18.20(2 \times 7.69 + 7.83) + 6.65 \times 7.69$$

$$+ 12 \times 1.082] - 18.65 \frac{10.00}{6} \left(13.26 + 12 \frac{1.352}{7.83}\right)$$

$$+ 2 \frac{10.00}{7.83^2} \left(\frac{35{,}075}{628} + \frac{35{,}075}{328}\right) \frac{1{,}672}{12^2} = +5{,}892$$

$$\delta_{30} = 18.65 \frac{1.0}{6} \frac{10.00}{7.83} [18.20(2 \times 7.83 + 7.83) + 13.26 \times 7.83$$

$$+ 12 \times 1.352] + 18.65 \frac{10.00}{6} \left(31.40 + 12 \frac{1.803}{7.83}\right)$$

$$+ 2 \frac{10.00}{7.83^2} \left(\frac{35{,}075}{628} + \frac{35{,}075}{328}\right) \frac{1{,}858}{12^2} = +3{,}918$$

$$\delta_{40} = -18.65 \frac{1.0}{6} \frac{10.00}{7.83} [18.20(2 \times 7.83 + 7.83) + 31.40 \times 7.83$$

$$+ 12 \times 1.803] + 58.45 \frac{0.966}{6} 10.00 \left(13.26 + 12 \frac{1.352}{7.83}\right)$$

$$+ 2 \frac{10.00}{7.83^2} \left(\frac{35{,}075}{628} + \frac{35{,}075}{328}\right) \frac{1{,}672}{12^2} = -703$$

$$\delta_{50} = -58.45 \frac{0.966}{6} \frac{10.00}{7.69} [16.30(2 \times 7.83 + 7.69) + 13.26 \times 7.83$$

$$+ 12 \times 1.352] + 98.85 \frac{11.00}{6} \left(6.65 + 12 \frac{1.082}{7.69}\right)$$

$$+ 2 \frac{10.00}{7.69^2} \left(\frac{35{,}075}{628} + \frac{35{,}075}{400}\right) \frac{1{,}087}{12^2} = -4{,}230$$

$$\delta_{60} = -98.85 \frac{1.0}{6} \frac{11.00}{7.40} [12.45(2 \times 7.69 + 7.40) + 6.65 \times 7.69$$

$$+ 12 \times 1.082] = -8{,}550$$

Theory of Structures

Matrix

	X_1	X_2	X_3	X_4	X_5	X_6	δ_{k0}
1	39.120	−2.720	−	−	−	−	+9,620
2	−2.720	44.250	−5.110	−	−	−	+5,892
3	−	−5.110	53.640	−11.390	−	−	+3,918
4	−	−	−11.390	53.640	−5.120	−	−703
5	−	−	−	−5.120	41.920	−2.890	−4,230
6	−	−	−	−	−2.890	31.290	−8,550

Iteration results:

$X_1 = +256.0\,k'$ $X_2 = +158.8\,k'$ $X_3 = +86.9\,k'$
$X_4 = -6.23\,k'$ $X_5 = -121.5\,k'$ $X_6 = -285.0\,k'$

Moments:

$M_{01}^{(T)} = -285.0 \dfrac{35,075}{31,036} = -322\,k'$ $M_{21}^{(T)} = 158.8 \dfrac{35,075}{19,277} = +299\,k'$

$M_{01}^{(B)} = -285.0 \dfrac{27,000}{31,036} = -248\,k'$ $M_{21}^{(B)} = 158.8 \dfrac{3,482}{19,277} = +29\,k'$

$M_{10}^{(T)} = 256.0 \dfrac{35,075}{21,553} = +416\,k'$ $M_{23}^{(T)} = -6.23 \dfrac{35,075}{19,277} = -11.4\,k'$

$M_{10}^{(B)} = 256.0 \dfrac{8,034}{21,553} = +96\,k'$ $M_{23}^{(B)} = -6.23 \dfrac{3,482}{19,277} = -1.1\,k'$

$M_{12}^{(T)} = -121.5 \dfrac{35,075}{21,553} = -198\,k'$ $M_{32}^{(T)} = 86.9 \dfrac{35,075}{19,277} = +158\,k'$

$M_{12}^{(B)} = -121.5 \dfrac{8,034}{21,553} = -45\,k'$ $M_{32}^{(B)} = 86.9 \dfrac{3,482}{19,277} = +16\,k'$

Normal force:

$-N_{1T} = N_{1B} = \dfrac{1,087 - 2 \times 256}{7.69}$ $-N_{3T} = N_{3B} = \dfrac{1,858 - 2 \times 86.9}{7.83}$

$\qquad\qquad\quad = 75\,k$ $\qquad\qquad\quad = 215\,k$

$$-N_{2T} = N_{2B} = \frac{1{,}672 - 2 \times 158.8}{7.83} \qquad -N_{4T} = N_{4B} = \frac{1{,}672 + 2 \times 6.23}{7.83}$$
$$= 173\,k \qquad\qquad\qquad\qquad = 215\,k$$

Zero points and shear:

0-1:

$$x_{T1} = \frac{322 \times 11.0}{322 + 416} = 4.8', \quad x'_T = 6.2':$$

$$V_{1T} = \frac{322}{4.8} = 67.3\,k$$

$$x_{B1} = \frac{248 \times 11.0}{248 + 96} = 7.94', \quad x'_B = 3.06':$$

$$V_{1B} = \frac{248}{7.94} = 31.3\,k$$

$$z_0 = \frac{248 \times 7.4}{248 + 322} = 3.22', \quad z'_0 = 4.18':$$

$$V_{0v} = \frac{322}{4.18} = 77.0\,k \sim N_1 = 75.0\,k,$$

$$\sum_B^T V_1 = 98.6\,k \sim V_d$$
$$= 98.85\,k$$

$$(\Delta 0V = 2.0\,k)$$

1-2:

$x_{T2} = 3.98', \quad x'_{T2} = 6.02': \quad V_{2T} = 50.0\,k$
$x_{B2} = 6.08', \quad x'_{B2} = 3.98': \quad V_{2B} = 7.4\,k$

$$\sum_B^T V_2 = 57.4\,k < V_{02}$$

$z_1 = 1.43', \quad z'_1 = 6.26': \quad V_{1v} = 98.0\,k = N_2 - N_1 = 173 - 75.0 = 98.0\,k$

2-3:

$x_{T3} = 0.68', \quad x'_{T3} = 9.32': \quad V_{3T} = 17.0\,k$
$x_{B3} = 0.65', \quad x'_{B3} = 9.35': \quad V_{3B} = 1.7\,k$

$$\sum_B^T V_3 = 18.7\,k = V_{03}$$

$z_{2v} = 7.14', \quad z'_{2v} = 0.69': \quad V_{2v} = 43.5\,k = N_3 - N_2 = 215 - 173 = 42.0',$

$$(\Delta 2V = 1.3\,k)$$

$$V_{3v} = 0.0\,k$$

Moment and force diagram:

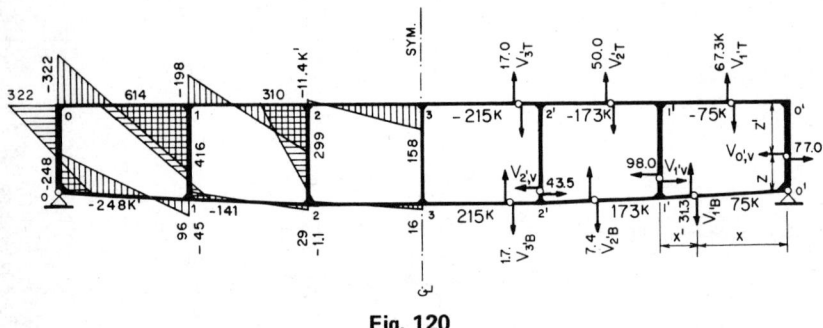

Fig. 120

The moments M_k and forces V_k, N_k obtained are shown in Fig. 120. The reason the horizontal forces between $V_k v$ and N_k do not check is that the chords were taken parallel in computing the final shear V_k and normal forces N_k. However, in cases in which the ζ values are relatively small, the differences are negligible.

Deflections of Vierendeel Girder

The influence of shear upon the deflection of a beam is commonly negligible. However, in a Vierendeel girder the deflections of chords and verticals due to the shear V_k and moments M_k are considerable and cannot be disregarded in computing the deflections. The exact computation of deflections for a Vierendeel girder in accordance with elastic theory is rather complex and time-requiring. For this reason, a qualitatively acceptable approximation is used.

The differential equation of the elastic curve of a beam is [Eqs. (108) and (109)]

$$\frac{d^2\delta}{dx^2} = \frac{M_0}{EI_c} + \frac{\kappa}{AG}\frac{dV}{dx}$$

In this equation, the quantity $\kappa/AG \cdot dV/dx$ denotes the shear influence upon deflection in panel members only; therefore, it must be extended to compensate also for deformations caused by moments M_k in the panels.

For $\varphi_K = \varphi_J = 0$ and infinitely rigid joints, the angular rotation of a member c_k in the Vierendeel panel is

$$\vartheta_k = \frac{\Sigma M_k c_{k0}}{12 EI_k} = \frac{V_k c_{k0}^2}{12 EI_k} \qquad (288)$$

and thus

$$\Sigma \vartheta_k = \frac{1}{12E}\left(\frac{c_{k0}^2}{I_k} + \frac{24\kappa}{A_k}\right) \qquad (289)$$

Assuming that the elastic curve due to V_k, M_k is proportional to the bending deflections M_0, the differential equation for the elastic curve of a Vierendeel girder becomes

$$\frac{d^2\delta}{dx^2} = \frac{M_0}{EI_c} + \omega\frac{dV}{dx}$$

where

$$\omega\frac{dV}{dx} = \frac{1}{12E}\left[\left(\frac{c_{k0}^2}{I_T + I_B} + \frac{24\kappa}{A_T + A_B}\right)\frac{dV_c}{dx} + \left(\frac{h_{k0}^2}{I_v} + \frac{24\kappa}{A_v}\right)\frac{dV_v}{dx}\right] \qquad (290)$$

The total deflection of the girder is

$$\Sigma\delta = \delta_0 + \delta_v = \int dx \int \frac{M_0}{EI} dx + \int \omega V\, dx \qquad (291)$$

Considering that for joint loads P_k the shear V_k is constant along a panel, the deflection due to shear is

$$\delta_{kv} = \frac{1}{12E}\sum_{k=0}^{k} c_{k0}\left[\left(\frac{c_{k0}^2}{I_T + I_B} + \frac{24\kappa}{A_T + A_B}\right)V_k + \left(\frac{h_{k0}^2}{I_{kv}} + \frac{24\kappa}{A_{kv}}\right)V_{kv}\right] \qquad (292)$$

The deflection δ_{k0} due to bending M_0 is computed as for any beam.
Using the data from the previous example (Figs. 119 and 120), the deflection of the girder will be

$I_c = kI_0 = 1.03 \times 1.69 \times 10^6 = 1.74 \times 10^6$ in.^{4}
$E = 3{,}200$ ksi $EI_c = 5.57 \times 10^9$ ksi $12E = 3.84 \times 10^4$

$c_{10} = 8.25$ ft $= 99$ in. $c_{20} = 7.75$ ft $= 93$ in. $c_{30} = 8.25$ ft $= 99$ in.
$h_0 = 5.0$ ft $= 60$ in. $h_{10} = 5.67$ ft $= 68$ in. $h_{20} = h_{30} = 6.0$ ft
$= 72$ in.

$24\kappa = 28.8$

$\delta_{0,\max} = \dfrac{1}{EI_c} \int M_0 M_1 \, dx$

$\phantom{\delta_{0,\max}} = \dfrac{1}{5.57 \times 10^9} \dfrac{5}{12} 20{,}220 \times 186 \times 744$

$\phantom{\delta_{0,\max}} = 0.210$ in.

$M_0 = N_3 h_3 = 215 \times 7.83$
$ = 1{,}685 \, k' = 20{,}220 \, k''$

$M_1 = \dfrac{1.0 L}{4} = \dfrac{1.0 \times 62.0 \times 12}{4}$
$ = 186 \, k''$

$L = 62.0 \times 12 = 744$ in.

$\delta_{v,\max} = \dfrac{1}{3.84 \times 10^4} \sum\limits_{k=0}^{k=3} 99 \left[\left(\dfrac{99^2}{52{,}920} + \dfrac{28.8}{1{,}100} \right) 98.6 + \left(\dfrac{60^2}{70{,}000} + \dfrac{28.8}{648} \right) 77 \right]$

$\phantom{\delta_{v,\max} =} + 93 \left[\left(\dfrac{93^2}{40{,}825} + \dfrac{28.8}{992} \right) 57.3 + \left(\dfrac{68^2}{40{,}500} + \dfrac{28.8}{540} \right) 98.0 \right]$

$\phantom{\delta_{v,\max} =} + 99 \left[\left(\dfrac{99^2}{38{,}587} + \dfrac{28.8}{956} \right) 18.7 + \left(\dfrac{72^2}{20{,}700} + \dfrac{28.8}{432} \right) 43.5 \right]$

$\phantom{\delta_{v,\max}} = 0.1945$ in. direct shear $\simeq 0.041$ in.

$\sum \delta_{\max} = 0.210 + 0.1945 = 0.4045$ in. $\sim 13/32$ in.

The deflections computed by Eqs. (291) and (292) are rather close to the actual deflections of Vierendeels measured in the field.

22. BEAM-SHEAR ACTION

In high-rise buildings, mostly shear walls must be used to resist high wind and seismic forces. The walls usually have openings dividing them into strips connected horizontally at floor levels by floor structure. To simplify statical computation, usually the horizontal connections are disregarded or the openings are overlooked. In the first case, the shear wall consists of a series of cantilevers, and in the second case, of a single cantilever. Both of these simplifications lead to overdesign or underdesign of the walls, which may have serious consequences, especially in the event of seismic and hurricane loading.

Estimating the stress conditions and deformations in the wall and horizontal connections with acceptable accuracy is a complex and time-requiring procedure. To overcome this difficulty, Rosman* has developed a rather simple method, the accuracy of which is, in most conceivable cases, well within acceptable limits.

Let us consider a shear wall subjected to horizontal loading $2w$, as illustrated in Fig. 121. The stiffness of the wall is decreased because of the openings by a ratio

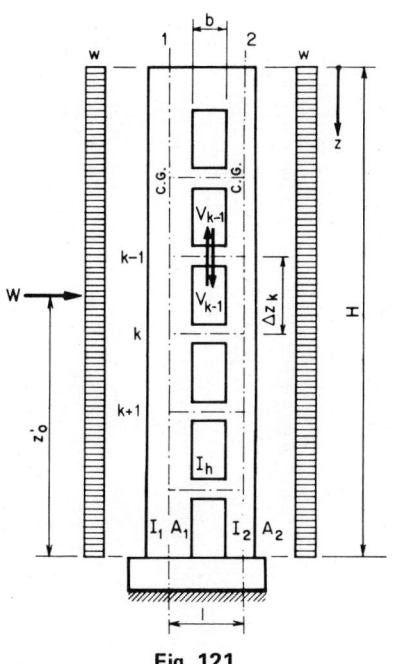

$$\epsilon'_k = \frac{\Delta z_k \, b^3}{12 E I_{Hk}}$$

$$k = 0, 1, 2, \ldots, n \qquad (293)$$

$$\epsilon = \frac{\Delta z \, b^3}{12 I_H}$$

Fig. 121

The shear force V_k in the horizontals resisting the deformation of wall elements 1 and 2 can be computed by the principle of minimum deformation energy. The shear force expressed in shear stresses is

$$V = \int_0^z \tau \, dz \qquad (294)$$

*Riko Rosman, Beitrag zur statischen Berechnung waagerecht belasteter Querwaende bei Hochbauten, Der Bauingenieur, no. 4, 1960.

206 Theory of Structures

and thus the total deformation energy of the horizontals is

$$U_H = \frac{1}{2}\int_0^H \epsilon' \tau^2 dz = \frac{1}{2E}\int_0^H \epsilon V'^2 dz \tag{295}$$

Assuming that Hooke's law applies, the external moment as a function of z is

$$M_z = wz^2 - V_z l \tag{296}$$

and the moments carried by walls 1 and 2 are

$$\begin{aligned} M_{1z} &= M_z \frac{I_1}{I_1 + I_2} = M_z \frac{I_1}{\Sigma I_w} \\ M_{2z} &= M_z \frac{I_2}{I_1 + I_2} = M_z \frac{I_2}{\Sigma I_w} \end{aligned} \tag{297}$$

The deformation energy of the wall strips (walls 1 and 2) is

$$U_w = \frac{1}{2E}\int_0^H \left(\frac{M_z^2}{\Sigma I_w} + \frac{1}{\Sigma A_w} V^2\right) dz \quad \Sigma A_w = \frac{1}{A_1} + \frac{1}{A_2} \tag{298}$$

The total energy of horizontals and walls is

$$U = \frac{1}{2E}\int_0^H \left(\epsilon V'^2 + \frac{M_z^2}{\Sigma I_w} + \frac{1}{\Sigma A_w} V^2\right) dz \tag{299}$$

Thus, in accordance with the minimum-energy principle,

$$\partial U = \frac{1}{E}\int_0^H \left[\left(\frac{l^2}{\Sigma I_w} + \frac{1}{\Sigma A_w}\right)V - \frac{wlz^2}{\Sigma I_w} - \epsilon V''\right] \partial V \, dz = 0 \tag{300}$$

The boundary conditions are $z = 0: V = 0$ and $z = H: \partial V = 0$. Denoting

$$\lambda^2 = \left(\frac{l^2}{\Sigma I_w} + \frac{1}{\Sigma A_w}\right)\frac{1}{\epsilon}$$

$$\beta = \frac{wl}{\Sigma I_w}\frac{1}{\epsilon}$$

(301)

the differential equation for shear V becomes

$$V'' - \lambda^2 V + \beta z^2 = 0 \tag{302}$$

The solution of this linear differential equation can most simply be obtained by converting the differential equation into a difference equation [Eq. (12)]:

$$-\frac{V_{k-1} - 2V_k + V_{k+1}}{\Delta z_k} + \lambda^2 V_k = \beta z_k^2$$

$$-V_{k-1} + (2 + \lambda^2 \Delta z_k^2)V_k - V_{k+1} = \beta \Delta z_k^2 z_k^2$$

(303)

In accordance with Eq. (14b), for fixed end $k = n$: $V_{n+1} = V_{n-1}$ and for free end $k = 0$: $V_0 = 0$. Denoting

$$\delta_{kk} = 2 + \lambda^2 \Delta z_k^2 \quad \delta_{k-1,k} = -1 \quad \delta_{k,k+1} = -1$$

$$\delta_{k0} = \beta \Delta z_k^2 z_k^2$$

(304)

the difference equations are

$$k = 1: V_1 \delta_{11} + V_2 \delta_{12} = \delta_{10}$$

$$k: V_{k-1}\delta_{k,k-1} + V_k \delta_{kk} + V_{k+1}\delta_{k,k+1} = \delta_{k0}$$

$$k = n: 2V_{n-1}\delta_{n,n-1} + V_n \delta_{nn} = \delta_{n0}$$

The matrix of the linear equations for the wall points $k = 1, \ldots, 6$ illustrated in Fig. 121 is

	V_1	V_2	V_3	V_4	V_5	V_6	δ_{k0}
1	δ_{11}	-1	–	–	–	–	δ_{10}
2	-1	δ_{22}	-1	–	–	–	δ_{20}
3	–	-1	δ_{33}	-1	–	–	δ_{30}
4	–	–	-1	δ_{44}	-1	–	δ_{40}
5	–	–	–	-1	δ_{55}	-1	δ_{50}
6	–	–	–	–	-2	δ_{66}	δ_{60}

The convergence of these equations is very good. Thus they can be solved easily by iteration.

After the V_k values are computed, the actual moments in the walls and horizontals can be determined:

$$\sum M_k = wz_k^2 - lV_k \quad M_{1k} = \sum M_k \frac{I_1}{\Sigma I_w} \quad M_{2k} = \sum M_k \frac{I_2}{\Sigma I_w} \quad (305)$$

$$M_{Hk} = \pm (V_{k-1} - V_k)\frac{1}{2}b$$

The deflection of the wall is given by Eq. (111):

$$\delta_z = \frac{1}{E\Sigma I_w} \int \sum M_z \bar{M}_{z1}\, dz + \sum \frac{1}{E} \int N_z \bar{N}_{z1} \frac{dz}{A_{1,2}}$$

$$+ \frac{\kappa}{G\Sigma A_w} \int \sum V_z \bar{V}_{z1}\, dz \quad \kappa \simeq 1.2 \qquad (306)$$

where $\Sigma M_z = wz^2 - lV_z$, $\bar{M}_{z1} = z$, $N_z = \pm wz^2/l$, $\bar{N}_{z1} = \pm z^2/l$, and $V_z = 2wz$, $\bar{V}_{z1} = 1$. Generally, the influence of shear upon deflection is negligible. In the case of poor soil conditions, the deflection of the wall must be increased in the amount caused by the rotation of the foundation.

Analysis of a Six-story Tower Wall

For illustration of this method, a wall of a six-story service tower (Fig. 121) will be analyzed.

Combined Carrying Actions

$A_1 = 10.30 \text{ ft}^2 \quad A_2 = 8.14 \text{ ft}^2 \quad \Sigma A_w = 18.44 \text{ ft}^2 \quad b = 4.00 \text{ ft.}$
$I_1 = 12.53 \text{ ft}^4 \quad I_2 = 9.27 \text{ ft}^4 \quad \Sigma I_w = 21.80 \text{ ft}^4 \quad I_H = 3.95 \text{ ft}^4$

$\Delta z_{1\ldots 5} = 10.75 \text{ ft} \quad \Delta z_{6,7} = 5.37 \text{ ft} \quad \sum_0^H \Delta z = 64.50 \text{ ft} \quad l = 8.86 \text{ ft}$

$$\epsilon = \frac{\Delta z b^3}{12 I_H} = \frac{10.75 \times 4.00^3}{12 \times 3.95} = 14.50$$

$$\lambda^2 = \left(\frac{l^2}{\Sigma I_w} + \frac{1}{\Sigma A_w}\right) \frac{1}{\epsilon}$$

$$= \left(\frac{8.86^2}{21.80} + \frac{1}{18.44}\right) \frac{1}{14.50} = 0.249$$

$$\beta = \frac{wl}{\Sigma I_w} \frac{1}{\epsilon} = \frac{0.5 \times 8.86}{21.80} \frac{1}{14.50} = 0.014$$

$\delta_{11\ldots 55} = 2 + \lambda^2 \Delta z_k^2 = 2 + 0.249 \times 10.75^2 = 30.80,$

$$\delta_{66\ldots 77} = 9.20$$

$\delta_{10} = \beta \Delta z_1^2 z_1^2 = 0.0140 \times 10.75^2 \times 10.75^2 = 187$
$\delta_{20} = 748, \quad \delta_{30} = 1{,}680, \quad \delta_{40} = 2{,}985, \quad \delta_{50} = 4{,}660$
$\delta_{60} = \beta \Delta z_6^2 z_6^2 = 0.0140 \times 5.37^2 \times 59.12^2 = 1{,}405$

$$\delta_{70} = 1{,}675$$

MATRIX:

	V_1	V_2	V_3	V_4	V_5	V_6	V_7	δ_{k0}
1	30.80	−1	−	−	−	−	−	187
2	−1	30.80	−1	−	−	−	−	748
3	−	−1	30.80	−1	−	−	−	1,680
4	−	−	−1	30.80	−1	−	−	2,985
5	−	−	−	−1	30.80	−	−1	4,660
6	−	−	−	−	−1	9.20	−1	1,405
7	−	−	−	−	−	−2	9.20	1,675

$V_k = $ 7.0 26.3 59.0 104.0 160.0 194.0 225.0 k
$\Delta V_k = 7.0 \quad 19.3 \quad 32.7 \quad 45.0 \quad 56.0 \quad 65.0 \quad \Sigma \Delta V_k = 225.0\,k$

$$\Sigma M_7 = wH^2 - V_7 l = 0.5 \times 64.5^2 - 225 \times 8.86 \simeq 90\,k'$$
$$M_{H5} = \pm(V_5 - V_7)b/2 = \mp 65.0 \times 4.0/2 = \mp 130\,k'$$

When several walls are combined into a unit by floors and transversal walls (Fig. 122) subjected to horizontal loading W, w, each of the combined walls must carry the external load in accordance with its stiffness. If the center of resistance of the walls does not coincide with the center of gravity of the external loads, a torsional moment M_T will be present, which also must be counteracted by the individual walls.

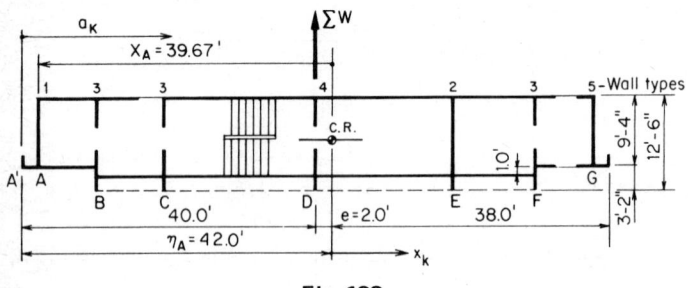

Fig. 122

The stiffness of a wall or element is determined by its deflection under load $(P = 1.0)$:

$$S_{0k} = \frac{1.0}{\delta_{k,\max}} \tag{307}$$

Thus the relative stiffnesses of the service-tower walls, illustrated in Fig. 122, will be

$$S_k = \frac{\delta_{k,\max}}{\sum_{1}^{n} \delta_{k,\max}} = \frac{S_{0k}}{\sum_{1}^{n} S_{0k}} \qquad k = 1, 2, \ldots, n \tag{308}$$

Using the data from the previous example for wall type 3, its deflection $\delta_{3,\max} = 0.342$ in., and thus its stiffness is

$$S_{03} = \frac{1.0}{0.342} = 2.92$$

Computing the stiffnesses of the other wall types, we obtain $S_{01} = 3.14$, $S_{02} = 9.35$, $S_{04} = 4.44$, and $S_{05} = 1.91$. Thus the total stiffness of the walls is

$$\sum_{A}^{G} S_{0k} = 27.60$$

The relative stiffnesses of the walls, in accordance with Eq. (308), are

Wall A:	S_A	= 0.114
Wall B, C, F:	$S_{B,C,F}$	= 0.106
Wall D:	S_D	= 0.161
Wall E:	S_E	= 0.339
Wall G:	S_G	= 0.069

$$\sum_{A}^{G} S_k = 1.0$$

The center of gravity of the stiffness of the service tower is

$$\eta_{A'} = \frac{\Sigma S_k a_k}{\Sigma S_k} \simeq 42.0 \text{ ft} \tag{309}$$

and the eccentricity is

$$e = \eta_{A'} - b = 42.0 - 40.0 = 2.0 \text{ ft}$$

The torsional moment is

$$M_T = e \Sigma W = 2.0 \Sigma W \tag{310}$$

Assuming a linear distribution for the loading of the walls due to torsional moment M_T, the load to be carried by the individual walls, expressed for example in terms of wall A, will be

$$W_k = \frac{x_k}{x_A} \frac{S_k}{S_A} W_A, \quad x_A = \eta_{A'} - a_A = 39.67 \text{ ft} \tag{311}$$

$$\sum_{A}^{G} \Delta M_k = \sum_{A}^{G} \left(\frac{x_k^2}{x_A} \frac{S_k}{S_A} \right) W_A = M_T$$

$$W_A = \frac{M_T}{\Sigma[(x_k^2/x_A)(S_k/S_A)]}$$

(311)
(Cont.)

Assuming $\Sigma W = 420\,k$, the torsional moment $M_T = +2.0 \times 420 = 840\,k'$

$$\sum_{A}^{G} \left(\frac{x_k^2}{x_A} \frac{S_k}{S_A} \right) = 134.29$$

Thus the wall loadings are

$$\Sigma W_A = 0.114 \times 420 + 1.0 \frac{840}{134.29} = 47.88 + 6.25 = 54.13\,k$$

$$\Sigma W_G = 0.069 \times 420 - \frac{35.67}{39.67} \cdot \frac{0.069}{0.114} \cdot 6.25 = 25.60\,k$$

For uniform load distribution, the load density for wall k will be

$$2w_k = \frac{\Sigma W_k}{H} \tag{312}$$

The final results of the W_k values or moments are obtained by multiplying the unit-load ($2w = 1.0\,k/\text{ft}$) values with the ratio w_k/W_k.

6

CIRCULAR BEAMS
Beam–Torsion Action*

In contemporary architecture and engineering, simple and continuous circular beams are very often unavoidable. Because the methods available to describe the stress and deformation condition in such beams are inadequate, engineers in practice are forced to apply rough approximation or to replace the circular curvature entirely with regular polygons. This leads to architecturally unacceptable, forced solutions and many structural uncertainties. In bridge construction, especially at intersections where the construction depth is limited and the structure has to follow the alignment of the road, such polygon substitutions force the use of short spans and numerous piers, resulting in obstructed and annoying views. In cases in which circular beams have been used, the

---

*Kurt Beyer, "Die Statik im Eisenbetonbau," Springer-Verlag OHG, Berlin, 1934.
Werner Vreden, "Die Berechnung des gekrümmten Durchlaufträgers," Wilhelm Ernst & Sohn KG, Berlin, 1964.

214 Theory of Structures

torsional moments due to the curvature commonly have been balanced by piers, which results in unbalance and heaviness of the structure and leads usually to uneconomical solutions, especially where large temperature changes must be accounted for.

In the following, a usable method for analyzing circular beams with acceptable accuracy will be discussed.

The equilibrium of external forces at section ds requires that (Fig. 123)

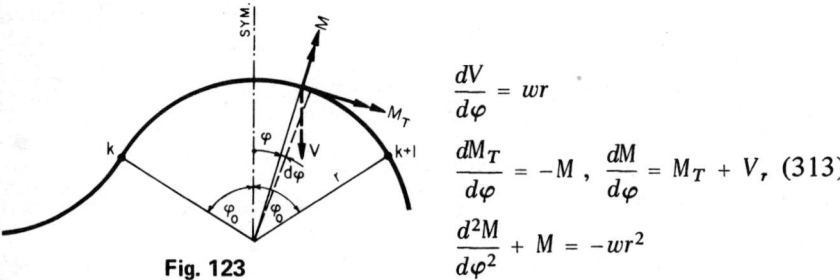

Fig. 123

$$\frac{dV}{d\varphi} = wr$$

$$\frac{dM_T}{d\varphi} = -M, \quad \frac{dM}{d\varphi} = M_T + V_r \quad (313)$$

$$\frac{d^2M}{d\varphi^2} + M = -wr^2$$

The solution of this differential equation for a statically determined system for uniform loading w is

$$M = A \sin\varphi + B \cos\varphi - wT^2$$

Thus, by direct integration [Eq. (313)], we obtain

$$M_T = -\int M \, d\varphi$$

$$V = \frac{1}{r}\left(\frac{dM}{d\varphi} - M_T\right)$$
(314)

The integration constants A, B can be determined from the boundary conditions for M and $dM/d\varphi$. For example, at symmetry line $\varphi = 0$:

For symmetric loading: $V = 0$ $M_T = 0$
For antisymmetric loading: $M = 0$ $\delta = 0$
and at support $\varphi = \varphi_0$:

Simple support: $M = 0$ $M_T = 0$ $\delta = 0$
Fixed-end: $dM/d\varphi = 0$ $\delta = 0$
Continuous beam: $dM_z/d\varphi = -dM_r/d\varphi$

In general, statically indeterminate systems are analyzed as any indeterminate structural system, that is, by cutting the beam at intermediate supports or any other suitable place and thus establishing a statically determinate principal system. Usually, the moments M, M_T and shear V acting in the cut faces are the redundants X_i ($i = 1, \ldots, n$). They will be computed from the continuity equations [Eqs. (7) and (8)]. However, for certain loadings, the redundants are zero or known, and thus the M, M_T and V can be computed directly by the equilibrium equations [Eqs. (313) and (314)]. The number of redundants of a circular continuous beam on simple supports is $n - 2$, where n denotes the number of spans. Thus a two-span beam is statically determinate and a three-span beam is only once indeterminate. A continuous circular beam with torsion-stiff end supports is n times indeterminate and is $2n - 1$ times when torsion is stiff at all supports.

Statically Determinate Circular Beams

As the two-span statically determinate circular beam serves mostly as the principal system, a nonsymmetric two-span beam (Fig. 124) as a general case will be analyzed first.

Geometric data:

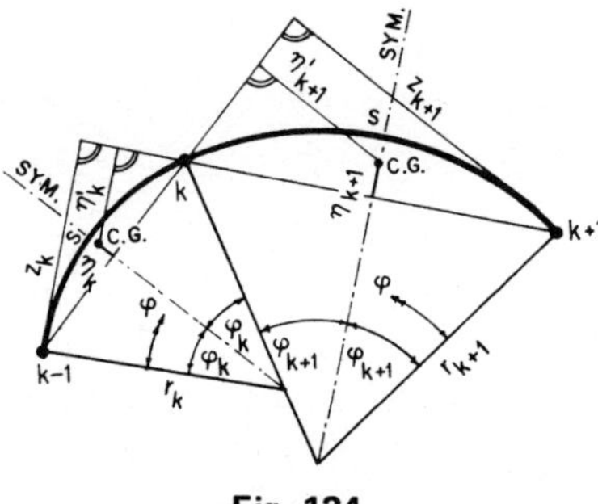

Fig. 124

$$z_k = r_k[\cos\varphi_{k+1} - \cos(\varphi_{k+1} + 2\varphi_k)]$$
$$z_{k+1} = r_{k+1}[\cos\varphi_k - \cos(\varphi_k + 2\varphi_{k+1})]$$
$$\eta_k = r_k\left(\frac{\sin\varphi_k}{\varphi_k} - \cos\varphi_k\right)$$
$$\eta'_k = r_k\left[\cos\varphi_{k+1} - \cos(\varphi_k + \varphi_{k+1})\frac{\sin\varphi_k}{\varphi_k}\right]$$
$$\eta_{k+1} = r_{k+1}\left(\frac{\sin\varphi_{k+1}}{\varphi_{k+1}} - \cos\varphi_{k+1}\right) \quad (315)$$
$$\eta'_{k+1} = r_{k+1}\left[\cos\varphi_k - \cos(\varphi_k + \varphi_{k+1})\frac{\sin\varphi_{k+1}}{\varphi_{k+1}}\right]$$

Uniform load w:

Total load W_k on span k or, respectively, on span $k+1$:

$$W_k = \int_{\varphi=0}^{2\varphi_k} wr_k\,d\varphi = wr_k 2\varphi_k \qquad W_{k+1} = \int_{\varphi=0}^{2\varphi_{k+1}} wr_{k+1}\,d\varphi \qquad (316)$$

$$= wr_{k+1} 2\varphi_{k+1}$$

The flexural moment M, torsional moment M_T, and shear V are, in accordance with Eqs. (314),

$$M_k = A\sin\varphi + B\cos\varphi - w_k r_k^2$$

Boundary conditions (span k loaded) are

$\varphi = 0:$ $\qquad\qquad M = 0 \quad M_T = 0$

$\varphi = 2\varphi_k:$ $\qquad M_{T,k} = W_k \eta_k, \quad M_{T,k+1} = W_{k+1}\eta_{k+1}$

Thus:

$$\begin{aligned} A &= V_k r_k \quad B = w_k r_k^2 \\ M &= V_k r_k \sin\varphi - (1 - \cos\varphi) w_k r_k^2 \\ M_T &= \mp\int M\,d\varphi = \mp V_k r_k (1 - \cos\varphi) \pm (\varphi - \sin\varphi) w_k r_k^2 \end{aligned} \qquad (317)$$

Uniform load w_k on span k:

$$V_{k-1} = \frac{W_k \eta'_k}{z_k} \qquad V_{k+1} = -\frac{W_k \eta_k}{z_{k+1}} \qquad (318)$$

Uniform load w_{k+1} on span $k+1$:

$$V_{k-1} = \frac{W_{k+1}\eta_{k+1}}{z_k} \qquad V_{k+1} = \frac{W_{k+1}\eta'_{k+1}}{z_{k+1}} \qquad (319)$$

Circular Beams

The flexural moments M in unloaded span k or $k+1$ are

$$M_k = V_{k-1} r_k \sin\varphi$$
$$M_{k+1} = V_{k+1} r_{k+1} \sin\varphi \qquad (320)$$

and the torsional moments are

$$M_{Tk} = V_{k-1} r_k (1 - \cos\varphi)$$
$$M_{Tk+1} = V_{k+1} r_{k+1} (1 - \cos\varphi)$$

$$(321)$$

The moments due to individual span loading, in accordance with Eq. (317), are illustrated in Fig. 125.

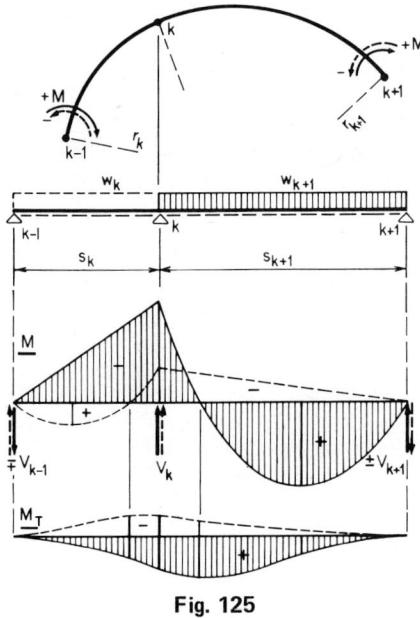

Fig. 125

Concentrated load:

Geometric data:

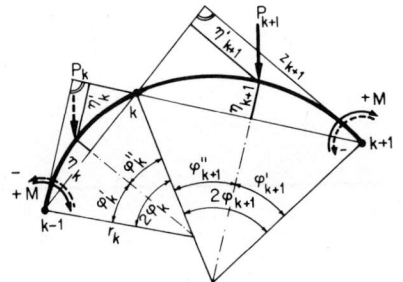

$$\eta_k = r_k [\cos(\varphi'_k - \varphi_k) - \cos\varphi_k]$$
$$\acute{\eta}_k = r_k [\cos\varphi_{k+1} - \cos(\varphi''_k + \varphi_{k+1})]$$

$$\eta_{k+1} = r_{k+1}[\cos(\varphi_{k+1} - \varphi'_{k+1}) - \cos\varphi_{k+1}]$$
$$\acute{\eta}_{k+1} = r_{k+1}[\cos\varphi_k - \cos(\varphi''_{k+1} + \varphi_k)] \qquad (322)$$

Load P_k on span k and load P_{k+1} on span $k+1$:

$$V_{k-1} = \frac{P_k \eta'_k}{z_k} \quad V_{k+1} = -\frac{P_k \eta_k}{z_{k+1}}$$

$$V_{k-1} = \frac{P_{k+1} \eta_{k+1}}{z_k} \quad V_{k+1} = \frac{P_{k+1} \eta'_{k+1}}{z_{k+1}} \tag{323}$$

$$M = Vr \sin\varphi - Pr \sin\varphi''$$
$$M_T = \mp Vr(1 - \cos\varphi) \pm Pr(1 - \cos\varphi') \tag{324}$$

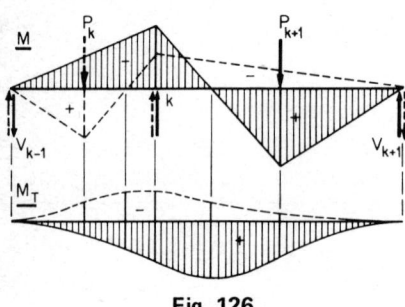

Fig. 126

The moments, in accordance with Eqs. (324), are illustrated in Fig. 126.

Torsional moment M_{T0} on k or $k+1$:

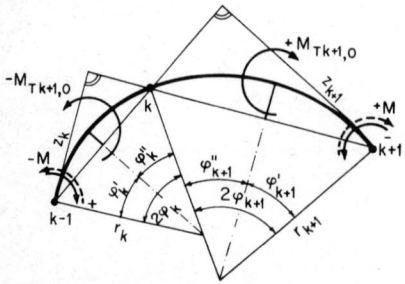

In case a circular beam is subjected to a torsional moment $M_{T0} = Py_\varphi$, where y_φ is the distance of a concentrated load P from the centerline of the beam, the shears (V_{k-1}, V_{k+1}) due to M_{T0} on span k or $k+1$ are

Moment M_{Tk0} on span k or moment M_{Tk+1} on span $k+1$:

$$V_{k-1} = \pm \frac{M_{Tk0} \cos(\varphi_{k+1} + \varphi_k'')}{z_k}$$

$$V_{k+1} = \pm \frac{M_{Tk0} \cos(\varphi_k - \varphi_k')}{z_{k+1}}$$

$$V_{k-1} = \mp \frac{M_{Tk+1,0} \cos(\varphi_{k+1} - \varphi_{k+1}')}{z_k} \quad (325)$$

$$V_{k+1} = \mp \frac{M_{Tk+1,0} \cos(\varphi_k + \varphi_{k+1}'')}{z_{k+1}}$$

Moments:

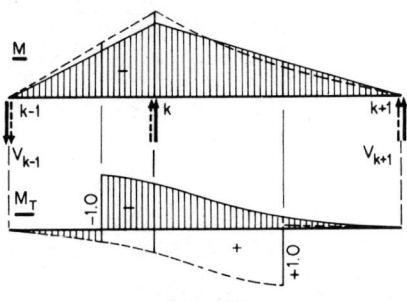

Fig. 127

$$M = \mp V_k r_k \sin\varphi \pm M_{Tk0} \sin\varphi''$$

$$M_T = \mp V_k r_k (1 - \cos\varphi) \pm M_{Tk0} \cos\varphi'' \quad (326)$$

Flexural moment at beam end:

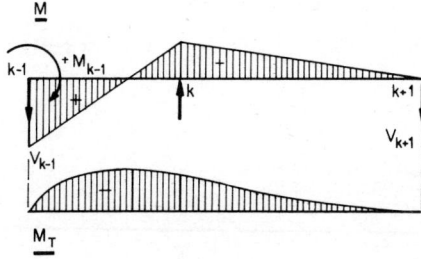

Fig. 128

$$V_{k-1} = -\frac{M_{k-1} \sin(2\varphi_k + \varphi_{k+1})}{z_k}$$

$$V_{k+1} = -\frac{M_{k-1} \sin\varphi_k}{z_{k+1}} \quad (327)$$

$$M_k = V_{k-1} r_k \sin\varphi + M_{k-1} \cos\varphi$$
$$M_{k+1} = V_{k+1} r_{k+1} \sin\varphi$$

$$M_{Tk} = -V_{k-1} r_k (1 - \cos\varphi) - M_{k-1} \sin\varphi$$
$$M_{Tk+1} = V_{k+1} r_{k+1} (1 - \cos\varphi) \quad (328)$$

Torsional moment at beam end:

It is customary to denote the moments M and M_T at the plan (or elevation) by a double arrow, where the moment acting in the clockwise direction is considered positive and in the counterclockwise direction, negative.

Moments:

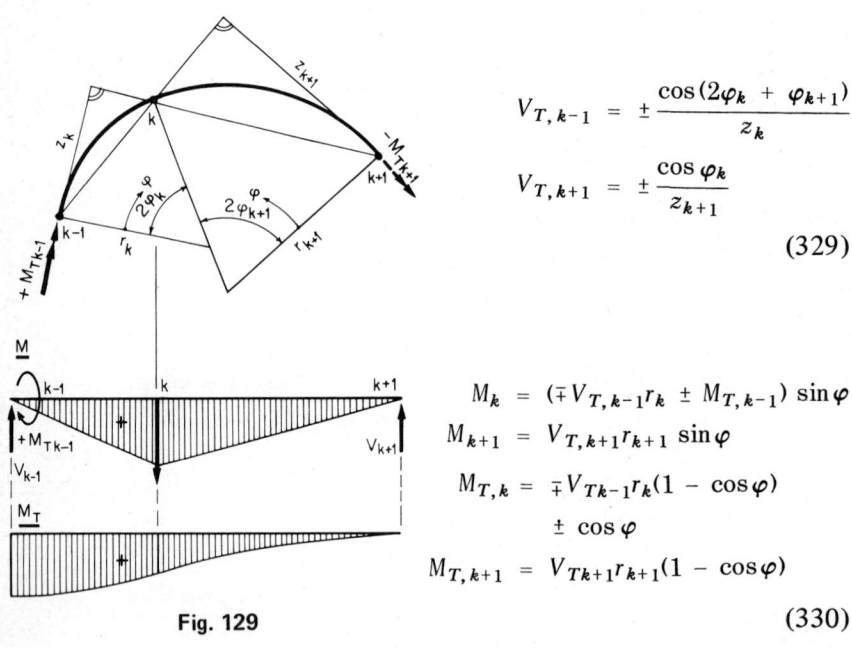

$$V_{T,k-1} = \pm \frac{\cos(2\varphi_k + \varphi_{k+1})}{z_k}$$

$$V_{T,k+1} = \pm \frac{\cos \varphi_k}{z_{k+1}}$$

(329)

$$M_k = (\mp V_{T,k-1} r_k \pm M_{T,k-1}) \sin \varphi$$
$$M_{k+1} = V_{T,k+1} r_{k+1} \sin \varphi$$
$$M_{T,k} = \mp V_{Tk-1} r_k (1 - \cos \varphi)$$
$$\pm \cos \varphi$$
$$M_{T,k+1} = V_{Tk+1} r_{k+1} (1 - \cos \varphi)$$

(330)

Fig. 129

Statically Indeterminate Circular Beams

Principal system:

Two-span circular beams (Fig. 130)

Redundants: $X_{1,3,\ldots}$ —bending moments
$X_{2,4,\ldots n}$ —torsional moments

Circular Beams

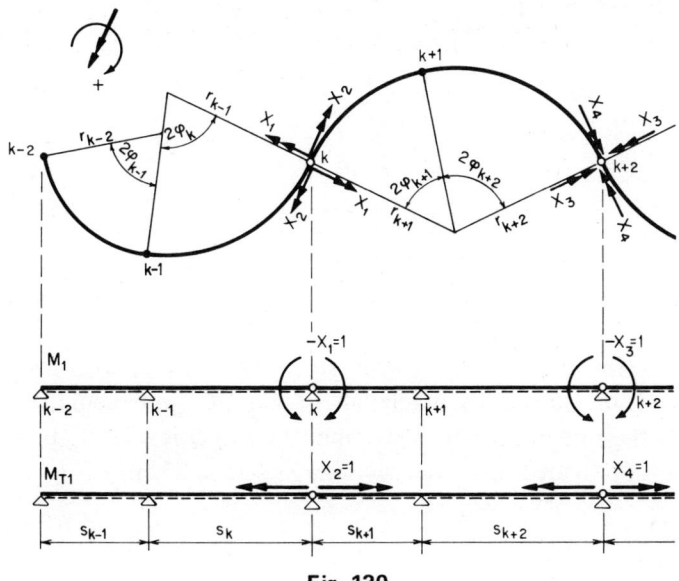

Fig. 130

Continuity equations (7) written in matrix form:

	X_1	X_2		X_{n-1}	X_n	
1	δ_{11}	δ_{12}		$\delta_{1,n-1}$	δ_{1n}	δ_{10}
2	δ_{21}	δ_{22}		$\delta_{2,n-1}$	δ_{2n}	δ_{20}
$n-1$	$\delta_{n-1,1}$	$\delta_{n-1,2}$		$\delta_{n-1,n-1}$	$\delta_{n-1,n}$	$\delta_{n-1,0}$
n	δ_{n1}	$\delta_{n,2}$		$\delta_{n,n-1}$	δ_{nn}	$\delta_{n,0}$

δ_{ii}, $\delta_{i,i+1}$, and δ_{i0} values:

$(k, i = 1, 2, \ldots, n)$

$$EI_c \delta_{ii} = \sum_{0}^{k=n} \int_{0}^{2\varphi_k} \overline{M}_i^2 r_k \, d\varphi + \rho \sum_{0}^{k=n} \int_{0}^{2\varphi_k} \overline{M}_{Ti}^2 r_k \, d\varphi \qquad \rho = \frac{EI}{GI_T},$$

$$r_k \, d\varphi = ds_k \qquad (331)$$

$$EI_c\delta_{i,i+1} = \sum_0^{k=n} \int_0^{2\varphi_k} \overline{M}_i \overline{M}_{i+1} r_k \, d\varphi + \rho \sum_0^{k=n} \int_0^{2\varphi_k} \overline{M}_{Ti} \overline{M}_{T,i+1} r_k \, d\varphi$$

$$EI_c\delta_{i0} = \sum_0^{k=n} \int_0^{2\varphi_k} M_0 \overline{M}_i r_k \, d\varphi + \rho \sum_0^{k=n} \int_0^{2\varphi_k} M_{T0} \overline{M}_{Ti} r_k \, d\varphi \qquad \text{(331)} \atop \text{(Cont.)}$$

The moments $\overline{M}_i$ and torsional moments $\overline{M}_{Ti}$ due to $X_{i(i=1,2,\ldots,n)} = 1$ and bending moments M_0 and torsional moments M_{T0} due to external loadings (wP, M_{T0}) in the principal system will be computed in accordance with Eqs. (317) to (330). In cases in which $I_k I_{Tk}$ vary, the $\delta_{ii}, \delta_{i,i+1}, \delta_{i0}$ integrals can best be solved by using an analytical approximation. Simpson's rule is the most accurate. The application of this rule requires that the segments are divided into even sections $n\Delta s = s$. Then

$$\int y \, ds = \frac{\Delta s}{3} (y_0 + 4y_1 + 2y_2 + \cdots + 4y_{n-1} + y_n) \qquad (332)$$

where $\Delta s = r \, d\varphi$ and y is the function to be integrated ($\overline{M}_k^2, \overline{M}_{Tk}^2, \overline{M}_i \overline{M}_k \ldots$). Direct integration is more simple for constant I, I_T, especially for uniform loading conditions.

The matrix can be solved by iteration because the convergence is rather good. The actual shear V and moments M are obtained by superposition:

$$\begin{aligned} \Sigma V_k &= V_{k0} - X_1 \overline{V}_k - X_2 \overline{V}_{Tk} + \cdots \\ \Sigma M_k &= M_{k0} - X_1 \overline{M}_k - X_2 \overline{M}_{Tk} + \cdots \end{aligned} \qquad (333)$$

For example, a four-span circular beam (Fig. 131) will be computed for $w_{0-1} = w_{2-3} = 0$, $w_{1-2} = w_{3-4} = 1.0 \, k/\text{ft}$.

Redundants: X_1, X_2 at support $k = 2$.

Geometric data:

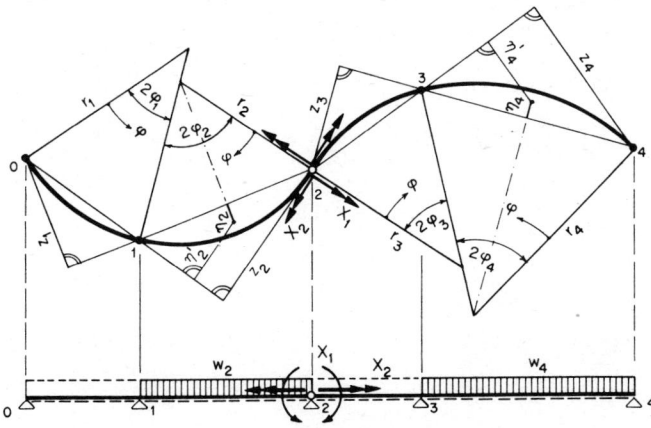

Fig. 131

$r_1 = 100$ ft $2\varphi_1 = 44°$, $\varphi_1 = 22°$ $s_1 = 76.80$ ft
$r_2 = 85$ ft $2\varphi_2 = 72°$, $\varphi_2 = 36°$ $s_2 = 106.60$ ft
$r_3 = 100$ ft $2\varphi_3 = 44°$, $\varphi_3 = 22°$ $s_3 = 76.80$ ft
$r_4 = 125$ ft $2\varphi_4 = 56°$, $\varphi_4 = 28°$ $s_4 = 122$ ft

In accordance with Eq. (315)

$z_1 = r_1[\cos\varphi_2 - \cos(\varphi_2 + 2\varphi_1)] = 63.5$ ft
$z_2 = r_2[\cos\varphi_1 - \cos(\varphi_1 + 2\varphi_2)] = 84.8$ ft
$z_3 = r_3[\cos\varphi_4 - \cos(\varphi_4 + 2\varphi_3)] = 57.4$ ft
$z_4 = r_4[\cos\varphi_3 - \cos(\varphi_3 + 2\varphi_4)] = 90.0$ ft

$\eta_2 = r_2\left(\dfrac{\sin\varphi_2}{\varphi_2} - \cos\varphi_2\right) = 10.8$ ft

$\eta_2' = r_2\left[\cos\varphi_1 - \cos(\varphi_2 + \varphi_1)\dfrac{\sin\varphi_2}{\varphi_2}\right] = 36.6$ ft

$\eta_4 = r_4\left(\dfrac{\sin\varphi_4}{\varphi_4} - \cos\varphi_4\right) = 9.75$ ft

$\eta_4' = r_4\left[\cos\varphi_3 - \cos(\varphi_4 + \varphi_3)\dfrac{\sin\varphi_4}{\varphi_4}\right] = 38.7$ ft

V_k and M_k, M_{Tk} values due to w_k in principal system, (Eqs. (317) to (321)):

$$W_2 = w_2 s_2 = 1.0 \times 106.6 = 106.6 \, k$$

$$V_2 = \frac{W_2 \eta_2'}{z_2} = \frac{106.6 \times 36.6}{84.8} = 46.0 \, k$$

$$V_0 = -\frac{W_2 \eta_2}{z_1} = -\frac{106.6 \times 10.8}{63.5} = -18.15 \, k$$

$$W_4 = w_4 s_4 = 1.0 \times 122.0 = 122.0 \, k$$

$$V_2 = -\frac{W_4 \eta_4}{z_3} = -\frac{122.0 \times 9.75}{57.4} = -20.8 \, k$$

$$V_4 = \frac{W_4 \eta_4'}{z_4} = \frac{122.0 \times 38.7}{90.0} = 52.5 \, k$$

$$M_{01} = V_0 r_1 \sin \varphi = -1{,}815 \sin \varphi$$
$$M_{21} = V_2 r_2 \sin \varphi - (1 - \cos \varphi) w_2 r_2^2$$
$$= 3{,}910 \sin \varphi - 7{,}225 (1 - \cos \varphi)$$
$$M_{23} = V_2 r_3 \sin \varphi = -2{,}080 \sin \varphi$$
$$M_{43} = V_4 r_4 \sin \varphi - (1 - \cos \varphi) w_4 r_4^2$$
$$= 6{,}563 \sin \varphi - 15{,}625 (1 - \cos \varphi)$$

$$M_{T01} = V_0 r_1 (1 - \cos \varphi) = -1{,}815 (1 - \cos \varphi)$$
$$M_{T21} = V_2 r_2 (1 - \cos \varphi) + (\varphi - \sin \varphi) w_2 r_2^2$$
$$= -3{,}910 (1 - \cos \varphi) + 7{,}225 (\varphi - \sin \varphi)$$
$$M_{T23} = -V_2 r_3 (1 - \cos \varphi) = +2{,}080 (1 - \cos \varphi)$$
$$M_{T43} = V_4 r_4 (1 - \cos \varphi) - (\varphi - \sin \varphi) w_4 r_4^2$$
$$= 6{,}563 (1 - \cos \varphi) - 15{,}625 (\varphi - \sin \varphi)$$

$\overline{V}_k$ and $\overline{M}_k$, $\overline{M}_{Tk}$ due to X_1 and X_2 in principal system, (Eqs. (327) to (330)):

$X_1 = -1$:

$$\overline{V}_0 = -\frac{\sin \varphi_2}{z_1} = -0.00926 \, k$$

$$\overline{V}_2 = -\frac{\sin (2\varphi_2 + \varphi_1)}{z_2} = -0.01176 \, k$$

$$\overline{V}_2 = -\frac{\sin (2\varphi_3 + \varphi_4)}{z_3} = -0.0166 \, k$$

$$\bar{V}_4 = -\frac{\sin\varphi_3}{z_4} = -0.00416\,k$$

$$\bar{M}_{01} = \bar{V}_0 r_1 \sin\varphi = -0.926\sin\varphi$$
$$\bar{M}_{21} = \bar{V}_2 r_2 \sin\varphi + \cos\varphi = -1.00\sin\varphi + \cos\varphi$$
$$\bar{M}_{23} = \bar{V}_2 r_3 \sin\varphi + \cos\varphi = -1.66\sin\varphi + \cos\varphi$$
$$\bar{M}_{43} = \bar{V}_4 r_4 \sin\varphi = -0.521\sin\varphi$$

$$\bar{M}_{T01} = \bar{V}_0 r_1 (1-\cos\varphi) = -0.926(1-\cos\varphi)$$
$$\bar{M}_{T21} = -\bar{V}_2 r_2 (1-\cos\varphi) - \sin\varphi = 1.00(1-\cos\varphi) - \sin\varphi$$
$$\bar{M}_{T23} = -\bar{V}_2 r_3 (1-\cos\varphi) - \sin\varphi = 1.66(1-\cos\varphi) - \sin\varphi$$
$$\bar{M}_{T43} = \bar{V}_4 r_4 (1-\cos\varphi) = -0.521(1-\cos\varphi)$$

$X_2 = -1$:

$$\bar{V}_{T0} = \frac{\cos\varphi_2}{z_1} = 0.01273\,k$$

$$\bar{V}_{T2} = \frac{\cos(2\varphi_2 + \varphi_1)}{z_2} = 0.000824\,k$$

$$\bar{V}_{T2} = \frac{\cos(2\varphi_3 + \varphi_4)}{z_3} = 0.0054\,k$$

$$V_{T4} = \frac{\cos\varphi_3}{z_4} = 0.0103\,k$$

$$\bar{M}_{01} = \bar{V}_{T,0} r_1 \sin\varphi = 1.273\sin\varphi$$
$$\bar{M}_{21} = (\bar{V}_{T,2} r_2 - 1)\sin\varphi = 0.925\sin\varphi$$
$$\bar{M}_{23} = (\bar{V}_{T2} r_3 + 1)\sin\varphi = 1.540\sin\varphi$$
$$\bar{M}_{43} = \bar{V}_{T4} r_4 \sin\varphi = 1.290\sin\varphi$$

$$\bar{M}_{T01} = \bar{V}_{T,0} r_1 (1-\cos\varphi) = 1.273(1-\cos\varphi)$$
$$\bar{M}_{T21} = \bar{V}_{T2} r_2 (1-\cos\varphi) - \cos\varphi = 0.075(1-\cos\varphi) - \cos\varphi$$
$$\bar{M}_{T23} = \bar{V}_{T2} r_3 (1-\cos\varphi) - \cos\varphi = 0.540(1-\cos\varphi) - \cos\varphi$$
$$\bar{M}_{T43} = \bar{V}_{T4} r_4 (1-\cos\varphi) = 1.290(1-\cos\varphi)$$

$EI_c \delta_{ii}$, $EI \delta_{i,i+1}$ values:

$$EI_c \delta_{ii} = \sum_0^{k=4} \int \bar{M}_i^2 r_k\, d\varphi + \rho \sum_0^{k=4} \int \bar{M}_{Ti}^2 r_k\, d\varphi,$$

$$\rho = \frac{EI}{GI_T} \sim 0.55, \quad I = 1$$

226 Theory of Structures

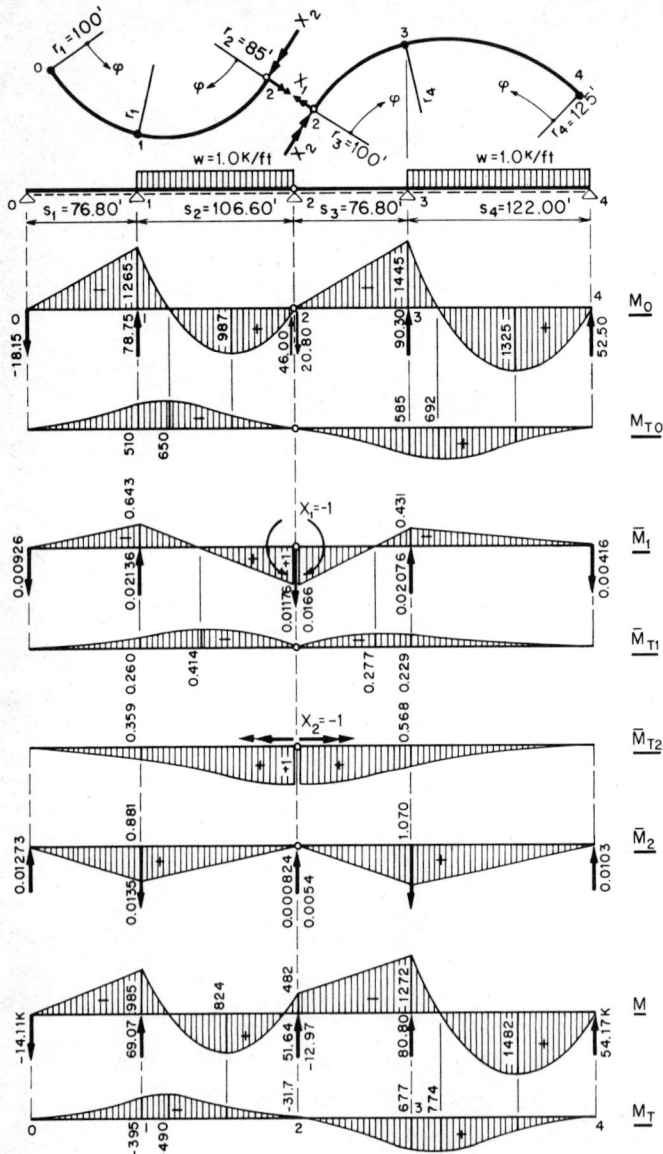

Fig. 132

$$= \int_0^1 \overline{M}_{11}^2 r_1 \, d\varphi + \int_2^1 \overline{M}_{21}^2 r_2 \, d\varphi + \int_2^3 \overline{M}_{31}^2 r_3 \, d\varphi$$

$$+ \int_4^3 \overline{M}_{41}^2 r_4 \, d\varphi$$

$$+ \rho \left[\int_0^1 \overline{M}_{T11}^2 r_1 \, d\varphi + \int_2^1 \overline{M}_{T21}^2 r_2 \, d\varphi + \int_2^3 \overline{M}_{T31}^2 r_3 \, d\varphi \right.$$

$$\left. + \int_4^3 \overline{M}_{T41}^2 r_4 \, d\varphi \right]$$

$$= 64.530 + 0.55 \times 16.686 = 73.690$$

$$EI_c \delta_{22} = \sum_0^{k=4} \int_0^{2\varphi_k} \overline{M}_2^2 r_k \, d\varphi + \rho \sum_0^{k=4} \int_0^{2\varphi_k} \overline{M}_{T2}^2 r_k \, d\varphi$$

$$= 123.600 + 0.55 \times 119.440 = +189.350$$

$$EI_c \delta_{12} = \sum_0^{k=4} \int_0^{2\varphi_k} \overline{M}_1 \overline{M}_2 r_k \, d\varphi + \rho \sum_0^{k=4} \int_0^{2\varphi_k} \overline{M}_{T1} \overline{M}_{T2} r_k \, d\varphi$$

$$= -35.855 - 0.55 \times 64.550 = -71.350$$

$EI_c \delta_{i,0}$ values:

$$EI_c \delta_{10} = \sum_0^{k=4} \int_0^{2\varphi_k} M_0 \overline{M}_1 r_k \, d\varphi + \rho \sum_0^{k=4} \int_0^{2\varphi_k} M_{T0} \overline{M}_{T1} r_k \, d\varphi$$

$$= +32{,}958 + 0.55 \times 2{,}585 = +33{,}380$$

$$EI_c \delta_{20} = \sum_0^{k=4} \int_0^{2\varphi_k} M_0 \overline{M}_2 r_k \, d\varphi + \rho \sum_0^{k=4} \int_0^{2\varphi_k} M_{T0} \overline{M}_{T2} r_k \, d\varphi$$

$$= -30{,}120 + 0.55 \times 3{,}000 = -28{,}470$$

Redundants:

$$X_1 = \frac{\delta_{10} \delta_{22} - \delta_{20} \delta_{12}}{\delta_{11} \delta_{22} - \delta_{12}^2} = \frac{33{,}380 \times 189.35 - 28{,}470 \times 71.35}{73.69 \times 189.35 - 71.35^2} = 482 \, k'$$

$$X_2 = \frac{\delta_{20}\delta_{11} - \delta_{10}\delta_{12}}{\delta_{11}\delta_{22} - \delta_{12}^2} = \frac{-28{,}470 \times 73.69 + 33{,}380 \times 71.35}{73.69 \times 189.35 - 71.35^2} = 31.7\,k$$

The actual shear and moments are obtained by superposition:

$$\Sigma V = V_0 - X_1 \overline{V}_1 - X_2 \overline{V}_2$$
$$\Sigma M = M_0 - X_1 \overline{M}_1 - X_2 \overline{M}_2$$
$$\Sigma M_T = M_{T0} - X_1 \overline{M}_{T1} - X_2 \overline{M}_{T2}$$

The shear-force and flexural and torsional moment diagrams in the principal and actual system are illustrated in Fig. 132.

For a closed circular beam at n supports subjected to uniform loading w, as illustrated in Fig. 133, the shear V and M, M_T can be computed directly from the equations of equilibrium (314):

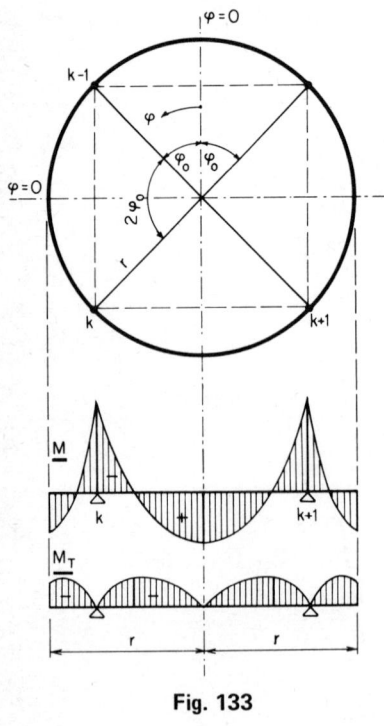

Fig. 133

$$2\varphi_0 = \frac{2\pi}{n} \qquad \varphi_0 = \frac{\pi}{n}$$

$$V_{k,k+1} = -V_{k,k-1} = \frac{\pi}{n} rw = \varphi_0 rw$$

$$V = \varphi rw \qquad R = 2\varphi_0 rw \qquad (334)$$

Because of symmetry, $M_T = 0$ at the supports and midspan. Thus we obtain by direct integration [Eqs. (314)],

$$M_k = wr^2 \left(\frac{\pi}{n} \cot \varphi_0 - 1 \right)$$

$$M = wr^2 \left(\frac{\pi}{n} \frac{\cos \varphi}{\sin \varphi_0} - 1 \right) \qquad (335)$$

$$M_T = -wr^2 \left(\frac{\pi}{n} \frac{\sin \varphi}{\sin \varphi_0} - \varphi \right)$$

Circular Beams

A fixed-end circular beam subjected to symmetric loading w, P, illustrated in Fig. 134, is once statically indeterminate because at the crown

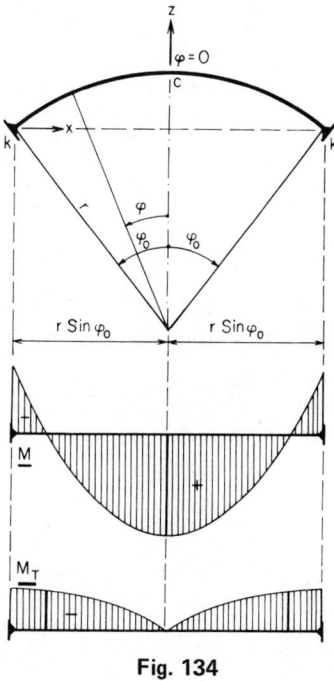

Fig. 134 $\qquad M_{TC} = V_C = 0$

Taking the bending moment M_C at the crown as redundant (X_1), the principal system consists of two cantilevers:

$$X_1 = \frac{\delta_{10}}{\delta_{11}} \qquad M_c = -X_1 \qquad \rho = \frac{EI_c}{GI_T I_c / I}$$

$$EI\delta_{11} = 2r \int_{\varphi=0}^{\varphi=\varphi_0} (\cos^2 \varphi + \rho \sin^2 \varphi) d\varphi$$

$$= \frac{r}{2}[2(\rho + 1)\varphi_0 - (\rho - 1)] \sin 2\varphi_0 \tag{336}$$

$$EI\delta_{10} = 2r \int_{\varphi=0}^{\varphi=\varphi_0} (M_0 \cos \varphi - \rho M_{T0} \sin \varphi) d\varphi$$

230 Theory of Structures

Uniform load:

$w = 1.0 \, k/ft$ (Fig. 134):

$$X_1 = -wr^2 \frac{\dfrac{\rho+1}{\rho-1}(4\sin\varphi_0 - 2\varphi_0) - \dfrac{4\rho}{\rho-1}\varphi_0\cos\varphi_0 + \sin 2\varphi_0}{2\dfrac{\rho+1}{\rho-1}\varphi_0 - 2\varphi_0} \qquad (337)$$

$$M = -wr^2(1 - \cos\varphi) - X_1 \cos\varphi$$

$$M_T = wr^2(\varphi - \sin\varphi) + X_1 \sin\varphi$$

P_c at crown c (Fig. 135):

$$X_1 = -P_c r \frac{2\dfrac{\rho+1}{\rho-1}(\cos\varphi_0 - 1) + \sin^2\varphi_0}{2\dfrac{\rho+1}{\rho-1}\varphi_0 - \sin 2\varphi_0}$$

$$M = -\frac{P_c}{2} r \sin\varphi - X_1 \cos\varphi \qquad (338)$$

$$M_T = \frac{P_c}{2} r(1 - \cos\varphi) + X_1 \sin\varphi$$

Symmetric loads in Span: (Fig. 135)

Two concentrated loads P' and P'' symmetrically located from crown c:

$P' = P'' = P$

$$X_1 = 2Pr \frac{\dfrac{2\rho}{\rho-1}(\cos\varphi_0 - \cos\varphi'') + \dfrac{\rho+1}{\rho-1}\varphi' \sin\varphi'' + \sin\varphi_0 \cdot \sin\varphi'}{2\dfrac{\rho+1}{\rho-1}\varphi_0 - \sin 2\varphi_0}$$

$$M = -Pr \sin(\varphi - \varphi'') - X_1 \cos\varphi$$

$$M_T = Pr[1 - \cos(\varphi - \varphi'')] + X_1 \sin\varphi \qquad (339)$$

For example:

$r = 100$ ft $\varphi_0 = 36°$ $2\varphi_0 = 72°$ $\rho = 0.55$

$P_c = 1:$ $X_1 = -15.40 \, k'$ $M_c = +15.40 \, k'$

$M_k = -16.90 \, k'$ $M_{kT} = +0.51 \, k'$

$P' = P'' = P = 1:$ $\quad \varphi' = 15°$ $\quad \varphi'' = 21°$
$$X_1 = -5.20\,k' \quad M_c = +5.20\,k'$$
$$M_k = -21.68\,k' \quad M_{kT} = +3.65\,k'$$

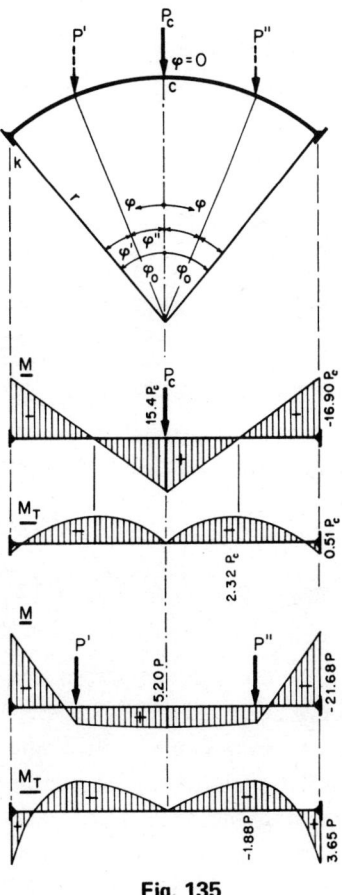

Fig. 135

Prestressing

The moment M_{pr} due to prestressing in any section x of a beam or girder is

$$M_{xpr} = e_{0x} H_{pr\,x} \tag{340}$$

where e_{0x} is the distance of the tendons from the center of gravity of the section considered x, and H_{prx} is the prestressing force acting on the section (Fig. 136).

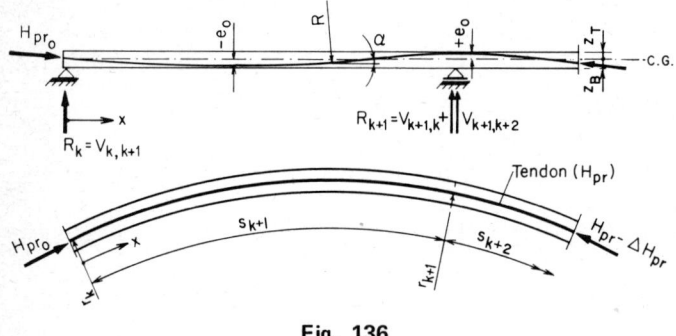

Fig. 136

Due to the curvature of the tendons, the prestressing force H_{pr0} is counteracted by friction during the application of prestressing. Frictional reduction in the prestressing force is commonly expressed by the following exponental function:

$$H_{prx} = H_{pr0} e^{-\beta} \qquad \beta = \mu\alpha + kx = \mu\alpha + kr\varphi, \qquad x = r\varphi \qquad (341)$$

where k is the friction-wobble coefficient, x the distance from the jacking end, μ the coefficient of friction, and α the total angular change of tendon from the jacking end. As average values, $k = 0.0020$ and $\mu = 0.25$. For long tendons with small radius, the frictional loss can be considerable. However, by proper design, prestressing application, and workmanship, the magnitude of frictional loss can be reduced almost entirely.

In circular beams, the tendons are curved vertically as well as horizontally. The radius r_k of the horizontal curvature is relatively large, and therefore its influence upon frictional losses is usually negligible.

The reactions R_{kpr} of a statically determinate circular beam due to prestressing are zero, as is the case with a straight beam. Due to this fact, the torsional moment M_{Tkpr} is also zero. Only in cases in which the tendons are not parallel to the centerline of the beam are torsional moments developed due to prestressing and must be considered.

The horizontal moment due to prestressing is

$$M_{ypr} = e_{0y} H_{pr\,x} \tag{342}$$

As a result of this moment, the curvature of the beam changes. The change is counteracted by torsional moment M_{Tpr}. In general, the M_{ypr} is related to M_{Tpr} as the vertical moment M_k to M_{Tk}. For statically indeterminate circular beams, the reactions $R_{kpr} = V_{kpr}$ and torsional moments M_{Tkpr} are not zero and must be determined in the same manner as discussed before. However, as the torsional moment in the principal system is zero, the δ_{k0} values consist of the first part of Eqs. (331) only. Thus,

$$EI\delta_{10} = \sum_0^n \int_{\varphi=0}^{\varphi=\varphi_k} M_{kpr}\overline{M}_{ki}\, r\, d\varphi \tag{343}$$

$$EI\delta_{T0} = 0$$

The $EI\delta_{ik}$ values are the same as those given by Eqs. (331). The M_{kpr} does not follow an easily defined mathematical curvature, mainly because of the frictional losses in H_{pr0}. Introducing the frictional losses into Eq. (342), we obtain

$$M_{kpr} = H_{pr0} e_0 \cdot e^{-(\mu a + k r \varphi)} \tag{344}$$

For relatively small radius r, R and straight tendons in Eq. (344), $\alpha \to \varphi$. Thus,

$$M_{kpr} = H_{pr0} e_0 \cdot e^{-(\mu + k r)\varphi}$$
and
$$EI\delta_{k0} = H_{pr0} e_0 \cdot r \int_{\varphi=0}^{\varphi=\varphi_k} e^{-(\mu + k r)\varphi} \overline{M}_{k1}\, d\varphi \tag{345}$$

The $EI\delta_{k0}$ value for either case is best obtained by numerical integration [Simpson's rule, Eq. (332)].

Equations (345) apply also in determining the redundants X_n for the case in which the centerline of the prestressing force is not parallel to the centerline curvature of the beam.

The final moments and shear are obtained by superposition:

$$\begin{aligned} M_{pr} &= M_{pr0} - X_1\overline{M}_1 - X_2\overline{M}_2 - \cdots - X_n\overline{M}_n \\ M_{Tpr} &= \phantom{M_{pr0}} - X_1\overline{M}_{T1} - X_2\overline{M}_{T2} - \cdots - X_n\overline{M}_{Tn} \\ V_{pr} &= \phantom{M_{pr0}} - X_1\overline{V}_1 - X_2\overline{V}_2 - \cdots - X_n\overline{V}_n \end{aligned} \quad (346)$$

The losses in prestressing force due to plastic flow and shrinkage as well as time deflection are obtained by Eqs. (80) and (128).

7

SPACE STRUCTURES

Shells are relatively thin, curved or prismatic surface structures. As a rule, shells are described by their transversal, longitudinal, and horizontal sections and by their degree of axial or rotational symmetry.

For example, a spherical shell (dome) is rotationally symmetrical; its meridional (transversal) section can be a circle, ellipse, cone, or any other symmetric curvature (Fig. 137).

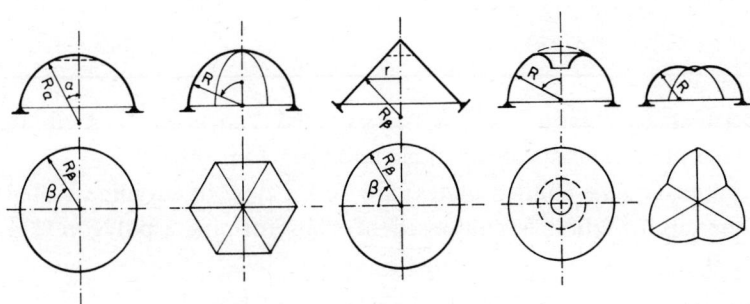

Fig. 137

The horizontal sections are always circles or regular polygons (polygonal domes).

Curvilinear cylindrical shells (barrels) have single or two-axial symmetry. The curvature of the transversal section can be circular, elliptical, cycloidal, or any curvature which remains higher than the pressure line of the loading the shell is subjected to (Fig. 138). As the parabola is the pressure line for uniform loading, no shell action occurs and the load is transmitted directly by arch action into the supports.

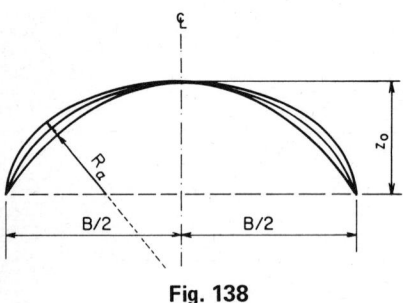

Fig. 138

The longitudinal and horizontal sections are rectangular. The curvilinear shells are supported by rigid frames (Fig. 139).

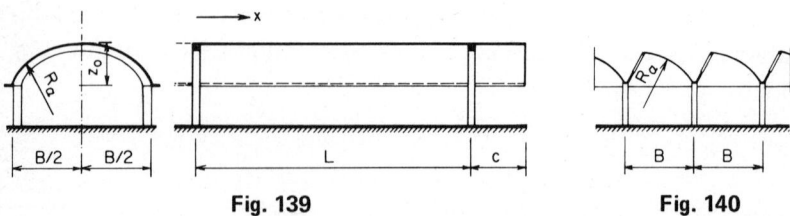

Fig. 139   Fig. 140

A particular curvilinear shell is the so-called "north-light" shell. Its transversal section consists of a partial curvature (Fig. 140).

Polygonlinear shells (folded plates) are also a type of curvilinear shell. Their transversal section is composed of slabs forming a polygon (Fig. 141).

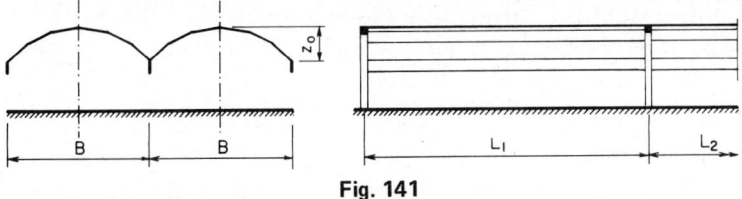

Fig. 141

When the longitudinal section also consists of a curvature, a two-directional curved shell is obtained. In this category of shells belong hyperbolic paraboloids and translational shells (Figs. 142 and 143.)

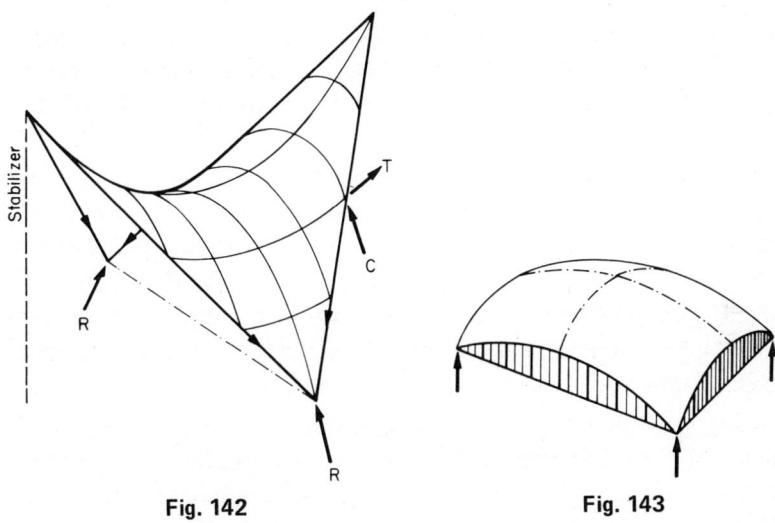

Fig. 142   Fig. 143

A series of combinations of these shell types can be used to obtain dramatic unit structures.

Carrying action for the rotationally symmetric and the translational shells is basically a combination of the arch and suspension action. Only the curvilinear cylindrical shells carry by true shell action.

The basis of shell analysis is the theory of elasticity, in which the relationship between stresses and deformations is mathematically formalized. The stresses, as a rule, are expressed as linear functions of the deformations (Hooke's law). The stress-strain functions and the equations of equilibrium for external forces, when combined, lead to nonhomogeneous simultaneous differential equations of high order. The successive integration of these differential equations allows us to determine the internal forces, moments, and deformations in the shell. Commonly, the integration of such nonhomogeneous differential equations is a complex problem. To overcome the integration difficulties and to obtain clarity, the nonhomogeneous differential equation in shell analysis is solved by two independent operations: the particular solution (particular integrals) and the homogeneous solution. The final results are obtained by application of the principle of superposition. This means, in its broader aspect, that shell analysis is similar to the analysis of indeterminate structures.

First: The shell is made statically determinate by allowing unrestricted rotation and displacement of the shell's edge at its support. In this condition, the external loads are transmitted to the supports by normal stresses or by so-called *membrane stresses*. The membrane-stress condition of such a shell is described by particular integrals of the nonhomogeneous differential equation.

Second: The strains due to membrane stresses in the shell yield rotations, displacements, and reactions along the supported edges of the shell that do not comply with the actual boundary conditions. However, there are cases in which boundary conditions can be created, especially by means of posttensioning, that comply with membrane deformations. In such cases, the membrane theory describes the stress and deformation conditions in the shell with qualitatively acceptable accuracy. Unfortunately, the boundary conditions in most cases cannot be satisfied and use must be made of the homogeneous part of the general nonhomogeneous differential equation to satisfy the boundary condition of the shell. Because the membrane theory or the particular integral satisfy the equilibrium equations, the shell for the homogeneous part of the differential equation is free from external surface loads and is acted upon by boundary loadings only to satisfy the geometric support conditions.

The boundary loading causes normal as well as flexural stresses in the shell which diminish in a dampened-wave-like manner with distance from the boundary. Generally, such so-called *boundary disturbances* are limited to a relatively small part of the shell; however, in a particularly unfavorable condition (cylindrical shell), they can extend over the entire shell and may be of such a magnitude that they control the design.

23. SPHERICAL SHELLS*

To derive the equations for determining the internal forces (N_α, N_β, and $N_{\alpha\beta}$) in a statically determinate (membrane theory) spherical shell acted upon by external vertical rotationally symmetric loading w, a differential ring element cut out of a shell (Fig. 144) will be considered. The shell is characterized by

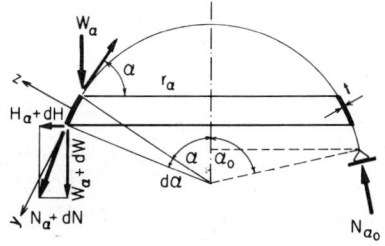

$$R_\alpha = R_\beta = R \quad t = \text{const}$$
$$r_\alpha = R \sin\alpha$$
$$dA = r_\alpha\, d\beta\, R\, d\alpha$$

For rotationally symmetric loading, the N_α, N_β, and $N_{\alpha\beta}$ values are not dependent on β, and thus all derivatives with respect to β are zero. For nonrotationally symmetric loading,

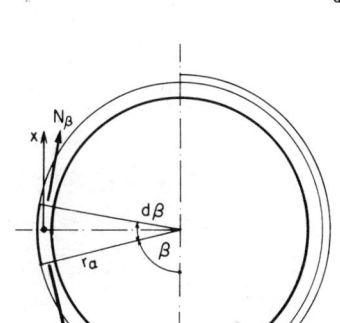

$$\frac{\partial N}{\partial \beta} \neq 0$$

Fig. 144

The axis of the differential element is denoted by x, y, and z, where x is tangent to the horizontal curve, y is tangent to meridian or transversal curvature, and z is normal to the element. Equilibrium requires that all forces in the x, y, and z direction be zero. Thus we obtain

---

*Kurt Beyer, "Die Statik im Eisenbetonbau," Springer-Verlag OHG, Berlin, 1934.
S. Timoshenko, "Theory of Plates and Shells," McGraw-Hill Book Co., Inc., 1940.

Theory of Structures

$$\Sigma X = 0: \quad \frac{\partial N_\beta}{\partial \beta} d\beta R\, d\alpha + N_{\alpha\beta} R\, d\alpha\, d\beta + \frac{\partial}{\partial \alpha}(N_{\alpha\beta} r_\alpha d\beta)\, d\alpha$$
$$+ w_x r_\alpha d\beta R\, d\alpha = 0$$

$$\Sigma Y = 0: \quad \frac{\partial}{\partial \alpha}(N_\alpha r_\alpha d\beta)\, d\alpha - N_\beta R\, d\alpha\, d\beta \cos\alpha + \frac{\partial N_{\alpha\beta}}{\partial \beta} d\beta R\, d\alpha \quad (347)$$
$$+ w_y r_\alpha d\beta R\, d\alpha = 0$$

$$\Sigma Z = 0: \quad N_\alpha r_\alpha d\beta\, d\alpha + N_\beta R\, d\alpha\, d\beta \sin\alpha + w_z r_\alpha d\beta R\, d\alpha = 0$$

These equations can be simplified to

$$\frac{1}{\sin\alpha}\frac{\partial N_\beta}{\partial \beta} + 2 N_{\alpha\beta} \cot\alpha + \frac{\partial N_{\alpha\beta}}{\partial \alpha} + R w_x = 0$$
$$\frac{\partial N_\alpha}{\partial \alpha} + (N_\alpha - N_\beta) \cot\alpha + \frac{1}{\sin\alpha}\frac{\partial N_{\alpha\beta}}{\partial \beta} + R w_y = 0 \quad (348)$$
$$N_\alpha + N_\beta + R w_z = 0$$

For vertical rotationally symmetric loading, $w_x = 0$ and $N_{\alpha\beta} = 0$. Substituting N_β from the third equation into the second, we obtain by direct integration

$$N_\beta = -(N_\alpha + w_z R)$$
$$N_\alpha = -\frac{R}{\sin^2\alpha} \int (w_y + w_z \cot\alpha) \sin^2\alpha\, d\alpha + C \quad (349)$$

The differential coefficient $C = f(\alpha)$ can be determined from the boundary conditions for $\alpha = 0$: $N_\alpha = N_\beta$.

The N_α value can also be obtained by Guldin's rule:

$$N_\alpha = \frac{\Sigma W_\alpha}{2\pi r_\alpha \sin\alpha} = \frac{\Sigma W_\alpha}{2\pi R \sin^2\alpha} \quad (350)$$

Considering that the shell surface area A_α from $\alpha = 0$ to α is (Fig. 145)

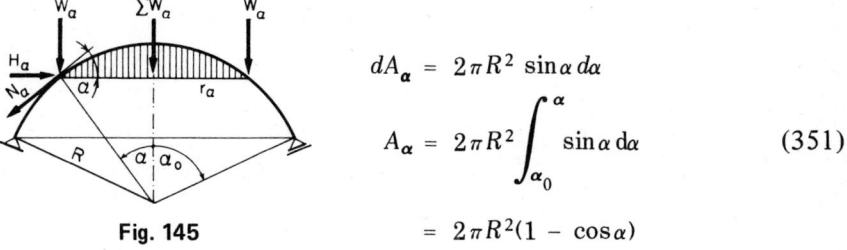

Fig. 145

$$dA_\alpha = 2\pi R^2 \sin\alpha\, d\alpha$$

$$A_\alpha = 2\pi R^2 \int_{\alpha_0}^{\alpha} \sin\alpha\, d\alpha \qquad (351)$$

$$= 2\pi R^2 (1 - \cos\alpha)$$

Thus, for example, N_α due to dead load is

$$\Sigma W_\alpha = 2\pi R^2 (1 - \cos\alpha) w_D$$
$$W_\alpha = R(1 - \cos\alpha) w_D$$
$$N_\alpha = -\frac{R w_D (1 - \cos\alpha)}{\sin^2\alpha} = -\frac{R w_D}{1 + \cos\alpha} \qquad (352)$$
$$N_\beta = \frac{R w_D}{1 + \cos\alpha}(1 - \cos\alpha - \cos^2\alpha)$$
$$N_{\alpha\beta} = 0$$

Deformations

As a rule, the horizontal displacement Δ_r and the rotation of meridian tangent ϑ of the shell are adequate to describe the deformations of the spherical shell. They can be simply expressed by strain or directly by N_α, N_β values. Thus,

$$\Delta r_\alpha = r_\alpha \epsilon_\beta = -\frac{r_\alpha}{Et}(N_\beta - \nu N_\alpha)$$

$$\vartheta_\alpha = (\epsilon_\alpha + \epsilon_\beta)\cot\alpha - \frac{\partial \epsilon_\beta}{\partial \alpha} \qquad (353)$$

$$= \frac{1}{Et}\left[(N_\beta - N_\alpha)(1 + \nu)\cot\alpha + \frac{\partial}{\partial \alpha}(N_\beta - \nu N_\alpha)\right]$$

Particular Integrals

The normal forces N_α, N_β, shear $N_{\alpha\beta}, V_\alpha$, deformations Δr_α, and rotation of the meridian tangent ϑ_α in a spherical shell ($R_\alpha = R_\beta = R$) subjected to various loadings encountered in practice are:

Dead load (Fig. 146):

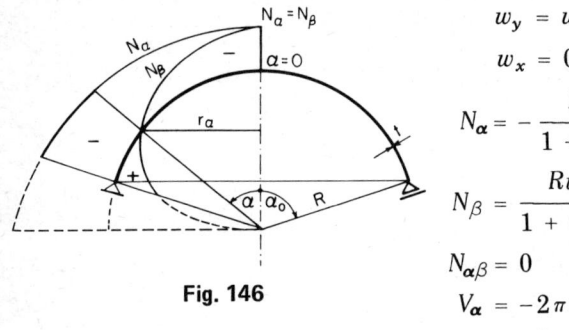

Fig. 146

$$w_y = w_D \sin\alpha \qquad w_z = w_D \cos\alpha$$
$$w_x = 0$$
$$N_\alpha = -\frac{Rw_D}{1 + \cos\alpha}$$
$$N_\beta = \frac{Rw_D}{1 + \cos\alpha}(1 - \cos\alpha - \cos^2\alpha)$$
$$N_{\alpha\beta} = 0$$
$$V_\alpha = -2\pi r_\alpha N_\alpha \sin\alpha$$
$$\Delta r_\alpha = \frac{R^2 w_D}{Et} \sin\alpha \left(\frac{1 + \nu}{1 + \cos\alpha} - \cos\alpha\right)$$
$$\vartheta_\alpha = -\frac{Rw_D}{Et}(2 + \nu) \sin\alpha$$

(354)

Surcharge (Fig. 147):

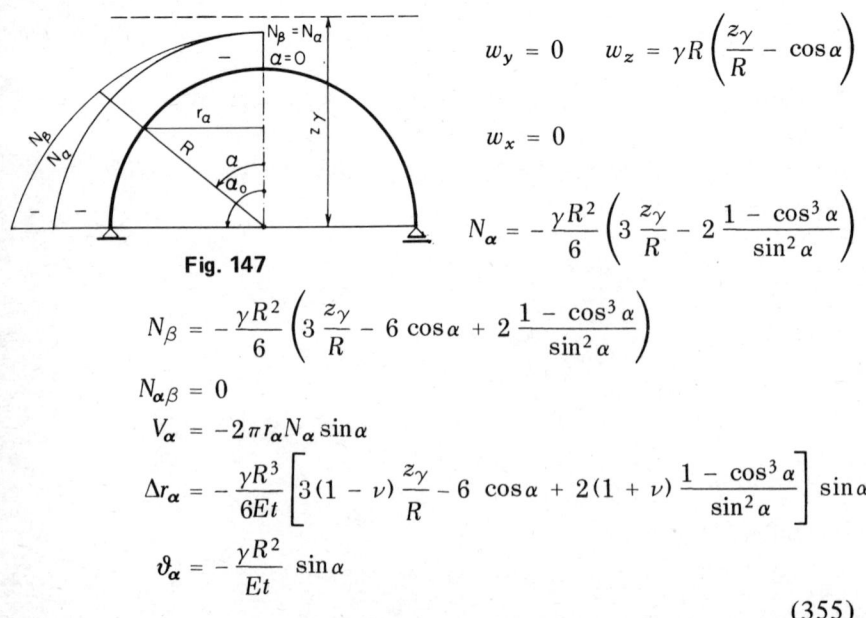

Fig. 147

$$w_y = 0 \qquad w_z = \gamma R\left(\frac{z_\gamma}{R} - \cos\alpha\right)$$
$$w_x = 0$$
$$N_\alpha = -\frac{\gamma R^2}{6}\left(3\frac{z_\gamma}{R} - 2\frac{1 - \cos^3\alpha}{\sin^2\alpha}\right)$$
$$N_\beta = -\frac{\gamma R^2}{6}\left(3\frac{z_\gamma}{R} - 6\cos\alpha + 2\frac{1 - \cos^3\alpha}{\sin^2\alpha}\right)$$
$$N_{\alpha\beta} = 0$$
$$V_\alpha = -2\pi r_\alpha N_\alpha \sin\alpha$$
$$\Delta r_\alpha = -\frac{\gamma R^3}{6Et}\left[3(1 - \nu)\frac{z_\gamma}{R} - 6\cos\alpha + 2(1 + \nu)\frac{1 - \cos^3\alpha}{\sin^2\alpha}\right]\sin\alpha$$
$$\vartheta_\alpha = -\frac{\gamma R^2}{Et}\sin\alpha$$

(355)

Live load (Fig. 148):

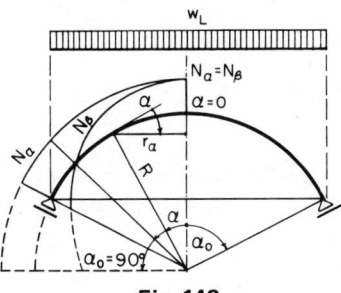

Fig. 148

$$w_y = w_L \sin\alpha \cos\alpha \qquad w_z = w_L \cos^2\alpha$$

$$w_x = 0$$

$$N_\alpha = -\frac{Rw_L}{2}$$

$$N_\beta = -\frac{Rw_L}{2} \cos 2\alpha$$

$$N_{\alpha\beta} = 0 \qquad (356)$$

$$V_\alpha = -2\pi r_\alpha N_\alpha \sin\alpha$$

$$\Delta r_\alpha = \frac{R^2 w_L}{Et}\left(\frac{1+\nu}{2} - \cos^2\alpha\right)\sin\alpha$$

$$\vartheta_\alpha = -\frac{Rw_L}{Et}(3+\nu)\sin\alpha \cos\alpha$$

Wind loading (Fig. 149):

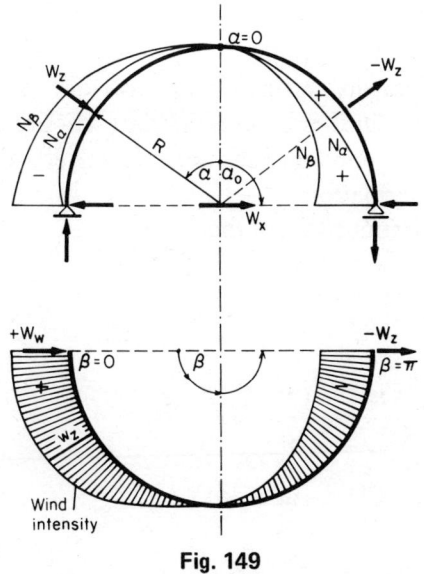

Fig. 149

$$w_y = 0 \qquad w_z = w_w \sin\alpha \cos\beta$$

$$w_x = 0$$

$$N_\alpha = -\frac{w_w R \cos\alpha}{3 \sin^3\alpha}(2 - 3\cos\alpha$$

$$+ \cos^3\alpha)\cos\beta$$

$$N_\beta = \frac{w_w R}{3\sin^3\alpha}(2\cos\alpha - 3\sin^2\alpha$$

$$- 2\cos^4\alpha)\cos\beta$$

$$N_{\alpha\beta} = -\frac{w_w R}{3\sin^3\alpha}(2 - 3\sin\alpha$$

$$+ \cos^3\alpha)\sin\beta \qquad (357)$$

For wind pressure, see Eq. (245).

Temperature change and shrinkage:

$$N_\alpha = N_\beta = N_{\alpha\beta} = 0 \quad \begin{aligned} \Delta r_{\alpha\Delta T} &= \epsilon_{\Delta T} \Delta TR \sin\alpha & \vartheta_{\alpha\Delta T} &= 0 \\ \Delta r_{\alpha\epsilon} &= \epsilon_{sr} R \sin\alpha & \vartheta_{\alpha\epsilon} &= 0 \end{aligned} \qquad (358)$$

The membrane theory, as stated before, is based upon the assumption that the shell is supported continuously and uniformly at its edge and so that the reaction is tangent to the meridian of the shell. This can be achieved without the use of a marginal member only if the tangent to the meridian is vertical at the edge of the shell. If this is not the case, use must be made of the homogeneous solutions to satisfy existing boundary conditions.

Homogeneous Solution

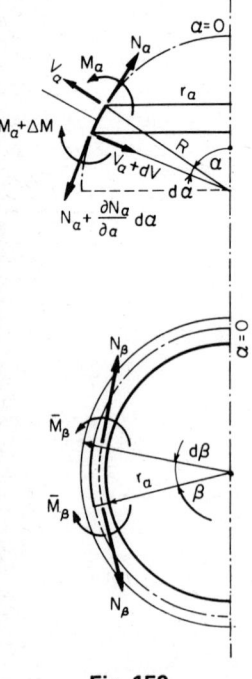

Fig. 150

The shell will be characterized as before for membrane forces by $R_\alpha = R_\beta = R =$ constant and $r_\alpha = R \sin\alpha$. The unknowns due to the boundary loading $\bar{w} = 0$ are $\bar{N}_\alpha, \bar{N}_\beta, \bar{V}_\alpha, \bar{M}_\alpha,$ and $\bar{M}_\beta$ (Fig. 150). Three of the five unknowns can be computed from the three equilibrium equations, but two remain as redundants and must be computed from deformations.

Equilibrium requires that

$$\frac{d(\bar{N}_\alpha \sin\alpha)}{d\alpha} - \bar{N}_\beta \cos\alpha - \bar{V}_\alpha \sin\alpha = 0$$

$$\frac{d(\bar{V}_\alpha \sin\alpha)}{d\alpha} \frac{1}{\sin\alpha} + \bar{N}_\alpha + \bar{N}_\beta = 0$$

$$\frac{d\bar{M}_\alpha}{d\alpha} - (\bar{M}_\beta - \bar{M}_\alpha) \cot\alpha - \bar{V}_\alpha R = 0$$

(359)

Denoting the flexural stiffness of the shell by $D = Et^3/12(1 - \nu^2)$, the

$\bar{M}$ values expressed by rotation of the meridian tangent $\bar{\vartheta}$ are

$$\bar{M}_\alpha = -\frac{D}{R}\left(\frac{d\bar{\vartheta}}{d\alpha} + \nu\bar{\vartheta}\cot\alpha\right)$$

$$\bar{M}_\beta = -\frac{D}{R}\left(\bar{\vartheta}\cot\alpha + \nu\frac{d\bar{\vartheta}}{d\alpha}\right) \qquad (360)$$

$$\frac{d\bar{M}_\alpha}{d\alpha} = -\frac{D}{R}\left[\frac{d^2\bar{\vartheta}}{d\alpha^2} + \nu\left(\frac{d\bar{\vartheta}}{d\alpha}\cot\alpha - \frac{\bar{\vartheta}}{\sin^2\alpha}\right)\right]$$

Introducing these values into the third equilibrium equation (359), we obtain

$$\frac{d^2\bar{\vartheta}}{d\alpha^2} + \frac{d\bar{\vartheta}}{d\alpha}\cot\alpha - \bar{\vartheta}\cot^2\alpha - \nu\bar{\vartheta} = \frac{R^2}{D}\bar{V}_\alpha \qquad (361)$$

Considering that for an unloaded shell $w_y = w_z = w_x = 0$,

$$\bar{N}_\alpha \sin\alpha = \bar{V}_\alpha \cos\alpha \qquad (362)$$

Thus:

$$\bar{N}_\alpha = \bar{V}_\alpha \cot\alpha$$

Introducing this value for N_α into the second equation (359), we obtain

$$\bar{V}_\alpha \cos\alpha + \bar{N}_\beta \sin\alpha - \frac{d\bar{V}_\alpha}{d\alpha}\sin\alpha - \bar{V}_\alpha \cos\alpha = 0 \qquad (363)$$

Thus:

$$\bar{N}_\beta = \frac{d\bar{V}_\alpha}{d\alpha}$$

In accordance with Eqs. (353),

$$Et\bar{\vartheta} = (\bar{N}_\beta - \bar{N}_\alpha)(1 + \nu) \cot\alpha + \frac{\partial}{\partial\alpha}(\bar{N}_\beta - \nu\bar{N}_\alpha)$$

Substituting the N values from Eqs. (362) and (363), we have the second required differential equation

$$\frac{d^2\bar{V}_\alpha}{d\alpha^2} + \frac{d\bar{V}_\alpha}{d\alpha}\cot\alpha - \bar{V}_\alpha \cot^2\alpha + \nu\bar{V}_\alpha = Et\bar{\vartheta} \qquad (364)$$

for determining the two unknowns.

In these two homogeneous differential equations [Eqs. (361) and (364)], $\bar{\vartheta}$, $d\bar{\vartheta}/d\alpha$, and $\bar{V}_\alpha$, $d\bar{V}_\alpha/d\alpha$ are negligible in comparison with $d^2\bar{\vartheta}/d\alpha^2$ and $d^2\bar{V}_\alpha/d\alpha^2$. Thus,

$$\frac{d^2\bar{\vartheta}_\alpha}{d\alpha^2} = -\frac{R^2}{D}\bar{V}_\alpha$$

$$\frac{d^2\bar{V}_\alpha}{d\alpha^2} = Et\bar{\vartheta}_\alpha \qquad (365)$$

By differentiating twice and eliminating $\bar{V}_\alpha$ or $\bar{\vartheta}_\alpha$, we obtain

$$\frac{d^4\bar{\vartheta}_\alpha}{d\alpha^4} + \frac{R}{D}Et\bar{\vartheta}_\alpha = 0$$

or $\qquad\qquad\qquad\qquad\qquad\qquad\qquad\qquad\qquad\qquad\qquad\qquad$ (366)

$$\frac{d^4\bar{V}_\alpha}{d\alpha^4} + \frac{R^2}{D}Et\bar{V}_\alpha = 0$$

Taking

$$4L^4 = \frac{R^2}{D}Et = \frac{12(1-\nu^2)R^2}{t^2}$$

$$L = \sqrt{\frac{R}{t}}\sqrt[4]{3(1-\nu^2)} \qquad (367)$$

L is the characteristic constant of the spherical shell.

The differential equations (365) in the final form will be

$$\frac{d^4\bar{\vartheta}_\alpha}{d\alpha^4} + 4L^4\bar{\vartheta}_\alpha = 0$$
$$\frac{d^4\bar{V}_\alpha}{d\alpha^4} + 4L^4\bar{V}_\alpha = 0 \tag{368}$$

For convenience, to determine the values of the deformations (or forces) in the shell, take $\omega_B = \alpha_0 - \alpha$ and $d\omega = -d\alpha$, thus changing the coordinates from which angular measurements are taken (Fig. 151). With this modification, the solution of the differential equations is

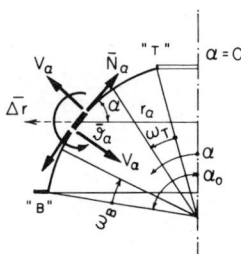

Fig. 151

$$\bar{\vartheta}_\alpha = e^{-L\omega}(A_1 \cos L\omega + B_1 \sin L\omega)$$
$$\bar{V}_\alpha = e^{-L\omega}(A_2 \cos L\omega + B_2 \sin L\omega) \tag{369}$$

Substituting $C \cos\psi$ for the differential coefficient A_2 and $C \sin\psi$ for B_2 in the second equation (369), we obtain

$$\bar{V}_\alpha = C e^{-L\omega} \cos(L\omega + \psi) \tag{370}$$

Thus the other functions, expressed in terms of $\bar{V}_\alpha$, are

$$\bar{\vartheta}_\alpha = \frac{d^2 \bar{V}_\alpha}{Et\, d\alpha^2} = C\frac{2L^2}{Et} e^{-L\omega} \sin(L\omega + \psi)$$

$$\bar{N}_\alpha = -\bar{V}_\alpha \cot\alpha$$

$$\bar{N}_\beta = -\frac{d\bar{V}_\alpha}{d\alpha} = \pm CL\sqrt{2}\, e^{-L\omega} \sin\left(L\omega + \psi + \frac{\pi}{4}\right)$$

$$\bar{M}_\alpha = -\frac{D}{R}\left(\frac{d\bar{\vartheta}}{d\alpha} + \bar{\vartheta} \cot\alpha\right)$$

$$\simeq \mp C \frac{D}{Et R} 2L^3 \sqrt{2}\, e^{-L\omega} \cos\left(L\omega + \psi + \frac{\pi}{4}\right) \tag{371}$$

$$\overline{M}_\beta = \nu \overline{M}_\alpha - \frac{D}{R} \bar{\vartheta} \cot \alpha$$

$$\overline{\Delta r} \simeq \frac{r_\alpha}{Et} CL\sqrt{2}\, e^{-L\omega} \sin\left(L\omega + \psi + \frac{\pi}{4}\right)$$

(371) (Cont.)

The differential coefficients C and ψ are determined from the following boundary conditions of the shell.

Hinged edge:

$$\alpha = \alpha_0 \qquad \omega = 0$$
$$M_{\alpha_0} = \overline{M}_{\alpha_0} = 0 \quad \text{and} \quad \Delta r_{\alpha_0} = \Delta r_0 + \overline{\Delta r} = 0 \tag{372}$$

To satisfy the first boundary condition requires that [Eqs. (371)]

$$\overline{M}_{\alpha_0} = C\frac{D}{EtR} 2L^3 \sqrt{2} \cos\left(\psi + \frac{\pi}{4}\right) = 0 \tag{372a}$$

This condition is satisfied only when $\cos(\psi + \pi/4) = 0$; thus $\psi = \pi/4$. The second boundary condition (Δr_0 computed by membrane theory),

$$\Delta r_{\alpha_0} = \Delta r_0 - \frac{r_{\alpha_0}}{Et} CL\sqrt{2}\, \sin\left(\frac{\pi}{4} + \frac{\pi}{4}\right) = 0 \tag{372b}$$

is satisfied when

$$C = \frac{\Delta r_0}{r_{\alpha_0} L \sqrt{2}} Et$$

Fixed edge:

$$\alpha = \alpha_0 \qquad \omega = 0$$
$$\vartheta_{\alpha_0} = \bar{\vartheta}_{\alpha_0} = 0 \quad \text{and} \quad \Delta r_{\alpha_0} = \Delta r_0 + \overline{\Delta r}_{\alpha_0} = 0 \tag{373}$$

The first boundary condition requires that

$$\bar{\vartheta}_{\alpha_0} = C\frac{2L^2}{Et} e^{-L\omega} \sin(L\omega + \psi) = 0 \qquad \vartheta_{\alpha_0} \sim 0 \tag{373a}$$

This condition is satisfied when $\sin(L\omega + \psi) = 0$; therefore, $\psi = 0$. The second boundary condition requires that

$$\Delta r_{\alpha_0} = \Delta r_0 - \frac{r_{\alpha_0}}{Et} CL\sqrt{2}\, e^{-L\omega} \sin\frac{\pi}{4} = 0 \qquad (373b)$$

and

$$C = \frac{\Delta r_0}{r_{\alpha_0} L} Et$$

The final forces and deformations in the shell are obtained by superposition of the values (particular integrals) from Eqs. (354) to (359) and (371).

Example:

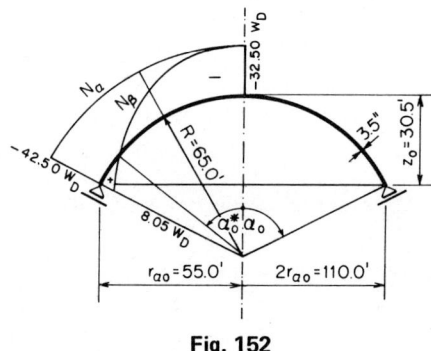

Fig. 152

Geometric data:

$t = 3.5$ in. $= 0.292$ ft

$$D = \frac{E\, 0.292^3}{12(1 - 0.167^2)}$$

$\quad = 0.00212 E$

$R = 65.00$ ft $\qquad \alpha_0 = 58°0'$

$r_{\alpha_0} = R \sin\alpha_0 \simeq 55.00$ ft

$2 r_{\alpha_0} = 110.00$ ft

$z_0 = R(1 - \cos\alpha_0) \simeq 30.50$ ft

Loading:

$w_0 = 1.0 w_D$

Particular integrals (membrane theory):

$$N_{\alpha_0} = -\frac{R w_D}{1 + \cos\alpha_0} = -42.50\, w_D$$

$$N_{\beta(\alpha_0)} = \frac{R w_D}{1 + \cos\alpha_0}(1 - \cos\alpha_0 - \cos^2\alpha_0) = +8.05\, w_D$$

$$N_{\alpha=0} = N_{\beta(\alpha=0)} = -\frac{Rw_D}{2} = -32.50\,w_D$$

$$Et\Delta r_{\alpha_0} = R^2 \sin\alpha_0 \left(\frac{1+\nu}{1+\cos\alpha_0} - \cos\alpha_0 \right) = 0.197\,R^2 w_D$$

Homogeneous solution
Boundary conditions (hinged edge):

$$\alpha = \alpha_0 \qquad \omega = 0: \quad M_{\alpha_0} = \Delta r_0 = 0$$

$$L = \sqrt{\frac{R}{t}\sqrt{3(1-\nu^2)}} = 19.50$$

$$\psi = \frac{\pi}{4}, \quad C = \frac{\Delta r_{\alpha_0}}{RL\sqrt{2}} Et = \frac{0.197\,Rw_D}{L\sqrt{2}} = 0.464\,w_D$$

Introducing the ψ and C values into Eqs. (371), we obtain the $\overline{N}_\alpha$ and $\overline{M}_\alpha$ values caused by boundary disturbances:

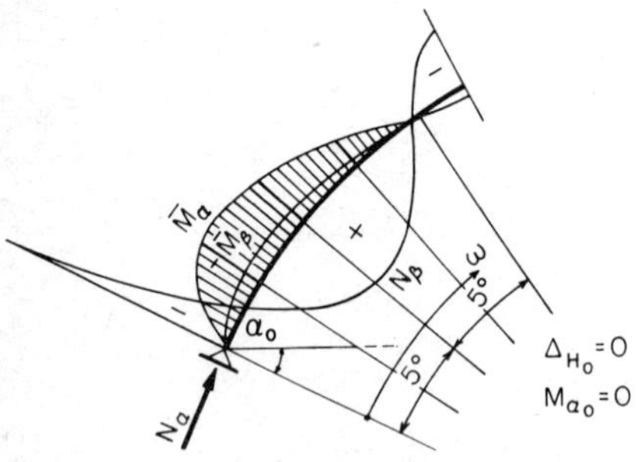

$$\overline{N}_\alpha = -0.464\,w_D\,e^{-L\omega} \cos\left(L\omega + \frac{\pi}{4}\right)$$

$$\overline{N}_\beta = -4.225\,w_D\,e^{-L\omega} \cos(L\omega)$$

$$\overline{M}_\alpha = -1.085\,w_D\,e^{-L\omega} \cos(L\omega)$$

$$\overline{M}_\beta = 0.167\,\overline{M}_\alpha - 0.0396\,w_D$$

$$\cot\alpha\,e^{-L\omega} \sin\left(L\omega + \frac{\pi}{4}\right)$$

Fig. 153

The N_α, N_β values by membrane theory are illustrated in Fig. 152, and the boundary disturbances $\overline{N}, \overline{M}$ in Fig. 153. The final results are obtained by superposition of both values, respectively.

Marginal Member

The magnitude of the shell thrust at $\alpha = \alpha_0$ is

$$H_{\alpha_0} = N_{\alpha_0} \cos\alpha_0 = Rw\,\frac{\cos\alpha_0}{1+\cos\alpha_0} \qquad (374)$$

The thrust H_{α_0} causes tension in the marginal member (Fig. 154):

$$T = H_{\alpha_0} R \sin\alpha_0 = \frac{R^2 w}{2} \frac{\sin 2\alpha_0}{1 + \cos\alpha_0} \tag{375}$$

Since the marginal member is subjected to deformations (due to T, w_m), the shell itself is no longer statically determinate. The boundary disturbances due to the interaction of the shell and the edge member are approximately zero when the difference of strain $\Delta\epsilon$ between shell and edge member is zero.

Denoting the area of the edge member by $A_m = bd_0$ and its unit strain by ϵ_m, we can write

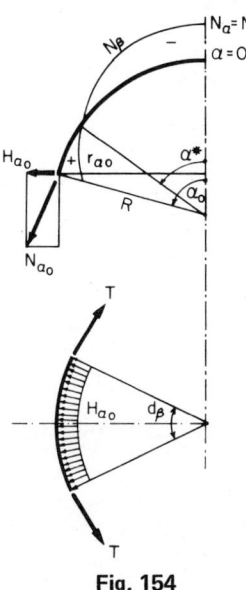

Fig. 154

$$\Delta\epsilon = \epsilon_m - \epsilon_\beta = \frac{T}{EA_m} - \frac{N_\beta - \nu N_{\alpha_0}}{Et}$$

$$= \frac{Rw}{E(1 + \cos\alpha_0)} \left[\frac{R}{2A_m} \sin 2\alpha_0 - \frac{1}{t}(1 - \cos\alpha_0 - \cos^2\alpha_0 + \nu) \right] \tag{376}$$

To balance the strains of shell and marginal member, two possibilities are available. First, considering that the N_β value is a compressive force from $\alpha = 0$ to $\alpha_D^* = 51°50'$ for dead load and to $\alpha_L^* = 45°$ for live load, then for $\alpha > \alpha^*$, N_β is a tensile force (Fig. 154). Due to this fact, the location of the edge beam can be chosen where the condition $\Delta\epsilon = 0$ is satisfied. Second, the $\epsilon_m > \epsilon_\beta$ and the condition $\Delta\epsilon = 0$ commonly can be satisfied by posttensioning of the edge member to the extent that $\epsilon_\beta = \epsilon_{pr}$.

In cases in which a spherical shell is supported by an elastic circular ring structure, the balancing of stresses is a complex problem because the deformations of the supporting member as well as the shell are interacting to a greater degree. The analysis is here strikingly similar to the analysis of ordinary indeterminate structures.

The principal system comprises two independent structures—a statically determinate shell and a circular beam-column system (Fig. 155).

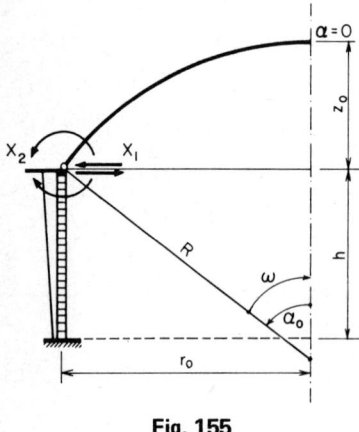

Redundants:

X_1 — thrust (H_{α_0})

X_2 — moment (M_{α_0})

$$\delta_{10} + X_1 \delta_{11} + X_2 \delta_{12} = 0$$
$$\delta_{20} + X_1 \delta_{21} + X_2 \delta_{22} = 0$$

$$X_1 = \frac{\delta_{10}\delta_{22} - \delta_{20}\delta_{12}}{\delta_{11}\delta_{22} - \delta_{12}^2}$$

$$X_2 = \frac{\delta_{20}\delta_{11} - \delta_{10}\delta_{12}}{\delta_{11}\delta_{22} - \delta_{12}^2}$$

Fig. 155

$X_1 = 1$:

Boundary condition:

$$\alpha = \alpha_0, \quad \omega = 0: \quad \bar{V}_{\alpha 01} = -\sin\alpha_0$$
$$\bar{M}_{\alpha 01} = 0$$
$$\psi = \frac{\pi}{4} \quad C = -\sqrt{2}\sin\alpha_0 \tag{377}$$

Thus, in accordance with Eqs. (371),

$$\bar{N}_{\alpha,1} = \sqrt{2}\sin\alpha_0\, e^{-L\omega} \cos\left(L\omega + \frac{\pi}{4}\right)$$

$$\bar{N}_{\beta,1} = 2L\sin\alpha_0\, e^{-L\omega}\cos(L\omega)$$

$$\bar{M}_{\alpha,1} = \frac{Lt}{\sqrt{3(1-\nu^2)}}\sin\alpha_0\, e^{-L\omega}\sin(L\omega), \quad (\bar{M}_{\beta 1} = 0) \tag{378}$$

$$\overline{V}_{\alpha,1} = -\sqrt{2}\,\sin\alpha_0\,e^{-L\omega}\cos\left(L\omega + \frac{\pi}{4}\right)$$

$$\overline{\vartheta}_{\alpha,1} = -\frac{2L^2\sqrt{2}}{Et}\sin\alpha_0\,e^{-L\omega}\sin\left(L\omega + \frac{\pi}{4}\right)$$

$$\Delta\overline{r}_{\alpha,1} = \frac{2LR}{Et}\sin^2\alpha_0\,e^{-L\omega}\cos(L\omega)$$

(378) (Cont.)

$X_2 = 1$:

Boundary condition:

$$\alpha = \alpha_0,\ \omega = 0: \quad \overline{M}_{\alpha 2} = -1$$
$$\overline{V}_{\alpha 2} = 0$$
$$\psi = \frac{\pi}{2},\ C = \frac{Et\,R}{2L^3 D}$$

(379)

Thus, in accordance with Eqs. (371),

$$\overline{N}_{\alpha 2} = \frac{EtR}{2L^3 D}\cot\alpha\,e^{-L\omega}\sin(L\omega)$$

$$\overline{N}_{\beta 2} = -\frac{EtR}{\sqrt{2}\,L^2 D}\,e^{-L\omega}\cos\left(L\omega + \frac{\pi}{4}\right)$$

$$\overline{M}_{\alpha 2} = -\sqrt{2}\,e^{-L\omega}\sin\left(L\omega + \frac{\pi}{4}\right),\quad (\overline{M}_{\beta 2} = 0)$$

$$\overline{V}_{\alpha 2} = -\frac{EtR}{2L^3 D}\,e^{-L\omega}\sin(L\omega)$$

$$\overline{\vartheta}_{\alpha 2} = \frac{R}{LD}\,e^{-L\omega}\cos(L\omega)$$

$$\Delta\overline{r}_{\alpha 2} = -\frac{R^2}{\sqrt{2}\,L^2 D}\sin\alpha\,e^{-L\omega}\cos\left(L\omega + \frac{\pi}{4}\right)$$

(380)

$\delta_{kk},\ \delta_{ik}$ values:

$$\overline{\delta}_{11} = \overline{\Delta r}_{\alpha 0 1} = \frac{2LR}{Et}\sin^2\alpha_0$$

$$\overline{\delta}_{22} = \overline{\vartheta}_{\alpha 0 2} = \frac{R}{LD}$$

(381)

$$\bar{\delta}_{21} = \bar{\vartheta}_{\alpha_0 1} = -\frac{2L^2}{Et}\sin\alpha_0$$

$$\bar{\delta}_{12} = \overline{\Delta r}_{\alpha_0 2} = -\frac{R^2}{2L^2 D}\sin\alpha_0$$

$$\bar{\delta}_{21} = \bar{\delta}_{12} : \frac{2L^2}{Et} = \frac{R^2}{2L^2 D} \qquad (381) \text{ (Cont.)}$$

To these $\bar{\delta}_{ik}$ values must be added the δ_{ik} values due to the deformation of the supporting system. Thus,

$$\Sigma\delta_{11} = \bar{\delta}_{11} + \delta_{11}$$
$$\Sigma\delta_{22} = \bar{\delta}_{22} + \delta_{22} \qquad (382)$$
$$\Sigma\delta_{12} = \bar{\delta}_{12} + \delta_{12}$$

δ_{k0} *values:*

The δ_{k0} values due to loading $(w, \Delta T, H_{pr})$ in the principal system are obtained from Eqs. (354) to (358). Thus,

$$\begin{array}{ll|l} \Sigma\delta_{10} = \delta_{10} + \bar{\delta}_{10} & \delta_{10} = -H\cdot\bar{\delta}_{11} & X_1 = -H \\ \Sigma\delta_{20} = \delta_{20} = \bar{\delta}_{20} & \delta_{20} = -H\cdot\bar{\delta}_{12} & X_2 = 0 \end{array} \qquad (383)$$

where $H = N_\alpha \cos\alpha$ is the thrust at $\alpha = \alpha_0$ and $N_\alpha = N_{\alpha 0} + \bar{N}_{\alpha 0}$.

The final values are obtained by superposition:

$$\Sigma N = N + X_1\bar{N}_1 + X_2\bar{N}_2$$
$$\Sigma M = -X_1\bar{M}_1 + X_2\bar{M}_2 \qquad (384)$$

For rotationally symmetric shells having a different meridian curvature than circle the $R_\alpha \neq R_\beta$ and are not constant, which makes their analysis more time-requiring, but practically the procedure is the same as for circular curvature. For example, the equations of equilibrium are obtained for a conical shell (Fig. 156) when, in Eqs. (348), $d\alpha/dy$ is substituted for R_β ($R_\beta = \infty$), $r_\alpha = y \cos\alpha$ for r_α, and $N_y, N_{y\beta}$ for $N_\alpha, N_{\alpha\beta}$.

For rotationally symmetric loading, $w_x = 0$ and $N_{y\beta} = 0$; and making use of Guldin's rule [Eq. (350)], we obtain

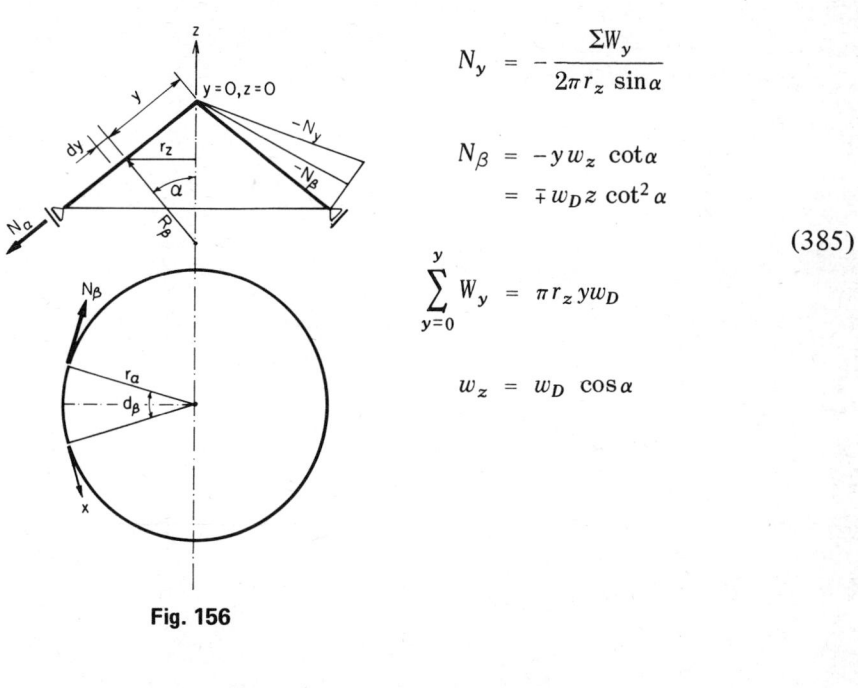

$$N_y = -\frac{\Sigma W_y}{2\pi r_z \sin\alpha}$$

$$N_\beta = -y w_z \cot\alpha$$
$$= \mp w_D z \cot^2\alpha \qquad (385)$$

$$\sum_{y=0}^{y} W_y = \pi r_z y w_D$$

$$w_z = w_D \cos\alpha$$

Fig. 156

$$\Delta\vartheta_z = \mp \frac{w_D z \cot\alpha}{Et \sin^2\alpha}\left[\frac{1}{2} + \nu - (2 + \nu)\cos^2\alpha\right]$$

$$\Delta r_z = \pm \frac{w_D z^2 \cot\alpha}{2Et \sin^2\alpha}(2\cos^2\alpha - \nu)$$

24. POLYGONAL DOMES

Domes composed of cylindrical sections have an advantage over circular domes because the cylindrical sections act more or less as curvilinear shells (barrels) and transmit their surface load over the ridges into the supports ($N_{\alpha_0} \simeq 0$). The shell carrying action allows the polygonal dome to be supported at the corners without heavy edge members, which commonly is not the case with spherical shells.

The meridian (transversal) curvatures generally preferred are the circle, cycloid, or ellipse (Fig. 138). The horizontal sections are polygons with an even (Fig. 157) or uneven number of sides. The even number of sides is structurally preferable because it gives an even number of intersecting cylindrical shells with entirely balanced ridge loading. In the membrane-stress condition, the normal forces are N_x, N_α and shear is $N_{x\alpha}$.

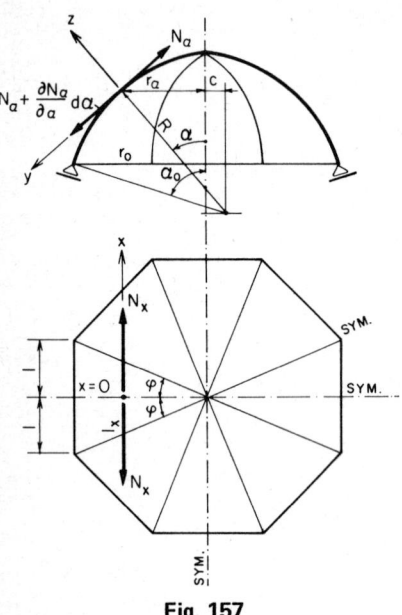

$$w_y = w_D \sin\alpha \qquad w_z = w_D \cos\alpha$$

$$w_x = 0$$

$$M_x = M_\alpha \sim 0$$

$$r_\alpha = R \sin\alpha - c$$

Fig. 157

The equilibrium equations are

$$\Sigma X = 0: \quad \frac{\partial N_x}{\partial x} dx\, R\, d\alpha + \frac{\partial N_{x\alpha}}{\partial \alpha} d\alpha\, dx + w_x\, dx\, R\, d\alpha = 0$$

$$\Sigma Y = 0: \quad \frac{\partial N_\alpha}{\partial \alpha} d\alpha\, dx + \frac{\partial N_{x\alpha}}{\partial x} dx\, R\, d\alpha + w_y\, dx\, R\, d\alpha = 0 \qquad (386)$$

$$\Sigma Z = 0: \quad N_\alpha dx\, d\alpha + w_z\, dx\, R\, d\alpha = 0$$

Dividing these equations by $dx\, R\, d\alpha$, we obtain

$$\frac{\partial N_x}{\partial x} + \frac{\partial N_{x\alpha}}{R\, \partial \alpha} + w_x = 0 \qquad N_\alpha + R w_z = 0$$

$$\frac{\partial N_{x\alpha}}{\partial x} + \frac{\partial N_\alpha}{R\, \partial \alpha} + w_y = 0 \qquad N_\alpha = -R w_z \qquad (387)$$

The N_α value, as can be seen from Eqs. (387), is dependent only on R and the component of surface load w_z normal to the shell's surface, as is characteristic for all curvilinear cylindrical shells:

$$N_\alpha = -Rw_z \qquad \alpha = 90°: \; N_\alpha = 0 \tag{388}$$

The shear $N_{x\alpha}$ and normal force N_x are obtained by direct integration from the equations of equilibrium (387):

$$N_{x\alpha} = -\int \frac{1}{R} \frac{\partial N_\alpha}{\partial \alpha} dx - \int w_y \, dx + C_1(\alpha) \tag{389}$$

$$N_x = -\int \frac{1}{R} \frac{\partial N_{x\alpha}}{\partial x} dx - \int w_x \, dx + C_2(\alpha)$$

The integral constants C_1 and C_2 are independent of x and are functions of α. As a symmetric loading $w_x = 0, w_y = w \sin\alpha, w_z = w \cos\alpha$ is also independent of x and a function of α, we can solve the integrals [Eqs. (389)] directly in the general way:

$$N_{x\alpha} = -\left(\frac{\partial N_\alpha}{R \, \partial \alpha} + w_y\right) x + C_1(\alpha)$$

$$= -w_y^* x + C_1(\alpha) \qquad w_y^* = \frac{\partial N_\alpha}{R \, \partial \alpha} + w_y \tag{390}$$

$$N_x = \frac{\partial w_y^*}{2R \, \partial \alpha} x^2 - \frac{\partial C_1(\alpha)}{R \, \partial \alpha} x + C_2(\alpha)$$

At the symmetry of the shell $x = 0$, $N_{x\alpha} = 0$; therefore, the differential coefficient $C_1 = 0$. The differential coefficient C_2 will be determined for the condition that the ridge is not subjected to bending stresses but is subjected mainly to normal stresses. This is possible only when the forces from the shell are transmitted to the ridge tangentially and all components in the direction of the binormal of the ridge are zero.

Thus, in accordance with Dischinger (Fig. 158), for $x = l$:

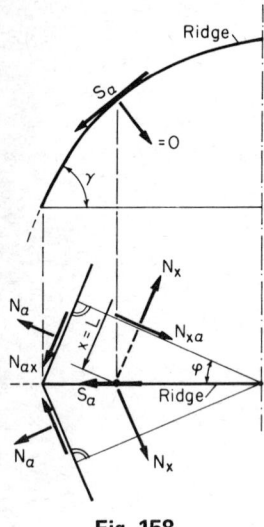

Fig. 158

$$(N_x - 2N_{\alpha x} \cos\alpha \tan\varphi + N_\alpha \cos^2\alpha \tan^2\varphi)$$
$$\times R\, d\alpha \cos\varphi = 0 \tag{391}$$

$$N_x = \frac{\partial w_y^*}{2R\, \partial\alpha} l^2 + C_2(\alpha)$$

$$N_{\alpha x} = -w_y^* l \qquad C_1 = 0$$

Substituting the N_x and $N_{x\alpha}$ values into Eq. (391) and solving for $C_2(\alpha)$, we obtain

$$C_2(\alpha) = -\frac{l^2}{2} \frac{\partial w_y^*}{R\, \partial\alpha} - 2w_y^* \cos\alpha \tan\varphi + N_\alpha \cos^2\alpha \tan^2\varphi \tag{392}$$

Thus the normal forces and shear in the shell are

$$N_\alpha = -w_z R$$
$$N_{x\alpha} = -w_y^* x \tag{393}$$
$$N_x^{(s)} = -\frac{l^2 - x^2}{2} \frac{\partial w_y^*}{R\, \partial\alpha} - (2w_y^* l \cos\alpha \tan\varphi + N_\alpha \cos^2\alpha \tan^2\varphi)$$

Ridge loading:

For $x = l$ components of the forces from the shell sections transmitted tangentially into the ridges, N_α and the normal force of ridges S_α must be in equilibrium with the external loads $\sum^{2n} W_\alpha$. Therefore, the S_α force can be computed most simply from this condition (Fig. 159):

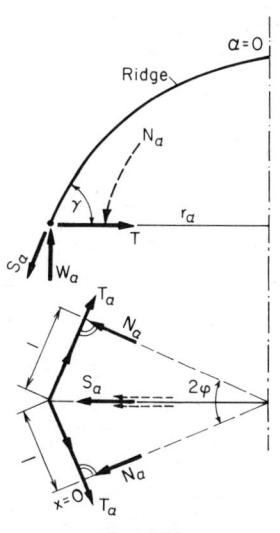

$$\sum^{2n} W_\alpha + 4nlN_\alpha \sin\alpha + 2nS_\alpha \sin\gamma = 0$$

$$\sin\gamma = \frac{1}{\sqrt{1 + \cot^2\alpha/\cos^2\varphi}}$$

$\sum^{2n} W_\alpha$ = total load from $\alpha = 0$ to $\alpha = \alpha_0$

$2n$ = number of sides of polygon

Fig. 159

$$S_\alpha = -\left(\frac{\sum^{2n} W_\alpha}{2n} - 2N_\alpha r_\alpha \sin\alpha \tan\varphi\right)\sqrt{1 + \cot^2\alpha/\cos^2\varphi} \qquad (394)$$

Also, the equilibrium requires that all horizontal components of any section be zero:

$$T_\alpha + \frac{S_\alpha \cos\gamma}{2\sin\varphi} + N_\alpha r_\alpha \cos\alpha$$

$$+ \int_0^\alpha (2w_y^* l \cos\alpha \tan\varphi + N_\alpha \cos^2\alpha \tan^2\varphi)R\,d\alpha = 0 \qquad (395)$$

The derivative of T_α with respect to α produces an $N_x^{(d)}$ value:

$$N_x^{(d)} = \frac{\partial T_\alpha}{R\,\partial\alpha} = \frac{\partial}{R\,\partial\alpha}\left(\frac{S_\alpha \cos\gamma}{2\sin\varphi} + N_\alpha r_\alpha \cos\alpha\right) \tag{396}$$
$$+ (2w_y^* l \cos\alpha \tan\varphi + N_\alpha \cos^2\alpha \tan^2\varphi)$$

This $N_x^{(d)}$ must be added to the $N_x^{(s)}$ value from Eqs. (393) to obtain the final value of N_x. Thus,

$$N_x = N_x^{(s)} + N_x^{(d)}$$
$$= -\frac{l^2 - x^2}{2}\frac{\partial w_y^*}{R\,\partial\alpha} + \frac{\partial}{\partial\alpha R}\left(\frac{S_\alpha \cos\gamma}{2\sin\varphi} + N_\alpha r_\alpha \cos\alpha\right) \tag{397}$$

As a conclusion, the polygonal dome's carrying action consists of two parts: curvilinear cylindrical (s) shell action and spherical (d) shell action.

Example:

For demonstration of the theory, a circular polygonal dome will be considered (Fig. 160).

$w_D = 1.0\,\text{ksf}$
$w_y = w_D \sin\alpha \quad w_z = w_D \cos\alpha \quad w_x = 0$
$2n = 8 \quad n = 4 \quad \alpha_0 = 72° \quad R = 70.00\,\text{ft}.$

$\varphi = \dfrac{2\pi}{16} = \dfrac{\pi}{8} = 22.5°$

$h = 21.63\,\text{ft} \quad z_0 = 48.37\,\text{ft} \quad B = 66.57\,\text{ft}.$
$l = 21.50\,\text{ft}$

$N_\alpha = -w_D R \cos\alpha = -70.00 \cos\alpha$

$w_y^* = \dfrac{\partial N_\alpha}{\partial\alpha} + w_y = 2w_D \sin\alpha$

$N_{\alpha x} = -w_y^* x = -2w_D x \sin\alpha \quad x = l = R \tan\varphi \sin\alpha$
$\quad\quad = -2Rw_D \tan\varphi \sin^2\alpha = -57.96 \sin^2\alpha$

$$\sum_{\alpha}^{2n} w_\alpha = 4n \tan\varphi \int_{\alpha=0}^{\alpha=\alpha} w_D R r_\alpha \, d\alpha \qquad r_\alpha = R \sin\alpha$$

$$= 4n \tan\varphi \, w_D R [R(1 - \cos\alpha)] = 32{,}359(1 - \cos\alpha)$$

$$w_\alpha = \frac{\sum_{\alpha}^{2n} w_\alpha}{2n} = 4{,}045(1 - \cos\alpha)$$

$$S_\alpha = -\left(\frac{\sum_{\alpha}^{2n} w_\alpha}{2n} - 2N_\alpha r_\alpha \sin\alpha \tan\varphi\right)\sqrt{1 + \cot^2\alpha/\cos^2\varphi}$$

$$= 2w_D R^2 \tan\varphi[(1 + \sin^2\alpha)\cos\alpha - 1]\sqrt{1 + \cot^2\alpha/\cos^2\varphi}$$

$$\simeq 4{,}057[(1 + \sin^2\alpha)\cos\alpha - 1]\sqrt{1 + \cot^2\alpha/0.854}$$

$$N_x \atop (x=0) = \frac{w_D R}{\cos^2\varphi}\left[\frac{1 - \cos\alpha - \cos^2\alpha}{1 + \cos\alpha} + (1 + 4\sin^2\alpha)\cos\alpha \sin^2\varphi\right]$$

$$\simeq 82.0\left[\frac{1 - \cos\alpha - \cos^2\alpha}{1 + \cos\alpha} + 0.147(1 + 4\sin^2\alpha)\cos\alpha\right]$$

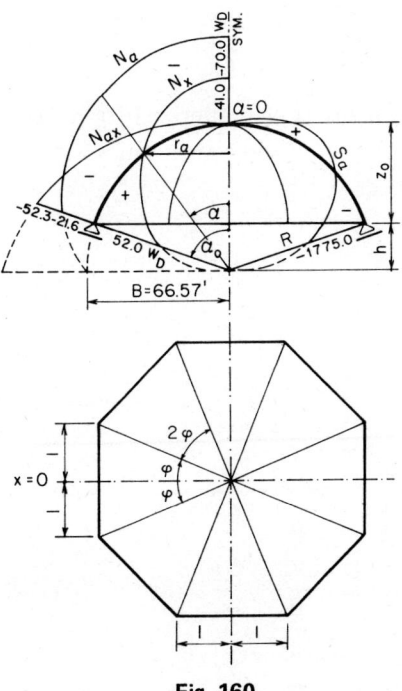

Fig. 160

25. ROTATIONALLY SYMMETRIC CYLINDRICAL SHELLS

Shells belonging to this category are commonly used for silos, tanks, etc. (Fig. 161). The horizontal pressure $p_y = f(z)$ the shell is subjected to is rotationally symmetric, but its intensity is a function of z. Because cylindrical shells have a single curvature (a circle), membrane theory is inadequate to describe the stress and deformation conditions in the shells used in practice. Any restriction at the boundaries ($\Delta r, \vartheta$) leads to boundary disturbances which do not disappear within a short distance from the boundary, as is the case with spherical shells, but which usually penetrate from boundary to boundary and have magnitudes that very often control the design.

To derive general equations for the internal forces, moments, and deformations in the shell wall, an infinitely small element cut out of the wall will be considered (Fig. 162).

The element is characterized by $r = $ constant, $\alpha = 90°$, and wall thickness $t = $ constant.

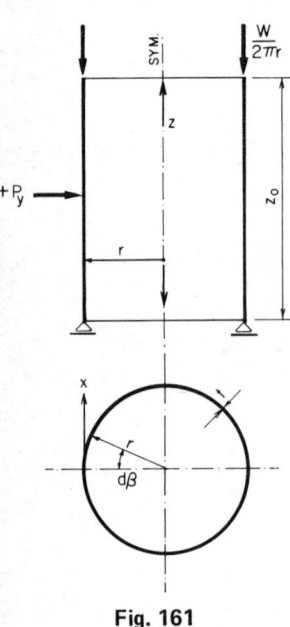

Fig. 161

Equilibrium requires that

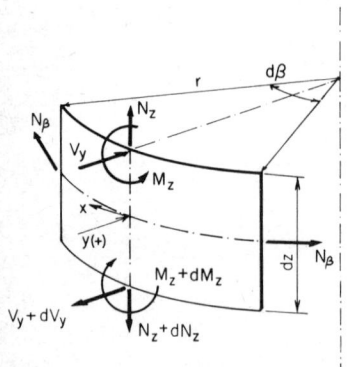

Fig. 162

$$\Sigma Z = 0: \quad \frac{dN_z}{dz} + pz = 0$$

$$\Sigma Y = 0: \quad \frac{dV_y}{dz} + \frac{N_\beta}{r} + py = 0 \quad (398)$$

$$\Sigma M_z = 0: \quad \frac{dM_z}{dz} - V_y = 0$$

Eliminating V_y from the third equation (398) and introducing it into the second, we obtain

$$\frac{d^2 M_z}{dz^2} + \frac{N_\beta}{r} + p_y = 0 \tag{399}$$

Denoting the radial deflection of the wall by ω, the unknowns M_z and N_β expressed in terms of ω are

$$D = \frac{Et^3}{12(1 - \nu^2)}$$

$$\epsilon_\beta = \frac{1}{Et}(N_\beta - \nu N_z) = -\frac{\omega}{r}$$

$$\epsilon_z = \frac{1}{Et}(N_z - \nu N_\beta) \tag{400}$$

$$N_\beta = Et\left(-\frac{\omega}{r} + \nu \epsilon_z\right)$$

$$M_z = -D\frac{d^2\omega}{dz^2}$$

Introducing the N_β and M_z values into Eq. (399), we obtain the differential equation of the cylindrical shell:

$$\frac{d^2\omega}{dz^2}\left(D\frac{d^2\omega}{dz^2}\right) + \frac{Et}{r^2}\omega = p_y + \frac{\nu}{r}N_z \tag{401}$$

For $t =$ constant, this equation simplifies to

$$\frac{d^4\omega}{dz^4} + \frac{4}{L^4}\omega = \frac{1}{D}\left(p_y + \nu\frac{N_z}{r}\right) \tag{402}$$

$$L^4 = \frac{r^2 t^2}{3(1 - \nu^2)} \qquad L = \sqrt[4]{\frac{r^2 t^2}{3(1 - \nu^2)}} = \frac{r}{1.31\sqrt{r/t}}$$

where L is the characteristic constant (length) of a cylindrical shell and $\nu = 0.167$ is Poisson's ratio. The N_z value is obtained by direct integration from Eqs. (398):

$$N_z = -\int p_z \, dz + C \tag{403}$$

264 Theory of Structures

The solution of this nonhomogeneous differential equation [Eq. (401)] may be obtained by expressing the radial deflection ω of the shell wall in terms of the particular integral of the nonhomogeneous equation for ω_0 (that is, the deflection of the statically determinate shell—boundary conditions ignored—loaded by the pressures p_y and p_z) and in terms of the expression for $\bar{\omega}$ (that is, the general solution of the homogeneous equation for the deflection of the wall when it is loaded only by the redundants $X_1, \ldots, i$ acting at the boundaries of the cylinder).

The final radial deflection ω is obtained by applying the principle of superposition:

$$\omega = \omega_0 + \sum_{1}^{i} X_i \bar{\omega}_i \tag{404}$$

The ω_0 value is computed by membrane theory, and $\bar{\omega}$ is obtained from the homogeneous equation with $i = 4$ differential coefficients determined from the boundary conditions of the shell.

Particular Integrals

Particular integrals of the nonhomogeneous equation for the loadings most often encountered in practice are the following.

Dead load and surcharge:

$$\begin{aligned} p_y &= 0 & p_z &= w_D z + \frac{w_z}{2r\pi} \\ N_\beta &= 0 & N_z &= p_z \\ \omega_0 &= -\frac{\nu r}{Et}\left(w_D z + \frac{w_z}{2\pi}\right) \\ \vartheta_0 &= -\frac{d\omega_0}{dz} = -\frac{\nu r}{Et} w_D \end{aligned} \tag{405}$$

Liquid pressure:

$$\begin{aligned} p_y &= -\gamma z & p_z &= 0 & \gamma &= \text{unit weight} \\ N_\beta &= \gamma r z & N_z &= 0 \\ \omega_0 &= -\frac{\gamma r^2}{Et} z \\ \vartheta_0 &= -\frac{\gamma r^2}{Et} \end{aligned} \tag{406}$$

Earth pressure:

$$p_y = \gamma_E \tan^2\left(45° - \frac{\Phi}{2}\right)\left(z + \frac{w_s}{\gamma_E}\right) \qquad p_z \simeq 0$$

$$N_\beta = -p_y r \qquad N_z \simeq 0 \qquad \Phi = \text{angle of int. friction} \tag{407}$$

$$\omega_0 = \frac{r^2}{Et} p_y \qquad\qquad w_s = \text{surcharge}$$

$$\vartheta_0 = \frac{r^2 \gamma_E}{Et} \tan^2\left(45° - \frac{\Phi}{2}\right)$$

Silo pressure:

$$p_y = \frac{\gamma A}{\mu u}(1 - e^{-z/z_0}) \qquad z = 0, \; p_z = \mu p_y$$

$$z_0 = \frac{A}{\mu u \tan^2(45° - \Phi/2)} \qquad A = \pi r_i^2, \; u = \pi 2 r_i$$

$$N_\beta = r p_y \qquad N_z \simeq 0 \qquad \mu = \text{coefficient of friction} \tag{408}$$

$$\omega_0 = -\frac{r^2}{Et} p_y$$

$$\vartheta_0 = -\frac{\gamma A}{\mu u}\frac{r^2}{z_0} e^{-z/z_0}$$

Wind pressure:

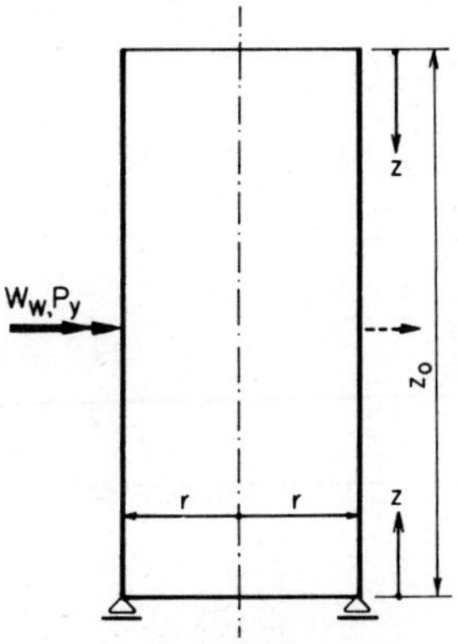

Fig. 163

$$p_y = \Sigma Z_n \cos n\beta$$
$$= w_w(-0.655 + 0.280 \cos\beta + 1.115 \cos 2\beta$$
$$+ 0.400 \cos 3\beta - 0.113 \cos 4\beta$$
$$- 0.027 \cos 5\beta)$$

$$N_\beta = -r \Sigma Z_n \cos n\beta$$

$$N_z = \frac{z^2}{2r} \Sigma n^2 Z_n \cos n\beta \qquad Z_n = \text{const}$$

$$N_{z\beta} = -z \Sigma n Z_n \cos n\beta \tag{409}$$

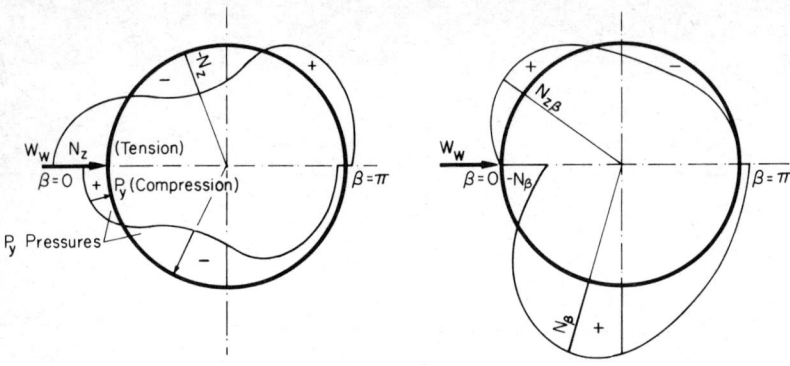

Fig. 164

The wind pressure p_y and related N_x, N_β, and $N_{z\beta}$ values are illustrated in Fig. 164.

Temperature and shrinkage:

$$N_\beta = N_z = N_{z\beta} = 0$$
$$\omega_{\Delta r} = -\alpha_{\Delta T} \Delta Tr \quad \vartheta = 0 \tag{410}$$
$$\omega_{sr} = \epsilon_{sr} \quad \vartheta = 0$$

Homogeneous Solution

The homogeneous differential equation to satisfy the boundary conditions of the shell has the form [Eqs. (402)]

$$\frac{d^4 \bar{\omega}}{d(\lambda\eta)^4} + 4\bar{\omega} = 0 \tag{411}$$

where $\lambda = z_0/L$ is the shell coefficient and is a constant for any given shell and $\eta = z/z_0$. The solution of this equation is

$$-\bar{\omega} = C_1 \cosh\lambda\eta \cos\lambda\eta + C_2 \cosh\lambda\eta \sin\lambda\eta \\ + C_3 \sinh\lambda\eta \cos\lambda\eta + C_4 \sinh\lambda\eta \sin\lambda\eta \tag{412a}$$

As a rule, when $\lambda > 7$, the hyperbolic functions can be expressed by exponential functions. Thus,

$$\bar{\omega} = e^{-\lambda\eta}(A_1 \cos\lambda\eta + B_1 \sin\lambda\eta) \\ + e^{-\lambda\eta}(A_2 \cos\lambda\eta' + B_2 \sin\lambda\eta') \quad \eta' = \frac{z_0 - z}{z_0} \tag{412b}$$

For this case, the shell is considered infinitely long, and therefore the redundants acting at $z = 0$ ($\eta = 0$) do not have appreciable influence upon the redundants acting at $z = z_0$ ($\eta' = 0$), and vice versa. Thus the differential coefficients A_1, B_1 and A_2, B_2 are independent of each other and can be determined from the boundary conditions of the shell at $\eta = 0$ and $\eta' = 0$, respectively. However, if the characteristic length L is great and the height z_0 relatively small, the forces at opposite boundaries of the shell influence each other, and therefore both portions of Eq. (412b) must be included in the computation of the deflection $\bar{\omega}$ at the boundary under consideration. In prestressed designs, L is usually small; thus $\lambda > 7$, and the computation reduces to either the first or the second part of Eq. (412b), depending on which boundary is under consideration:

$$\frac{d\bar{\omega}}{d(\lambda\eta)} = -e^{-\lambda\eta}[A_1(\sin\lambda\eta + \cos\lambda\eta) + B_1(\sin\lambda\eta - \cos\lambda\eta)]$$

$$\frac{d^2\bar{\omega}}{d(\lambda\eta)^2} = 2e^{-\lambda\eta}(A_1 \sin\lambda\eta - B_1 \cos\lambda\eta) \qquad (413)$$

$$\frac{d^3\bar{\omega}}{d(\lambda\eta)^3} = -2e^{-\lambda\eta}[A_1(\sin\lambda\eta - \cos\lambda\eta) - B_1(\sin\lambda\eta + \cos\lambda\eta)]$$

$$\bar{N}_\beta = -\frac{Et}{r}\bar{\omega} \qquad \sum N_\beta = N_\beta + \sum_{1}^{K} X_k \bar{N}_{\beta k}$$

$$\bar{M}_z = -\frac{D}{L^2}\frac{d^2\bar{\omega}}{d(\lambda\eta)^2} \qquad \sum M_z = M_z + \sum_{1}^{K} X_k \bar{M}_{zk} \qquad (414)$$

$$\bar{V}_y = -\frac{D}{L^3}\frac{d^3\bar{\omega}}{d(\lambda\eta)^3} \qquad \sum V_y = V_y + \sum_{1}^{K} X_k \bar{V}_{yk}$$

Redundants (Fig. 165):

$X_1 = 1.0k$:

Boundary conditions for $\lambda = z_0/L > 7$, $\eta = 0$:

$\bar{V}_{y1} = -1 \qquad M_{z1} = 0$

268   Theory of Structures

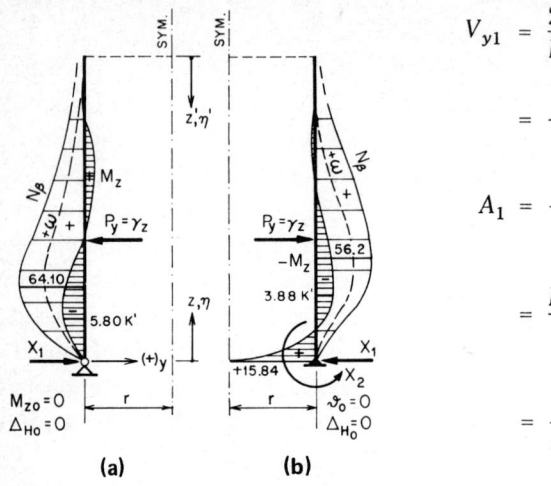

$$V_{y1} = \frac{2D}{L^3} e^{-\lambda\eta} A_1(\sin\lambda\eta - \cos\lambda\eta)$$

$$= -1.0$$

$$A_1 = +\frac{L^3}{2D}$$

$$= \frac{L^4}{L} \frac{12(1-\nu^2)}{2Et^3} \frac{r^2 t^2}{3(1-\nu^2)}$$

$$= \frac{2r^2}{LEt}$$

Fig. 165

$$M_{z1} = 0: B_1 = 0$$

$$\bar{\omega}_1 = \frac{2r^2}{LEt} e^{-\lambda\eta} \cos\lambda\eta \qquad (415)$$

$$\bar{\vartheta}_1 = -\frac{2r^2}{L^2 Et} e^{-\lambda\eta}(\sin\lambda\eta + \cos\lambda\eta)$$

$$\bar{N}_{\beta 1} = -\frac{2r}{L} e^{-\lambda\eta} \cos\lambda\eta$$

$$\bar{M}_{z1} = -L e^{-\lambda\eta} \sin\lambda\eta$$

$$\bar{V}_{y1} = e^{-\lambda\eta}(\sin\lambda\eta - \cos\lambda\eta)$$

$$X_2 = 1.0 k'$$

Boundary conditions for η' or $\eta = 0$: $\bar{V}_{y2} = 0$, $M_{z2} = -1.0$

$$A_1 = 0, \quad B_1 = -\frac{L^2}{2D}$$

$$\bar{\omega}_2 = -\frac{2r^2}{L^2 Et} e^{-\lambda\eta}(\cos\lambda\eta - \sin\lambda\eta)$$

$$\bar{\vartheta}_2 = \frac{4r^2}{L^3 Et} e^{-\lambda\eta} \cos\lambda\eta \qquad (416)$$

$$\bar{N}_{\beta 2} = \frac{2r}{L^2} e^{-\lambda\eta}(\cos\lambda\eta - \sin\lambda\eta)$$

$$\bar{M}_{z2} = e^{-\lambda\eta}(\sin\lambda\eta + \cos\lambda\eta)$$

$$\bar{V}_{z2} = -\frac{2}{L} e^{-\lambda\eta} \sin\lambda\eta$$

$$\delta_{10} + X_1\delta_{11} + X_2\delta_{12} = 0$$
$$\delta_{20} + X_1\delta_{21} + X_2\delta_{22} = 0 \tag{416 Cont.}$$

δ_{kk}, δ_{ik}, and δ_{k0} values:

$$\delta_{11} = \bar{\omega}_{01} = \frac{2r^2}{LEt} \quad \delta_{22} = \frac{4r^2}{L^3Et} \tag{417}$$

$$\delta_{12} = \delta_{21} = \bar{\omega}_{02} = -\frac{2r^2}{L^2Et}$$

$$\left.\begin{array}{l}\delta_{10} = \omega_0 \\ \delta_{20} = \vartheta_0\end{array}\right| \begin{array}{l}\text{Particular integrals}\\ \text{Eqs. (405) to (410)}\end{array} \tag{418}$$

(The ω values in the direction of load are taken positive.)

$$r = 45.0 \text{ ft} \quad z_0 = 30.0 \text{ ft} \quad t = 0.75 \text{ ft}$$
$$E = 5.6 \times 10^6 \text{ psi} \quad \nu = 0.167$$
$$L = \frac{r}{1.31\sqrt{r/t}} = 4.45 \quad \lambda = \frac{z_0}{L} = 6.74$$
$$\eta = \frac{z}{z_0} \quad z_0 = \lambda L \quad z = \eta\lambda L$$

$$Et \sum_D^\gamma \omega_0 = -r(\nu N_z + rN_\beta) \quad N_z = w_D z + w_z$$
$$= w_D \lambda L(1 - \eta) + w_z$$
$$w_z = 1.205 \text{ k/ft}$$
$$N_\beta = \gamma r\lambda L(1 - \eta)$$

$$Et \sum_D^\gamma \omega_0 = \delta_{10} = +[(\nu w_D + \gamma r^2)\lambda L + \nu w_z]$$
$$= +[(0.167 \times 0.112 + 0.063 \times 45^2)30.0 + 0.167 \times 1.205] = 3{,}796$$

$$Et \sum_D^\gamma \vartheta_0 = \delta_{20} = +(\nu r w_D + \gamma r^2) = 127.34$$

$$Et\delta_{11} = \frac{2r^2}{L} = \frac{2 \times 45^2}{4.45} = 910 \qquad Et\delta_{22} = \frac{4r^2}{L^3} = 91.8$$

$$Et\delta_{12} = -\frac{2r^2}{L^2} = -204$$

$$X_1 = \frac{3{,}796 \times 91.8 - 127.34 \times 204}{910 \times 91.8 - 204^2} = 7.74\, k/ft = -V_y$$

$$X_2 = \frac{127.34 \times 910 - 3{,}796 \times 204}{910 \times 91.8 - 204^2} = -15.84\, k\text{-}ft/ft = -M_z$$

The rotationally symmetric cylindrical and spherical shells combined are most suitable for atomic-energy applications for use as containers.

26. SEGMENTAL CYLINDRICAL SHELLS*

Principally the forces, moments, and deformations in a segmental cylindrical shell (Fig. 166) for rotationally symmetric loading are determined in the same way as for a closed cylindrical shell. However, additional forces ΔN_β and moments M_y in the horizontal direction must be considered because the radial deflection ω [$\omega = f(z)$] at the vertical boundary of the shell is zero. Also, the rotation of the tangent $\vartheta_y = f(z)$, depending upon the vertical boundary condition, may or may not be zero. The final stress and deformation condition is obtained by application of the principle of superposition.

The strain caused by normal force $N_\beta(\epsilon_e)$, shrinkage ϵ_{sr}, change of temperature $\epsilon_{\Delta T, T}$, and displacement of supports Δ results in a reduction (ΔN_β) in N_β (computed for a closed cylindrical shell). Denoting the change of span L by $\Sigma\delta_{a0}$ due to the strain (e, sr, ΔT, T, and Δ) and δ_{aa} due to the thrust 1.0, we obtain

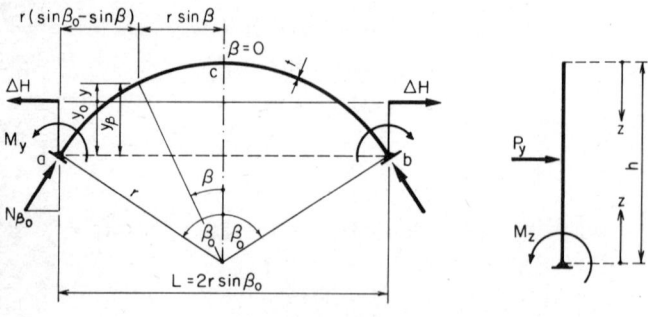

Fig. 166

*A. E. Komendant, Economy and Safety of Different Types of Concrete Dams, *ASCE Proc.*, vol. 81, 1955.

$$\Delta H = N_\beta \cos\beta = -\frac{\Sigma \delta_{a0}}{\delta_{aa}} \qquad A = 1.0\,t \qquad (419)$$

$$\delta_{a0} = \frac{2r}{Et} \int_{\beta=0}^{\beta=\beta_0} N_\beta \cos\beta\, d\beta + 2r(\epsilon_s \mp \epsilon_T)\sin\beta$$

$$- 2\frac{\epsilon_{\Delta T}}{t} \int_{\beta=0}^{\beta=\beta_0} \left[(\cos\beta - \cos\beta_0) \right. \qquad (420)$$

$$\left. - \rho\left(\frac{\sin\beta_0}{\beta_0} - \cos\beta_0\right) \right] r\, d\beta + 2\Delta$$

$$\delta_{aa} = \frac{2}{EI} \int_{\beta=0}^{\beta=\beta_0} (Y_\beta^2 - Y_\beta Y_0) r\, d\beta + \frac{2}{Et} \int_{\beta=0}^{\beta=\beta_0} \cos^2\beta\, r\, d\beta$$

where ρ is the degree of restraint at support (for fixed support $\rho = 1.0$) and the ordinates of the circle are

$$\begin{aligned}
Y_\beta &= r(\cos\beta - \cos\beta_0) \\
Y_0 &= \rho r\left(\frac{\sin\beta_0}{\beta_0} - \cos\beta_0\right) \\
Y &= Y_\beta - Y_0
\end{aligned} \qquad (421)$$

After integration, rearranging the expressions, and considering the influence of plastic flow φ upon the shrinkage and displacement of supports Δ, we obtain

$$\begin{aligned}
\Delta H_e &= -N_\beta C_H \\
\Delta H_{sr} &= -\epsilon_{sr} Et\, C_H \frac{1-e^{-\varphi}}{\varphi} \qquad e = 2.7183 \\
\Delta H_T &= \mp \epsilon_T Et\, C_H \\
\Delta H_{\Delta T} &= \epsilon_{\Delta T} r\, C_H C_{\Delta T} \\
\Delta H_\Delta &= -\frac{\Delta Et}{r\sin\beta_0} C_H e^{-\varphi}
\end{aligned} \qquad (422)$$

where

$$C_H = \frac{\sin\beta_0}{\tfrac{1}{4}(\sin 2\beta_0 + 2\beta_0) + 12r^2/t^2[\beta_0(\cos^2\beta_0 + \tfrac{1}{2}) - \tfrac{3}{4}\sin 2\beta_0 - \rho r(\sin\beta_0/\beta_0 - \cos\beta_0)^2]}$$

$$C_{\Delta T} = (1 - \beta_0 \cot\beta_0)(1 - \rho)$$

The horizontal moment M_y due to ΔH is

$$M_y = \Delta H r \left[(\cos\beta - \cos\beta_0) - \rho\left(\frac{\sin\beta_0}{\beta_0} - \cos\beta_0\right) \right] \tag{423}$$

The stresses and deflection are obtained by superposition:

$$\sum N_\beta = N_\beta + \Delta N_\beta \qquad \Delta N_\beta = \frac{\Delta H}{\cos\beta}$$

$$\sigma_y = \mp \frac{\sum N_\beta}{t} \pm \frac{M_y}{2I} t \tag{424}$$

$$\omega = \frac{r}{Et}(\sum N_\beta - \nu N_z)$$

For a relatively large center angle ($2\beta_0$) and r/t ratio, the horizontal flexural stresses (due to M_y) are negligible in comparison with the normal stresses due to $\sum N_\beta$. When the N_β is a compressive force, the design is commonly controlled by stability of the shell. In accordance with Euler's formula, the buckling strength of a straight element is

$$P_B = \frac{\pi EI}{L_B^2} \qquad \sigma_\beta = \frac{P_\beta}{A} = \frac{\pi\epsilon}{L_\beta^2}\frac{I}{A} \tag{425}$$

Introducing $I = At^2/12$, the buckling stress will be

$$\sigma_B = \frac{\pi E t^2}{12 L_B^2} \tag{426}$$

For a restrained shell having center angle $2\beta_0$, the buckling length is approximately

$$L_B = \frac{2\beta_0 r}{3} \tag{427}$$

Introducing the L_B value into Eq. (426), we obtain the limiting ratio for buckling stability:

$$\frac{r}{t} \cong \frac{0.77}{\beta_0} \sqrt{\frac{E}{\sigma_B}} \qquad (428)$$

For example, assuming $\beta_0 = 60°$, F.S. = 3.5, and $E/\sigma_d \sim 4{,}000$, the minimum thickness of the shell, as controlled by buckling, is

$$t_{min} = \frac{r\beta_0}{0.77\sqrt{E/\sigma_d}} \sim \frac{r}{45}$$

27. POLYGONLINEAR SHELLS (FOLDED PLATES)

The carrying action of this type of space structure is most clearly explained by polygonlinear shells. Therefore, they will be discussed and analyzed first.

Membrane and Bending Theories

The coordinates x, y, and z and polygon angles α, γ are chosen as indicated in Fig. 167.

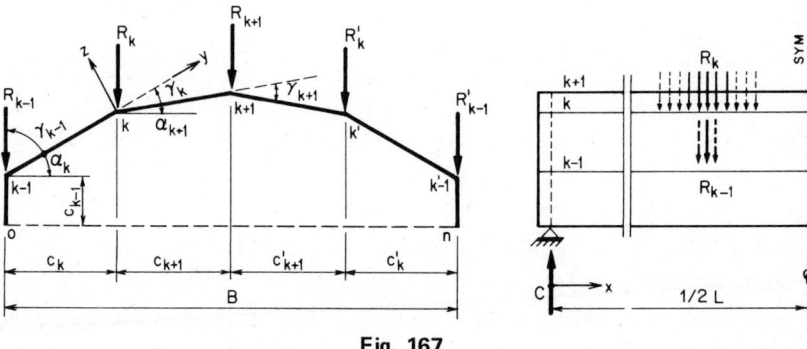

Fig. 167

The external uniform surface loads

$$\left(\sum_0^n w_k C_k = \sum_0^n R_k \right)$$

will be transmitted by beam action in the y direction into the ridges k. Thus the polygon panels act transversally, like continuous slabs supported by ridges. The reaction of the slab (R_k) produces uniform line loads from $x = 0$ to $x = L$ for ridges which are transmitted along the polygon panels in the x direction into supporting elements C.

The structural ability of the ridges to provide support for the panels in the y direction is best illustrated as shown in Fig. 168.

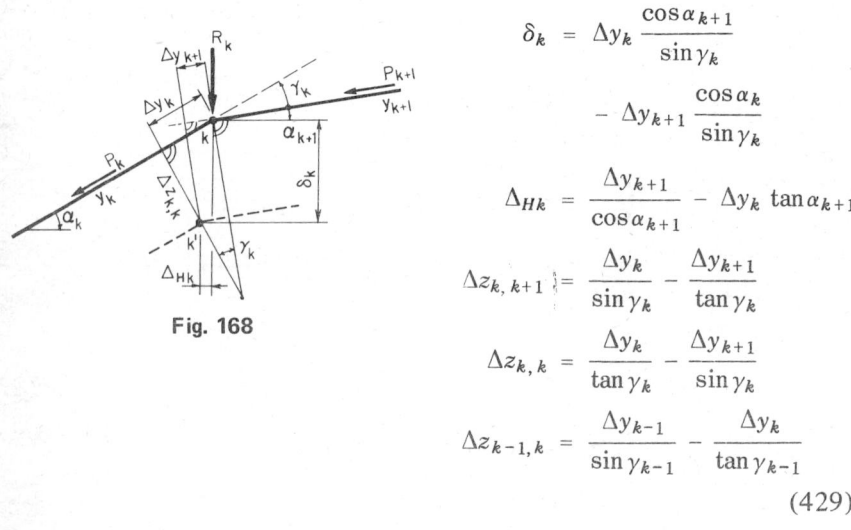

Fig. 168

$$\delta_k = \Delta y_k \frac{\cos \alpha_{k+1}}{\sin \gamma_k} - \Delta y_{k+1} \frac{\cos \alpha_k}{\sin \gamma_k}$$

$$\Delta_{Hk} = \frac{\Delta y_{k+1}}{\cos \alpha_{k+1}} - \Delta y_k \tan \alpha_{k+1}$$

$$\Delta z_{k,k+1} = \frac{\Delta y_k}{\sin \gamma_k} - \frac{\Delta y_{k+1}}{\tan \gamma_k}$$

$$\Delta z_{k,k} = \frac{\Delta y_k}{\tan \gamma_k} - \frac{\Delta y_{k+1}}{\sin \gamma_k}$$

$$\Delta z_{k-1,k} = \frac{\Delta y_{k-1}}{\sin \gamma_{k-1}} - \frac{\Delta y_k}{\tan \gamma_{k-1}}$$

(429)

The ridge loads R_k are resolved into panel loads P_k [Fig. 174, Eq. (458)]. The panels acted upon in their y plane by P_k and having span L deform and yield deflections Δy_k, resulting in vertical and normal deflections $(\Delta_k, \Delta z_k)$ of the polygon.

From these equations (429), it can be seen that if $\Delta y_k = 0$, $\Delta_k = \Delta_{Hk} = \Delta z_k = 0$. Considering that the depth of panels y_k is relatively large in comparison with the span L, the Δy values are rather small. But for relatively flat polygons, $\sin \gamma_k$ is relatively small in comparison with $\cos \alpha_k$ or $\cos \alpha_{k+1}$. As a result, the Δz_k values can be considerable regardless of the small values of Δy.

The classic folded-plate theory assumes $\Delta_k = 0$ and computes the ridge load R_k as reactions for continuous slabs on nonyielding supports. Now, considering that any yielding of support k of a continuous slab results in a redistribution of reactions R_k, on which the deflections Δ_k and moments M_{ky} depend, the problem is rather complex and in a high degree statically indeterminate.* In practice, successive approximations are commonly used to avoid the difficulties of the analysis. This means that after the Δ_k values are obtained, the new corresponding reactions R'_k as functions of Δ_k [Eqs. (12) and (14)] and improved panel loading P'_k are computed. The procedure is carried on until qualitatively acceptable results are obtained. But regardless of the effort involved, the solution seldom converges for folded plates in which the M_y moments affect the shell action appreciably, because this method considers only the displacements ($\Delta y, \Delta z$) and disregards the rotation (φ_k) of the ridges. Making use of basically the same method as that applied in the analysis of rotationally symmetric shells—that is, a membrane solution superimposed by a homogeneous solution—Friedemann Auberlen† has shown an elegant method of obtaining the unknowns σ_x and M_{ky} simply by one operation from a set of simultaneous equations written for each ridge k of the polygon.

Membrane Theory

Individual panels in a polygon subjected to panel load P_k differ from ordinary structural beams only by boundary conditions for the top and bottom fibers. At the common boundary (ridge k), the adjacent panels must meet the condition

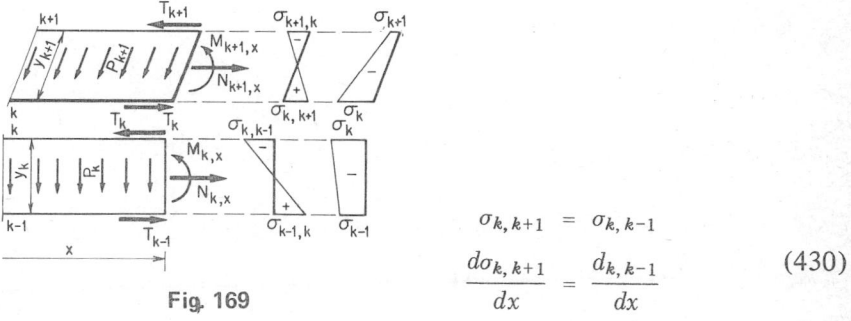

Fig. 169

$$\sigma_{k,k+1} = \sigma_{k,k-1}$$

$$\frac{d\sigma_{k,k+1}}{dx} = \frac{d_{k,k-1}}{dx} \qquad (430)$$

---

*W. Fluegge, "Statik und Dynamik der Schalen," Springer-Verlag OHG, Berlin, 1934.
†Friedemann Auberlen, Praktische Berechnung steifknotiger Faltwerke, *Beton-Stahlbetonbau*, no. 2, February, 1968.

This condition can be satisfied only by shear force T_k. Equilibrium requires that (Fig. 169)

$$\sum N_x = 0: \quad N_k = T_k - T_{k-1}$$
$$\sum M_x = 0: \quad M_k = M_{k0} - \frac{1}{2} y_k (T_k + T_{k-1}) \tag{431}$$

The beam moment for uniform loading P_k is

$$M_{k0,\max} = \frac{P_k L^2}{8} \qquad M_{k0,x} = 4 M_{k0,\max} \xi \xi'$$

and for sinusoidal loading is

$$M_{k0,\max} = \frac{P_k L^2}{\pi^2} \qquad M_{k0,x} = M_{k0,\max} \sin \frac{\pi x}{L}$$

For other types of moment diagrams, $\Delta_k / \Delta f(x)$ can be used as a correction factor.

The requirement that the stresses σ_{kx} in adjacent panels along their common ridge must be equal allows the determination of the shear force T_k. Considering that the cross-sectional area (A) and section modulus (S) are

$$A_k = t y_k$$
$$S_k = \frac{t y_k^2}{6}$$

the fiber stresses $\sigma_{k,k+1}$ and $\sigma_{k,k-1}$ of the panels at ridge k are

$$\sigma_{k,k+1} = \frac{N_{k+1}}{A_{k+1}} + \frac{M_{k+1,0} - \frac{1}{2}(T_k + T_{k+1})}{S_{k+1}}$$
$$= \sigma_{k+1,0} - \frac{4}{A_{k+1}} T_k - \frac{2}{A_{k+1}} T_{k+1} \tag{432}$$
$$\sigma_{k,k-1} = -\frac{N_k}{A_k} - \frac{M_{k0} - \frac{1}{2}(T_k + T_{k-1})}{S_k}$$
$$= -\sigma_{k0} + \frac{4}{A_k} T_k + \frac{2}{A_k} T_{k-1}$$

Equating these equations we obtain

$$\frac{\sigma_{k+1,0} + \sigma_{k0}}{2} = \frac{1}{A_k} T_{k-1} + 2\left(\frac{1}{A_k} + \frac{1}{A_{k+1}}\right) T_k + \frac{1}{A_{k+1}} T_{k+1} \qquad (433)$$

Such a linear equation can be written for any ridge, and the solution of these simultaneous equations delivers the shear T_k ($k = 1, 2, \ldots, n$).

Considering that the lever arm y_k of the T_k-force couple is relatively large, the influence of beam action (M_{kB}) in comparison with shell action [$M_{kT} = \tfrac{1}{2} y_k (T_k + T_{k-1})$] is relatively small:

$$M_{kB} = \int_{k-1}^{k} \sigma_k y\, dA = M_{k0} - \tfrac{1}{2} y_k (T_k + T_{k-1}) \to 0 \qquad (434)$$

This means that stress distribution over the depth y_k of the panel is almost linear.

Bending Theory (Friedemann Auberlen)

Considering that stress distribution over the depth y_k of the panels is almost linear and the transversal moments due to deformation (M_{ky}) are influencing the P_k loading in considerable degree, it is appropriate to use $\sigma_x = f(T_k)$ and M_{ky} as unknowns. The final ridge moments are obtained by superposition $\Sigma M_{ky} = M_{ky0} - M_{ky}$. The M_{ky0} moments are computed for fixed and nonyielding ridges.

For simplification of the solutions, the moments M_{x0} due to the panel loading P_k and corresponding T_k values are computed at $x = L/2$ and it is assumed that they follow longitudinally the function $f(x) = \sin(\pi x/L)$. Also, the transversal moments due to deformation (M_{ky}) are considered affined to the elastic line of the corresponding ridge. By this well-justified assumption,* the relationship between P_k, T_k, σ_k, and Δ_k is established. In cases in which the moment deviates considerably from the sinusoidal [$f(x) = \sin(\pi x/L)$] moment diagram, correction must be made in the $M_{x,\max}$ values. For example; for the triangular moment diagram,

---

*David Yitzhaki, "Prismatic and Cylindrical Shell Roofs," Haifa Science Publishers, Israel, 1958.

$\Delta_k/\Delta f(x) = 0.822$; thus, $M'_{x,\max} = 0.822 M_{x,\max}$. For the rectangular moment diagram, $M_x =$ constant; $M'_{x,\max} = 1.23 M_x$. For the parabolic moment diagram, $\Delta_k/\Delta f(x) = 1.0$.

Introducing in the principal system the $\sigma_k = f(T_k)$ as unknowns, the T_k values become (Fig. 170)

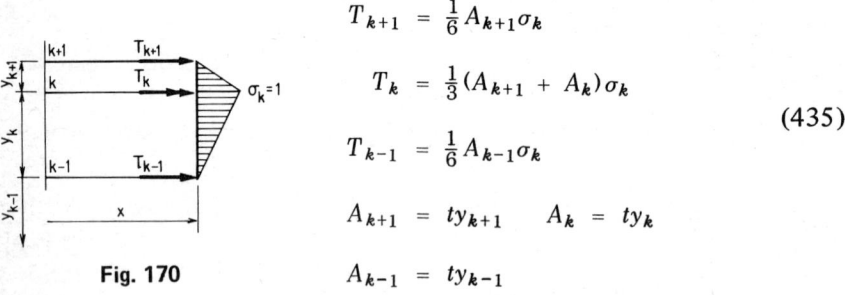

Fig. 170

$$T_{k+1} = \tfrac{1}{6} A_{k+1} \sigma_k$$
$$T_k = \tfrac{1}{3}(A_{k+1} + A_k)\sigma_k \tag{435}$$
$$T_{k-1} = \tfrac{1}{6} A_{k-1} \sigma_k$$
$$A_{k+1} = t y_{k+1} \qquad A_k = t y_k$$
$$A_{k-1} = t y_{k-1}$$

Equilibrium requires that at any ridge $\Sigma X = 0$. Thus we obtain

$$\tfrac{1}{6} A_{k+1}\sigma_{k+1} + \tfrac{1}{3}(A_{k+1} + A_k)\sigma_k + \tfrac{1}{6} A_{k-1}\sigma_{k-1} = T_{k0} + \sum_{k=0}^{k=n} M_{ky} T_{ki} \tag{436}$$

where $T_{k0} = P_k L^2/8 y_k$ and $T_{ki} = P_{ki} L^2/\pi^2 y_k$ due to $M_{ky} = 1.0$ $(k, i = 1, \ldots, n-1)$.

The requirement to satisfy the boundary conditions $k = 0$ to $k = n$ is that after the deformations of the panels the ridge angles (Fig. 168) remain unchanged. Thus,

$$\Delta \alpha_k = \Delta \gamma_k = 0$$

This condition is satisfied when

$$E \sum_{k=1}^{k=n-1} \sigma_k \vartheta_{ki} + E \sum_{k}^{k+2} M_{ky} \delta_{ki} = 0$$

For sinusoidal loading,

$$E \vartheta_{ki} = T_{ki}$$

Thus:

$$\sum_{k=1}^{k=n-1} \sigma_k T_{ki} + E \sum_{k}^{k+2} M_{ky} \delta_{ki} = 0 \tag{437}$$

where σ_k is the fiber stress due to M_{ky}, ϑ_{ki} the rotation of the panel k due to $\sigma_k = 1$, and δ_{ki} the panel end rotations due to $M_{ky} = 1$.

The general equations (436) and (437) for the polygon linear shell illustrated in Fig. 167 are written in matrix form [Eq. (438)].

Matrix:

	σ_0	σ_{k-1}	σ_k	σ_{k+1}	$\overline{M_{k-1,y}}$	$\overline{M_{k,y}}$	$\overline{M_{k+1,y}}$	T_{k0}
0	$\frac{1}{3}A_{k-1}$	$\frac{1}{6}A_{k-1}$	–	–	$-T_{0,k-1}$	$-T_{0,k}$	–	T_0
$k-1$	$\frac{1}{6}A_{k-1}$	$\frac{1}{3}(A_{k-1}+A_k)$	$\frac{1}{6}A_k$	–	$-T_{k-1,k-1}$	$-T_{k-1,k}$	$-T_{k-1,k+1}$	$T_{k-1,0}$
k	–	$\frac{1}{6}A_k$	$\frac{1}{3}(A_k+A_{k+1})$	$\frac{1}{6}A_{k+1}$	$-T_{k,k-1}$	$-T_{k,k}$	$-T_{k,k+1}$	$T_{k,0}$
$k+1$	–	–	$\frac{1}{6}A_{k+1}$	$\frac{1}{3}A_k$	$-T_{k+1,k-1}$	$-T_{k+1,k}$	$-T_{k+1,k+1}$	$T_{k+1,0}$
$k-1,y$	–	$-T_{k-1,k-1}$	$-T_{k-1,k}$	$-T_{k-1,k+1}$	$E\delta_{k-1,k-1}$	$E\delta_{k-1,k}$	–	0
k,y	–	$-T_{k,k-1}$	$-T_{k,k}$	$-T_{k,k+1}$	$E\delta_{k,k-1}$	$E\delta_{k,k}$	$E\delta_{k,k+1}$	0
$k+1,y$	–	$-T_{k+1,k-1}$	$-T_{k+1,k}$	$-T_{k+1,k+1}$	–	$E\delta_{k+1,k}$	$E\delta_{k+1,k+1}$	0

(438)

Solution of the matrix gives directly the stresses σ_k in ridges and transversal moments due to deformations M_{ky}.

P_{ki}, δ_{ki} and T_{ki} values:

The redundants $X_k = 1$ ($k, i = 1, 2, \ldots, n$) in the principal system cause panel loads P_{ki} and panel end rotations δ_{ki}, resulting in T_{ki} values. These values all follow the function $f(x) = \sin(\pi x/L)$. In the following, they will be determined for $x = L/2$.

The moments M_{ki} and shear T_{ki} due to panel load P_{ki} and the ridge rotation δ_{ki} due to the redundants $X_{k,i}$ have to be affined to the elastic line of the ridges. Thus,

$$M_{ki} = \pm \frac{L^2}{\pi^2} P_{ik} \qquad m = \frac{L^2}{\pi^2} = \text{const}$$

$$T_{ki} = \pm \frac{M_{ki}}{y_k} = \pm m \frac{\Sigma P_{ik}}{y_k} \tag{439}$$

280   Theory of Structures

In accordance with Eqs. (8) and Table 3 for E and I_s = constant,

$$EI_s \delta_{ki} = \sum \int X_i X_k \, dy \tag{440}$$

$$EI_s \delta_{kk} = \sum \int X_k^2 \, dy$$

For the end panel (0–1) the moment due to X_1 must be balanced by torsion (see "Torsion," Sec. 7):

$$\frac{d^2 \delta_{11}}{dx^2} = -\frac{1}{GI_T} \sin \frac{\pi x}{L} \tag{441}$$

The δ_{11} is obtained by direct integration:

$$E\delta_{01} = \frac{L^2}{\pi^2} \frac{1}{GI_T} = 0.235 \frac{L^2}{I_T} \quad G = \frac{E}{2(1+\nu)} \,;\; \nu = 0.167$$

$$EI_s \delta_{11} = EI_s \delta_{01} + \int_1^2 X_1^2 \, dy \tag{442}$$

$X_{k+1} = 1k'$ (symmetry point):

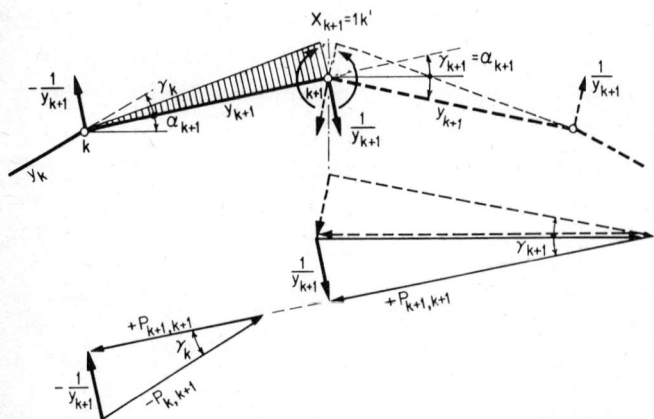

Fig. 171

$$P_{k+1,k+1} = \frac{1}{y_{k+1}\tan\gamma_{k+1}}$$
$$P_{k+1,k+1} = \frac{1}{y_{k+1}\tan\gamma_k} \quad \bigg| \ \Sigma P_{k+1,k+1}$$
$$P_{k,k+1} = -\frac{1}{y_{k+1}\sin\gamma_k}$$
(443)

$$EI_s \delta_{k+1,k+1} = \tfrac{1}{3} y_{k+1}$$
$$EI_s \delta_{k+1,k} = \tfrac{1}{6} y_{k+1}$$
(444)

$X_k = 1\,k'$:

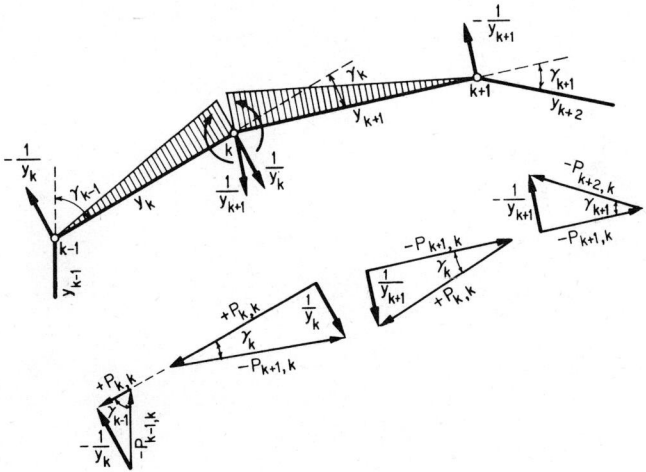

Fig. 172

$$P_{k+2,k} = -\frac{1}{y_{k+1}\sin\gamma_{k+1}}$$
$$P_{k+1,k} = -\frac{1}{y_{k+1}\tan\gamma_{k+1}}$$
$$P_{k+1,k} = -\frac{1}{y_{k+1}\tan\gamma_k} \quad \bigg| \ \Sigma P_{k+1,k}$$
$$P_{k+1,k} = -\frac{1}{y_k \sin\gamma_k}$$
(445)

$$P_{k,k} = \frac{1}{y_{k+1}\sin\gamma_k}$$

$$P_{k,k} = \frac{1}{y_k \tan\gamma_k} \quad \bigg| \quad \Sigma P_{k,k}$$

$$P_{k,k} = \frac{1}{y_k \tan\gamma_{k-1}}$$

$$P_{k-1,k} = -\frac{1}{y_k \sin\gamma_{k-1}} \tag{445}$$
(Cont.)

$$EI_s \delta_{k,k+1} = \tfrac{1}{6} y_{k+1}$$

$$EI_s \delta_{kk} = \tfrac{1}{3}(y_k + y_{k+1}) \tag{446}$$

$$EI_s \delta_{k,k-1} = \tfrac{1}{6} y_k$$

$X_1 = 1 k'$ (boundary):

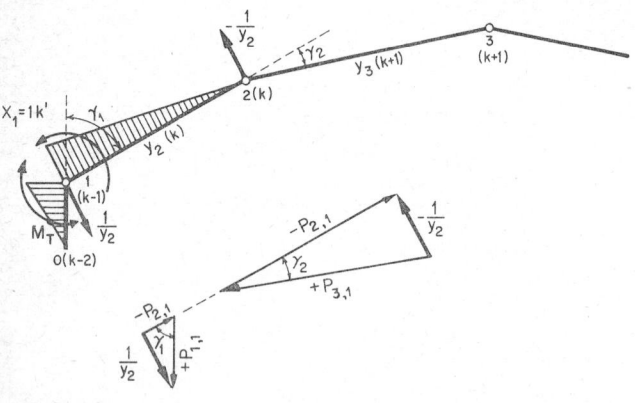

Fig. 173

$$P_{31} = \frac{1}{y_2 \sin\gamma_2}$$

$$P_{2,1} = -\frac{1}{y_2 \tan\gamma_2} \quad \bigg|$$

$$P_{2,1} = -\frac{1}{y_2 \tan\gamma_1} \quad \bigg| \quad \Sigma P_{2,1} \tag{447}$$

$$P_{1,1} = \frac{1}{y_2 \sin\gamma_1}$$

$$EI_s\delta_{21} = \tfrac{1}{6}y_2$$

$$EI_s\delta_{11} = \tfrac{1}{3}y_2 + 0.235 L^2 \frac{I_s}{I_T} \tag{448}$$

T_{ki} *values:*

   – compression fiber of panel
   + tension fiber of panel

$$X_{k+1} = 1k': \quad \left|\begin{array}{l} T_{k+1,k+1} = -m\,\dfrac{\Sigma P_{k+1,k+1}}{y_{k+1}} \\[4pt] T_{k,k+1} = m\left(-\dfrac{P_{k,k+1}}{y_k} + \dfrac{P_{k+1,k+1}}{y_{k+1}}\right) \\[4pt] T_{k-1,k+1} = m\,\dfrac{P_{k,k+1}}{y_k} \end{array}\right| \quad \sum_{i=k-1}^{i=k+1} T_{i,k+1} = 0 \tag{449}$$

$$X_k = 1k': \quad \left|\begin{array}{l} T_{k+2,k} = m\,\dfrac{P_{k+2,k}}{y_{k+2}} \\[4pt] T_{k+1,k} = m\left(-\dfrac{\Sigma P_{k+1,k}}{y_{k+1}} - \dfrac{P_{k+2,k}}{y_{k+2}}\right) \\[4pt] T_{k,k} = -m\left(\dfrac{\Sigma P_{k,k}}{y_k} - \dfrac{P_{k+1,k}}{y_{k+1}}\right) \\[4pt] T_{k-1,k} = m\left(-\dfrac{P_{k,k}}{y_k} - \dfrac{P_{k-1,k}}{y_{k-1}}\right) \\[4pt] T_{k-2,k} = m\,\dfrac{P_{k-1,k}}{y_{k-1}} \end{array}\right| \quad \sum_{i=k-2}^{i=k+2} T_{i,k} = 0 \tag{450}$$

$$X_1 = 1k': \quad \left|\begin{array}{l} T_{3,1} = -m\,\dfrac{P_{3,1}}{y_3} \\[4pt] T_{2,1} = m\left(-\dfrac{\Sigma P_{2,1}}{y_2} + \dfrac{P_{3,1}}{y_3}\right) \\[4pt] T_{1,1} = -m\left(\dfrac{\Sigma P_{2,1}}{y_2} + \dfrac{P_{1,1}}{y_1}\right) \\[4pt] T_{0,1} = m\,\dfrac{P_{1,1}}{y_1} \end{array}\right| \quad \sum_{i=0}^{3} T_{i,1} = 0 \tag{451}$$

$EI_2 \delta_{k0}$ *values:*

All uniform loads that the panels are subjected to are computed as acting normal to the panel (y_k). Thus,

Dead load: $\quad w_{kz} = w_D \cos\alpha_k \quad w_{ky} = w_D \sin\alpha_k \quad\quad w_{kx} = 0$

Live load: $\quad w_{kz} = w_L \cos^2\alpha_k \quad w_{ky} = w_L \sin\alpha_k \cos\alpha_k \quad w_{kx} = 0$

$$\tag{452}$$

In accordance with Eqs. (8) and making use of the integration Table 3, the ridge rotations due to external loading w, P are

$$EI_s \delta_{k0} = \sum_k^{k+1} \int_0^{y_k} M_{k0} X_k \, dy \quad M_{k0} = \frac{w_{kz} y_k^2}{8} \tag{453}$$

$$= \tfrac{1}{3}(M_{k0} y_k + M_{k+1,0} y_{k+1})$$

For the case $\Delta_k = 0$ (nonyielding ridges), the first part of Eq. (437) is zero ($\Delta_{ki} = 0$). Continuity then requires that the second part of Eq. (437), $M_{ky} \to X_k$, satisfy the condition

$$\sum X_{ki} \delta_{ik} = \delta_{k0} \tag{454}$$

These equations, written in matrix form [Eq. (455)] yield the $M_{ky0} = -X_k$, as obtained in accordance with membrane theory.

Matrix:

	X_{k-2}	X_{k-1}	X_k	X_{k+1}	X_{k+2}	δ_{k0}
$k-2$	$-\delta_{k-2,k-2}$	$-\delta_{k-2,k-1}$	—	—	—	$\delta_{k-2,0}$
$k-1$	$-\delta_{k-1,k-2}$	$-\delta_{k-1,k-1}$	$-\delta_{k-1,k}$	—	—	$\delta_{k-1,0}$
k	—	$-\delta_{k,k-1}$	$-\delta_{k,k}$	$-\delta_{k,k+1}$	—	$\delta_{k,0}$
$k+1$	—	—	$\delta_{k+1,k}$	$-\delta_{k+1,k+1}$	$-\delta_{k+1,k+2}$	$\delta_{k+1,0}$
$k+2$	—	—	—	$-\delta_{k+2,k+1}$	$-\delta_{k+2,k+2}$	$\delta_{k+2,0}$

$$\tag{455}$$

The ridge loads $\bar{R}_k$ as a function of the redundants X_k are

$$\bar{R}_k = \pm \frac{X_k - X_{k-1}}{c_k} + \frac{X_k - X_{k+1}}{c_{k+1}}$$

$$\bar{R}_{kx} = R_k \sin \frac{\pi x}{L}$$
(456)

where the minus sign applies for edge and symmetry-point ridge loads.

To satisfy the equilibrium as well as the boundary conditions [Eqs. (436) and (437)] in a simple operation, the $\bar{R}_k$ and ridge loads R_{k0} due to w_{kz} must be added to obtain the T_{k0} values for Eq. (438):

$$\sum R_k = R_{k0} + \bar{R}_k$$

$$\sum_0^n \bar{R}_k = 0$$
(457)

ΣT_k values:

The panel loads P_k due to ΣR_k are (Fig. 174)

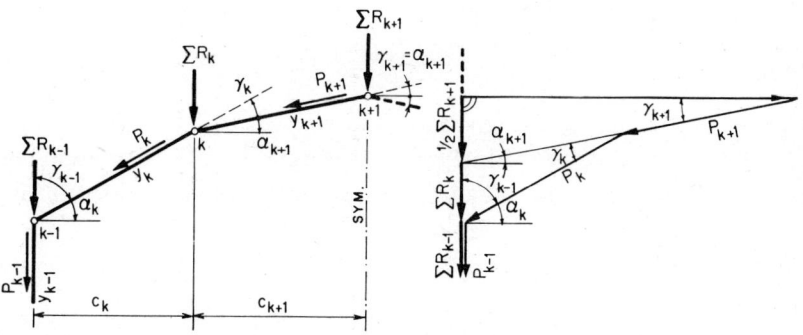

Fig. 174

$$P_{k+1} = \frac{1}{2}\sum R_{k+1} \frac{1}{\sin \gamma_{k+1}} - \sum R_k \frac{\cos \alpha_k}{\sin \gamma_k}$$

$$P_k = \sum R_k \frac{\cos \alpha_{k+1}}{\sin \gamma_k} - \sum R_{k-1} \frac{\cos \alpha_{k-1}}{\sin \gamma_{k-1}}$$
(458)

$$P_{k-1} = \sum R_{k-1}$$

The shear forces T_{k0} are computed from the moment M_{k0} due to the panel loading P_k

$$T_{k0} = \frac{M_{k0}}{y_k} = \frac{P_k L^2}{8 y_k} \qquad (459)$$

$$T_{k0,x} = T_{k0} \sin \frac{\pi x}{L}$$

at symmetry points $\Sigma T_k = 0$.

Introducing T_{k0} into Eq. (438), the solution of the matrix gives directly the normal stresses σ_k in ridges and transversal moments due to deformations M_{ky}.

Deflections

The displacements of ridges are computed by Eqs. (429), whereas the Δy_k values are computed by Eq. (111):

$$\Delta y_k = \frac{1}{EI_s} \int_{x=0}^{x=L} M_{k0} \overline{M}_{k1} \, dx \qquad (460)$$

The panel moments M_{k0} are given by the stresses σ_k [Eq. (438)] and $\overline{M}_{k1} = 1.0 L/4$. Making use of integration formulas (Table 3) or applying numerical integration, the obtained Δy_k values will be of sufficient accuracy.

The horizontal (Δ_{H0}) and vertical (δ_0) displacements of the free edge $(0, n)$ will be

$$\Delta_{H0} = \frac{\Delta y_2}{\cos \alpha_2} - \frac{\Delta y_1}{\tan \alpha_2} + \delta_{1y_1} \qquad (461)$$

$$\delta_0 = \Delta \delta_1$$

At the valley ridge of adjoining polygonlinear shells, the vertical and horizontal displacements are (Fig. 175)

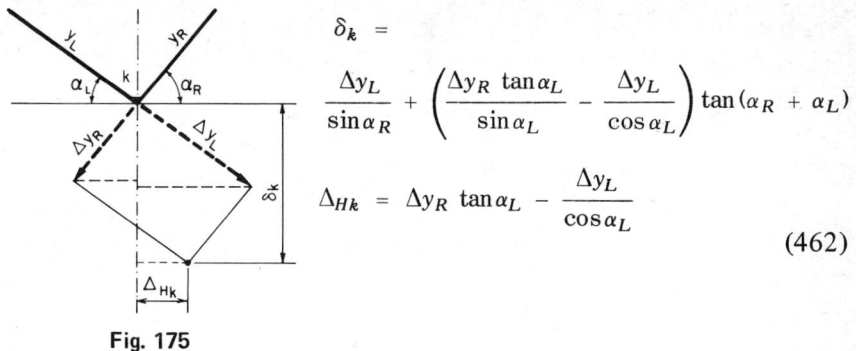

$$\delta_k = \frac{\Delta y_L}{\sin \alpha_R} + \left(\frac{\Delta y_R \tan \alpha_L}{\sin \alpha_L} - \frac{\Delta y_L}{\cos \alpha_L} \right) \tan(\alpha_R + \alpha_L)$$

$$\Delta_{Hk} = \Delta y_R \tan \alpha_L - \frac{\Delta y_L}{\cos \alpha_L} \tag{462}$$

Fig. 175

For symmetric shells $\Delta_{Hk} = 0$ and $\alpha_L = \alpha_R$, Eqs. (462) simplify to

$$\delta_k = \frac{\Delta y_L}{\sin \alpha_L} \tag{463}$$

These equations apply also for the crown of the individual shells.

For illustration of this method, the polygonlinear shell illustrated in Fig. 176 will be analyzed.

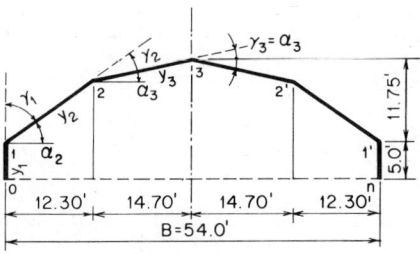

Fig. 176

Geometric data:

$y_1 = 5.0$ ft $y_2 = y_3 = 15.0$ ft
$\alpha_1 = 90°$ $\alpha_2 = 35°$ $\alpha_3 = 12°$ $L = 120.0$ ft
$\gamma_1 = 55°$ $\gamma_2 = 23°$ $\gamma_3 = \alpha_3 = 12°$ $B = 54.0$ ft
$c_2 = 12.30$ ft $c_3 = 14.70$ ft

$t_1 = 10$ in. $t_2 = t_3 = 4$ in. $I_s = 0.0031$ ft^4
$A_1 \simeq 4.2$ ft^2 $A_2 = A_3 \simeq 5.0$ ft^2 $I_T = 0.865$ ft^4

Principal system:

Hinged polygon.
Redundants: $\sigma_{0x}, \sigma_{1x}, \sigma_{2x}, \sigma_{3x}$ and M_{1y}, M_{2y}, M_{3y}

Forces, moments, and deformations in principal system:

In accordance with Eqs. (443) to (451), the panel load P_{ik}, moments, shear T_{ik}, and ridge rotation δ_{ik} due to $X_1 = 1$, $X_2 = 1$ and $X_3 = 1$ are as follows.

P_{ik} values:

$$y_2 = y_3 = y = 15.0 \text{ ft}$$

$$y\Sigma P_{33} = \frac{1}{\tan \gamma_3} + \frac{1}{\tan \gamma_2} = 4.710 + 2.358 = 7.068$$

$$yP_{23} = -\frac{1}{\sin \gamma_2} = -2.558$$

$$y\Sigma P_{32} = -\left(\frac{1}{\tan \gamma_3} + \frac{1}{\tan \gamma_2} + \frac{1}{\sin \gamma_2}\right) = -7.068 - 2.558 = -9.626$$

$$y\Sigma P_{22} = \left(\frac{1}{\sin \gamma_2} + \frac{1}{\tan \gamma_2} + \frac{1}{\tan \gamma_1}\right) = 2.558 + 2.358 + 0.702$$
$$= 5.618$$

$$yP_{12} = -\frac{1}{\sin \gamma_1} = -1.221$$

$$yP_{31} = \frac{1}{\sin \gamma_2} = 2.558$$

$$y\Sigma P_{21} = -\left(\frac{1}{\tan \gamma_2} + \frac{1}{\tan \gamma_1}\right) = -2.358 - 0.702 = -3.060$$

$$y_1 P_{11} = \frac{1}{\sin \gamma_1} = 1.221$$

T_{ik} values:

$$m = \frac{L^2}{\pi^2} = \frac{120.0^2}{3.14^2} = 1{,}455 \qquad \frac{m}{y^2} = \frac{1{,}455}{15.0^2} = 6.48$$

$$T_{33} = -6.48(y\Sigma P_{33}) = -45.90$$
$$T_{23} = 6.48(-y\Sigma P_{23}) - T_{33} = +16.58 + 45.90 = 62.48$$
$$T_{13} = 6.48(y\Sigma P_{23}) = -16.55$$

$T_{32} = 6.48(-y\Sigma P_{32}) = +62.45$

$T_{22} = -6.48(y\Sigma P_{22}) - T_{32} = -36.50 - 62.45 = -98.95$

$T_{12} = -6.48(y\Sigma P_{22}) - T_{02} = 36.50 + 23.70 = 60.20$

$T_{02} = 1{,}455 \dfrac{y\Sigma P_{12}}{yy_1} = -23.70$

$T_{31} = -6.48(y\Sigma P_{31}) = -16.55$

$T_{21} = 6.48(-y\Sigma P_{21}) - T_{31} = 19.85 + 16.55 = 36.40$

$T_{11} = -6.48(-y\Sigma P_{21}) - T_{01} = -19.85 - 23.70 = -43.55$

$T_{01} = 1{,}455 \dfrac{y\Sigma P_{11}}{yy_1} = 23.70$

$EI_s \delta_{ik}$ values:

$EI_s\delta_{33} = -\tfrac{1}{3}y_3 = -\tfrac{1}{3}15.0 = -5.00 \quad\quad E\delta_{33} = -1.610 \times 10^3$

$EI_s\delta_{32} = -\tfrac{1}{6}y_3 = -\tfrac{1}{6}15.0 = -2.50 \quad\quad E\delta_{33} = -0.805 \times 10^3$

$\quad\quad\quad = EI_s\delta_{23} \quad\quad\quad\quad\quad\quad\quad = E\delta_{33}$

$EI_s\delta_{22} = -\tfrac{1}{3}(y_3 + y_2) = -\tfrac{1}{3}(15.0 + 15.0) = -10.00$

$\quad\quad\quad\quad\quad\quad\quad\quad\quad\quad\quad\quad E\delta_{22} = -3.230 \times 10^3$

$EI_s\delta_{21} = -\tfrac{1}{6}y_2 = -\tfrac{1}{6}15.0 = -2.50 \quad\quad E\delta_{21} = -0.805 \times 10^3$

$\quad\quad\quad = EI_s\delta_{12} \quad\quad\quad\quad\quad\quad\quad = E\delta_{12}$

$EI_s\delta_{11} = -\tfrac{1}{3}y_2 + 0.235 L^2 \dfrac{I_s}{I_T}$

$\quad\quad\quad = -\tfrac{1}{3}15.0 - 0.235 \times 120.0^2 \dfrac{0.0031}{0.865}$

$\quad\quad\quad = -5.00 - 12.13 = -17.13 \quad\quad E\delta_{11} = -5.525 \times 10^3$

$EI_2 \delta_{k0}$ values:

$w_D = 0.060 \text{ ksf}$

$w_{3z} = w_D \cos\alpha_3 = 0.060 \times 0.978 = 0.0587 \text{ ksf}$

$w_{2z} = w_D \cos\alpha_2 = 0.060 \times 0.819 = 0.0491 \text{ ksf}$

$w_{1D} = (0.83 \times 0.150 + 0.005)5.00 \simeq 0.65 k$

290 Theory of Structures

$$EI_s\delta_{30} = \int X_3 M_{30}\, dy = \frac{1.0}{3} M_{30}\, y_3 \quad M_{30} = \frac{0.0587 \times 15.0^2}{8} = 1.65\, k'$$

$$= \frac{1.65 \times 15.0}{3} = 8.25$$

$$EI_s\delta_{20} = \sum_{k=0}^{k=3} \int X_2 M_{k0}\, dy = \frac{1.0}{3}(M_{30}y_3 + M_{20}y_2) \quad M_{20} = 1.38\, k'$$

$$= \frac{1.65 + 1.38}{3}\, 15.0 = 15.15$$

$$EI_s\delta_{10} = \int X_1 M_{10}\, dy = \frac{1.0}{3}\, 1.38 \times 15.0 = 6.90 \quad M_{10} = 0$$

Matrix:

	X_1	X_2	X_3	$-\delta_{k0}$	
1	17.13	2.50	—	6.90	$X_1 = 0.228\, k'$
2	2.50	10.00	2.50	15.15	$X_2 = 1.194\, k'$
3	—	2.50	5.00	8.25	$X_3 = 1.053\, k'$

Ridge loads [Eqs. (456) and (457)]:

$$\Sigma R_k = R_{k0} + \overline{R}_k$$

$$\Sigma R_1 = 0.65 + \tfrac{1}{2} \times 0.060 \times 15.0 - \frac{1.194 - 0.228}{12.30} = 1.022\, k/\text{ft}$$

$$\Sigma R_2 = 0.060 \times 15.0 + \frac{1.194 - 0.228}{12.30} + \frac{1.194 - 1.053}{14.70}$$

$$= 0.988\, k/\text{ft}$$

$$\Sigma R_3 = \tfrac{1}{2} \times 0.060 \times 15.00 - \frac{1.194 - 1.053}{14.70} = 0.440\, k/\text{ft}$$

Panel loads [Eqs. (458)]:

$$P_1 = \Sigma R_1 = 1.022\, k/\text{ft}$$

$$P_2 = \Sigma R_2\, \frac{\cos \alpha_3}{\sin \gamma_2} = 0.988\, \frac{0.978}{0.391} = 2.465\, k/\text{ft}$$

$$P_3 = \Sigma R_3\, \frac{1}{\sin \gamma_3} - \Sigma R_2\, \frac{\cos \alpha_2}{\sin \gamma_2} \simeq 0.050\, k/\text{ft}$$

ΣT_k values:

$$T_0 = \frac{P_1 L^2}{8 y_1} = \frac{1.022 \times 120.0^2}{8 \times 5.00} = 368 k$$

$$T_1 = -\Sigma T_0 + \frac{P_2 \times L^2}{8 \times y_2} = -368 + \frac{2.465 \times 120.0^2}{8 \times 15.0} = -72 k$$

$$T_2 = -\frac{2.465 \times 120.0^2}{8 \times 15.0} + \frac{0.05 \times 120.0^2}{8 \times 15.0} \simeq -291 k$$

$$\Sigma T_3 = 0$$

Matrix [Eq. (438)]:

	σ_0	σ_1	σ_2	σ_3	M_{1y}	M_{2y}	M_{3y}	ΣT_k
0	1.400	0.700	–	–	−23.70	+23.70	–	368
1	0.700	3.067	0.833	–	+43.55	−60.20	+16.55	−72
2	–	0.833	3.333	0.833	−36.40	+98.95	−62.48	−291
3	–	–	0.833	1.667	+16.55	−62.45	+45.90	0
1y	–	+43.55	−36.40	+16.55	-5.525×10^3	-0.805×10^3	–	0
2y	–	−60.20	+98.95	−62.48	-0.805×10^3	-3.230×10^3	-0.805×10^3	0
3y	–	+16.55	−62.45	+45.90	–	-0.805×10^3	-1.610×10^3	0

The solution of the equation in matrix results:

$\sigma_0 = +299.454$ ksf $\sigma_1 = -81.144$ ksf $\sigma_2 = -48.533$ ksf

$\sigma_3 = -3.191$ ksf

$M_{1y} = -0.3174 k'$ $M_{2y} = -0.0830 k'$ $M_{3y} = +0.9980 k'$

$$\sum M_{ky} = -(X_k + M_{ky})$$

The σ_x stresses and transversal moment at $x = L/2$ for yielding and nonyielding ridges are illustrated in Fig. 177.

As can be seen from the moment and stress diagrams (Fig. 177), the influence of yielding of the ridges upon transversal moments as well as on normal stresses in the span direction is considerable.

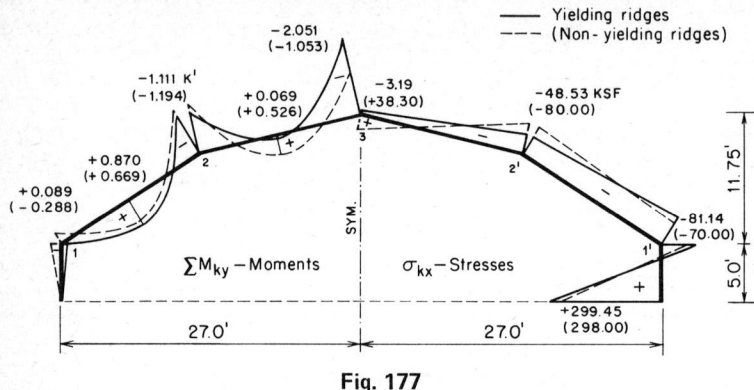

Fig. 177

Continuous Polygonlinear Shells

For a rough estimation of the stresses in panels, the intermediate supports C_m are computed as for continuous beams acting in the principal system as reverse ridge loads (Fig. 178). If the moment diagram deviates from the parabolic of sinusoidal moment, correction factors $\Delta/\Delta f(x)$, as stated before, must be used. Further, the $\bar{R}_k$ values can be taken zero. The panel loads P_k (concentrated loads) will be computed by Eqs. (458), and the shear T_{k0} by Eqs. (459). The T_{k0} values (see the example) introduced into the matrix [Eq. (438)] give the σ_{kx} stresses and M_{ky} moments for C_m loading. For example,

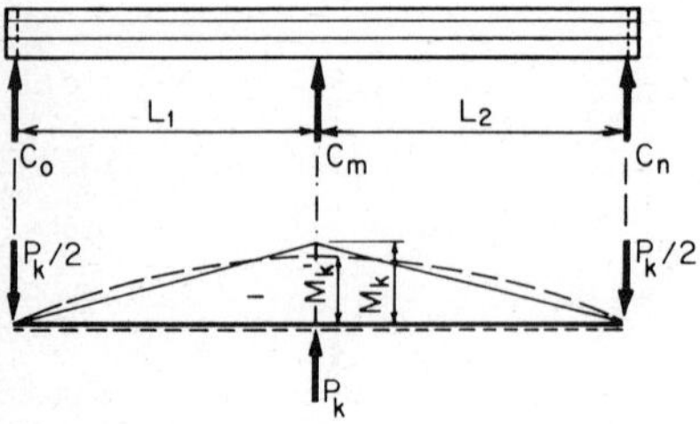

Fig. 178

$$L_1 = L_2$$

$$C_m \simeq 0.60 \sum_0^n wc_k(L_1 + L_2)$$

$$P_k = c_{mk}\frac{\cos\alpha_{k+1}}{\sin\gamma_k} - c_{m,k-1}\frac{\cos\alpha_{k-1}}{\sin\gamma_{k-1}}$$

$$M_{k,\max} = \frac{P_k(L_1 + L_2)}{4}$$

$$M'_{k,\max} = 0.822\, M_{k,\max}$$

$$T_k = \frac{M'_k}{\pi^2 y_k}$$

(464)

The final results—a single shell acted upon by R_k and C_m—are obtained by superposition.

In the case of multibay polygonlinear shells, the only difference from single shells is in the δ_{11} values, because for the intermediate marginal panels $(0-n)$, the torsion is zero. Otherwise, the values of adjacent panels will be dealt with as symmetry points.

Prestressing

Normally the marginal panels may require prestressing to avoid crack formation, to obtain economy, and to control deflections. Here also, the $\overline{R}_k$ values can be taken zero. The panel moments due to the prestressing force (H_{pr0}) produce T_{1pr} and T_{0pr} values (Fig. 179).

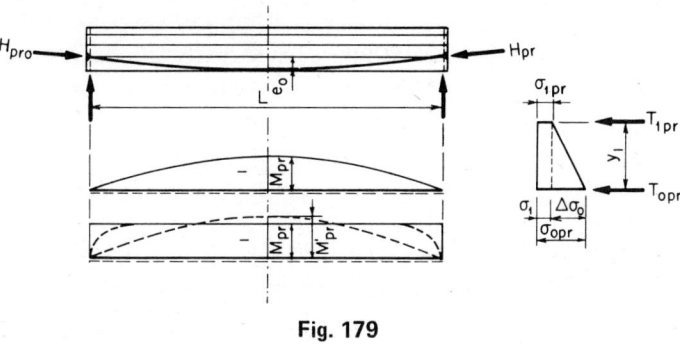

Fig. 179

Parabolic tendons:
Straight tendons:

$$M_{pr} = H_{pr} e_0$$
$$M'_{pr} = 1.23 H_{pr} e_0 \quad \bigg| \quad M_{prx} = M_{pr} \sin \frac{\pi x}{L}$$

$$T_{1pr} = -\tfrac{1}{2} A_1 \sigma_1 + \tfrac{1}{3} \Delta\sigma_0 A_1$$
$$T_{0pr} = -\tfrac{1}{2} A_1 \sigma_1 + \tfrac{2}{3} \Delta\sigma_0 A_1 \quad \bigg| \quad A_1 = t y_1$$

(465)

The T_{kpr} values will be introduced into Eq. (438). The solution of the matrix gives the stresses σ_{xpr} and transversal moments due to deformations M_{kypr}. The final stresses and transversal moments will again be obtained by superposition.

28. CURVILINEAR CYLINDRICAL SHELLS

The peculiar feature of curvilinear shells is the transmission of loads mainly by direct stresses with relatively small flexural stresses. As the material is most efficiently used when only direct stresses are involved in carrying action, relatively large areas can be bridged economically by thin shells without any intermediate supports.

In general, the analysis of curvilinear shells is similar to the analysis of ordinary beams having a curved cross section. The main differences in structural behavior between the beam and the cylindrical shell lie in the normal stress distribution over the cross section in the vertical direction and in the distinctive action of the shell in the transversal direction. These two basic differences are explained by the fact that the cross section in the transversal direction is rigid in the case of a beam but rather elastic in the case of a shell. However, there are cases in which these deviations are of secondary order and can be disregarded. In these cases, the reactions and internal forces in the shell due to uniform external loads can be determined with sufficient accuracy by the three equations of equilibrium or so-called membrane theory.

The basic requirements for applicability of membrane theory are:

1. The curvature of the cross section of the shell (meridian curvature) should have vertical or almost vertical tangents at its longitudinal edge ($\alpha_0 \simeq 90°$).
2. The boundary conditions must be satisfied.

In cases in which these two basic requirements cannot be met—for architectural, practical, or any other reasons—relatively large flexural stresses are developed in the shell, which very often penetrate from one longitudinal boundary to the other and entirely control the design. In such cases, use must be made of the bending theory of shells.

The bending theory of cylindrical shells is basically simple. First, the shell is analyzed in accordance with membrane theory, which satisfies the equilibrium but does not comply with the boundary requirements. The second phase of the analysis is to apply line loads as redundants along the longitudinal edges of the shell to satisfy the boundary conditions. The line loads commonly cause direct as well as flexural stresses in the shell. The final stresses are obtained by superposition of the two phases of the analysis. Regardless of how simple the bending theory is in principle, its application in actual design is rather complicated and time-requiring, mainly because the redundants are not constant but are functions of x which must be expressed in Fourier series to equalize and balance the normal force N_{α_0} and longitudinal shear force $N_{x\alpha_0}$ given by the membrane theory. In the case of multiple shells, the edge rotation ϑ_0 and lateral displacement Δ_H are zero. To satisfy these conditions, a moment and horizontal force normal to the edge must be applied as redundants in addition to the $\overline{N}_{\alpha_0} = -N_{\alpha_0}$ and $\overline{N}_{x\alpha_0} = -N_{x\alpha_0}$ redundants.

Membrane Theory

As a rule, the thickness t of the shell is taken constant and the loading in the longitudinal direction of the shell is assumed to be zero ($w_x = 0$). The vertical load to which the shell is subjected can be represented by the components w_y and w_z, which are assumed to be continuous functions of α (Fig. 180). The equilibrium requires that [see Eqs. (386) to (390)]

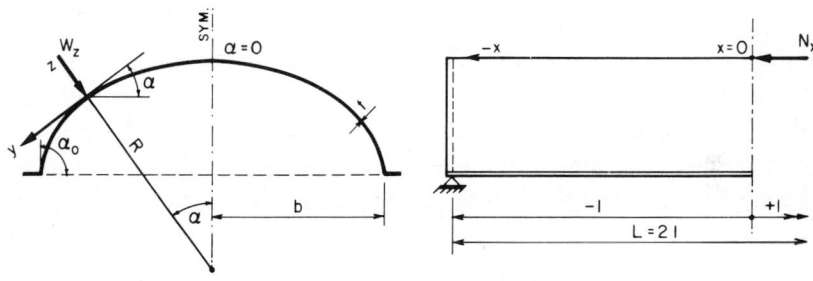

Fig. 180

$$\frac{\partial N_x}{\partial x} + \frac{\partial N_{x\alpha}}{R\partial \alpha} + w_x = 0$$

$$\frac{\partial N_{x\alpha}}{\partial x} + \frac{\partial N_\alpha}{R\partial \alpha} + w_y = 0$$

$$N_\alpha + Rw_z = 0$$

For symmetric steady loading $w_x = 0$, w_y and w_z are functions of α and independent of x. The shear forces $N_{x\alpha} = N_{\alpha x}$ and normal force N_x can be obtained by direct integration with respect to x. Thus,

$$N_{x\alpha} = -\left(\frac{\partial N_\alpha}{R\partial \alpha} + w_y\right)x + C_1(\alpha)$$

$$= -w_y^* x + C_1(\alpha) \qquad w_y^* = \frac{\partial N_\alpha}{R\partial \alpha} + w_y \qquad (466)$$

$$N_x = \frac{1}{2R}\frac{\partial w_y^*}{\partial \alpha}x^2 - \frac{1}{R}\frac{\partial C_1(\alpha)}{\partial \alpha} + C_2(\alpha) \qquad (466)$$
$$\text{(Cont.)}$$
$$N_\alpha = -Rw_z$$

The values of the integral constants $C_1(\alpha)$ and $C_2(\alpha)$ can be determined from the boundary conditions. Consider, for example, a shell simply supported at each end. In this shell at $x = 0$, $N_{x\alpha} = 0$ (by symmetry), and the value of $C_1(\alpha) = 0$. At $x = \pm l$, $N_x = 0$, so that $C_2(\alpha)$ can now be computed. Thus,

$$C_2(\alpha) = -\frac{1}{R}\frac{\partial}{\partial \alpha}w_y^* \frac{l^2}{2}$$
$$N_x = -\frac{1}{R}\frac{\partial}{\partial \alpha}w_y^*\left(\frac{l^2 - x^2}{2}\right) \qquad (466a)$$

It should be noted that at $x = \pm l$, $N_{x\alpha} \neq 0$ and should be offset by the addition of marginal members placed along these two end sections.

In the case of a shell supported at only one end as a cantilever, at $x = l$ (the free end of the cantilever), $N_x = N_{x\alpha} = 0$ and no marginal member is required. However, at $x = 0$ (the fixed end of the cantilever), $N_x \neq 0$ and $N_{x\alpha} \neq 0$. Obviously, a marginal member is required at the fixed end of the cantilever to offset these forces.

Since N_α is not expressed by a differential equation and is dependent only upon R and local loading w_z, it cannot be influenced by boundary conditions. Roofs are subjected primarily to vertical loading (snow and dead load). Thus, in a shell having vertical tangents at its longitudinal edges (that is, $\alpha_0 = 90°$), $w_z = 0$ and therefore $N_{\alpha_0} = 0$. In this case, the edge of the shell at $\alpha_0 = 90°$ is subjected only to $N_{x\alpha_0}$ forces, which must be offset by longitudinal edge members. Following from this fact, the stress condition can be described adequately by membrane theory only for curvature shapes, such as ellipses, circles, and cycloids, which may have vertical end tangents.

In the following, the shear forces $N_{x\alpha}$ and normal forces N_x, N_α for the three curvature shapes will be given.

Ellipse:

$$R = \frac{a^2 b^2}{(b^2 \sin^2 \alpha + a^2 \cos^2 \alpha)^{3/2}}$$

Dead load:

$$w_x = 0 \quad w_y = w_D \sin\alpha \quad w_z = w_D \cos\alpha \qquad (467)$$

$$N_\alpha = -w_D a^2 b^2 \frac{\cos\alpha}{(b^2 \sin^2\alpha + a^2 \cos^2\alpha)^{3/2}}$$

$$N_{x\alpha} = -w_D x \frac{2b^2 + (b^2 - a^2)\cos^2\alpha}{b^2 \sin^2\alpha + a^2 \cos^2\alpha} \sin\alpha$$

$$N_x = -\frac{w_D l^2}{2}\left(1 - \frac{x^2}{l^2}\right)$$

$$\frac{a^2 b^2 (b^2 \sin^2\alpha + a^2 \cos^2\alpha)^{1/2}}{3a^2 b^2 + 3b^2(a^2 - b^2)\sin^2\alpha - (b^2 \sin^2\alpha + a^2 \cos^2\alpha)^2}$$

Live load:

$$w_x = 0, \quad w_y = w_L \sin\alpha \cos\alpha, \quad w_z = w_L \cos^2\alpha \qquad (468)$$

$$N_\alpha = -w_L a^2 b^2 \frac{\cos^2\alpha}{(b^2 \sin^2\alpha + a^2 \cos^2\alpha)^{3/2}}$$

$$N_{x\alpha} = w_L 3x \frac{a^2 - 2(b^2 \sin^2\alpha + a^2 \cos^2\alpha)}{(b^2 \sin^2\alpha + a^2 \cos^2\alpha)} \sin\alpha \cos\alpha$$

$$N_x = -\frac{w_L l^2}{2} 3\left(1 - \frac{x^2}{l^2}\right)$$

$$\frac{a^2(b^2 \sin^2\alpha - a^2 \cos^2\alpha) + 2(b^2 \sin^2\alpha + b^2 \cos^2\alpha)^2(\cos^2\alpha - \sin^2\alpha)}{a^2 b^2 (b^2 \sin^2\alpha + a^2 \cos^2\alpha)^{1/2}}$$

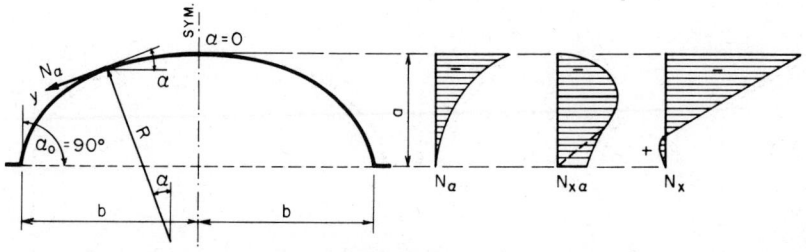

Fig. 181

Normal forces and shear distribution are illustrated in Fig. 181.

Circle:

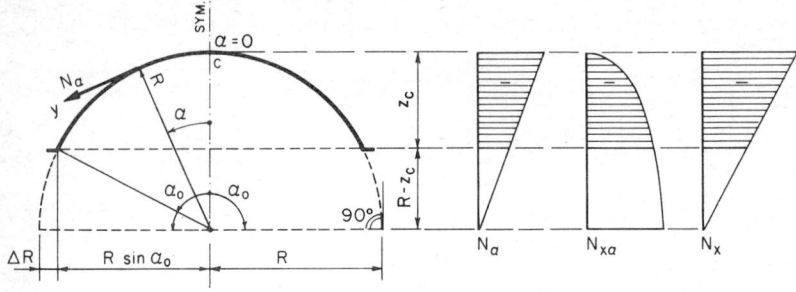

Fig. 182

$\alpha_0 \ne 90°$
$\Delta R = R(1 - \sin\alpha_0)$
$z = R(\cos\alpha - \cos\alpha_0)$

Dead load:

$$w_y = w_D \sin\alpha \qquad w_z = w_D \cos\alpha$$
$$N_\alpha = -w_D R \cos\alpha$$
$$N_{x\alpha} = -w_D 2x \sin\alpha \tag{469}$$
$$N_x = -\frac{w_D l^2}{R}\left(1 - \frac{x^2}{l^2}\right)\cos\alpha$$

Live load:

$$w_y = w_L \sin\alpha \cos\alpha \qquad w_z = w_L \cos^2\alpha$$
$$N_\alpha = -w_L R \cos^2\alpha$$
$$N_{x\alpha} = -w_L 3x \sin\alpha \cos\alpha \tag{470}$$
$$N_x = -\frac{w_L l^2}{2}\frac{3}{R}\left(1 - \frac{x^2}{l^2}\right)(1 - 2\sin^2\alpha)$$

Cycloid:

$$z_c = 2r = \frac{a}{2}$$
$$a = 2z_c = 4r \tag{471}$$

$$y_0 = \frac{z_c}{2}(\Theta - \sin\Theta)$$

$$z_0 = \frac{z_c}{2}(1 - \cos\Theta)$$

$$R = a\cos\alpha$$

$$\alpha = 90°: \quad R = 0$$

(471)
(Cont.)

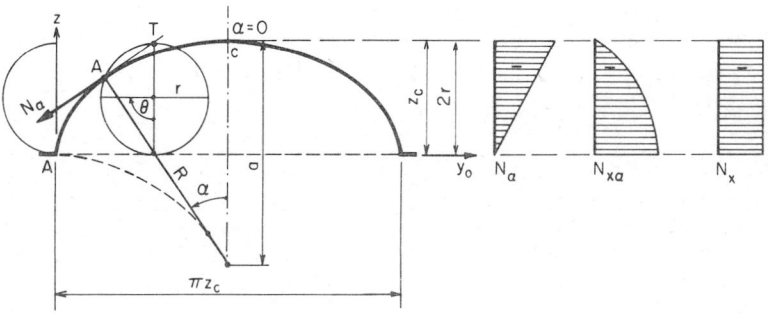

Fig. 183

Dead load:

$$w_y = w_D \sin\alpha \quad w_z = w_D \cos\alpha$$
$$N_\alpha = -w_D a \cos^2\alpha$$
$$N_{x\alpha} = -w_D 3x \sin\alpha \qquad (472)$$
$$N_x = -\frac{w_D l^2}{2}\frac{3}{a}\left(1 - \frac{x^2}{l^2}\right)$$

Live load:

$$w_y = w_L \sin\alpha \cos\alpha \quad w_z = w_L \cos^2\alpha$$
$$N_\alpha = -w_L a \cos^3\alpha$$
$$N_{x\alpha} = -w_L 4x \sin\alpha \cos\alpha \qquad (473)$$
$$N_x = w_L l^2 \frac{2}{a}\left(1 - \frac{x^2}{l^2}\right)\frac{1 - 2\cos^2\alpha}{\cos\alpha}$$

As can be seen from the above analyses and plotted diagrams, the equilibrium, in accordance with membrane theory, is possible only in

300 Theory of Structures

connection with an edge member. The cross section of the shell in the x direction is entirely in compression; only in the case of an ellipse is a relatively small tensile area present at the margin ($\alpha \to \alpha_0$) of the shell. In the case of cycloids, the N_x normal force is not influenced by α and is therefore constant over the entire cross-sectional area. The shear force $N_{x\alpha}$ is zero at the crown and increases toward the edges ($\alpha \to \alpha_0$), except in ellipses, where under dead-load condition it lightly decreases and becomes zero for live load. The unbalanced N_{α_0}, $N_{x(\alpha_0)}$, and $N_{x\alpha_0}$ forces at the shell edge must be carried and balanced by edge members.

For symmetric cross section of the shell, the tensile force T in the edge member at any point x is obtained by integration of the $N_{x\alpha}$ force over the half of the shell with respect to x. Thus,

$$T = -\int_l^x \left(w_y + \frac{1}{R}\frac{\partial N_\alpha}{\partial \alpha}\right) x\, dx$$

$$= -w_y^* \left(\frac{l^2 - x^2}{2}\right) \quad w_y^* = \left(w_y + \frac{1}{R}\frac{\partial N_\alpha}{\partial \alpha}\right)$$

(474)

Applying the equation of equilibrium $\Sigma x = 0$, we obtain

$$T + \int_0^{\alpha_0} N_x R\, d\alpha = 0$$

$$T = -\int_0^{\alpha_0} N_x R\, d\alpha$$

(475)

Thus the total compressive force (ΣN_x) in the cross section of the shell is in equilibrium with the tension (ΣT) in the edge members caused by shear. The main difficulty with membrane theory lies in balancing the differences in strain at the junction of the shell (ϵ_s) and edge members (ϵ_m). The difference in strain is especially pronounced in the case of cycloids. The constant normal force N_x over the entire cross section is balanced by equal tension T in the edge members. The difference in strain is

$$\Delta\epsilon = -\epsilon_s - (+\epsilon_m) \qquad (476)$$
$$= \epsilon_s + \epsilon_m$$

In addition to the "equal-strain" requirement ($\Delta\epsilon = 0$), to maintain membrane-stress condition, the deflection of the shell (δ_s) and edge member (δ_m) must also be equal. To balance somewhat the strain at the junction requires large edge members. As the shell is relatively rigid, its deflection in comparison with that of the edge members is negligible. Thus the shell has to carry almost the total weight of the edge members, which the shell is not able to do without developing large flexural stresses.

In the case of ellipses and semicircles, the strains and deflection can be somewhat balanced and the membrane-stress condition maintained by prestressing of the edge member. But this is practically impossible in cycloids and multiple shells or shells having $\alpha_0 \neq 90°$, regardless of curvature shape.

Prestressing:

Besides the most economical use of material, the membrane theory offers a clarity and simplicity that justify all the efforts involved in finding ways to satisfy the boundary condition of the shell, at least to the extent that the membrane stresses remain the controlling stresses for design.

Basically, there are two efficient means available to accomplish this:

1. Relatively light prestressed edge members
2. Prestressing tendons arranged in the direction of principal tensile stresses (tension trajectories) in the shell

The light edge member is required to avoid increasing the unbalanced forces ($N_\alpha, N_{x\alpha}$) at the edge of the shell by the weight of the marginal member. The arrangement of the tendons along the tensile trajectories is required to carry the relatively large portion of the loads by suspension action to the end supports (Fig. 184). For shells with vertical end tangent ($\alpha_0 = 90°$, $\cos\alpha_0 = 0$) in accordance with Eqs. (469) to (473) $N_{\alpha_0} = 0$, $N_{x\alpha_0} \neq 0$, and N_x—dependent on the curvature—may or may not be zero at the edge. Thus for equilibrium an edge member is required to balance $N_{x\alpha_0}$ and $N_x \neq 0$ values at the margin. If $\alpha_0 \neq 0$, V_{α_0} and H_{α_0} must also be balanced by the edge member (see Fig. 184).

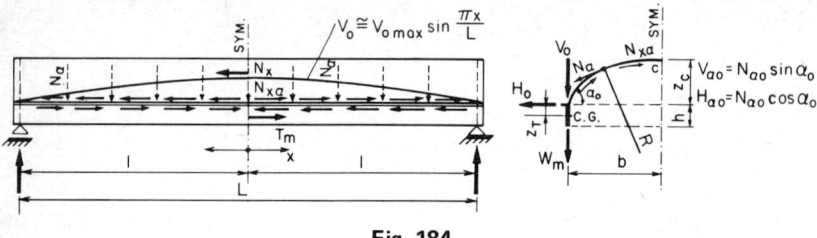

Fig. 184

When an edge beam is not desirable (not recommended because of deflection), the unbalanced shear and normal forces must be fed back to the shell as opposite-sign equivalent live loads, and the membrane state of stresses is lost for reasons already discussed. As a conclusion, to maintain the membrane-stress condition, the unbalanced edge loads, as given by membrane theory, must be carried by the edge beam and prestressing tendons.

As the T force acts at the top of the edge beam, the maximum moment at $x = 0$ in the beam will be

$$\overline{M}_{max} = Te_0 + \frac{w_m L^2}{8} \qquad N_\alpha = 0$$
$$\overline{M}_x = \overline{M}_{max} \sin \frac{\pi x}{L} \tag{477}$$

Also, the edge beam acted upon by a sinusoidal moment $\overline{M}_x$ and T must be designed mainly to satisfy the equal strain conditions at the junction between the edge beam and the shell. Using only the first term of the Fourier series to express the longitudinal distribution of the boundary forces is well justified, as experience with prestressed statically indeterminate systems indicates.

For the first trial, the prestressing force H_{pr} will be selected approximately equal to tensile force T_m. Thus,

$$H_{pr} = -T_m = -w_y^* \int_l^x x\,dx = w_y^* \left(\frac{l^2 - x^2}{2}\right)$$
$$w_y^* = \left(w_y + \frac{1}{R}\frac{\partial N_{\alpha 0}}{\partial \alpha}\right) \tag{478}$$

or

$$H_{pr} = -\int_0^{\alpha_0} N_x R \, d\alpha$$

Balancing the moment $\overline{M}_x$ and deflection δ_x determines the curvature of the tendons (e_{0x}).

The strain of the shell at its junction with the edge member or beam may be estimated for any point x (α = constant) by the following formulas:

$$\epsilon_x = \frac{1}{Et}(N_x - \nu N_\alpha)$$

$$\epsilon_\alpha = \frac{1}{Et}(N_\alpha - \nu N_x) \tag{479}$$

$$\gamma_{x\alpha} = \frac{(1 + \nu)}{Et} N_{x\alpha}$$

The strain of the edge beam is computed from the stresses due to the loading of the beam $(\overline{M}_m T)$ and prestressing (H_{pr}, M_{pr}):

$$\epsilon_x = \frac{1}{E}(\Sigma \sigma_x - \nu \Sigma N_\alpha) \tag{480}$$

The deflection δ_m of the edge beam is computed for the times $t = 0$ and $t = n$ as for ordinary prestressed beams. It is recommended that the size and prestressing of the beam be determined so that its time deflection will be upward or at least zero (see "Deformations," Sec. 11).

Commonly, the edge beams that are desirable are rather slender (L/d_0), and the strain at the junction cannot be equalized by posttensioning of the edge beam alone. In this case, use must be made of suspension action to reduce the values of unbalanced edge forces $(N_{x\alpha}, N_x)$ and tensile stresses in the shell.

To make the most effective use of prestressing, the tendons have to follow the trajectory of the principal tension forces in the shell. The principal forces $(N_{\varphi_{1,2}})$ are defined as forces acting on an angle φ to the x axis on which the shear S_φ is zero and which are in equilibrium with the membrane forces N_x, N_α and shear force $N_{x\alpha}$ (Fig. 185). These forces are

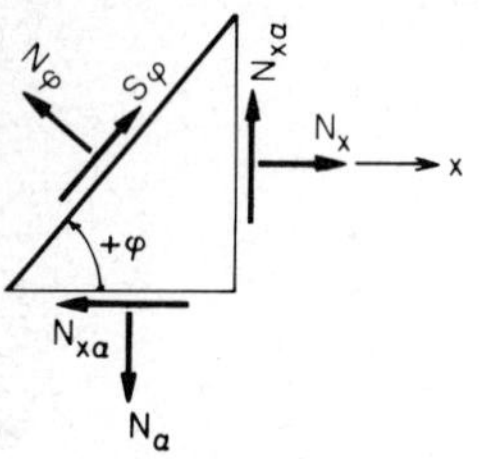

$$N_\varphi = N_x \sin^2\varphi + N_\alpha \cos^2\varphi - 2N_{x\alpha} \sin\varphi \cos\varphi$$
$$S_\varphi = (N_\alpha - N_x) \sin\varphi \cos\varphi + N_{x\alpha}(\cos^2\varphi - \sin^2\varphi)$$
(481)

Fig. 185

In accordance with these equations, the limiting values for $N_{\varphi 1,2}$ and $S_{\varphi 1,2}$ are:

$N_{\varphi 1,2} \to$ maximum when $S_\varphi = 0$; thus,

$$N_{\varphi 1,2} = \frac{N_x + N_\alpha}{2} \pm {}^1\!/_2 \sqrt{(N_x - N_\alpha)^2 + 4N_{x\alpha}^2}$$
$$\tan 2\varphi = \frac{2N_{x\alpha}}{N_x - N_\alpha}$$
(482)

$S_{\varphi 1,2} \to$ maximum when $N_\varphi = 0$; thus,

$$S_{\varphi 1,2} = \pm {}^1\!/_2 \sqrt{(N_x - N_\alpha)^2 + 4N_{x\alpha}^2}$$
$$\tan 2\bar\varphi = \frac{N_x - N_\alpha}{2N_{x\alpha}}$$
$$\bar\varphi - \varphi = 45°$$
(483)

The second principal forces N_{φ_2} and shear S_{φ_2} will be at right angles to the first principal forces $N_{\varphi_1}, S_{\varphi_1}$. The trajectories of the principal forces $N_{\varphi 1,2}$ are shown in Fig. 186.

As can be seen (Fig. 186), relatively large tensile forces caused by shear $N_{x\alpha}$ occur at the lower part toward the end support of the shell. These relatively large tensile stresses can be counteracted only by prestressing. Since the tendons have to follow the two-way curvature, for practical reasons they cannot be arranged so that only membrane forces are induced into the shell. Thus transversal bending moments are commonly required to obtain equilibrium.

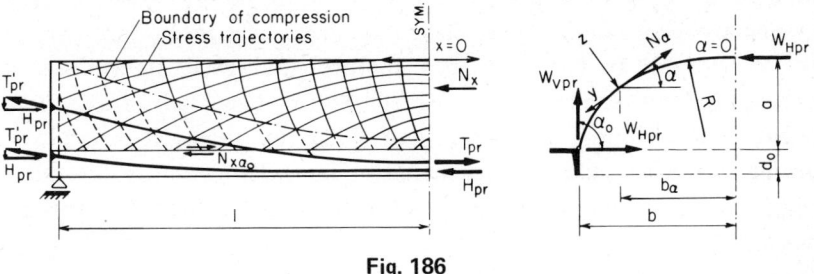

Fig. 186

In accordance with Eqs. (466), the $N_\alpha, N_{x\alpha},$ and N_x membrane forces due to prestressing will be

$$w_{v_{pr}} = \frac{8T_{pr} z_0}{L^2} \qquad w_{H_{pr}} = \frac{8T_{pr} y_0}{L^2}$$
$$w_{z_{pr}} = w_{H_{pr}} \sin\alpha - w_{v_{pr}} \cos\alpha \qquad (484)$$
$$w_{y_{pr}} = w_{H_{pr}} \cos\alpha + w_{v_{pr}} \sin\alpha$$

$$N_{\alpha_{pr}} = R w_{z_{pr}} = R(w_{H_{pr}} \sin\alpha - w_{v_{pr}} \cos\alpha)$$
$$N_{x\alpha_{pr}} = w_{y_{pr}}^* x = \left(w_{y_{pr}} + \frac{1}{R}\frac{\partial N_\alpha}{\partial \alpha}\right) x \qquad (485)$$
$$N_{x_{pr}} = \frac{1}{R}\frac{\partial w_{y_{pr}}^*}{\partial \alpha}\left(\frac{l^2 - x^2}{2}\right) + \sigma_{pr} t$$

The $w_{v_{pr}}$ loading is carried directly by suspension action to the end diaphragms of the shell; thus the membrane forces due to external loading are greatly reduced. The final membrane forces are obtained by superposition.

Since the location of the curved tendons is given by α_{pr}, the membrane forces are a function of x. Therefore, the N values must be computed for at least $x = 0$, $x = 0.5 l$, and $x = 0.25 l$ to obtain acceptable accuracy.

The tendons anchored at the transversal end support of the shell induce a direct compressive stress $\sigma_{x_{pr}}$ into the shell, as is the case in any other posttensioned beam. The distribution of these stresses depends mainly on the location of the end anchorage and also on the transversal rigidity of the end framing.

The nonbonded posttensioning force (ΣT_{pr}) is practically not influenced by live load because the rigidity of the shell, in comparison with a suspended system, is relatively large. Thus any increase in T_{pr} requires large deformations of the shell, which are not possible.

In the case of bonded posttensioning, the stress in tendons can be increased notably by live load. Also, the principal tensile stresses will be superimposed by prestress to a larger extent. These advantages with bonded posttensioning are more than offset by practical difficulties: grouting, a greater space requirement for large-diameter flexible metal tubing, and especially extensive crack formation along the large-diameter tendons.

Since the w_{pr} loading due to posttensioning is a rough estimation only and does not follow a simple mathematical function, flexural stresses are required to obtain continuity of the deformations in the shell.

As a conclusion, regardless of the degree and type of posttensioning, the flexural stresses in the shell cannot be eliminated entirely; they can only be reduced. In multiple shells, the edge rotation ϑ and horizontal displacement Δ_H must be zero. These conditions cannot be satisfied by membrane theory, and they induce additional bending in the shell. The extent of bending present in any shell can be evaluated only by the bending theory of shells, which will be discussed in the following.

Bending Theory

Since the membrane theory has limited usefulness in describing the stress and deformation conditions in shell types required in practical application, extensive work has been carried out to develop formulas and tables for various load-stress and deformation relationships. This material has been made readily available for practical use. Most understandable and complete are the excellent works by the ASCE Subcommittee on Thin Shell Design* and by Aas-Jakobsen.† But regardless of the amount of data available, the mathematical background of most practicing engineers is inadequate to make use of it. In addition, almost all the work available is concentrated on circular cylindrical shells. This is understandable because of the mathematical difficulties connected with other curvatures having variable radii.

The circle is architecturally the dullest and least functional of curvatures. It requires considerable height to obtain a structurally acceptable edge tangent (α_0). Commonly, to reduce the construction height, segments of a circle have to be used. This results in a large radial

---

*Design of Cylindrical Concrete Shell Roofs, *ASCE Manuals Eng. Pract.* 31, 1952.
†"Die Berechnung der Zylinderschalen," Springer-Verlag OHG, Berlin, 1958.

normal force N_α because of the relatively large w_z ($w_z = w_D \cos\alpha$). To balance the N_α force at the edge and to counteract the displacements requires a rather high and complicated edge member. In the case of relatively flat segments, such a shell acts very much like a T section. The advantages of such a shell girder over a T beam lie in the fact that the shell has a curved compressive area participating across its full width in the carrying action, which is not the case with a slab in a T section. In T sections, the slab participation in the carrying action is forced only by shear stresses between the beam and slab, but in shells, the N_x force is already present because of equilibrium requirements without the edge beam. Therefore, the interaction of the edge beam and shell segment is complete regardless of the arch span or spacing of the edge members. In addition, the load transmission to the edge member is considerably reduced due to shell action and occurs in a more economical way (arch action) than is the case with T beams (entirely by beam action).

Taking advantage of the above considerations, Lundgren has developed an acceptable and simple so-called "beam-method" to describe the stress and deformation condition in curvilinear cylindrical shells. The Lundgren method gives qualitatively acceptable results for long shells (lowest limit is $L/B > 1.5$), but it is not applicable for short shells, which still have to be analyzed by mathematical shell theory.

To explain the Lundgren method, a shell element illustrated in Fig. 187 will be considered.

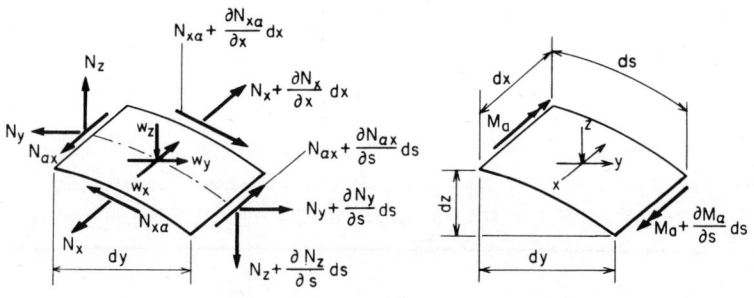

Fig. 187

$$M_x = M_{x\alpha} = V_x = 0$$
$$N_{x\alpha} = N_{\alpha x}$$
$$\Delta A = t\,ds$$

Using rectangular coordinates, the equilibrium between external loads and internal forces requires that

$$\frac{\partial N_x}{\partial x} + \frac{\partial N_{x\alpha}}{\partial s} + w_x = 0$$

$$\frac{\partial N_y}{\partial s} + \frac{\partial N_{x\alpha}}{\partial x}\frac{dy}{ds} + w_y = 0$$

$$\frac{\partial N_z}{\partial s} + \frac{\partial N_{x\alpha}}{\partial x}\frac{dz}{ds} + w_z = 0 \qquad (486)$$

$$\frac{\partial M_\alpha}{\partial s} + N_y \frac{dz}{ds} - N_z \frac{dy}{ds} = 0$$

By direct integration of these equations (486), we obtain

$$N_{x\alpha} = N_{x\alpha(0)} - \int_0^s \frac{\partial N_x}{\partial x} ds - \int_0^s w_x\, ds$$

$$\frac{\partial N_{x\alpha}}{\partial x} = \frac{\partial N_{x\alpha(0)}}{\partial x} - \int_0^s \frac{\partial^2 N_x}{\partial x^2} ds - \int_0^s \frac{\partial w_x}{\partial x} ds$$

$$N_y = N_{y(0)} - \int_0^s \frac{\partial N_{x\alpha}}{\partial x} dy - \int_0^s w_y\, ds \qquad (487)$$

$$N_z = N_{z(0)} - \int_0^s \frac{\partial N_{x\alpha}}{\partial x} dz - \int_0^s w_z\, ds$$

$$M_\alpha = M_{\alpha(0)} + \int_0^s N_z\, dy - \int_0^s N_y\, dz$$

Further, considering that the external loads are carried by the shell to the transversal end-supporting elements, the N_y, N_z, and $N_{x\alpha}$ are zero at the longitudinal edges of the shell (including the edge beam). Thus the equilibrium in the transversal (y) and vertical (z) directions of the shell from $s = 0$ to $s = n$ at any section x = constant requires that

$$\int_{s=0}^{s=n} \frac{\partial N_{x\alpha}}{\partial x} dy = \int_{s=0}^{s=n} w_y\, ds$$

$$\int_{s=0}^{s=n} \frac{\partial N_{x\alpha}}{\partial x} dz = \int_{s=0}^{s=n} w_z\, ds \qquad (488)$$

In accordance with Eqs. (466) and (466a), the N_x and $N_{x\alpha}$ are a function of x and can be evaluated by beam theory. For $w_x = 0$, $\partial^2 N_x/\partial x^2$, $\partial N_{x\alpha}/\partial x$, and N_α are independent of x. If the N_x-value distribution over the cross section $(0 \to n)$ of the shell is known, the $(\partial N_{x\alpha}/\partial x)(s)$, $N_{y(s)}$, $N_{z(s)}$, and $M_{\alpha(s)}$ can be computed by Eqs. (487) and (488).

To obtain a general equation valid for symmetric and nonsymmetric shells, the two-axial linear normal stress $(N_x = t\sigma_x)$ distribution is assumed to follow the function

$$\sigma_x = (a + by + cz) M(x) \tag{489}$$

The parameters a, b, c are constant and are determined from the relationship of stress resultants and loading.

The numerical integration over the entire cross section of the shell (including edge members) gives

$$\begin{aligned} N &= \Sigma \sigma_x \Delta A = (aA + bQ_y + cQ_z) M(x) \\ M_y &= \Sigma \sigma_x y \Delta A = (aQ_y + bI_y + cI_{yz}) M(x) \\ M_z &= \Sigma \sigma_x z \Delta A = (aQ_z + bI_{zy} + cI_z) M(x) \end{aligned} \tag{490}$$

where

$$A = \sum_0^n \Delta A \quad n = 1, 2, 3, \ldots \tag{491}$$

is the total cross-sectional area of shell (including edge members) and

$$\begin{aligned} Q_{y(s)} &= \sum^s y \Delta A \\ Q_{z(s)} &= \sum^s z \Delta A \end{aligned} \tag{492}$$

are the static moments of the cross section above or below the section s about the z and y axes passing through the center of gravity of the shell (centroidal axis) for which the $\sigma_{x(s)}$ stress is evaluated.

$$I_y = \sum_0^{s=n} y^2 \Delta A$$

$$I_z = \sum_0^{s=n} z^2 \Delta A \qquad (493)$$

$$I_{yz} = \sum_0^{s=n} yz \Delta A$$

are moments of inertia and polar moments of inertia about the centroidal axes (y_0, z_0).

The centroid of the shell cross section will be computed as for any section:

$$y_0 = \frac{\bar{Q}_y}{A} = \frac{\sum_0^n \bar{y} \Delta A}{\sum_0^n \Delta A}$$

$$z_0 = \frac{\bar{Q}_z}{A} = \frac{\sum_0^n \bar{z} \Delta A}{\sum_0^n \Delta A} \qquad (494)$$

where $\bar{y}, \bar{z}$ are the coordinates of arbitrarily chosen axes about which the initial static moments $\bar{Q}_y$ and $\bar{Q}_z$ are taken.

The coordinates of the sections ΔA from the centroidal axis are

$$\begin{aligned} y &= \bar{y} - y_0 \\ z &= \bar{z} - z_0 \end{aligned} \qquad (495)$$

The parameters a, b, and c are determined to satisfy Eqs. (490). For simple bending, Q_y and Q_z about the centroidal axis of the shell is zero. Therefore, $a = 0$ and

$$N_x = \Sigma \sigma_x \Delta A = (bQ_y + cQ_z) M(x) \qquad (496)$$

In case of prestressing, $N = H_{pr}$ and

$$a = \frac{H_{pr}}{A} \qquad (497)$$

For vertical loading $\sum_0^n w_z = 1$ k/ft and $w_y = 0$, $\partial^2 M/\Delta x^2 = -\sum_0^n w_z = 1$, we obtain

$$bI_y + cI_{yz} = 0$$
$$bI_{yz} + cI_z = 1 \tag{498}$$

From these two equations the parameters b and c can be computed.

The M_y, M_z moments can be caused by any external loads, including prestressing.

After the b and c values are known, we obtain [Eqs. (490)]

$$N_{x(s)} = -\sum_{s-1}^{s+1} \sigma_x \Delta A = (bQ_{y(s)} + cQ_{z(s)})M(x)$$

$$N_{x\alpha(s)} = -\sum_{s-1}^{s+1} \frac{\Delta \sigma_x}{\Delta x} \Delta A = (bQ_{y(s)} + cQ_{z(s)})\frac{\Delta M(x)}{\Delta x}$$

$$= (bQ_{y(s)} + cQ_{z(s)})V(x)$$

$$\frac{\Delta N_{x\alpha(s)}}{\Delta x} = \sum_{s-1}^{s+1} \frac{\Delta^2 \sigma_x}{\Delta x} \Delta A = (bQ_{y(s)} + cQ_{z(s)})\frac{\Delta V_z(x)}{\Delta x}$$

$$= (bQ_{y(s)} + cQ_{z(s)})\sum_a^c w_z \tag{499}$$

$$N_{y(s-1)} = \sum_{s-1}^{s+1} \frac{\Delta N_{x\alpha(s)}}{\Delta x}\Delta y \quad \Delta y = \Delta s \cos\alpha$$

$$N_{z(s-1)} = \sum_{s-1}^{s+1} \frac{\Delta N_{x\alpha(s)}}{\Delta x}\Delta z \quad \Delta z = \Delta s \sin\alpha$$

$$M_{\alpha(s)} = -\sum_{s-1}^{s} w_{z(s-1)}\Delta y + \sum_{s-1}^{s} N_{z(s-1)}\Delta y - \sum_{s-1}^{s} N_{y(s-1)}\Delta z$$

For convenience and simplicity, it is recommended that the normal forces and shear be computed separately for each loading (w_y, w_z). The final stresses are obtained by superposition.

For shells having symmetric cross section $I_{yz} = 0$, in accordance with Eqs. (498) for loadings $w_y = 0$ and $w_z = 1$ the parameters are: $b = 0$, $c = 1/I_z$ or for $w_y = 1$ and $w_z = 0$: $c = 0$, $b = 1/I_y$. Introducing these parameters into Eqs. (499), we obtain the general beam stress-loading relationship equations. Thus,

$$N_{x(s)} = \sum_{s-1}^{s+1} \sigma_x \Delta A = \frac{Q_{(s)}}{I} M(x)$$

$$N_{x\alpha(s)} = -\sum_{s-1}^{s+1} \frac{\Delta \sigma_x}{\Delta x} \Delta A = \frac{Q_{(s)} V_z(x)}{I}$$

$$\frac{\Delta N_{x\alpha(s)}}{\Delta x} = -\sum_0^s \frac{\Delta^2 \sigma_x}{\Delta x^2} = \frac{Q_{(s)} \Sigma w_z}{I} \qquad (500)$$

$$M_y = \sum_0^n \sigma_x y \Delta A$$

$$M_z = \sum_0^n \sigma_x z \Delta A$$

Note that for symmetric cross section of the shell, only half of the shell need be analyzed because at the symmetry point (crown) the shear $N_{x\alpha} = 0$ and $\partial N_{x\alpha}/\Delta x = 0$.

The vertical shear stress at section s of the shell is

$$\tau_{z(s)} = \frac{Q_{(s)} V_{z(x)}}{t I_z} \cos \alpha \qquad (501)$$

and the tangential shear stress is

$$\tau_{\alpha(s)} = \frac{\tau_{z(s)}}{\cos \alpha} = \frac{Q_{(s)} V_{z(x)}}{t I_z} \qquad (502)$$

For multiple shells, the boundary condition at the longitudinal edges requires that the rotation $\vartheta = 0$ and horizontal displacement $\Delta_H = 0$ be known. As the transversal moment M_α does not influence appreciably the stresses in the longitudinal direction, the constant redundants $X_1 X_2$, required to satisfy the boundary conditions, can be computed independently as for a simple arch (see Chap. 3), where the single shell serves as the principal system. The final results are obtained, as in any indeterminate system, by the principle of superposition.

The external load and redundants acting upon the shell and the internal resisting forces in the transversal direction are illustrated in Fig. 188.

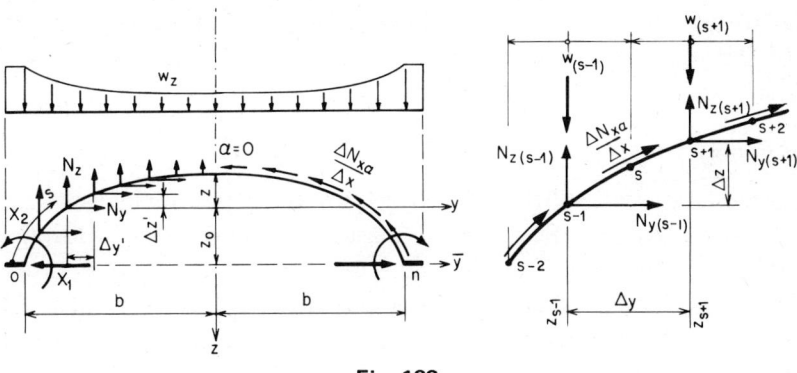

Fig. 188

Alternates:
1. ΔA = const
   Δy = variable
2. ΔA = variable
   Δy = const

As can be seen, the w_z loading and N_y loading, as well as the X_2 loading, develop negative moments (M_α) and N_z and X_1 positive moments counteracting the negative moments. In so-called "statically determinate" shells ($X_1 = X_2 = 0$), the transversal moments (M_α) along the entire cross section of the shell are negative.

Prestressing

In accordance with the beam method, the prestressing can be analyzed as for ordinary prestressed beams. The arrangement of posttensioning tendons in principle is the same as that discussed under membrane theory, above. The prestressing force is

$$\sum H_{pr} = \frac{\Sigma \sigma_B A}{\psi_B} \qquad (503)$$

and the moments due to prestressing are (Fig. 186)

$$-M_{zpr} = H_{pr}e_z + T_{pr}\sin\alpha\, e_{Tz}$$
$$-M_{ypr} = T_{pr}\cos\alpha\, e_y \tag{504}$$

Computation of the transversal moments $M_{\alpha pr}$ due to prestressing is similar to that of the w_z and w_y loadings. The losses in prestressing force due to shrinkage and plastic flow are computed, as for any prestressed beam, by using Eqs. (72) to (80).

Deformations

The deformations in a shell occur mainly by temperature change $T, \Delta T$ and shrinkage ϵ_{sr}. Elastic (ϵ_e) and plastic (ϵ_{pl}) shortenings, with the exception of prestressed shells, are considered of secondary order because of relatively small normal stresses. However, for long spans (L) and relatively wide ($2b$) shells, it is necessary to evaluate the vertical deformation δ_k and horizontal displacements Δ_H. The difference in edge deflections ($\Delta\delta_0$) of nonsymmetric shells may cause torsional moments M_T, especially in relatively flat north-light shells, which cannot be overlooked.

In statically determinate shells, the deformations occur without change of stresses. However, in statically indeterminate shells, appreciable stresses in the transversal direction are caused by temperature change ΔT and shrinkage ϵ_{sr}, and these must be considered if crack formation is to be avoided or at least reduced.

In longitudinal direction (L), the deflections δ_k of a curvilinear cylindrical shell can be estimated by beam theory [Eq. (111)]. Thus,

$$EI_z\delta_k = \int_0^L M_z\overline{M}_k\, dx + \frac{EI_z}{GA}\kappa\int_0^L V_{z(x)}\overline{V}_k\, dx$$
$$EI_z\delta_{kpr} = -\int_0^L M_{zpr}\overline{M}_k\, dx \tag{505}$$

where M_z and V_z are the moment and shear, respectively, due to external load $\sum_0^n w_z$, M_{zpr} is the moment due to prestressing force ΣH_{pr}, and $\overline{M}_k$ and

$\bar{V}_k$ are the moment and shear, respectively, caused by dummy load $P_k = 1$ acting at section k at the point at which the deflection is determined.

The deflection due to temperature difference between the crown and edges of the shell can be considerable, especially in shells having inadequate heat insulation. Denoting the temperature difference by ΔT and the thermal expansion of concrete by α_t the edge deflection is

$$\delta_{k\Delta T} = \pm \int_0^L \frac{\alpha_t \Delta T}{z_c} \bar{M}_k dx \tag{506}$$

The rising and sagging of the crown of the shell can be roughly estimated by Eqs. (142) and (144). These are due to atmospheric temperature change ΔT, shrinkage ϵ_{sr}, and elastic and plastic deformations $\epsilon_e + \epsilon_{pl}$ of the shell in the transversal direction:

$$\sum \Delta z_c = \left[\pm \alpha_t \Delta T - \epsilon_{sr} - \sum_0^c \frac{N_\alpha}{Et}(1 + \varphi_n) \right] \frac{b^2 + z_c^2}{z_c} \tag{507}$$

If there is any lateral displacement of the longitudinal edges ($\Sigma \Delta_H$) of the shell, the crown movement will be approximately

$$\Delta z_c = \pm \frac{\Sigma \Delta_H}{2} \frac{b}{z_c} \tag{508}$$

The temperature and shrinkage stresses in indeterminate shells (intermediate shells) in the transversal direction can be estimated by the segmental cylindrical shell equations [Eqs. (420) to (424), Fig. 166] when α is substituted for β in these equations and z for y. These equations are valid only for the circular cross section of the shell. For other curvatures, the integrals must be solved by numerical integrations (Simpson's rule).

The deformations and stresses obtained by the above equations are slightly higher than those determined by mathematical shell theory, so far as the deformations and secondary stresses of this nature have been treated. But considering that deformations increase and flexural stresses decrease when plastic flow takes place, and especially if crack formation occurs (the normal case), the results obtained by the given equations are adequate for practical use and at least give the limiting values for deformations and flexural stresses to evaluate the behavior and economy of curvilinear cylindrical shells.

End Supports

At its end ($x = \pm l$), the shell is commonly supported transversally by arched frames or diaphragms. The loads that the end support must carry are the shear $N_{x\alpha}$ acting tangent to the shell, the concentrated loads due to prestressing force T_{pr}, and the dead load of the end support itself.

The vertical and horizontal components of the tangential shear force $N_{x\alpha}$ at the end support ($x = \pm l$) are

$$\begin{aligned} V_s &= N_{x\alpha} \sin\alpha \\ H_s &= N_{x\alpha} \cos\alpha \end{aligned} \tag{509}$$

The normal force N_x due to the external loads w is zero at the end support. The prestressing force T'_{pr} anchored at the end supports has a horizontal component in the longitudinal direction of the shell. Denoting the angle between the tendon and x axis by β, the horizontal and tangential prestressing forces at the end support are (Figs. 73 and 186)

$$\begin{aligned} T_{prx} &= T'_{pr} \cos\beta \\ T_{pr\alpha} &= T'_{pr} \sin\beta \end{aligned} \tag{510}$$

Thus the vertical and horizontal components of the prestressing force at the end support are

$$\begin{aligned} V_{spr} &= T_{pr\alpha} \sin\alpha \\ H_{spr} &= T_{pr\alpha} \cos\alpha \end{aligned} \tag{511}$$

The boundary condition of the end frame must coincide with the shell (Δ_H, ϑ).

The stress distribution $N_{x\alpha}$ due to T_{prx} is computed as for an eccentrically loaded column. Thus N_{prx} can be positive or negative, depending on the location of the end anchorage in relation to the center of gravity of the shell.

Prestressing is seldom required for end supports. However, in the case of solid end diaphragms, straight tendons posttensioning of the bottom margin of the diaphragm is recommended. The main reasons for this are economy and to avoid extensive time deflection. The analysis of such arch frames was thoroughly discussed in Chap. 3.

Commonly the shell is fixed at the end support, and therefore the boundary conditions at its transversal edges are not satisfied. Due to this relatively large bending, stresses occur which may lead to crack formation if the bending is not considered in design.

The magnitude of the bending moments M_x and shear V_x can be estimated by Eqs. (414) to (418), developed for rotationally symmetric cylindrical shells.

The variation of the moment M_x in the transversal direction from $\alpha = 0$ to α_0 follows cosine function.

Analysis of a Shell

For illustration, an inner shell of a multiple-shell roof will be analyzed. The transversal section of the curvilinear shells is a cycloid. The shells are connected with each other by a channel accommodating mechanical ducts and electric wiring. The cycloids span 100 ft and are supported at the ends by two-hinged arches. The geometric data of the roof are given in Fig. 189.

Geometric data:

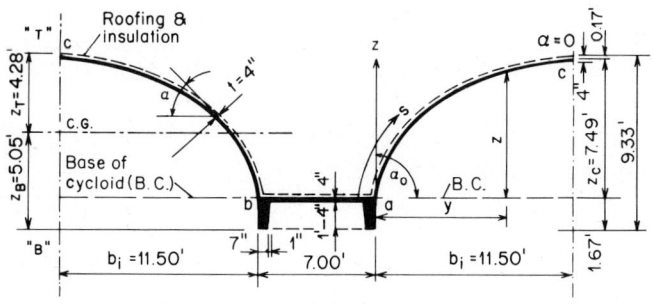

Fig. 189

$$t = 0.333 \text{ ft}$$
$$a_i = 14.64 \text{ ft} \qquad a_o = 14.98 \text{ ft}$$
$$R_i = a_i \cos\alpha \text{ (inside face)}$$
$$z_{ci} = \frac{a_i}{2} = 7.32 \text{ ft} \qquad z_{co} = 7.49 \text{ ft}$$
$$b_i = \frac{\pi a_i}{4} = 11.50 \text{ ft} \qquad b_o = 11.67 \text{ ft}$$

$$s = 2a_o = 2 \times 14.98 = 29.96 \text{ ft}$$

$$z = \frac{a}{4}(1 - \cos\Theta)$$

$$y = \frac{a}{4}(\Theta - \sin\Theta)$$

Center of gravity (C.G.):

Cycloid: $\eta_0 = \dfrac{\Sigma Q}{\Sigma \Delta A} = 4.96$ ft

Cycloid + channel: $\sum \eta_0 = \dfrac{\Sigma \Sigma Q}{\Sigma \Sigma \Delta A} = 3.38$ ft $z_T = 4.28$ ft

$z_B = 5.05$ ft

$$\sum I_0 = 2 \times 67.708 = 135.416 \text{ ft}^4$$

$$\sum_0^b A = 2(4.88 + 2.00) = 13.76 \text{ ft}^2$$

Loading:

Cycloid: 4 in. concrete + 2 in. insulation
Lead-covered copper roofing $w_D = 62$ psf
Live load $w_L = 30$ psf

Channel: 4-in. slab + ribs + L.W. concrete
Roofing and mechanical $P_D = 0.86$ k/ft
Live load $P_L = 0.20$ k/ft

$$\sum_a^b w_D = 0.062 \times 29.96 + 0.86 = 2.72 \text{ k/ft}$$

$$\sum_a^b w_L = 0.030 \times 23.33 + 0.20 = 0.90 \text{ k/ft}$$

$$k_D = \frac{\Sigma w_D}{\Sigma w} = \frac{2.72}{3.62} = 0.75_{.7}$$

$$k_L = 1.0 - 0.75 = 0.25$$

Method of analysis:

Considering that the relative lateral displacement of the edges of cycloids is zero and the span-to-width ratio is about 4.3 (long shell), the stress condition in the shell is described with sufficient accuracy by the Lundgren method.

Moments, shear, and stresses:
1. Longitudinal, x direction:

$$M_{max} = \frac{1}{8} \sum_a^b wL^2 = \frac{1}{8} \cdot 3.62 \times 102^2 = 4,700\,k'$$

$$V_{max} = \frac{1}{2} \sum_a^b L = \frac{1}{2} \cdot 3.62 \times 100 = 181.00\,k/\text{end}$$

$$f_{cT} = \frac{M_{max}}{\Sigma I_0} z_T = \frac{4,700}{135.416} \cdot 4.28 = -148.5 \text{ ksf} \sim 1,030 \text{ psi}$$

$$f_{cB} = \frac{M_{max}}{\Sigma I_0} z_B = \frac{4,700}{135.416} \cdot 5.05 = +175.5 \text{ ksf} \sim 1,220 \text{ psi}$$

$$v_{C.G.} = \frac{v_{max} Q}{2t/\cos\alpha \, \Sigma I_0} = \frac{181.00 \times 19.6}{0.84 \times 135.416} = 31.20 \text{ ksf} \sim 217 \text{ psi}$$

$$v_0 = \frac{v_{max} Q_0}{2t \, \Sigma I_0} = \frac{181.00 \times 15.56}{0.666 \times 135.416} = 31.00 \text{ ksf} \sim 215 \text{ psi}$$

2. Prestressing:

$$e_0 = \Sigma\eta_0 - 0.338 = 3.00 \text{ ft}$$

$$i^2 = \frac{\Sigma I_0}{\Sigma A} = \frac{135.416}{13.76} = 9.85 \text{ ft}^2$$

$$\psi_T = 1 - \frac{e_0 z_T}{i^2} = 1 - \frac{3.00 \times 4.28}{9.85} = -0.30$$

$$\psi_B = 1 + \frac{e_0 z_B}{i^2} = 1 + \frac{3.00 \times 5.05}{9.85} = 2.54$$

$$H_{prn} = \frac{\Sigma f_B \Sigma A}{\psi_B} = \frac{175.5 \times 13.76}{2.54} \simeq 950\,k$$

Use: 2×3 -6/0.5 cables-nonbonded

$$H_{pr0} = 6 \times 162 = 972\,k \; (0.75 f'_s)$$

3. Stresses:

$t = 0$:

$$\sigma_{0pr} = \frac{H_{pr0}}{\Sigma A} = -\frac{972}{13.76} = -70.7 \text{ ksf} \sim 490 \text{ psi}$$

$$\sigma_{prT} = \psi_T \sigma_{pr0} = +0.300 \times 490 = +147 \text{ psi}$$

$$\sigma_{prB} = \psi_B \sigma_{pr0} = -2.54 \times 490 = -1{,}245 \text{ psi}$$

$$\sum_{pr}^{D+L} f_T = -f_{cT} + \sigma_{prT} = -1{,}030 - 147 = -883 \text{ psi}$$

$$\sum_{pr}^{D+L} f_B = f_{cB} + \sigma_{pr0} = +1{,}220 - 1{,}245 = -25 \text{ psi}$$

$$\sigma_{\text{I,II}} = \frac{\sigma_{0pr}}{2} \mp {}^1\!/_2 \sqrt{\sigma_{0pr}^2 + 4v_{C.G.}^2} = -\frac{490}{2} \mp {}^1\!/_2 \sqrt{490^2 + 4 \times 217^2}$$

$$= + 79 \text{ psi}$$
$$- 569 \text{ psi}$$

The loss in prestress (ΔH) and the stresses for $t = n$ are computed as for beams described under prestressing (Sec. 8) and plastic flow and shrinkage (Sec. 9).

Bending in the y direction:

Since the shell is symmetric, only half of it is considered. The half cycloid is divided into six sections; the dividing points are designated by the numbers 0, 2, 4, 6, 8, 10, 12, and the centers of gravity of these sections are designated by 1, 3, 5, 7, 9, 11. The coordinates, weight (dead and live) of the sections $w_0, w_1, \ldots$, and static moments Q for points 0, 2, 4, . . . are determined and indicated in Fig. 190.

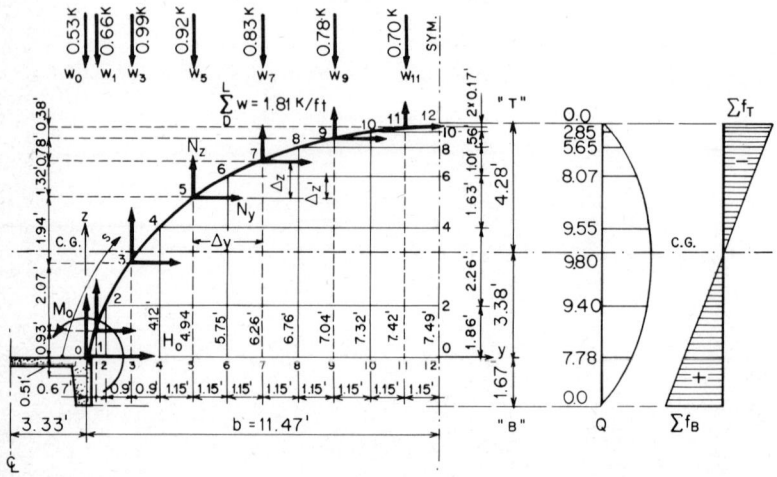

Fig. 190

1. Statically determinate cycloid ($H_0 = M_0 = 0$) (principal system):
External moments:

$$w_i = \Delta s_i w_0 + \Delta y w_L$$

$$M_{n0} = M_{n-1,0} + \sum_{0}^{n-1} w \Delta y + {}^1{}_2 w_i \Delta y$$

$$n = 0, 2, 4, \ldots \quad i = 1, 3, 5, \ldots$$

M_{n0}

$M_{10} = -0.355\,k'$ $M_{80} = -7.144\,k'$
$M_{20} = -0.404\,k'$ $M_{100} = -10.572\,k'$
$M_{40} = -1.839\,k'$ $M_{120} = -14.471\,k'$
$M_{60} = -4.226\,k'$

Internal moments:

$$c = \frac{\Sigma w}{I_0} = \frac{1.810}{67.708} = 0.0267$$

$$\Delta N_{x\alpha} = cQ$$
$$N_z = \Delta N_{x\alpha} \Delta z$$
$$N_y = \Delta N_{x\alpha} \Delta y$$

$$\sum_{0}^{c} N_z = \sum_{0}^{c} w = 1.810\,k$$

N_z	N_y
$N_{z0} = 0.366\,k$	$N_{y0} = 0.0\,k$
$N_{z1} = 0.519\,k$	$N_{y1} = 0.316\,k$
$N_{z3} = 0.494\,k$	$N_{y3} = 0.520\,k$
$N_{z5} = 0.285\,k$	$N_{y5} = 0.496\,k$
$N_{z7} = 0.119\,k$	$N_{y7} = 0.350\,k$
$N_{z9} = 0.027\,k$	$N_{y9} = 0.172\,k$
$N_{z11} = 0.0\,k$	$N_{y11} = 0.0\,k$

$$M_{nz} = M_{n-1,z} + \sum_{0}^{n-1} N_z \Delta y + {}^1{}_2 N_{iz} \Delta y$$

$$M_{ny} = M_{n-1,y} + \sum_{0}^{n-1} N_y \Delta z + {}^1{}_2 N_{iy} \Delta z$$

$$M_{\alpha 0} = M_{n0} + M_{zn} + M_{yn}$$

M_{zn}	M_{yn}
$M_{z2} = 0.572\,k'$	$M_{y2} = -0.294\,k'$
$M_{z4} = 2.458\,k'$	$M_{y4} = -1.596\,k'$
$M_{z6} = 5.936\,k'$	$M_{y6} = -3.364\,k'$
$M_{z8} = 9.893\,k'$	$M_{y8} = -4.887\,k'$
$M_{z10} = 14.024\,k'$	$M_{y10} = -5.876\,k'$
$M_{z12} = 18.184\,k'$	$M_{y12} = -6.191\,k'$

2. Statically indeterminate cycloid:

The redundants ($X_1 = H_0$, $X_2 = M_0$) to satisfy the boundary condition of the cycloid are determined from the continuity requirements:

$$\left. \begin{array}{l} \delta_{10} - X_1\delta_{11} - X_2\delta_{12} = 0 \\ \delta_{20} - X_1\delta_{21} - X_2\delta_{22} = 0 \end{array} \right| \quad \delta_{12} = \delta_{21}$$

$$I_c = \frac{I}{\cos\alpha} = 1$$

$X_1 = -1: M_1 = +z:$

$$EI_c\delta_{11} = 2\int_a^c M_1^2 dy = 2\int_a^c z^2 dy = 822$$

$X_2 = 1: M_2 = 1.0:$

$$EI_c\delta_{22} = 2\int_a^c M_2^2 dy = 23$$

$$EI_c\delta_{12} = 2\int_a^c M_1 M_2 dy = 129.5$$

$$EI_c\delta_{10} = 2\int_a^c M_{\alpha 0} M_1 dy = -254$$

$$EI_0\delta_{20} = 2\int_a^c M_{\alpha 0} M_2 dy = -39.2$$

Numerical integration by Simpson's rule [Eq. (112)]

$$X_1 = \frac{\delta_{10}\delta_{22} - \delta_{20}\delta_{12}}{\delta_{11}\delta_{22} - \delta_{12}^2} = -0.367\,k$$

$$X_2 = \frac{\delta_{20}\delta_{11} - \delta_{10}\delta_{12}}{\delta_{11}\delta_{22} - \delta_{12}^2} = +0.357\,k'$$

$M_\alpha = M_{\alpha 0} - X_1 z - X_2$

$M_{0\alpha} = 0.0 + 0.0 - 0.357 = -0.357\,k'$

$$M_{2\alpha} = -0.778 + 0.367 \times 1.86 - 0.357 = +0.047 \, k'$$
$$M_{4\alpha} = -0.977 + 0.367 \times 4.12 - 0.357 = +0.181 \, k'$$
$$M_{6\alpha} = -1.654 + 0.367 \times 5.75 - 0.357 = +0.109 \, k'$$
$$M_{8\alpha} = -2.138 + 0.367 \times 6.76 - 0.357 = -0.011 \, k'$$
$$M_{10\alpha} = -2.423 + 0.367 \times 7.32 - 0.357 = -0.100 \, k'$$
$$M_{12\alpha} = -2.478 + 0.367 \times 7.49 - 0.357 = -0.090 \, k'$$

3. Prestressing:

When the tendons in the vertical and horizontal projections follow a parabola, the w_{zpr}, w_{ypr} are constant:

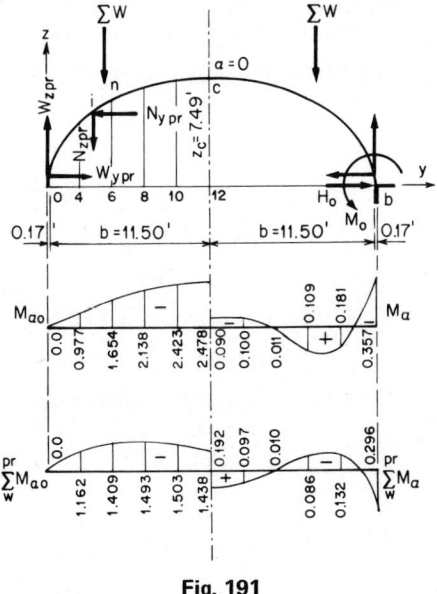

Fig. 191

$$w_{zpr} = \frac{8T_{pr}z_0}{L^2} = -\frac{8 \times 486 \times 3.14}{100^2} \simeq -1.219 \, k/\text{ft}$$

$$w_{ypr} = \frac{8T_{pr}y_0}{L^2} = +\frac{8 \times 486 \times 1.93}{100^2} \simeq +0.751 \, k/\text{ft}$$

External moments:

$$M_{pr0} = w_{zpr}\, y - w_{zpr}\, z$$

$$\begin{array}{rl} & M_{pr0} \\ M_{20} = & -0.588\,k' \\ M_{40} = & 0.395\,k' \\ M_{60} = & 1.970\,k' \\ M_{80} = & 4.005\,k' \\ M_{100} = & 6.380\,k' \\ M_{120} = & 9.090\,k' \end{array}$$

Internal moments:

It should be noted that the horizontal prestressing loading w_{ypr} produces simple bending. The internal moment due to vertical prestressing loading w_{zpr} is obtained from the relation of $w_{zpr}/\Sigma w = k$. Thus,

$$k = -\frac{1.219}{1.810} = -0.672$$

$$M_{npr} = k(M_{zn} + M_{yn})$$

$$M_{\alpha pr0} = M_{pr0} + M_{npr}$$

$$\begin{array}{rl} & M_{npr} \\ M_2 = & -0.085\,k' \\ M_4 = & -0.580\,k' \\ M_6 = & -1.725\,k' \\ M_8 = & -3.360\,k' \\ M_{10} = & -5.460\,k' \\ M_{12} = & -8.050\,k' \end{array}$$

$$\begin{array}{rl} & M_{\alpha pr0} \\ M_{\alpha 2 pr} = & -0.673\,k' \\ M_{\alpha 4 pr} = & -0.185\,k' \\ M_{\alpha 6 pr} = & +0.245\,k' \\ M_{\alpha 8 pr} = & +0.645\,k' \\ M_{\alpha 10 pr} = & +0.920\,k' \\ M_{\alpha 12 pr} = & +1.040\,k' \end{array}$$

$\sum_{w}^{pr}$-moments:

$$H_0 = M_0 = 0: \quad \sum_{w}^{pr} M_{\alpha 0} = M_{\alpha 0} + M_{\alpha pr0}$$

$$\Sigma M_{2\alpha} = -0.126 - 0.673 = -0.799\,k'$$
$$\Sigma M_{4\alpha} = -0.977 - 0.185 = -1.162\,k'$$
$$\Sigma M_{6\alpha} = -1.654 + 0.245 = -1.409\,k'$$
$$\Sigma M_{8\alpha} = -2.138 + 0.645 = -1.493\,k'$$
$$\Sigma M_{10\alpha} = -2.423 + 0.920 = -1.503\,k'$$
$$\Sigma M_{12\alpha} = -2.478 + 1.040 = -1.438\,k'$$

$H_0 \neq 0, M_0 \neq 0$:

$$EI_c \delta_{10} = 2 \int_0^c M_1 \Sigma M_\alpha dy = -183.0$$

$$EI_c \delta_{20} = 2 \int_0^c M_2 \Sigma M_\alpha dy = -29.6$$

$$X_1 = \frac{-183 \times 23 + 29.6 \times 129.5}{822 \times 23 - 129.5^2} = -0.178\,k$$

$$X_2 = \frac{-29.6 \times 822 + 183 \times 129.5}{822 \times 23 - 129.5^2} = -0.295\,k'$$

$\Sigma M_{\alpha 0} = 0.0 + 0.0 + 0.295 = 0.295\,k'$
$\Sigma M_{\alpha 2} = -0.799 + 0.178 \times 1.86 + 0.295 = +0.172\,k'$
$\Sigma M_{\alpha 4} = -1.162 + 0.178 \times 4.12 + 0.295 = -0.132\,k'$
$\Sigma M_{\alpha 6} = -1.409 + 0.178 \times 5.75 + 0.295 = -0.086\,k'$
$\Sigma M_{\alpha 8} = -1.493 + 0.178 \times 6.76 + 0.295 = +0.010\,k'$
$\Sigma M_{\alpha 10} = -1.503 + 0.178 \times 7.32 + 0.295 = +0.097\,k'$
$\Sigma M_{\alpha 12} = -1.438 + 0.178 \times 7.49 + 0.295 = +0.192\,k'$

The moment diagrams for both boundary conditions at midspan of the cycloid are plotted in Fig. 191. As the location of the tendons rises toward the crown, the $M_{p r 0}$ decreases; therefore, the transversal moments change from midspan to the end diaphragms.

29. PIPES

Equations (466) are general equations of curvilinear cylindrical shells, a category in which pipes belong also. Therefore, these equations can be directly applied in the design of pipes. For illustration, the forces in a circular pipe supported at the bell and spigot only and subjected to dead load and water pressure will be computed.

Dead Load

$w_x = 0 \quad w_y = w_D \sin\alpha \quad w_z = w_D \cos\alpha$
$N_\alpha = -R w_D \cos\alpha$

The integration constant $C_1(\alpha)$ is given by symmetry:
$x = 0$:

$$N_{x\alpha} = 0 \quad \text{also} \quad C_1 = 0$$
$$N_{x\alpha} = -w_y^* x = -2w_D x \sin\alpha \tag{512}$$
$$N_x = \frac{1}{2R} \frac{\partial w_y^*}{\partial \alpha} x^2 + C_2(\alpha)$$

The integration constant $C_2(\alpha)$ is determined from the condition $x = \pm l$:

$$N_x = 0 \quad \text{thus} \quad C_2 = -l^2$$
$$N_x = -\frac{w_D}{R}(l^2 - x^2)\cos\alpha$$

The normal force N_α, N_x and shear $N_{x\alpha}$ diagrams due to dead load are illustrated in Fig. 192.

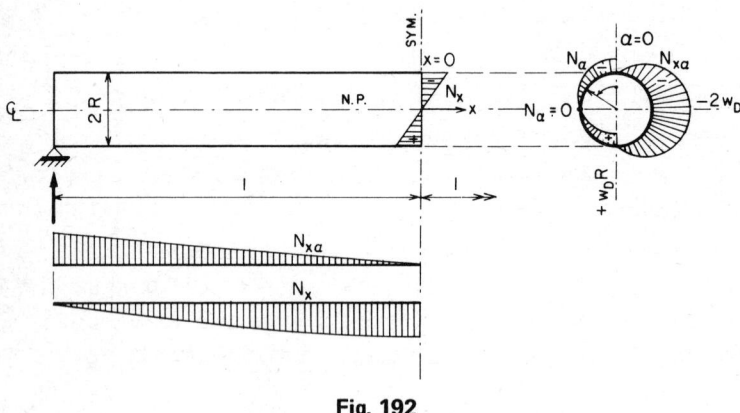

Fig. 192

As can be seen, the N_x and $N_{x\alpha}$ diagrams are the same as for a simply supported beam.

Liquid Pressure

$$w_x = w_y = 0$$
$$w_z = -\gamma(h - R\cos\alpha)$$
$$N_\alpha = R^2\gamma\left(\frac{h}{R} - \cos\alpha\right) \tag{513}$$

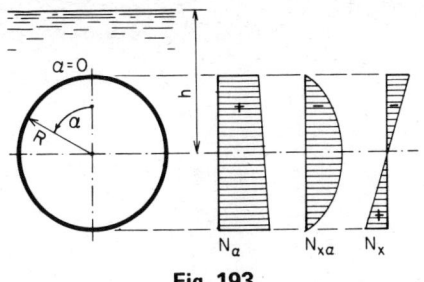

$$N_{x\alpha} = -R\gamma x \sin\alpha$$

$$N_x = -\gamma\left(\frac{l^2 - x^2}{2}\right)\cos\alpha \quad (513)$$
(Cont.)

Fig. 193

where γ is the weight per unit volume of the liquid and h is the pressure head from the center of the cross section of the pipe (Fig. 193).

As can be seen from the equations and normal force and shear diagrams, very high tensile stresses can occur in long-span pipes subjected to high liquid pressures. These stresses are frequently found to be of such a magnitude that they cannot be offset by conventional reinforcing. In such cases, prestressing must be applied to prevent the development of cracks.

Prestressing

The prestressing forces required to keep the pipe free of tensile stresses are computed from the N_α and N_x forces. The N_α forces determine the circumferential prestressing, and the N_x forces determine the longitudinal prestressing. The circumferential and longitudinal prestressings can be applied simultaneously by the basket-weave method developed by the Preload Company. When this method is used, the tension in the wires and the pitch of the spiral winding determine the magnitude of the longitudinal and circumferential components of the prestressing forces.

30. HYPERBOLIC PARABOLOIDS

The carrying action of a hyperbolic paraboloid is accomplished by two sets of identical parabolic curvatures translated to each other and intersecting at right angles. The first set of paraboloids with downward curvature represents the arch action, and the second set with upward curvature represents the suspension action.

The special characteristic of such a translation surface is that a straight line lying wholly in the surface may be drawn through any point on the surface. This property of the hyperbolic paraboloid makes it especially attractive for shell construction.

The general equation for a hyperbolic paraboloid with coordinates as shown on Fig. 195 and origin o in vertex is

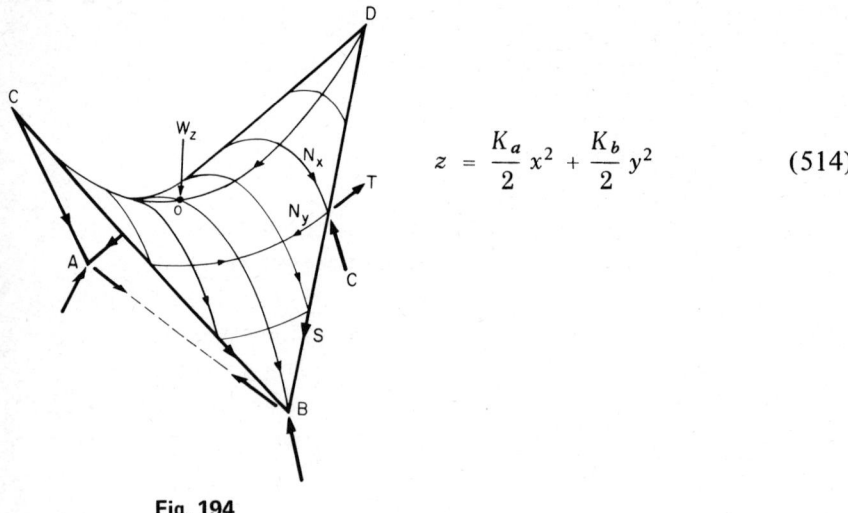

$$z = \frac{K_a}{2} x^2 + \frac{K_b}{2} y^2 \qquad (514)$$

Fig. 194

The parameters K_a, K_b of the parabolas having rises z_a, z_b and spans $2a, 2b$ in the x and y directions are determined from the conditions:

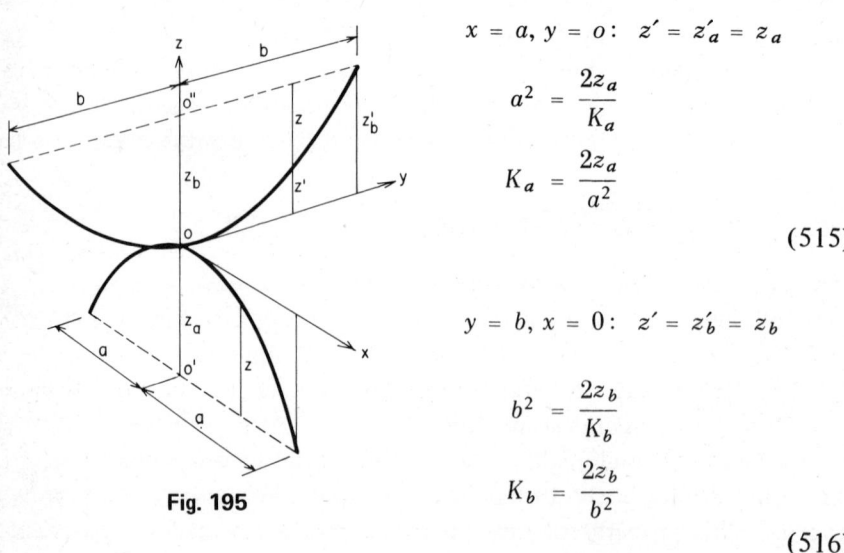

Fig. 195

$$x = a, y = 0: \quad z' = z'_a = z_a$$

$$a^2 = \frac{2z_a}{K_a}$$

$$K_a = \frac{2z_a}{a^2} \qquad (515)$$

$$y = b, x = 0: \quad z' = z'_b = z_b$$

$$b^2 = \frac{2z_b}{K_b}$$

$$K_b = \frac{2z_b}{b^2} \qquad (516)$$

Introducing the k values into Eq. (514), we obtain

$$z = \frac{z_a}{a^2} x^2 - \frac{z_b}{b^2} y^2 \tag{517}$$

For identical parabolas, $a^2/z_a = b^2/z_b$ and $K_a = K_b$.

In practice, it is more convenient to translate the origins o of the individual parabolas to o' and o''; then $z = z_a - z'$.

Also,

$$\begin{aligned} z_x &= z_a \left(1 - \frac{x^2}{a^2}\right) \\ z_y &= z_b \left(1 - \frac{y^2}{b^2}\right) \end{aligned} \tag{518}$$

The radius of a curvature is

$$\frac{1}{R_x} = \frac{d^2 z}{dx^2} = \frac{d\alpha}{ds} \tag{519}$$

For a relatively flat parabola $ds \simeq \Delta x$ and $\tan \alpha = \Delta z/\Delta x$, the radius at vertex becomes in the x and y directions, respectively,

$$\begin{aligned} R_x &= \frac{a^2}{2z_a} \\ -R_y &= \frac{b^2}{2z_b} \end{aligned} \tag{520}$$

Thus, for identical parabolas,

$$R_x = -R_y = R$$

and is practically constant.

The normal forces N_x and N_y, in accordance with Eqs. (348), must be in equilibrium with the external load w_z. Considering that for a flat curvature $\cos \alpha \simeq 1$, we obtain

$$N_y + N_x + w_z = 0 \tag{521}$$

and

$$N_y = -N_x = \frac{w_z}{2} R \simeq \text{const}$$

The equal compression and tension in the x and y directions are the principal stresses in the shell and, in accordance with Eqs. (483), are equivalent and identical to the principal shear $S_\varphi (N_x = N_y = 0)$ at $\varphi = 45°$ to the x and y axes (principal axis). Thus [Eqs. (483)],

$$S = \pm \tfrac{1}{2}\sqrt{(N_y - N_x)^2} = \pm \frac{w_z}{2} R \qquad (522)$$

To keep the edge members relatively light, the shell must be supported along straight lines at $\varphi = 45°$ to the principal axis (x, y), where the $N_x = N_y = 0$ and the edge beam is subject to the shear S only. When the shell is supported otherwise, the edge members must balance the resultant of the N_x, N_y forces at the shell boundary.

The straight edge member of hyperbolic paraboloids is compressed by accumulated shear from C to AB and D to AB and is also acted upon in its transversal direction by a moment

$$M_E = e \sum_{C,D}^{A,B} S \qquad (523)$$

where $e = b_0/2$ is the eccentricity of the shear in respect to the centerline of the edge member. Also, due to lack of curvature at the boundary, the shell is not able to carry the dead load of the edge member without developing bending stresses.

To keep the shell free of bending stresses, the edge-member deflection in the transversal direction due to M_E and in its plane due to w_{ED} must be avoided. This is most easily done by posttensioning (see "Deformations," Sec. 11).

Since the total external vertical load Σw_z is transmitted to the foundation or to supporting elements mainly at A and B, at least one more supporting element is required for stability of the shell and to resist wind pressure. This supporting element or stabilizer can be most effectively located at point C or D. If this is not possible for architectural or other reasons, then, to avoid torsion and undue stress concentration, two stabilizers are required, arranged symmetrically (if possible) along the edge members C-AB or D-AB.

31. ELLIPTIC PARABOLOIDS

The elliptic paraboloid is characterized by a parabola having a rise z_a and span $2a$ and sliding along of another parabola having a rise z_b and span $2b$. Both parabolas have downward curvatures and are supported on four boundaries by edge members which are capable of carrying loads only on their own plane (Fig. 196). The center surface of the shell is described by the equation

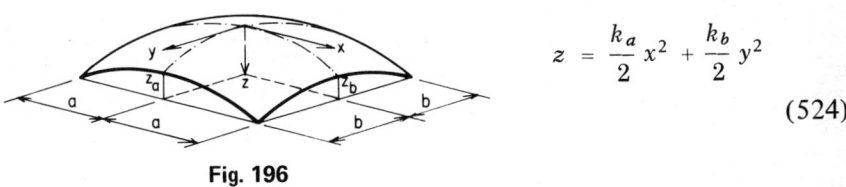

$$z = \frac{k_a}{2} x^2 + \frac{k_b}{2} y^2 \tag{524}$$

Fig. 196

The parameters k_a, k_b and radii R_a, R_b are [Eqs. (515), (516), and (520)]

$$\begin{aligned} k_a &= \frac{\partial^2 z}{\partial x^2} = \frac{1}{R_a} = \frac{2 z_a}{a^2} \\ k_b &= \frac{\partial^2 z}{\partial y^2} = \frac{1}{R_b} = \frac{2 z_b}{b^2} \\ k_c &= \frac{\partial^2 z}{\partial x \partial y} = 0 \quad \text{no twist} \end{aligned} \tag{525}$$

The loading in the direction of tangent is assumed to be zero. Considering that z_a and z_b are relatively small in comparison with a and b, the values of second order can be disregarded. Thus the normal forces N_x, N_y and shear N_{xy} projected to the x, y plane can be determined by the equations of equilibrium:

$$\begin{aligned} \frac{\partial N_x}{\partial x} + \frac{\partial N_{xy}}{\partial y} &= 0 \quad w_x = 0 \\ \frac{\partial N_y}{\partial y} + \frac{\partial N_{yx}}{\partial x} &= 0 \quad w_y = 0 \\ k_a N_x + k_b N_y + w_z &= 0 \end{aligned} \tag{526}$$

Klimov* expressed N_x, N_y, and N_{xy} by a stress function and simplified the analysis of relatively flat ($z_{max} = L/16$, $L = 2a$ or $2b$) elliptic paraboloids down to the solution of the Laplace equation. In the following, the Klimov method will be discussed.

$$N_x = \frac{1}{k_a}\left(N - \frac{w_z}{2}\right)$$

$$N_y = -\frac{1}{k_b}\left(N + \frac{w_z}{2}\right)$$

$$N = \frac{k_a}{2}N_x - \frac{k_b}{2}N_y \qquad (527)$$

$$= n\sqrt{k_a k_b}$$

$$= \frac{2n}{ab}\sqrt{z_a z_b} \qquad n = \frac{Nab}{2\sqrt{z_a z_b}}$$

To obtain general equations, dimensionless coordinates will be used:

$$\frac{x}{a} = \frac{\xi}{\sqrt{2z_a}} \qquad \frac{y}{b} = \frac{\eta}{\sqrt{2z_b}} \qquad (528)$$

The related derivatives are

$$dx = \frac{a}{\sqrt{2z_a}}d\xi \qquad dy = \frac{b}{\sqrt{2z_b}}d\eta \qquad (529)$$

By differentiation of Eqs. (527) with respect to x, y or ξ, η, we obtain

$$\frac{\partial N_x}{\partial x} = \frac{1}{k_a}\frac{\partial N}{\partial x}$$

$$\frac{\partial N_y}{\partial y} = -\frac{1}{k_b}\frac{\partial N}{\partial y}$$

$$\frac{\partial N}{\partial x} = \frac{2\sqrt{z_a z_b}}{ab}\frac{\sqrt{2z_a}}{a}\frac{\partial n}{\partial \xi} \qquad (530)$$

*Boris Klimov, Berechnung des flachen elliptischen Paraboloids, *Beton-Stahlbetonbau*, no. 1, January 1959.

$$\frac{\partial N}{\partial y} = \frac{2\sqrt{z_a z_b}}{ab} \frac{\sqrt{2z_b}}{b} \frac{\partial n}{\partial \eta}$$

$$\frac{\partial N_{xy}}{\partial x} = \frac{\sqrt{2z_a}}{a} \frac{\partial s}{\partial \xi}$$

$$\frac{\partial N_{yx}}{\partial y} = \frac{\sqrt{2z_b}}{b} \frac{\partial s}{\partial \eta}$$

$$N_{xy} = N_{yx} = s \qquad (530)$$
$$(\text{Cont.})$$

Introducing these values into Eqs. (526), we obtain the known Cauchy-Riemann equations:

$$\frac{\partial n}{\partial \xi} + \frac{\partial s}{\partial \eta} = 0$$
$$-\frac{\partial n}{\partial \eta} + \frac{\partial s}{\partial \xi} = 0 \qquad (531)$$

After differentiating the first equation (531) with respect to ξ and the second with respect to η and then subtracting the second from the first, and vice versa, we obtain the Laplace equation:

$$\frac{\partial^2 n}{\partial \xi^2} + \frac{\partial^2 n}{\partial \eta^2} = 0$$
$$\frac{\partial^2 s}{\partial \xi^2} + \frac{\partial^2 s}{\partial \eta^2} = 0 \qquad (532)$$

These entirely independent differential equations can be solved for given boundary conditions by the difference method [Eq. (12)].

Boundary Conditions

It is customary, mainly because of window arrangement, to use Vierendeel trusses for the edge members of elliptic paraboloid shells (Fig. 197). Long-span Vierendeels are relatively rigid in the vertical plane but rather flexible in the transversal direction. Due to this, the boundary conditions require that all horizontal forces perpendicular to the boundary must be zero. In accordance with Eqs. (527) and (528), we obtain:

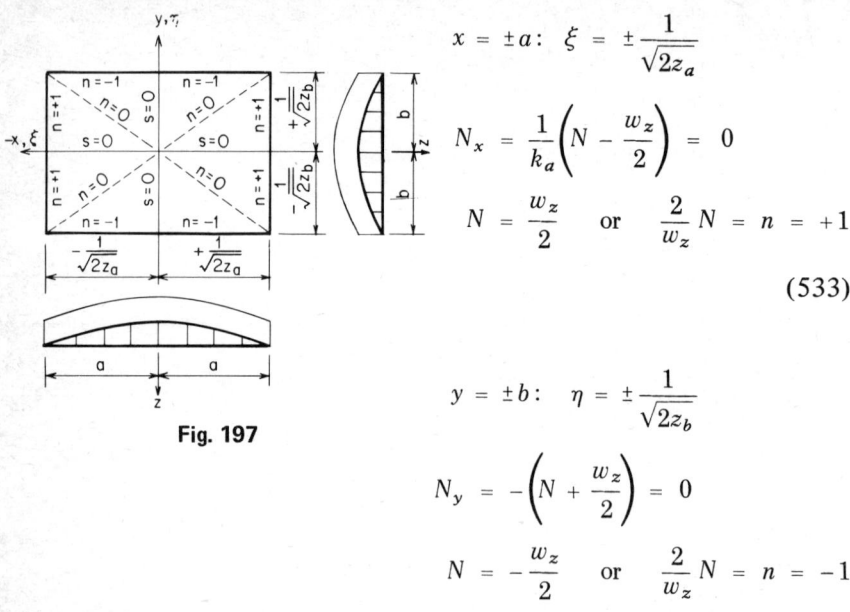

Fig. 197

$$x = \pm a: \quad \xi = \pm \frac{1}{\sqrt{2z_a}}$$

$$N_x = \frac{1}{k_a}\left(N - \frac{w_z}{2}\right) = 0$$

$$N = \frac{w_z}{2} \quad \text{or} \quad \frac{2}{w_z} N = n = +1$$

(533)

$$y = \pm b: \quad \eta = \pm \frac{1}{\sqrt{2z_b}}$$

$$N_y = -\left(N + \frac{w_z}{2}\right) = 0$$

$$N = -\frac{w_z}{2} \quad \text{or} \quad \frac{2}{w_z} N = n = -1$$

Equilibrium also requires the $n = (2/w_z)N$ values along the diagonals be zero. Further, by symmetry considerations, the shear N_{xy} along the x, ξ and y, η axis is zero.

n Values

To determine the n values, the first Laplace equation (532) would be most easily solved by replacing the partial differential equations with difference equations. To do this, the horizontal plane of the shell will be overlaid by a grid system (Fig. 198). The spacing of the grid depends on the accuracy required. We denote the spacing of the grid by m_ξ and m_η in the ξ, η directions, respectively, and we assume that $m_\xi^2 = \mu m_\eta^2$. Then, in accordance with Eq. (12), we obtain the following partial difference equation to be written for any grid point k:

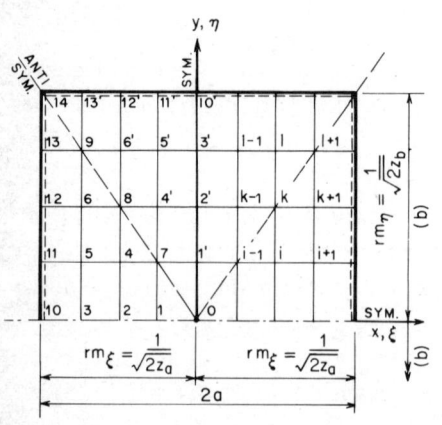

Fig. 198

$$\frac{\partial^2 n}{\partial \xi^2} + \frac{\partial^2 n}{\partial \eta} \approx \frac{n_{k+1} - 2n_k + n_{k-1}}{\mu m_\eta} + \frac{n_l - 2n_k + n_i}{m_\eta} = 0$$

$$\simeq n_{k+1} - 2n_k + n_{k-1} + \mu(n_l - 2n_k + n_i) = 0 \qquad (534)$$

$$n_k = \frac{1}{2(1 + \mu)} (n_{k+1} + n_{k-1} + \mu n_l + \mu n_i)$$

The unit n_k values for an 8×8 grid ($r = 4$) and the different μ values are given in Fig. 199 and Table 4.

To obtain $\mu > 1$ and justify the assumption $m_\xi^2 = \mu m_\eta^2$, the longer span $2b$ must be chosen in the y, η direction.

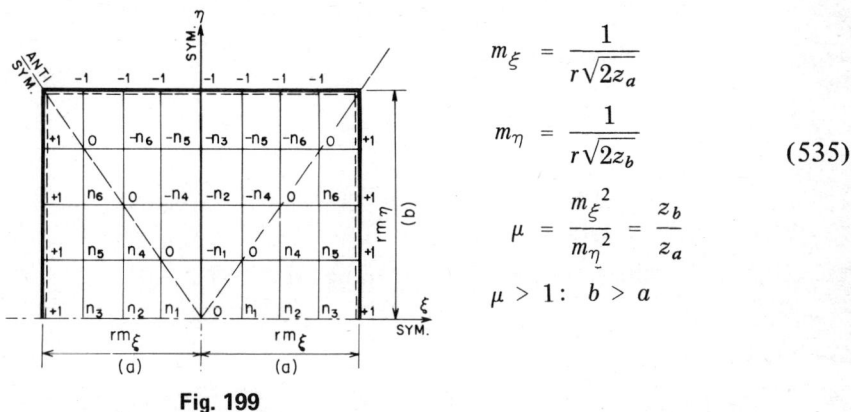

$$m_\xi = \frac{1}{r\sqrt{2z_a}}$$

$$m_\eta = \frac{1}{r\sqrt{2z_b}} \qquad (535)$$

$$\mu = \frac{m_\xi^2}{m_\eta^2} = \frac{z_b}{z_a}$$

$$\mu > 1: \quad b > a$$

Fig. 199

TABLE 4

μ \ k	1	1.2	1.4	1.6	1.8	2.0	2.5	3.0
n_1	0.066	0.057	0.045	0.038	0.035	0.032	0.024	0.018
n_2	0.264	0.246	0.218	0.205	0.196	0.189	0.162	0.143
n_3	0.587	0.572	0.552	0.534	0.523	0.515	0.485	0.465
n_4	0.201	0.188	0.169	0.157	0.150	0.143	0.123	0.108
n_5	0.542	0.528	0.510	0.490	0.481	0.473	0.444	0.425
n_6	0.386	0.371	0.355	0.342	0.333	0.325	0.304	0.286

s Values

For known unit n_k, the s_k values can be most simply computed from the Cauchy-Riemann equations (531) by the difference method. Thus,

$$\frac{\partial n}{\partial \xi} + \frac{\partial s}{\partial \eta} \simeq n_{k+1} + n_{l+1} - (n_k + n_l) + s_l + s_{l+1} - (s_k + s_{k+1}) = 0$$

$$\frac{\partial n}{\partial \eta} + \frac{\partial s}{\partial \xi} \simeq n_l + n_{l+1} - (n_k + n_{k+1}) + s_k + s_l - (s_{k+1} + s_{l+1}) = 0$$

(536)

Subtracting the second equation from the first, we obtain

$$n_{l+1} - n_k + s_l - s_{k+1} = 0$$

$$s_{l+1} = s_k + n_l - n_{k+1}$$

(537)

The s_k values at grid points on the ξ, η axis and the n_k values along the diagonals are zero. Starting from $s_{k=10} = 0$ $(x = -a, y = 0)$ in the diagonal direction, all s_k values can be computed using Eqs. (537).

Normal Forces and Shear

The N_x, N_y forces for any μ value and grid point k are computed using Eqs. (527):

$$N_x = \frac{1}{k_a}\left(N - \frac{w_z}{2}\right)$$

$$N_y = -\frac{1}{k_b}\left(N + \frac{w_z}{2}\right)$$

Multiplying these equations by $2/w_z$ and introducing k values, we obtain

$$\frac{4z_a}{a^2 w_z} N_{x,k} = \frac{2}{w_z} N - 1 = n_k - 1$$

(538)

$$N_{x,k} = (n_k - 1)\frac{a^2 w_a}{4z_a}$$

$$\frac{4z_b}{b^2 w_z} N_{y,k} = -\frac{2}{w_z} N - 1 = -(n_k + 1)$$

$$N_{y,k} = -(n_k + 1)\frac{b^2 w_z}{4z_b}$$

(538)
(Cont.)

The shear force $N_{xy} = N_{yx}$ for any grid point k equals s_k multiplied by the factor

$$n = \frac{N}{\sqrt{k_a k_b}} = \frac{ab}{\sqrt{z_a z_b}} w_z$$

$$N_{x,y,k} = N_{y,x,k} = s_k \frac{ab}{\sqrt{z_a z_b}} w_z$$

(539)

The total vertical shear V_z around the shell boundary must equal the total load. Thus,

$$\Sigma N_{xy} \sin\alpha = \Sigma V_z = 4abw \qquad (540)$$

where the $\tan \alpha$ value of a parabola is

$$\tan \alpha = \frac{\Delta z}{\Delta x} = \frac{2k_a}{a^2} x \qquad (541)$$

The membrane theory cannot describe the stress condition at the corners (singular point, $x = \pm a$, $y = \pm b$: $N_x = N_y = 0$). Considering that in accordance with the Eqs. (525) $k_c = 0$, $R_{xy} \to$ infinity; also $w_z = 0$. This means that perpendicular to the diagonal, the shell acts at a corner more or less as a slab to balance the Σw_z load. Due to this, the membrane stresses are superimposed by bending stresses. The magnitude of the bending involved at a corner region can be roughly evaluated from the difference in vertical shear $\Delta V_z = \Sigma V_z - 4abw_z$.

The principal forces $N_{\varphi_{1,2}}$ and $S_{\varphi_{1,2}}$ at the corner region, in accordance with Eqs. (482) and (483), are equal to N_{xy}. To keep the relatively high stresses, especially the diagonal tension (N_{φ_1}), within safe limits very often it is required to increase the shell thickness t at the corner and to apply posttensioning perpendicular to the diagonal of the shell.

As the shell is somewhat restrained at the Vierendeels, the bending stresses involved can be estimated by Eqs. (371) and (373).

Analysis of an Elliptic Paraboloid

For illustration, an elliptic paraboloid, shown on Fig. 200, will be analyzed.

Geometric data:

$a = 46.0$ ft $\qquad \mu = \dfrac{z_b}{z_a} \simeq 1.4$

$z_a = 5.75$ ft $\qquad r = 4$

$b = 64.0$ ft $\qquad \Delta x = 16.0$ ft

$z_b = 8.0$ ft $\qquad \Delta y = 11.5$ ft

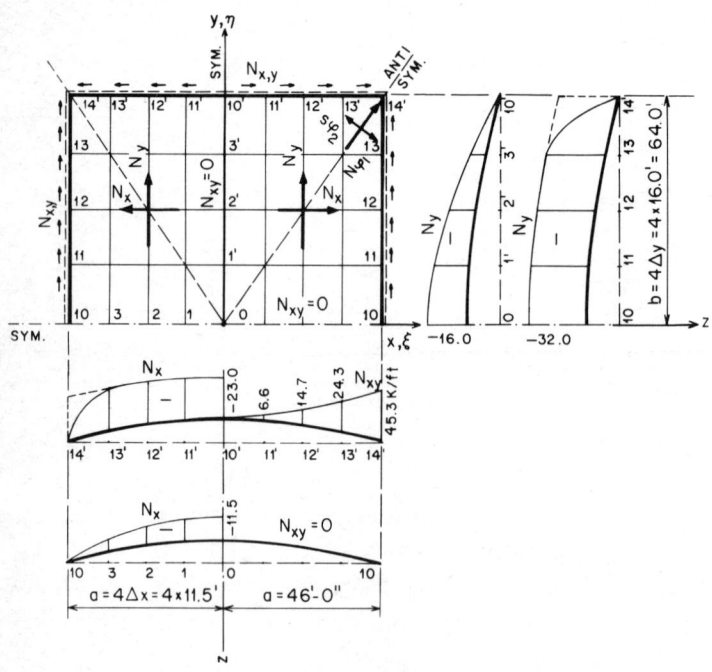

Fig. 200

Loading:

Dead load: $w_D = 0.085$ ksf
Live load: $\underline{w_L = 0.040 \text{ ksf}}$
$\quad\quad\quad\quad w_z = 0.125$ ksf

Normal forces:

$$N_{x,k} = (n_k - 1)\frac{a^2 w_z}{4 z_a} = (n_k - 1)\frac{46.0^2 \times 0.125}{4 \times 5.75}$$
$$= (n_k - 1) 11.50 \ k/\text{ft}$$

$k = 0: \ n_0 = 0.0 \quad\quad N_{x,0} = -1.00 \times 11.50 = -11.50 \ k/\text{ft}$
$k = 1: \ n_1 = 0.045 \quad N_{x,1} = -0.955 \times 11.50 = -11.00 \ k/\text{ft}$
$k = 2: \ n_2 = 0.218 \quad N_{x,2} = -0.782 \times 11.50 = -9.00 \ k/\text{ft}$
$k = 3: \ n_3 = 0.552 \quad N_{x,3} = -0.448 \times 11.50 = -5.25 \ k/\text{ft}$
$k = 10 \ n_{10} = 1.0 \quad\ N_{x,10} = 0.0$

$k = 10'\text{ to } 13' \quad n_k = -1.0$
$N_k = -2.0 \times 11.5 = 23.0 \ k/\text{ft}$
$n_{14'} = 0.0 \quad N_{14'} = 0.0$

$$N_{Yk} = -(n_k + 1)\frac{b^2 w_z}{4 z_b} = -(n_k + 1)\frac{64.0^2 \times 0.125}{4 \times 8.0}$$
$$= -(n_k + 1) 16.0$$

$k = 0: \quad\quad n_0 = 0.0 \quad\quad\quad N_{y,0} = -1.0 \times 16.00 = -16.0 \ k/\text{ft}$
$k = 1': \quad\quad n_1 = -0.045 \quad N_{y,1'} = -0.955 \times 16.00 = -15.30 \ k/\text{ft}$
$k = 2': \quad\quad n_2 = -0.218 \quad N_{y,2'} = -0.782 \times 16.00 = -12.50 \ k/\text{ft}$
$k = 3': \quad\quad n_3 = -0.552 \quad N_{y,3'} = -0.448 \times 16.00 = -7.15 \ k/\text{ft}$
$k = 10': \quad\ n_{10 \text{ ft}} = -1.0 \quad\ N_{y,10'} = 0.0$

$k = 10'\text{ to } 13' \quad n_k = +1.0$
$N_k = -2.0 \times 16.0 = -32.0 \ k/\text{ft}$
$n_{14} = 0.0 \quad N_{14} = 0.0$

Shear:

s_k values (Fig. 198, Table 4, $\mu = 1.4$):

$$s_{l+1} = s_k + n_l - n_{k+1}$$
$$s_5 = s_{10} + n_{11} - n_3 = 0 + 1.0 - 0.552 = 0.448$$
$$s_8 = s_5 + n_6 - n_4 = 0.448 + 0.355 - 0.169 = 0.634$$
$$s_9 = s_8 + n_6 - n_{6'} = 0.634 + 2 \times 0.355 = 1.344$$
$$s_{14} = s_9 + n_{13} - n_{13'} = 1.344 + 2 \times 1.0 = 3.344$$
$$s_{13} = s_6 + n_{12} - n_9 = 0.802 + 1.0 - 0.0 = 1.802$$
$$s_{12} = s_5 + n_{11} - n_6 = 0.448 + 1.0 - 0.355 = 1.092$$
$$s_{11} = s_3 + n_{10} - h_5 = 0 + 1.0 - 0.510 = 0.490$$
$$N_{x,y} = s_k \frac{ab}{4\sqrt{z_a z_b}} \quad w_z = s_k \frac{46 \times 64.0}{4\sqrt{8.0 \times 5.75}} \; 0.125 = 13.50 \; s_k \; k/\text{ft}$$

The $N_{x,k}$, $N_{y,k}$, and N_{xy_k} values are plotted for boundaries and x, y axis on Fig. 200.

32. STABILITY OF SHELLS

The factor of safety of a shell is based upon the stress condition and stability which exist under a particular external load the shell is subjected to. The stresses, as has been discussed, are determined from the stable equilibrium requirements between internal forces and external loads, where "stable equilibrium" is understood to mean that within certain limits any slight change of the loading condition does not produce a disproportionate increase in stresses and deformations of the assumed structural system. The limits are defined by allowable stress, which also determines the degree of safety as far as the strength of the shell in its stable state is concerned. The allowable limit of the deformations, until the stable equilibrium is possible, determines the degree of safety against instability of the system. As the thickness t of the shell, in comparison with its other dimensions, is relatively small, safety against instability is one of the most important features in shell design.

The theory and stress-stability analysis of shells are based on the assumption that concrete is a fully elastic material. As extensive research indicates, this assumption is well justified up to approximately 0.40

ultimate stress limit, but beyond this limit the plastic characteristics of concrete control and the linear stress-strain relationship are lost. In general, the stresses in concrete are relatively low because the thickness of the shell is commonly more controlled by practical necessities than by stresses. Therefore, the stresses in the shell are normally below the 0.4 f'_c limit, and classic shell theory describes the stress-strain condition in the shell with acceptable accuracy.

The prevailing stability analysis of shells also assumes concrete as a fully elastic material. The stability criterion is based upon the so-called *critical stress* (σ_{cr}) estimated by stability theories. The margin between actual stresses and stresses before breakdown of internal resistance of the material determines the factor of safety, regardless of whether the elastic limit is exceeded or not. The validity of this theory is not verified by experimental research because the critical stresses obtained by theory are considerably higher than those reached in actual testing. Such disagreement is understandable when we consider that no shell exists in which the internal force follows exactly the centerline curvature in the manner assumed by the theory and also, principally, when we consider that concrete is not a fully elastic material and the instability phenomenon is controlled by the entire elastic and plastic complex stress-strain relationship of the concrete under consideration.

The mathematical treatment of the stability phenomenon is usually based on the assumption that the shell is already deformed in a periodic-wave-type manner commonly described by the function

$$\bar{\omega} = \cos m\alpha \, \sin \lambda \xi \tag{542}$$

where $\bar{\omega}$ is the deflection in radial direction $\lambda = n\pi\xi$ and $\xi = x/L$ or $\xi = R/L$ ($m, n = 1, 2, 3, \ldots$).

The essence and nature of the stability phenomenon of a shell can be explained by considering a shell subjected to an extreme external load $w_{cr} = N_{cr}/R$. Due to this loading, the shell is deformed and its centerline is displaced, as illustrated in Fig. 201.

The shell section s ($s = d\alpha R$) is displaced in the x, y, z direction and rotated by ϑ. Maintaining stability and equilibrium requires the

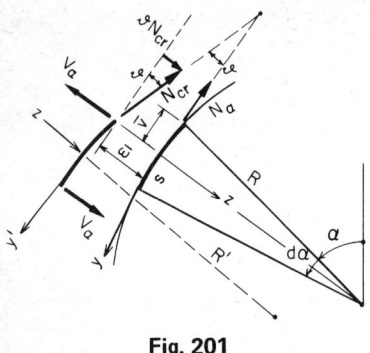

Fig. 201

development of a radial loading

$$z = \vartheta N_{cr} \tag{543}$$

and shear to counteract the rotation.

By neglecting the terms of higher order, the normal force N, shear V, and ϑ can be expressed in terms of $\bar{\omega}$ and N_{cr} can be determined from the condition that the sum of all components in the radial direction must be zero. Besides the trivial solution $\bar{\omega} = 0$ ($\lambda = 0$), the obtained differential equation has a system of so-called *characteristic solutions* ($\lambda \neq 0$; $m, n = 1, 2, 3, \ldots$) pertaining to the characteristic values of N_{cr} as a function of λ and m. The smallest N_{cr} value indicates the limit of elastic stability. It can be determined from the condition

$$\frac{dN_{cr}}{d(\lambda, m)} = 0 \tag{544}$$

Such rather complex analyses have been carried out most explicitly by Flügge. For a complete spherical shell, he obtains the critical normal force

$$\begin{aligned} N_{cr} &= \frac{2E}{\sqrt{3(1-\nu^2)}} \frac{t^2}{R} \quad \nu = \frac{1}{6} \\ &= 1.16 E \frac{t^2}{R} \end{aligned} \tag{545}$$

For curvilinear shells (one-direction curvature), the critical normal force is approximately half that of the spherical shell. Thus,

$$N_{cr} = 0.58 E \frac{t^2}{R} \tag{546}$$

The factor of safety is

$$v = \frac{N_{cr}}{N_{(\alpha x)}} \tag{547}$$

For example, the minimum thickness of a spherical shell is $t = 3.5$ in. $=$ 0.292 ft, the radius $R = 100$ ft, and $\sum_0^L w = 0.055 + 0.040 = 0.095$ ksf.

$$E_0 = 4.5 \times 10^6 \text{ psi} \sim 6.5 \times 10^5 \text{ ksf}$$

$$\Sigma N_\alpha = \Sigma wR = -0.095 \times 100 = -9.50 \text{ k/ft}$$

$$\Sigma \sigma_\alpha = \frac{\Sigma N_\alpha}{12t} = -226 \text{ psi} \qquad \sigma_D = 131 \text{ psi}$$

$$N_{cr} = 1.16 E \frac{t^2}{R} = -1.16 \times 6.5 \times 10^5 \frac{0.292^2}{100} = \sim 65 k$$

$$\sigma_{cr} = \frac{N_{cr}}{12t} = -1{,}550 \text{ psi}$$

$$v = \frac{N_{cr}}{\Sigma N_\alpha} = 6.85$$

Using the same data as for a spherical shell, the corresponding critical values for a curvilinear shell will be half. Thus,

$$\Sigma N_\alpha = -9.5 \text{ k/ft}$$
$$\Sigma \sigma_\alpha = -226 \text{ psi}$$
$$N_{cr} = -65/2 = -32.5 k$$
$$\sigma_{cr} = -1{,}550/2 = -775 \text{ psi}$$
$$v = 6.85/2 = 3.42$$

These critical forces and factor of safety would be very close to the actual ones if the assumption on which the stability theory has been based were true. Unfortunately, this is not the case.

Assuming that even a relatively small deviation from the theoretical centerline exists, a disturbing moment $\overline{M}_0$ will be present. Thus,

$$\overline{\omega} = \frac{\overline{M}_0}{N_{cr}} = \frac{\overline{M}_0}{v N_\alpha} \tag{548}$$

$$v = \frac{\bar{M}_0}{\bar{\omega} N_\alpha} = \frac{\bar{M}_0}{M_{n=1}} \quad n = 1, 2, 3, \ldots$$

$$M_{n=1} = \frac{\bar{M}_0}{v} \tag{548}$$
(Cont.)

The additional moment $M_{n=1}$ caused by the initial disturbing moment $\bar{M}_0$, tacitly assumed to be affined to the buckling waves, causes further additional moments (Dischinger):

$$M_{n=2} = \frac{M_{n=1}}{v} = \frac{\bar{M}_0}{v^2}$$

$$M_{n=3} = \frac{M_{n=2}}{v} = \frac{\bar{M}_0}{v^3}$$

$$\ldots \ldots \ldots \ldots$$

The total moment due to the elastic deformations of the centerline of the shell will be, for time $t = 0$,

$$\sum_{n=1}^{n=\infty} M = \bar{M}_0 \left(1 + \frac{1}{v} + \frac{1}{v^2} + \frac{1}{v^3} + \cdots \right)$$
$$= \bar{M}_0 \frac{v}{v-1} \tag{549}$$

The final moment due to elastic and plastic deformations, in accordance with Eq. (275), reaches its maximum value when plastic flow of the concrete is finished. Thus, for time $t = t_n$,

$$\Sigma M_{t_n} = \bar{M}_0 \frac{v}{v-1} e^{\varphi_n/(v-1)} \tag{550}$$

For example, assuming that $\bar{\omega} = 1/4$ in. $= 0.0208$ ft and $\varphi_n = 2.0$, the stresses and factor of safety for the same curvilinear shell discussed above ($\sigma_{cr} = 775$ psi) will be, for $t = 0$ and $t = t_n$:

Dead load ($N_\alpha = 5.5\,k$):

$$\bar{M}_0 = \bar{\omega} N_\alpha = 0.0208 \times 5.50 = 0.115\,k'$$

$$v = \frac{N_{cr}}{N_\alpha} = \frac{32.5}{5.5} = 5.9$$

$$\sum_{n=1}^{n=\infty} M_{t=0} = \overline{M}_0 \frac{v}{v-1} = \sim 0.138\,k'$$

$$\sum_{n=1}^{n=\infty} M_{t=t_n} = \sum M_{t=0}\, e^{\varphi_n/(v-1)} = 0.138 \times 2.781^{2.0/4.9} \simeq 0.210\,k'$$

Live load ($N_\alpha = 4.0\,k$):

$$v = 3.42$$
$$\overline{M}_0 = \overline{\omega} N_\alpha = 0.0208 \times 4.0 = 0.083\,k'$$

$$\sum_{n=1}^{n=\infty} M = \overline{M}_0 \frac{v}{v-1} = 0.083 \times \frac{3.42}{2.42} = 0.117\,k'$$

Stresses and factor of safety:

$t = 0$:

$$\sum \sigma_\alpha = \frac{N_\alpha}{12t} \pm \frac{\Sigma M}{S} \qquad S = \frac{bt^2}{6}$$

$$= -\frac{9500}{12 \times 3.5} \mp \frac{(0.138 + 0.117)\,12{,}000}{24.5} = \begin{array}{l} -351\text{ psi} \\ -101\text{ psi} \end{array}$$

$$v = \frac{\sigma_{cr}}{\Sigma \sigma_\alpha} \simeq \frac{775}{351} \simeq 2.2$$

$t = t_n$:

$$\sum \sigma_\alpha = -226 \mp \frac{(0.210 + 0.117)\,12{,}000}{24.5} = \begin{array}{l} -386\text{ psi} \\ -66\text{ psi} \end{array}$$

$$v \simeq \frac{775}{386} \simeq 2.0$$

In the above, the factor of safety has been computed from fiber stresses, which is not quite correct since the maximum fiber stress below the $0.4\,f'_c$ stress limit is not exactly decisive for instability of the shell. However, this rough analysis clearly proves that the critical stress computed by the elastic-stability theory can never be reached in actual condition.

PART 2

MATERIALS

Regardless of how exact and extensive the theoretical treatment of a design may be, the final work may lack significance unless more advanced ideas, theories, material technology, and construction methods are understood and applied.

From day to day what one "knows" changes, disappears, and gives way to new concepts, theories, and construction methods which are more rational and economical. The progress is so extensive and rapid that it appears to be limitless. But this advancement in technology is unavoidably accompanied by the steadily increasing specialization in all fields of engineering and also by a lack of understanding among specialized engineers working in offices, in laboratories, and in the field. The architect, who in almost any contemporary design has the leading role, is commonly the least up-to-date in terms of this vast technological achievement and thus is unable to make use of it for architectural benefit and is far less able to coordinate consciously the work of the team of specialists he is assumed to command. Allied fields in which the work of a project team is most closely related are materials technology and construction methods. Unfortunately, it is just in these particular fields that the gap of knowledge and understanding of one another's work is the widest and most profound.

In any contemporary structure, the basic construction materials are concrete and steel. For both of these materials our knowledge boundaries have been considerably extended by using special scientific instruments such as the electronic microscope, x-ray analysis, colloidal chemistry, and capillary physics. Scientific studies of materials in the manufacturing process as well as of the physical properties and behavior of materials under various possible conditions are largely responsible for our understanding of the true nature of today's materials. In the past, time-requiring testing and experimentation were the only means to obtain data about the strength, elasticity, etc., of materials. But regardless of how important such fact-finding tests are, they usually give only a two-dimensional picture of phenomena; they lack the depth, or third dimension, required for understanding the true nature of materials. The third dimension has been made available by basic scientific research. The scientific approach in contemporary technology is mainly responsible for its successes. It is at present already at such a level that the physicochemical characteristics of concrete and steel produced are sufficiently known that their behavior under any conceivable loading condition at any time can be predicted with accuracy. Thus, at present, any project can be carried out with required quality and reasonable safety.

The quality of a structure depends not only on design and material qualities but also to a great extent on construction methods and quality. Therefore, the construction methods should be determined in terms not only of economy but also of obtainable quality.

As the materials and their proper use in the final product (structure) are closely related, the architect, engineer, concrete technologist, and construction engineer should have at least a basic understanding of the materials and proper construction methods associated with a particular contemporary structure if maximum results at minimum costs are to be obtained. The following chapters are intended to give a basic knowledge of this subject and thus to help bridge the gap of knowledge and understanding among various specialists working as a team for the same end.

8
CONCRETE

33. GENERAL CHARACTERISTICS

Concrete is a physicochemical mixture of cement, water, and aggregate manufactured at or near the construction site. It contains a large number of voids and capillary tubes filled with water, moisture, and air. The volume of these voids and tubes, depending on the quality of concrete, may vary from 7 to 27 percent of the total volume of concrete. As such, concrete is not a solid body in the ordinary sense but is, rather, a pseudosolid and, for that reason, is subject to the capillary laws. This lends to concrete certain physical properties which change with time t, relative humidity of air ϵ_A, stresses σ due to loading, and temperature changes ΔT of the atmosphere.

The action of capillary forces in concrete results in internal stresses and deformations, the magnitudes of which depend on the volume and size of the capillary tubes and voids in the concrete. The capillary forces can be relatively high during the period of evaporation or condensation of water in the capillary tubes, as we shall see in the following paragraphs. They may lead to serious crack formation and reduction in strength and other qualities of concrete.

Due to this, the physical properties of concrete vary widely, depending on the qualities and properties of materials, the proportioning of mix, the gradation of aggregate, the method of batching, mixing, placing, and curing, and the weather conditions during the hardening period. The neglect of any of these essentials may seriously hamper attainment of the results desired, even though all the other requirements of good practice are properly satisfied.

To fully understand the behavior and physical characteristics of concrete and the factors which influence and affect them, one has to know the increments the concrete is made of. Therefore, in the following the constituents of concrete and the other essentials mentioned above will be discussed.

Cement*

Cement is a product of minerals of very complicated crystal structure. When mixed with water, the minerals react and turn into a colloidal state. The physicochemical nature of the colloidal mixture is responsible for the hydraulic hardening of the cement. The chemical compositions of portland cement minerals are:

Tricalcium silicate:	$3CaO \cdot SiO_2$	(C_3S)
Dicalcium silicate:	$2CaO \cdot SiO_2$	(C_2S)
Tricalcium aluminate:	$3CaO \cdot Al_2O_3$	(C_3A)
Tetracalcium aluminoferrite:	$4CaO \cdot Al_2O_3 \cdot Fe_2O_3$	(C_4AF)

Customary denoting of clinker minerals in industry is given in parentheses.

Tricalcium silicate contains 73.7 percent calcium oxide (CaO) and 26.3 percent silicon dioxide (SiO_2). When water is added, the tricalcium silicate transforms as follows:

$$2(3CaO \cdot SiO_2) + 6H_2O \rightarrow$$
$$3CaO \cdot 2SiO_2 \cdot 3H_2O + 3Ca(OH)_2$$

---

*Wolfgang Czernin, Cement Chemistry and Physics for Civil Engineers, Chemical Publishing Co., Inc., New York, 1962.

By molecular weight, the tricalcium silicate (100 percent) can bind 24 percent of water, resulting in 75 percent $3CaO \cdot 2SiO_2 \cdot 3H_2O$ (tobermorite) and 49 percent calcium hydrate [$Ca(OH)_2$]. Thus fully hydrated tricalcium silicate contains 40 percent by weight of calcium hydrate. The hydration is rather rapid and generates about 120 cal/gm of heat. Due to this, the tricalcium silicate is mainly responsible for the early strength of the cement (Fig. 202).

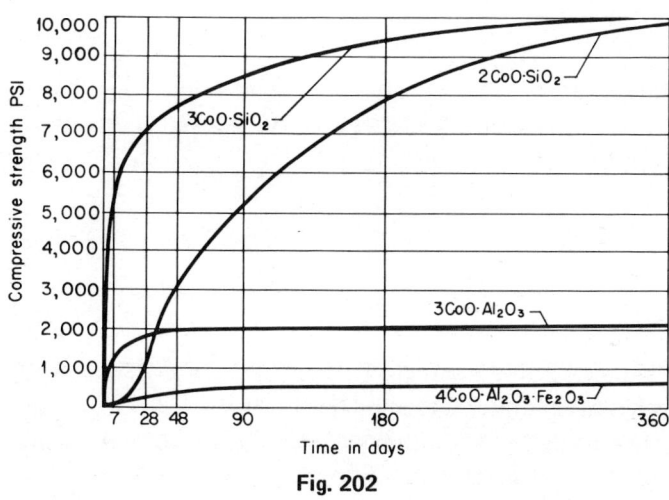

Fig. 202

Dicalcium silicate contains 65.1 percent calcium oxide (CaO) and 34.9 percent silicon dioxide (SiO_2). It hydrates by reaction:

$2(2CaO \cdot SiO_2) + 4H_2O \rightarrow$
$3CaO \cdot 2SiO_2 \cdot 3H_2O + Ca(OH)_2$

Thus 100 percent of dicalcium silicate combines 21 percent water, resulting in 100 percent of $3CaO \cdot 2SiO_2 \cdot 3H_2O$ (tobermorite) and 21 percent by weight of calcium hydrate [$Ca(OH)_2$]. Thus fully hydrated dicalcium silicate yields approximately 18 percent by weight of calcium hydrate. Its hydration is much lower than that of tricalcium silicate and generates only 62 cal/gm of heat. Dicalcium silicate is mainly responsible for the later increase of strength of the cement.

Tricalcium aluminate contains 62.3 percent calcium oxide and 37.7 percent aluminum oxide (Al_2O_3). It hydrates by reaction:

$$3CaO \cdot Al_2O_3 + 6H_2O \rightarrow$$
$$3CaO \cdot Al_2O_3 \cdot 6H_2O$$

Also, it hydrates without lime involvement.

Thus 100 percent of tricalcium aluminate combines 40 percent of water. Its hydration is very rapid and generates 207 cal/gm of heat. As a result, tricalcium aluminate contributes largely to the early strength of the cement.

Tetracalcium aluminoferrite is the least understood of the constituents in cement, and considerable differences exist as to its reaction mechanics. Commonly, it is presented as

$$4CaO \cdot Al_2O_3 \cdot Fe_2O_3 + 2Ca(OH)_2 + 10H_2O \rightarrow$$
$$3CaO \cdot Al_2O_3 \cdot 6H_2O + 3CaO \cdot Fe_2O_3 \cdot 6H_2O$$

In accordance with this reaction, 100 percent of tetracalcium aluminoferrite combines with 37 percent by weight of water. Thus about 30 percent of calcium hydrate set free during hydration of silicates becomes bound chemically by hydration of tetracalcium aluminoferrite. The rate of hydration is relatively steady and generates 100 cal/gm of heat.

Besides these main constituents, there are minor ones, such as magnesium oxide (MgO), sulphuric anhydride (SO_3) and alkalis (K_2O and Na_2O). But in spite of the fact that these constituents are present in small amounts, their effect on the properties of cement can be considerable.

The lime (CaO) content in cement is the highest; then follow silicon, alumina, and finally ferrite. The quantity of lime present in cement is of great importance to the physical characteristics and behavior of cement in the course of time. It is commonly controlled by the so-called *lime saturation factor* (LSF). The LSF represents the ratio of lime actually present in cement (in percentage of weight) to the quantity of lime chemically required to form the main constituents. The LSF in portland cement should not be higher than approximately 1.0 and lower than 0.66.

If the LSF is higher, free lime (CaO) unavoidably remains in the cement. During hydration this results in excessive heat and the formation of coarse-structure calcium hydroxide [Ca(OH)$_2$] crystals. Some of the free lime, trapped inside the crystalline structure of the clinker, is not easily accessible to mixing water. In the course of time, when the cement is already hardened, the free lime takes in moisture and transforms to calcium hydroxide. This process is connected with unavoidable volume change, which results in relatively high internal stresses and which, depending on the quantity present, may lead to considerable crack formation. As the calcium hydroxide is soluble in pure water (rainwater), it can be leached out from concrete (efflorescence on the concrete surface).

If the lime concentration drops below the permissible limit, and considering that all cement hydration products are stable only in aqueous solutions containing a certain minimum concentration of calcium hydrate, the chemically combined calcium oxide goes into solution, decomposing the calcium silicates, aluminates, and ferrites. Lime-poor hydraulic products have considerably less strength than products having proper lime saturation.

In general, the chemical composition of portland cement is approximately

CaO:	63–66%
SiO$_2$:	20–22%
Al$_2$O$_3$:	5.5–7.7%
Fe$_2$O$_3$:	3–4.5%
MgO:	1.5–5%
SO$_3$:	2.5–3%

In nature, magnesium oxide is commonly associated in small quantities with calcium carbonate (CaCO$_3$) and thus is unavoidably present in cement. It does not go into chemical reaction with acidic oxide but remains in cement as "free magnesia." As in the case of free lime, it gradually takes moisture in, which results in an increase of volume. The magnesia expansion is far more destructive than lime expansion because the rate of its reaction with water is relatively slow and the resulting crack formation may appear after many years. Due to this danger, the magnesia content is limited to maximum of 5 percent.

The sulphuric anhydride (SO$_3$) is added to cement as gypsum (CaSO$_4$) during its grinding to regulate the setting time of cement. If the addition of gypsum exceeds a certain percentage and the SO$_3$ content increases above the 3 percent limit, it may result in so-called *sulphate expansion*. Attempts have been made to use the sulphate expansion for balancing shrinkage of cement. The best known of this type of nonshrinking cement are the Lossier cements. Lossier intergrinds sulfoaluminate admixtures in the amount of 10 to 20 percent by weight with portland cement. These sulfoaluminates, commonly produced by a fusion process or even by mixing calcium sulfates and aluminates, react with the aluminates in portland cement to form hydrated sulfoaluminates during hydration and setting. For stabilizing of the expansion, blast-furnace slag is used.

Klein uses in his expansive cement an anhydrous calcium aluminosulfate component as the expansive agent. The calcium aluminosulfate is a fusion product formed by a mixture of CaO, Al_2O_3, and SO_3. The obtained clinker, having $CaSO_4$ not greater than 5 percent, a combined $CaO + SO_3$ content of about 37 percent, and free lime (CaO) about 19 percent, is ground and blended about 20 to 30 percent with 75 percent low-aluminum-content portland cement. It is claimed that no stabilizer is required. The magnitude and rate of expansion are controlled by the amount of expansive component blended with the portland cement and by the water-cement ratio.

Thus the physical characteristics and mineralogical composition of portland cement can be greatly affected even by a relatively small change in chemical composition or by intergrinding admixtures of pozzolanic materials or other additives such as blast-furnace slag.

For instance, hardening can be accelerated by increasing the rapidly hydrating tricalcium silicate and tricalcium aluminate content and also by finer grinding of the clinker. A reduction in hydration heat generation can be obtained by eliminating the tricalcium aluminate and reducing the tricalcium silicate content. Also, for example, white or other colored portland cement can be obtained by eliminating ferrite and selecting pure raw materials or by intergrinding suitable color pigments.

Hydration Process

As soon as cement is mixed with water, the cement grains attract the surrounding water and the surface softens. The speed of water attraction and surface softening is rapid with tricalcium silicate and somewhat slower with dicalcium silicate. The calcium silicates hydrate to tobermorite and release calcium hydroxide, which saturates the gauging water. The gauging water dissolves gypsum, which reacts with aluminates to form calcium sulphoaluminate. By this process, the specific surface of dissolved cement fragments increases considerably (about 1,000 times) and is approximately proportional to the chemically bound water. This is understandable when we consider that the cement-grain size varies from 100 to 1μ (1μ = 1 micron = $1/1,000$ mm) and that the size of the fully hydrated cement fragments in colloidal state varies from 10 to 1 $m\mu$ ($1m\mu = \mu/1,000 = 1/10^6$ mm). This transformation is associated with the corresponding increase in the specific surface of the hydrated products (colloids).

The hydration and cement gel formation starts from the outermost surface of the cement grains and is forced by swelling (the gel volume is more than twice the volume of the original cement, with a specific surface up to $2 \times 10^6 \, cm^2/gm$) into the water-filled capillary space between the grains. When the hydration proceeds, the gel gradually consolidates and becomes denser, penetrates the capillaries, and establishes close contact with the cement constituents. If the amount of mixing water or capillary space present is insufficient to allow spreading of the gel, clusters of slowly softening fragments are formed, hindering the hydration and transforming the fragments into colloidal solution.

During hydration, chemical heat is generated. As hydration proceeds from the surface of the grains or clusters, the heat builds up in the cores of larger grains (up to approximately $178°F$). Due to this, the hydration toward the interior is slowed down because the temperature of the surrounding water must be increased at least up to the core temperature level for the water to be able to penetrate the cores. The rate of hydration of the cores of larger grains and clusters is further slowed down by hindered water penetration through the surrounding gel and high vapor pressure due to the chemical heat. When the chemical heat is high (as in the case of cements rich in aluminum) and cannot escape (as in mass construction), artificial cooling is required to avoid evaporation and high vapor pressure and thus to accomplish sufficient hydration and reduce thermal expansion.

It is possible that after years the hydration of larger cement grains inside massive concrete is complete in the grains no deeper than a few microns from the surface, even in the presence of sufficient capillary water. This possibility is substantiated by the phenomenon of "healing" of concrete. When cracks occur in concrete, the unhydrated or not fully hydrated particles along the faces of the cracks will be exposed and thus made accessible to hydration. Gel is formed and forced into the cracks.

Tests show that when high-strength concrete is loaded at 3 days for approximately 10 min up to 75 percent and then reduced to 0.6 of its ultimate, the 28-day strength is considerably higher than that of the same concrete cured under similar conditions but not loaded. This can be explained principally by the more complete hydration due to forced capillary-water penetration of the unhydrated cement fragments and the reduced voids due to the volume increase of the hydrated products.

To obtain more complete hydration, fine-ground cement should be used. Cluster formation can be avoided by using a w/c ratio not lower than 0.4, by admixture, and by thorough mixing. The fineness of cement results in more complete hydration and high early strength. However, it must be realized that rapid hydration is associated with higher chemical heat and shrinkage. Both of these phenomena will be discussed under the physical characteristics of concrete.

The colloidal rigid cement gel is a crystalline homogeneous substance having relatively high strength. The crystals are extremely small and cannot be seen even by the electronic microscope but can be determined only by means of x-ray analysis. The electronic microscope reveals, however, a general picture of occurrences of the hydration and following hardening of cement. It is interesting to see that after weeks of hydration and hardening, the larger cement fragments are still present and visible.

The hydration process is schematically illustrated in Fig. 203.

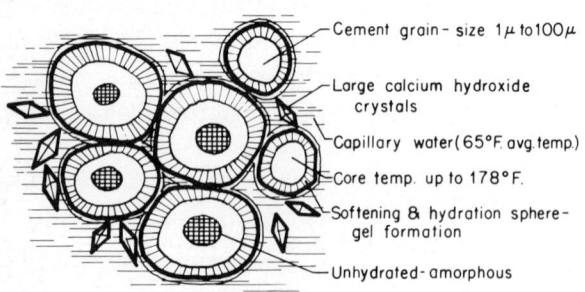

Fig. 203

The water in the cement paste is present in three states: chemically bound water, gel water, and capillary water. In fully hydrated cement, the chemically bound water is roughly estimated to be 25 percent of the weight of the cement and gel water to be 13 to 15 percent. The rest of the mixing water is capillary, or free, water.

In the crystallization state of the gel, the pressure is relatively high, and it resists the swelling due to hydration and causes compression of the chemically bound and gel water. The gel water is assumed to be only weakly chemically bound and can be evaporated, but it is not capable of any further chemical reaction. The capillary and absorbed (aggregate) water serves only for further hydration. Once air fills the capillaries and voids, hydration ceases. This phenomenon explains the importance of curing in the early stages of hardening. Furthermore, when evaporation of capillary water starts, very high capillary forces develop (see Sec. 42), resulting in serious surface-crack formation. As the surface quality of exposed concrete has great importance, the drying of the cement paste must be delayed by curing until the tensile strength has developed enough to resist the capillary forces.

The initial set of cement paste is largely dependent on temperature and is commonly considered to take 1 hr at 70°F. It characterizes the degree of hydration and gel formation to the extent that some stiffness is notable (Vicat needle test). The final set is theoretically meaningless; it indicates the time until the mix is still workable without causing mechanical damage to the already developed crystals.

The unit weight of cement is approximately 94 lb/ft^3. The specific weight is 3.1. Thus the solid mass is 0.487 ft^3 and the voids volume approximately 0.573 ft^3.

As the physical characteristics and properties of cement and concrete are essentially the same, they will be discussed together under "Physical Properties of Concrete," Sec. 41.

34. AGGREGATES

The aggregates represent about 75 percent of the volume of concrete. Therefore, the quality of the aggregates has an important bearing upon the lasting quality and behavior of concrete in service. In practice, the importance of aggregate quality is commonly overlooked except in connection with the 28-day strength of concrete. For contemporary concrete structures with reduced dimensions and large exposed areas, there are, besides strength, other, even more important qualities to consider, such as durability in all weather conditions, fire resistance, protection of reinforcing, deformations, defect-free appearance, non-staining. It is not enough to be satisfied with the standard requirements—such as clean, hard, strong, durable, uncoated particles free from injurious amounts of soft, thin, or laminated pieces, alkali, and organic or other deleterious matter and proper and uniform grading—because for high-quality concrete, the mineralogical structure, geological origin, and chemical composition of the natural mineral aggregates is equally important. Besides, for lightweight aggregates the manufacturing process is also significant.

In recent years, extensive research has been carried out by concrete research laboratories,* and valuable information is available concerning the aggregates and concrete as a whole. But there can be considerable difference in laboratory and field conditions. First, the laboratory works under controlled conditions, uses small specimens, and obtains isolated, basic scientific data. The field, on the other hand, works under actual conditions and must consider all factors, which usually are interrelated and complex. Due to this fact, the practicing engineer must use his own judgment and personal interpretation of the available research data and must relate them with existing conditions. Therefore, it is essential to know what the inherent characteristics and properties of aggregates are and the extent and degree of their significance, in relation with other factors involved, upon the final overall quality of concrete for a particular finished structure.

Fine Aggregates

The gradation of fine aggregates—natural and structural lightweight—is an important factor in controlling the volume of cement paste, density, and water content in mortar. The recommended gradation for natural sand is given in Fig. 204. For lightweight and architectural concrete, the percentage of fine grain sizes should be selected slightly higher, and thus closer to the fineness limit, than for purely structural concrete. The grading of aggregate very often has to be improved by adding or separating certain grain sizes.

As a rule, the strength of the aggregate material should be higher than the strength of the cement paste. However, tests indicate that high-strength concrete can be obtained even with aggregates not meeting this requirement. This is explained by the three-axial confinement of aggregate in the surrounding cement paste.

*"Manual of Concrete Practice," American Concrete Institute, 1968.

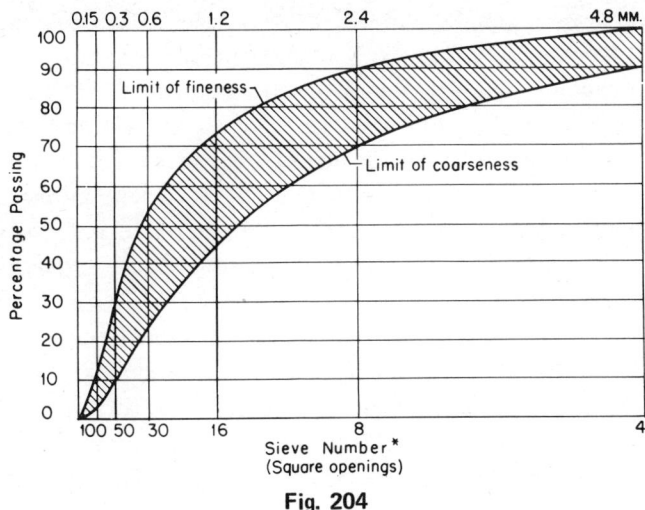

Fig. 204

The strength of natural quartz sand is very high, approximately 30,000 to 65,000 psi. The shapes of the grains are approximately round, and the moisture is present as film water surrounding the grains. Absorption capacity is about 0.7 percent by dry weight, and permeability is practically zero.

As to thermal expansion, there are two types of quartz in nature: α-quartz and β-quartz. The α-quartz (SiO_2) is the most common. At approximately 1000°F (575°C), it transforms into β-quartz (SiO_2) without any change in chemical composition but with a change in mineralogical structure, becoming more coarse. Its volume increases by approximately 24 percent. In volcanic areas, the change of α-quartz to β-quartz has already taken place. This phenomenon explains the damage in concrete containing α-quartz in fire. The thermal expansion of α-quartz is approximately 5.21×10^{-6}. Heat conductivity is 0.030. At normal temperatures and grain sizes, it is chemically unreactive. The unit weight of dry natural sand having good gradation is approximately 100 to 110 lb/ft^3. The amount of film water in bulk storage is approximately 2 to 4 percent by dry weight. Thus the voids are 33 to 40 percent. Specific weight is approximately 2.65.

Commercial structural lightweight aggregate is a cellular material

*Standard size designation is based on the number of meshes per linear inch.

processed of shales, clays, slates, or iron blast-furnace slags. The cells in particles may vary from microscopic to macroscopic size and may be closed or interconnected. The cellular structure is formed at high temperature, up to 2000°F, by entrapped gases or steam retained on cooling of the mass. The methods used for manufacturing commonly are the Klein process, which is similar to that used in cement clinker manufacturing, and the sintering process, using raw material mixed with fuel and burned. The gases formed expand the viscous mass and are entrapped and retained in cooling.

In both processes, the raw material can be pelletized prior to burning. The pelletized aggregate has a rounded shape, and the surface texture is smooth and closed. The surface texture of crushed materials is relatively rough, and the shapes are irregular. Proper grading is obtained by crushing and screening. The color is dark brown, almost black, or dark blue-gray.

Depending on the raw material used and the method of manufacturing, the physical characteristics of lightweight aggregate may vary greatly. Water absorption is relatively high, up to 25 percent by dry weight for crushed materials but far less for pelletized materials, and film water is practically zero. The durability of particles having no fractures is satisfactory, even in a high degree of saturation. The aggregate is chemically unreactive; thermal expansion and heat conductivity are relatively low in comparison with natural stone.

The strength is a function of the unit weight, depending on the size, volume of voids, and degree of sintering. The weight can be relatively high for fine grains, approximately 60 to 65 lb/ft^3. The specific weight is approximately 2.4.

Coarse Aggregates

The gradation of coarse aggregate is not as important as for fine aggregate because it influences the amount of voids in concrete only slightly. The minimum volume of voids of coarse aggregate depends on gradation and size of particles. The maximum particle size for high-quality concrete is 3/4 in. For exposed (architectural) and lightweight concrete, the maximum size of particles is 1/2 in. or even 3/8 in. (see "Mix Design," Sec. 37). The recommended gradation for coarse aggregate is given in Fig. 205.

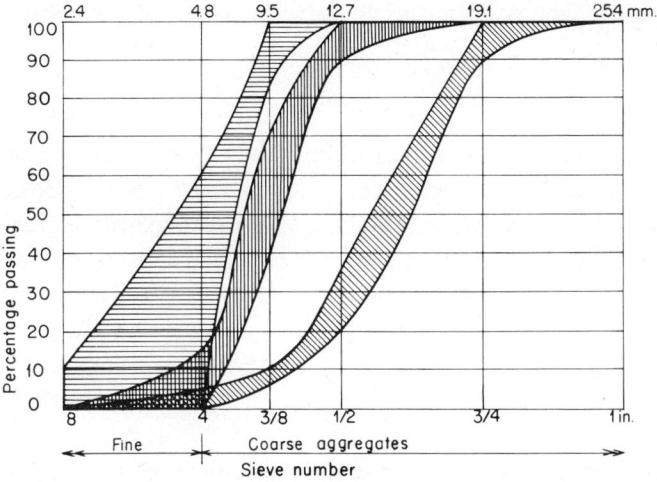

Fig. 205

The physical characteristics of coarse aggregate vary in accordance with geological origin, mineralogical structure, and chemical composition.

Water absorption of igneous rocks with coarse mineralogical structure is up to 0.5 by dry weight and is up to 0.3 for finer structures (traprock, especially basalt). For sediment stones used in aggregate, the water absorption can be as high as 5 percent by dry weight. In bulk storage, the film water present can be up to 2 percent by dry weight.

Unit dry weight is 95 to 105 lb/ft³. The average specific weight is approximately 2.65. Thus the solid volume V_s and voids per cubic foot are

$$V_s = \frac{w}{2.65 \times 62.4} \simeq 0.57 \text{ to } 0.63 \text{ ft}^3$$

$$V_v = 1 - V_s = 0.43 \text{ to } 0.37 \text{ ft}^3$$

The strength of stones (compressive and tensile) for coarse aggregate is more important than for fine aggregate. It is, for coarse-structure igneous stones,

Compressive strength: 23,000–42,000 psi
Tensile strength: 1,400–3,000 psi

and even far higher for traprock. The strength of sediment stones is generally less than for igneous stone. It is, for limestone, approximately

Compressive strength: 11,000–25,000 psi
Tensile strength: 850–2,150 psi

Thermal expansion for igneous stone is approximately 4.7×10^{-6} and is slightly less for limestone, approximately 4.4×10^{-6}. For igneous stones containing silicon constituents, the thermal expansion may increase considerably at elevated temperatures and, therefore, should not be used where extensive, long-duration heat (above 800°F) can be expected.

Heat conductivity, granite: 0.0045–0.0050
              limestone: 0.0025–0.0070

Generally, the stones used for coarse aggregate are chemically unreactive. However, there are stones, such as dolomite, dolomitic limestone, opal, opaline silica, and chalcedony, which react chemically with alkalis (K_2O, N_2O) in portland cement, resulting in deterioration of the cement (alkali-aggregate reaction expansion).

Gradation for lightweight structural coarse aggregate is the same as for stone aggregate. The water-absorption capacity of pelletized, uncrushed, uncracked, round particles is practically no more than 2 percent by dry weight. Also, for closed-cell crushed particles it is considerably less than for particles with interrelated void-cell structure. The unit dry weight is 45 to 50 lb/ft^3, and the specific weight is approximately 2.4 (apparent specific weight = 1.4).

35. ADMIXTURES

Admixtures are used to affect some of the characteristics of hydration and capillary forces and to reduce internal friction and intermolecular action of concrete constituents. The admixtures can be classified generally into four groups:

1. Retarders or accelerators of initial setting
2. Plasticising and water-reducing agents
3. Air-entrainment agents and lubricators
4. Fillers and coloring additives

Many of the admixtures on the market perform one or more of these functions simultaneously. The effect of some admixtures is significantly modified by cement type and amount present, hydration heat, mixing time, and water-cement ratio. Because of this, admixtures represent a complicated phenomenon and their effect and usefulness can be evaluated only by prior thorough testing under actual field conditions. This is especially true for the many agents whose chemical compositions are trade secrets.

Several organic and inorganic compounds are known as useful retarders of cement. Examples are sodium phosphates, calcium inorganic salts, sulphate lye, many types of sugar compounds, and lignosulfonates.

The best-known accelerator is calcium chloride ($CaCl_2$). It does not shorten the initial setting but decreases considerably the final setting time and increases the rate of hardening. If it is used in amounts of less than 1 percent by weight of cement, it is considered harmless to intermediate-grade reinforcing steel. It should not be used in the presence of high-carbon steel (prestressing tendons). Commonly, the retarding agents act to a certain degree as air-entraining and water-reducing agents (especially the lignosulfonates and sulphite lye).

The plasticising agents not affecting setting time are hydroxylated carboxylic salts and also some sugar compounds. Their effect is mainly caused by their ability to reduce the surface tension of water, resulting in a reduction of cohesive forces and thus making the film water around the aggregate particles more available for carrying and distributing the cement. In addition, the cluster formation of cement grains is reduced by the dispersive action of the agents.

Commonly, the plasticising agents contain calcium chloride to counteract the retarding effect of the agent. The quantity of calcium chloride present must be checked, especially when high-carbon steel is used.

Because of the water-reducing ability of plasticising agents and the more complete hydration due to their dispersive action, there is a slight increase in strength, especially at a later date.

The purpose of an air-entraining agent is to induce relatively small air bubbles (up to approximately 0.004 in. in diameter) into cement paste. Such agents consist of oils, fats, resins, etc. Addition of the agent up to 1 percent by weight of portland cement (type I) creates an air-bubble volume up to 4 percent of the mix (approximately 13 percent of the volume of paste). Such finely dispersed and entrapped air bubbles act as lubricants, reducing friction and the heat conductivity of the concrete. The mix is more plastic, and workability is slightly improved—provided that the mix is not oversanded to start with. However, any type of fine surface-finishing work is difficult because the interaction of expanding air due to hydration heat and capillary forces makes the mix instable. The main advantage of air entrainment is the substantial improvement in the resistance of concrete to freezing and thawing.

The amount of water-reducing and air-entraining agents must be very closely controlled to obtain the required dispersion of air bubbles. When the water-cement ratio is high, the diameter of bubbles and their spacing are increased, resulting in decreased effectiveness of the air-entraining agent. This may seriously reduce the strength of the concrete. The air-entraining agents on the market are often not properly processed and act as plain lubricants. The lubricating effect of such an agent reduces seriously the bond between aggregate particles and cement gels, resulting in reduction in strength and in extensive plastic deformations. Also, the original spacing and size of bubbles are considerably increased by extensive vibration. To reduce the uncertainty in field operations, air-entrained cement should be preferred and the air volume should be kept at the 4 percent level. This amount of air bubbles has been found adequate to secure the durability of quality concrete.

Mineral powders, chemically unreactive or cementitious, can be added to concrete mix to influence the physicochemical behavior of the cement paste during hydration and hardening, to improve workability without increasing the cement content, and to reduce hydration heat.

The minerals used in practice and added to the batch are pozzolanas, which are weathering products of volcanic deposits or diatomaceous earth (fossilized skeletons of diatoms); fly ash, which is recovered in dedusters of pulverized-coal–fired power stations; fine-ground limestone; hydrated lime; etc.

Chemically, the pozzolanas consist mainly of silica, varying amounts of alumina, ferric oxide, and traces of lime and magnesia. The silica reacts with the calcium hydroxide set free during hydration of portland cement, forming calcium silicate hydrates at ordinary temperatures. Alumina also reacts with calcium hydroxide, forming calcium aluminate hydrate. As the calcium hydroxide is soluble in pure water and thus primarily responsible for efflorescence on concrete surfaces, the reduction of its content in concrete is desirable, especially in exposed architectural concrete. The addition of finely divided minerals increases the durability and workability of concrete. However, the expected increase in strength is not always achieved, mainly because of the presence of other components in pozzolanas (sometimes up to 40 percent) which do not contribute but rather reduce the quality of cement paste. The amount of pozzolanas added should be determined by field testing. Generally, it should not be more than 10 percent by weight of cement.

Fly ash belongs in the class of pozzolanas. Chemically it consists mainly of silicon dioxide (SiO_2), small amounts of lime, magnesium, unburned coal particles, gypsum. As some of the constituents in fly ash, depending on the amount present, can be detrimental to portland cement in the mix, its suitability as an admixture must be thoroughly tested before it is used. The hydraulic activity and the effect of fly ash upon the quality of cement paste are practically the same as for natural pozzolanas.

Finely ground limestone is rather coarse in comparison with the pozzolanas. Its advantages are mainly that it improves workability and fire resistance and especially that it increases the tensile and bond strength of concrete. It must be borne in mind that the porosity of ground limestone may be up to 25 percent and its absorption up to 10 percent by dry weight. Therefore, it must be considered in mix design. Its advantages can be explained by the physicochemical properties of limestone. Water is held in fine lime particles by relatively weak bonds, and the absorbed water is pulled out by cement colloids when hydration proceeds. Thus the saturated limestone particles behave like small, evenly distributed water tanks; also, their presence seems to affect the capillary structure of the paste. As the lime particles are more flexible than the sand particles, the internal and surface stresses in concrete are considerably reduced, resulting in higher flexural, tensile, and bond strength. The voids, originally filled by absorbed water, will be filled by air, resulting in increased durability of the concrete. The higher fire resistance can be explained by the reduced heat conductivity at the surface layer caused by the decomposing of limestone ($CaCO_3$) into quicklime (CaO) and carbon dioxide (CO_2). The expelling of carbon dioxide results in considerable porosity (the weight loss is approximately 45 percent) and higher insulation against heat propagation toward the interior (heat conductivity: quartz, 0.03; lime, 0.00029). The amount added may run up to 30 percent by weight of cement.

Hydrated lime is added mainly to improve the workability and to lighten the color of concrete. The result is a reduction in strength. The expected reduced durability is believed to be counteracted by more closed surface due to the carbonization of hydrated free lime in the surface pores by the presence of carbon dioxide.

The natural color of cement is controlled by the magnesia, alumina, and iron content in cement. High magnesia content is mainly responsible for a light green-gray color. Magnesia-free cements are brownish. High alumina content commonly causes a darkened color, and high iron content a reddish-brown color. The shade of natural cement color can be influenced to a certain degree by fine aggregate and mineral fillers. If white quartz sand is used, the color of concrete obtained is relatively lighter than if yellow natural sand is used. Lightweight fine aggregate causes a muddy dark color. Coarse aggregate does not affect the shade of color of concrete unless it is exposed by acid treatment, sandblasting, or retarders. To obtain other desired colors for concrete (architectural concrete), color pigments are often added to the mixing water or are interground with the cement. Best results are obtained by interground inorganic or synthetic pigments. The pigments used should not affect setting time, strength, or other essential qualities of concrete. Besides thorough mixing, manufacturer's instructions should be carried out with the utmost care in order to obtain uniformity and stability of color.

36. MIX DESIGN

The qualities of concrete that are required for a particular structure are determined by the engineer and the architect. To obtain the specified qualities, mix design is commonly carried out and tested by materials testing laboratories. As the designers are responsible for the structure or structural elements the concrete is used for in field or plant condition, the final mix design must be checked and approved by the engineer for its structural qualities and by the architect for its architectural qualities.

Commonly, the 28-day compressive strength of concrete is used as the criterion for the other essential qualities: flexural tensile strength, bond and shear strength, durability, and watertightness (impermeability). As long as concrete is used for ordinary structural purposes only, using the strength criterion to determine overall concrete quality is justified. But for an advanced, contemporary structure with large surfaces exposed to severe climatic conditions, this criterion is not adequate because the other qualities mentioned above are not proportional to the 28-day compressive strength. The tensile strength of concrete used for structural purposes is considered of little importance and is usually disregarded.

This is true also for flexural strength, regardless of the fact that it is approximately twice as high as the direct tensile strength of concrete. However, tensile strength has the utmost value in terms of durability and, therefore, is extremely important for exposed concrete (structural and architectural). Its magnitude is not proportionally related to concrete strength. Also, bond strength and watertightness are not proportionally related to compressive strength. The engineer, to be able to properly judge a mix design, must know the factors related and contributing to certain qualities and rate of hydration and hardening of concrete.

As proved by numerous tests and practical experience, all the significant qualities of concrete are controlled primarily by the cement characteristics, by the porosity of the paste, and also by the strength of the bond between the paste and the aggregate particles. The strength of cement paste is controlled (besides by the porosity) not so much by its chemical composition as by the fineness of grinding, i.e., by the increased specific surface of cement grains exposed to hydration. However, the rate of hydration depends both on the fineness and on the chemical composition of cement. As already discussed, the grain sizes of portland cement (type I and III) may vary within a wide range—from 100 down to 1μ—and the specific surface may vary from 200 to 20,000 cm^2/gm, respectively. Therefore, the hydration and intermolecular forces are higher for fine-ground than for coarse-ground cement. The higher strength of high-early cement is especially pronounced in the early age—up to 3 days. After that, the diffusion of water, required for hydration, is slowed by the growing thickness of gel and by the heat-vapor barrier (Fig. 203). Freyssinet has proved that diffusion difficulties and the heat-vapor barrier can be overcome and complete hydration and reduced porosity of paste obtained in a few hours by the simultaneous application of high pressure and heat.

The porosity of cement paste (or mortar) depends on the quantity of water present, commonly expressed by the w/c ratio by weight and degree of consolidation. The drop in compressive and flexural strength of concrete versus the w/c ratio is illustrated in Fig. 206. If the consolidation of mix is imperfect, air-filled voids will be present and the porosity increased, resulting in lower strength values. The drop in flexural strength with the higher w/c ratio can be explained mainly by the increased bond between paste and aggregate and the reduced Poisson ratio of concrete in tension.

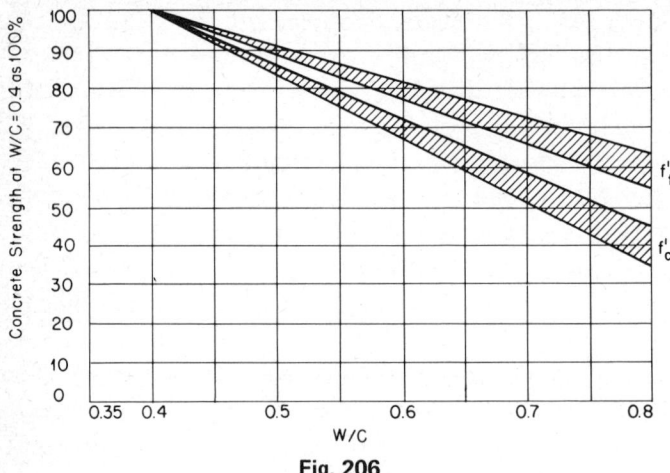

Fig. 206

As stated before, the minimum quantity of water for chemical reaction and gel water for complete hydration is about 0.38 (0.25 + 0.13) percent by weight of cement. The film water in aggregate joins in some degree the mixing water for carrying and distributing of cement. The absorbed water in aggregate particles contributes to later hydration but not to the workability. Also, it does not increase the absolute volume of water and is disregarded in mix design. If there is no absorbed water, the w/c ratio should not be less than 0.38. However, it is well known that the highest strength of concrete is obtained under field conditions by a w/c ratio as low as 0.35. This decrease can be possible only because of absorbed water. For works carried out in high climatic temperatures, the w/c ratio must be increased, sometimes up to 0.55, to compensate for the evaporation loss and incomplete absorption of aggregate. Commonly, these amounts of water are adequate to obtain acceptable workability, especially when high-frequency vibrators are used for compaction. In the case of relatively rough surfaces of aggregate particles, even the 0.55 w/c ratio may be inadequate to produce the necessary plasticity. In this case, instead of increasing the w/c ratio, water-reducing admixtures should be used. It must be borne in mind that keeping water content in the mix to a minimum increases in large degree all the qualities of concrete.

In addition to the water and cement content of the mix, a certain minimum amount of fine aggregate is required to obtain the desired plasticity. If there are not enough fine particles, the mortar will be harsh and watery. An excess of fine particles requires a high w/c ratio because a larger surface area has to be coated by cement paste. The excess water, required for complete hydration, will evaporate when the concrete is drying out, resulting in porosity, reduced strength, high shrinkage, increased permeability, and decreased durability. The degree of workability required for a particular mix is also controlled by the shape of the particles and the condition and method of placing. This is determined by the slump test.

The proper fine-coarse aggregate relationship depends on many factors. It has to be determined by trial mixes, carried out in the laboratory, which take into account all actual and practical considerations, such as materials available (including admixtures), time required for transport, climatic conditions, placing methods (regular, pumping, guniting, or slip-form operation), compaction equipment, and vibrator types.

There exists a quality limit for a given aggregate beyond which the increase in cement content in a mix only slightly increases the particular quality of concrete and may have a reverse effect on other qualities (see Sec. 41). Tests indicate that the quality limit for cement is reached for natural stone aggregate with approximately 8 bags/yd^3 and for crushed and lightweight aggregate with about 9 bags/yd^3.

The ratio of fine to coarse aggregate for all qualities lies between 35 and 55 percent by weight of total aggregate. Further, the quality depends in large degree on gradation and maximum size of aggregate. The mineralogical and chemical composition of stone affect mostly the flexural and bond strength of concrete.

The compressive strength increases with maximum grain size (maximum is 1 in.) and lowest fine-coarse aggregate ratio (0.35). For lightweight aggregate, particle size should not be more than 1/2 in. Beyond this size limit, the lateral displacement of particles (Poisson's ratio) increases beyond the capacity of tensile strength to provide adequate confinement for larger particles. Thus the maximum compressive strength for lightweight aggregate is controlled in considerable degree by the tensile strength of the paste.

The tensile, flexural, and bond strengths are closely related. Maximum strengths are obtained with fine aggregate gradation approximately midway between the two limits (Fig. 204), maximum aggregate size 1/2 in., and fine-coarse aggregate ratio up to 0.50. Particles, round in shape with coarse surface texture and moderate water-absorption capacity, are important contributing factors for both tensile and bond strength.

Replacing up to 20 percent of the fines passing sieve No. 50 by limestone powder increases considerably all qualities of concrete. Coarse aggregates having fine mineralogical structure, such as traprocks, are not suitable for concrete used for purposes in which tensile and bond strength are of importance.

Durability and watertightness of concrete are also closely related; one of the most important requirements for durable concrete, besides high tensile strength, is its watertightness. Hardened, fully hydrated cement paste with w/c = 0.40 is practically watertight, even though 1 ft^3 of the cement contains up to 0.513 ft^3 of void space. To fill this space requires an equal quantity (including the gel water) of gel. This is the case when w/c = 0.40, as proved by the following computation.

One bag (94 lb) of cement: Solid volume $= \dfrac{94}{3.1 \times 62.4} = 0.487$ ft^3

Water: $0.4 \times 94 = 37.5$ lb (5.3 gal)

Volume of chemically bound and gel water: $\dfrac{(0.25 - 0.06 + 0.15)\,94}{62.4} = 0.513$ ft^3

$$\Sigma V = 1.00 \text{ ft}^3$$

The $0.06 \times 94/62.4 = 0.091$ ft^3 is the reduction in volume of chemically bound water due to contraction (intrinsic shrinkage).

The capillary forces in gel are large and offer almost insurmountable resistance to the flow of water. In accordance with experience and test results, the capillary pores reach vulnerable size when w/c = 0.5 (6¼ gal) or when the paste is allowed to dry out prematurely (not complete hydration). In the case of air entrainment, the air bubbles, although far larger in size than capillary pores, commonly do not increase permeability because they are surrounded by impermeable cement paste and normally remain free of water until crack formation in the paste occurs.

However, concrete produced and cured under field conditions usually has considerable surface and internal cracking, as revealed by microscopic studies. As long as the width of such cracks remains below measurable limits by ordinary means they seem harmless, as far as permeability is concerned, but they may reduce the durability of concrete, especially under climatic conditions of repeated freezing and thawing. Through these cracks water will be drawn into the concrete by capillary action and will freeze in the larger capillaries when the temperature drops below freezing point. The phase change—water to ice—is associated with about 9 percent volume increase. The expanding ice-water system causes hydraulic pressure in concrete and, if not controlled, may exceed considerably the tensile strength of concrete. On renewed saturation, the hair cracks produced are filled with water, and thus frost action and damage are intensified with repeated cycles of freezing and thawing. The hydraulic pressure can be controlled to some degree (depending on the permeability of the paste, the rate of flow, and the temperature change) by providing isolated air voids as relief points, evenly and closely distributed in the paste.* If the spacing of air bubbles is larger than approximately 0.25 mm, the effectiveness of air entrainment is considerably reduced. The percentage of air voids required to compensate for the volume change of water to ice depends on the quantity of freezable water present in the mortar. The chemically combined water as well as the gel water, is nonfreezing; also, the water in small capillaries does not freeze under normal climatic conditions because of the high surface tension the water is subjected to. Thus water is freezable in large capillaries and in smaller capillaries only when the temperature drops considerably below the freezing temperature of ordinary water. Therefore, it is almost universally recognized that about 4 percent of the volume of concrete, as extra space, is required to secure durability against freezing and thawing.

Under normal conditions, full saturation of high-quality concrete is not probable, except in special cases such as pavement. Therefore, sufficient air voids normally will be present in concrete to accommodate the increase in volume resulting from some ice formation. It must be kept in mind that any increase in porosity reduces almost proportionally the compressive, flexural, and bond strength of concrete. Due to this consideration, air entrainment should be used only within the limits of its effectiveness and only in cases in which concrete may be fully saturated and subjected to repeated cycles of severe freezing and thawing.

Coarse aggregate, if it is not frost-resistant, may cause considerable surface damage, especially when particle size is more than 1/2 in. Large angular particles close to surface, which have a different thermal

---

*T. C. Powers, The Air Requirement of Frost-resistant Concrete, *Proc. Highway Res. Board,* vol. 29, 1949.

expansion from that of paste and smooth surfaces, can easily be plucked out of the matrix after the protective layer of mortar is cracked due to drying shrinkage and moisture has access to the particles.

Surface cracks and scaling usually also occur when, by improper vibration, the light, soft, erosive fine material and excess water are sucked to the surface (see Fig. 207). The inferior layer, having different physical properties from those of the concrete beneath, may even separate entirely from the surface.

The fire endurance of concrete can be considerably increased by the use of air entrainment and lightweight aggregate because of the lower heat conductivity and reduced volume change, especially of coarse particles, due to thermal expansion. Therefore, lightweight concrete retains almost its original strength at high temperature, which is not the case with regular concrete, especially when it contains siliceous coarse aggregate.

To prevent honeycombing and to obtain proper bonding and acceptable workability, the quantity of cement paste should be up to 15 percent greater than the volume of voids in fine aggregate and the quantity of mortar should be about an equal amount greater than the volume of voids in coarse aggregate.

Calculation of batch quantities per cubic yard of concrete is commonly based on the absolute-volume concept. For example:

Cement—8 bags: $\quad V_c = \dfrac{8 \times 94}{3.15 \times 62.4} = 3.82 \text{ ft}^3$
w/c ratio 0.4
  (incl. film water): $\quad V_w = \dfrac{0.4 \times 8 \times 94}{62.4} = 4.84 \text{ ft}^3$

Air entrainment 4 percent: $\quad V_A = 27 \times 0.04 = 1.08 \text{ ft}^3$

$$\Sigma V_s = 9.74 \text{ ft}^3$$

Aggregate: $27 - 9.74 = 17.26 \text{ ft}^3$
Assuming the fine-to-coarse aggregate ratio = 0.45:
  Fine aggregate = 7.77 ft^3
    $w_F = 7.77 \times 2.65 \times 62.4 = 1{,}296 \text{ lb}$

Coarse aggregate = 9.49 ft^3
    $w_C = 9.49 \times 2.65 \times 62.4 = 1{,}586 \text{ lb}$
    $\Sigma w = (8 \times 94)1.4 + 1{,}296 + 1{,}586 = 3{,}932 \text{ lb}$
    146 lb ft^3

Calculation of batch quantities, required for lightweight concrete, is more difficult because the space of pores inside the particles makes the absolute-volume method useless. Therefore, the proper proportion and required quantities are determined by trial mixes. If natural sand is used, the mortar mix is designed based on the absolute-volume method and coarse aggregate is considered mixed in as a filler in the possible highest quantity.

Usually several trial mixes are required to determine the best composition and workability of mix, taking into consideration the possible variations in field, to obtain and secure the desired qualities of concrete in the structure.

Mix Design for Pumpcrete and Shotcrete

Mix design for concrete to be pumped into forms is practically the same as that described above. However, to reduce friction and increase cohesiveness round and smaller aggregate particle shapes are preferred. Also, replacing fines passing sieve No. 50 by finely ground limestone or fly ash increases the stability of the mix, resulting in higher uniformity and reduced stoppage. Good flowability of the mix is especially important when concrete has to be pumped long distances, around sharp curvatures, and to considerable heights. Admixtures, such as intrusion aid and plasticizers, are commonly required to increase the flowability.

Pneumatically placed concrete (gunite) is used principally for thin, free-shaped shell construction and for repair work (see Part 3). There are two placing methods: (1) Dry-mix process, in which cement and damp sand are thoroughly mixed and conveyed by compressed air through a hose provided with a special nozzle at its end; water is added to the mix at the nozzle. (2) Wet-mix process, in which mix is prepared like any regular mix before being carried by compressed air through the hose and nozzle. For new construction the wet-mix process is preferred because it results in more uniform concrete and less dust and rebound.

The basic principles of mix design that apply for any high-quality concrete apply also for gunited concrete. In general, the gradation of sand passing sieve No. 4 must be excellent, that is, uniform and close to fineness limit (Fig. 204). The cohesiveness of the mix is most important to reduce sagging and rebound. It is obtained by using finely ground limestone (passing sieve No. 50) and high-early cement. Best results have been obtained by a w/c ratio approximately 0.4 to 0.45. Admixtures or additives of known types, except small amounts of asbestos fiber and accelerators (for cool, damp weather conditions), have been found not useful. As the mix design and workmanship control equally the quality of gunited concrete in place, numerous laboratory tests are required to determine the proper mix, equipment, and presssure intended to be used and the workmanship required. This is best done by studying the actual results gained by panels gunited in simulated job conditions. In general, the physical properties of quality gunite are in the same range as of conventional concrete having the same mix design and curing.

Grout Mix and Grouting Operation

Grout is used in the structural field to protect the posttensioning tendons, to establish bonds between the tendons and tubing after the posttensioning is applied, and to fill the joints of prefabricated elements. The space to be filled with grout is relatively narrow and uneven. The distances the grout is forced to flow under high pressure (up to 100 psi) may be up to 100 ft or even more. To meet these requirements, the grout must be carefully designed and tested for its suitability and quality before the grouting operation.

Two of the most important qualities of grout are sustained cohesiveness and flowability. Both of these qualities depend mainly on the fineness of mineral additives (large specific surface) and their ability to remain in suspension even under high pressure. In accordance with experience, best results are obtained by fly ash (SiO_2) and limestone powder ($CaCO_3$). Besides its fineness, fly ash has hydraulic properties. It remains in suspension for unlimited time and reacts with calcium hydroxide [$Ca(HO)_2$] in the solid state. However, fly ash may contain unburned coal particles and gypsum in quantities detrimental to portland cement. Also, its composition very often is not uniform. Due to these facts, fly ash must be thoroughly tested in these aspects before being used.

Limestone powder does not have hydraulic activity, but it has high bonding capacity with cement gel. Because of its water-absorption property, the w/c ratio is slightly higher than that required for fly ash to obtain equal flowability. But regardless of this, limestone-powder grout has far less bleeding or water separation (increased water dispersion) than grout containing fly ash.

The w/c ratio should be kept to the lowest possible for the condition of grouting. If w/c $>$ 0.5, water will separate from the mix, creating water pockets, sedimentation, and stoppages that result in frost damage, cracking, and corrosion. The flowability of grout is greatly improved by the use of intrusion aid. The intrusion aid reduces the surface tension of water. It commonly contains a small amount of aluminum powder, which creates gas bubbles during hydration that expand and counteract the intrinsic or chemical shrinkage. The small, evenly dispersed gas bubbles also decrease the frost damage, as already discussed (air entrainment).

Excellent results have been obtained with the following mix design:

High-early portland cement:	94 lb
Limestone powder:	40 lb
Intrusion aid:	1 lb
Water:	43 lb

The mixing operation should be performed with a high-speed mechanical mixer for at least 4 min. The quality of the grout should be tested at least for water separation and strength.

The water-separation test can be carried out in a glass tube ½ in. in diameter and 25 in. long with the bottom end corked. The tube is filled up to 24 in. with thoroughly mixed grout. Then the top is corked, and the tube is held in the vertical position. There should not be more than 1/16 in. of water at the top after initial set.

The strength of the grout, as tested on small cylinders (4 in./2 in.), should not be less than

3 days:	2,500 psi
28 days:	3,500 psi

Before actual grouting is started, the tubing should be checked for clearance and washed with water. In cold-weather operation, live steam should be used to melt the possible ice and to warm the tendons and tubing. When clear water flows out of the far end, grout is forced into the tubing slowly, under minimum pressure, until pure grout penetrates the tubing and flows out freely in sufficient amount. Then the far end is plugged and the pressure increased and held for a short period of time. If the grout does not penetrate, the tubing must be washed or blown out using compressed air. A small amount of water is then added to the mix, and the grouting operation is repeated from the same end.

Watertightness of Concrete

In general, no cracks develop in properly designed prestressed concrete because the concrete will remain in compression under any loading condition to which the structure would normally be subjected. But the prestressing does not prevent leakage if the concrete itself is not watertight. As the watertightness of concrete is of special interest in the design of architectural concrete, tanks, and pipes and is also of importance in other thin shells in which special precautions should be taken to prevent corrosion of the reinforcing steel, the design of watertight concrete will be discussed in this section.

There are two ways in which a watertight concrete can be obtained.

First, a waterproofing compound can be added to the mix. The waterproofing compound seals the voids after the surplus water is evaporated. Experience has proved, however, that only limited success can be achieved with this method since, even when the mix is properly designed, the compounds currently available are effective in preventing percolation of water through concrete for only a limited period of time. Further, most admixtures now in use adversely affect one or more of the other important properties of concrete. If the concrete mix is improperly designed, the effectiveness of the waterproofing compounds disappears very quickly.

Second, it is possible to design a mix in such a way that the concrete will remain watertight throughout the life of the structure, provided tensile cracks are not permitted to develop. By this method, concrete can be made permanently watertight, even under high pressure, without the use of compounds.

As discussed above under mix design, the watertightness of concrete depends mainly on the watertightness of the mortar and on the percentage of mortar in the concrete.

The mortar consists of fine sand, cement, and water. The percentage of each component may vary to a great extent. For complete chemical hydration, the weight of water required is approximately 13 percent of the weight of the cement used. By mixing neat cement with the exact amount of water required for hydration, *cement stone* is obtained. The specific gravity of cement stone depends upon the type of cement used and can be determined by laboratory tests. The apparent absolute volume of the cement stone is thus

$$V_{cs} = \frac{1.13\, w_c}{\gamma_{cs} \cdot 62.4} \tag{551}$$

where w_c is the density of loose dry cement in pounds per cubic foot and γ_{cs} is the apparent specific gravity of the cement stone.

The mortar will be watertight when all voids in the sand are filled by voidless cement stone. To achieve this, the sand must be sound and properly graded, the grains must be more or less rounded rather than angular in shape, and the ratio of the apparent absolute volume of cement stone to the volume of voids of the sand must be controlled within narrow limits. The properties of the sand to be used may be represented with sufficient accuracy by a factor k obtained by the following expression:

$$k = 3 - \frac{P_{70} + P_{18} + P_7}{P_{0.265}} \tag{552}$$

where P_{70}, P_{18}, P_7, and $P_{0.265}$ are the percentages by weight of sand passing standard sieve numbers 70 (210 μ), 18 (1,000 μ), 7 (2,830 μ), and 0.265 (6,730 μ). Generally, Eq. (552) gives a higher value of k for a coarse sand than for a fine sand.

The ratio of the volume V_{cs} of cement stone to the volume V_{vs} of voids in the sand gives the fill factor f_m:

$$\begin{aligned} f_m = \frac{V_{cs}}{V_{vs}} &= \frac{1.13\, w_c}{\gamma_{cs} \cdot 62.4[1 - (w_s/\gamma_s \cdot 62.4)]\, m_s} \\ &= \frac{1.13\, w_c}{(\gamma_s \cdot 62.4 - w_s)\gamma_{cs}/\gamma_s \cdot m_s} \end{aligned} \tag{553}$$

where w_s is the unit weight of dry, loose sand, γ_s is the specific gravity of the sand, and m_s is the ratio by volume of sand to cement in the mix.

The criterion for watertightness of the mortar is

$$k f_m > 1 \tag{554}$$

For example, suppose the mortar consists of 1 ft³ of cement weighing 90 lb and 3 ft³ of fine dry sand weighing 110 lb/ft³. The apparent specific gravity of cement stone is usually about 2.9, and that of fine sand about 2.65. Assume, in this case, that sieve analyses of the aggregates (sand and gravel) give the following results: $P_{70} = 2.2, P_{18} = 23.2, P_7 = 33.1, P_{0.265} = 49.8, m_s = 3$. Then

$$k = -\frac{2.2 + 23.2 + 33.1}{49.8} = 1.83$$

$$f_m = \frac{1.13 \cdot 90}{(2.65 \cdot 62.4 - 110) 2.9/2.65 \cdot 3} = 0.56$$

$$kf_m = 1.83 \cdot 0.56 = 1.025 > 1$$

Since the value of kf_m is greater than 1, a watertight mortar may be expected.

The concrete will be watertight when all voids in the coarse aggregate are filled by watertight mortar and when the stones of the coarse aggregate are resistant to the passage of water. Sufficient mortar will be present in the mix when the ratio f_c of the volume V_m of mortar to the volume V_{va} of the voids in the coarse aggregate is approximately

$$f_c = \frac{V_m}{V_{va}} \simeq 1.15$$

The amount of water in the mix should not exceed 6½ gal per sack of cement.

37. PLACING AND CONSOLIDATION

The most important rules for placing and consolidating concrete are:

1. No segregation. If, for a given mix, this is practically unavoidable under field conditions, the mix must be modified to increase the viscosity of the mortar.

2. Use of the right type of vibrators and proper vibrating.

There are three vibrator types available for consolidation of mix in the field: internal, surface, and form. The frequencies of internal vibrators are usually 6,000 to 9,000 rpm, with varying amplitudes and capacities. Surface vibrators commonly have lower frequencies, 3,000 to 6,000 rpm, and varying weights. The form vibrators in use are relatively powerful (centrifugal force 450 to 2,200 lb), with frequencies of 3,000 to 9,000 rpm. The range of efficiency depends to a large degree on mix design, degree of hydration, centrifugal force, and frequency used. For internal vibrators, as an average, the range is about 12 to 24 in. The efficient depth (consolidation of top and bottom fibers approximately equal) for surface vibrators in use is approximately 10 to 15 in. The weight of the vibrator should be adequate to keep it in steady contact with the surface of concrete. The effectiveness of form vibrators depends largely on the type of forms. It is smaller for heavy, rigid steel forms than for lighter wooden forms. In general, the area of efficient consolidation by a medium-size form vibrator (minimum frequency 3,000 rpm and amplitude approximately 1.0 mm) is vertically and longitudinally about 4.0 ft and in depth about 8 in.

Internal vibration should be carried out under specific overload (surcharge) to avoid acceleration of particles and to increase steady wave propagation. Experience has proved that the best results are obtained when the vibration range in depth is about 1.0 ft from the top of the last poured layer and 1.3 ft into the previously poured layer (Fig. 207). To avoid collection of water and light particles on the surface of the layer, commonly associated with weak bondage and disuniformity of surface (water traces, discoloration, etc.), the top part (1.0 ft) of the last poured layer (intermediate) should not be vibrated. The distance of vibration from any surface should not be less than 4 in. The water carrying light and weak particles should be drawn by vibration away from the surface

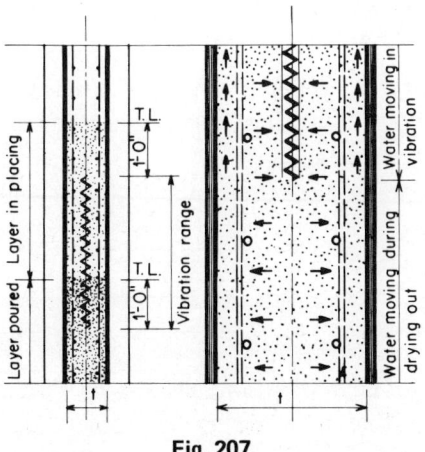

Fig. 207

toward the interior. Excess water collected in the interior is pulled back to the surface by capillary forces to balance water loss at the surface due to evaporation. Such water movement promotes further hydration (possible only when capillaries are filled with water) and efficiency of curing. As a result, the quality of the surface concrete is considerably improved, with less surface cracking and increased durability and uniformity. If the water movement is reversed by the use of a form vibrator or internal vibrator close to the forms, the fine, weak particles are collected not only on the surface but also below the reinforcing bars, resulting in corrosion and weak bonds. Furthermore, the pumping effect of the vibrating forms also results in air from the atmosphere being sucked between the forms and concrete, causing considerable bleeding traces and air holes at the surface.

To avoid the covibration of reinforcing bars, especially prestressed strands closely placed, internal plate vibrators must be used. The steel plate recommended is about 1/16 in. thick, the end approximately 2 in. and the top about 4 in. wide. The length is as required up to 4 ft. The plate can easily be attached or welded to any type of regular internal vibrator. Plate vibrators are easier to handle in cases of deep vibration and heavy reinforcing than regular internal vibrators, which stick, especially when the hydration has advanced for hours. Also, plate vibrators produce more uniform density and less bleeding than round-head vibrators. The water movement is along and perpendicular to the plate, and simultaneous uniform vibration of considerably larger depth is possible (Fig. 207).

Experience has shown that the best results are obtained when vibration has been carried out in two or even three steps:

1. Vibration during placing. Vibrator frequency is approximately 3,000 rpm, and timing is a short period (1 sec or so) to obtain uniform mass without honeycombs and to ease placing.

2. Vibration to consolidate concrete. A high-frequency (9,000 rpm) vibrator is used, and timing is earliest 1 hr and latest 3 to 4 hr after placing, depending primarily on the w/c ratio and climatic temperature. For relatively low w/c ratio (0.35 to 0.4) and high climatic temperature, 2 hr latest is recommended.

Experience also shows that even with later vibration, about 6 hr with a w/c ratio above 0.5 and a mix rich in fines, the quality of concrete was improved when a high-frequency vibrator (20,000 rpm) with relatively small amplitude (about 0.15 mm) was used. The latest time of vibration without physical damage to concrete is when the vibrator is easily pushed into the concrete under light pressure and when the hole closes completely and immediately after the vibrator is pulled out. The duration of vibration is a few seconds for concrete rich in coarse aggregate and slightly longer for concrete rich in fines. During later or second-step vibration, the forms should be hammered, starting from the bottom of the vibrated layer upward, to avoid horizontal cracking of the surface due to form friction. If the vibration is carried out in one operation, extensive segregation may occur, especially when 9,000-rpm frequency, powerful vibrators are used. After a certain degree of hydration, water is already saturated with a considerable amount of calcium hydroxide and the paste has gained density, resulting in more extensive wave propagation, uniformity of hydration, and hydration heat. The effect of later vibration also reduces and distributes more evenly the vapor pressure and capillary system, promotes crystallization, and reduces internal stresses.

38. CURING OF CONCRETE

As stated before, hydration can proceed only when capillaries are filled with water. The rate of hydration and hardening is accelerated by high temperature and reduced by low temperature. Hydration and hardening are practically zero at freezing temperature and are highest at approximately 200°F under atmospheric pressure. This highest-temperature-level curing is possible only by the use of low-pressure saturated live steam. The maximum temperature level at which moist or water curing is possible is approximately 100°F.

Test results show that if concrete is allowed to set after placing, even for a short period of time (about 4 hr), at a lower temperature, the obtained final strength will be somewhat higher than when starting immediately with a high temperature. This phenomenon is explained by the decreased solubility of calcium sulfate at higher temperatures, resulting in flash setting. When cement is gauged with water, the softening of the particles starts and water penetrates rather slowly toward the nucleus of the grains. If enough time is allowed for softening, proper arrangement of minerals, and leeching out of alkali-sulfate to form calcium sulfoaluminate (required to slow down the hydration of tricalcium silicate), the overall hydration will be more complete. This also largely explains the higher final strength if retarders are used. When flush or rather rapid setting occurs, the water, required for hydration of the coarser grains, must pass the heat barrier and the rapidly hydrated sphere of grains, resulting in incomplete softening and a colloidal state. For instance, the specific surface of normally (below 85°F) hydrated cement paste is approximately in the order of $2 \times 10^6 \, cm^2/gm$ and is only about one-tenth of this value if hydrated at elevated temperatures (200°F). This means that the crystal structure of gel is considerably more coarse which is associated with some degree of reduction in strength. But after the softening has occurred, any rise in temperature increases the chemical reaction between the cement particles and water as well as between the particles themselves. It must be understood that the cement minerals in the particles, regardless of how finely ground, are not exactly as required for crystallization. The proper arrangement of cement minerals and the most complete crystallization can take place only in the colloidal state. The delayed initial set of mortar with a high w/c ratio is explained by the fact that a certain degree of saturation of water with calcium hydroxide is required before aluminum enters into reaction or hydration process.

It can be understood from the above that the strength and other qualities of concrete largely depend on the conditions under which it is cured. But regardless of the importance of proper cure, there are considerable difficulties under field conditions in carrying it out, at least in the extent theoretically desirable, especially during cold- and hot-weather periods of construction. Since the reaction continues for years—so long as capillaries are filled, temperature is favorable, and hydrated cement is available—the strength of concrete may increase 1½ times or even more over its 28-day strength. One of the most inconvenient difficulties in field operation is the prolonged curing time. Therefore, some compromise between practical and theoretical considerations is necessary. In general, a sacrifice of quality is tolerable if concrete is cured effectively under normal conditions for at least 3 days when finely ground cement is used and 6 days for coarse-grained cement.

Extended curing time is required for low-temperature operation. The curing time, as a function of temperature, can be computed approximately by Saul's rule, based on the so-called *maturity concept:*

$$M = a_t(t + 10) \tag{555}$$

where M is the maturity of concrete, a_t the time in days, t the curing temperature in degrees Celsius, and 10 the temperature below freezing at which hydration is assumed to be zero. Thus, if the curing time at which a given concrete develops a certain strength or maturity at normal temperature (70°F) is known, the time required to reach equal strength or maturity at a lower curing temperature can be computed by this rule. For example, let us assume that the strength of a concrete at 3-day curing at 70°F (21°C) is 3,500 psi. Its maturity is

$$M = 3(21 + 10) = 93$$

The curing time required at 40°F (5°C) to obtain approximately the same strength is

$$x = \frac{93}{(5 + 10)} \simeq 6$$

or for $t = 86°F$ (30°C), curing time is approximately

$$x = \frac{93}{(30 + 10)} \simeq 2\frac{1}{4}$$

The strength development must be checked by testing the site-cured specimens.

A second difficulty is in preventing concrete from drying out during the curing period. If drying out is permitted, severe surface cracking occurs. Preventing evaporation of horizontal flat surfaces is not difficult; this can be done by protecting the surface against sun and wind and keeping it wet. More difficult is to avoid the drying out of vertical surfaces. Experience shows that the drying out of surfaces cannot be prevented simply by keeping the forms in place; the forms must be opened slightly, as soon as is possible without mechanical damage to the surfaces, and enough water must be supplied to keep the surface wet. Using this curing method in addition to proper vibration (discussed above), the surfaces can be kept completely free of shrinkage cracks, even when high-early cement is used.

The effect of low-pressure high-temperature steam curing is illustrated in Fig. 208. In accordance with thermodynamics, the core of a cement grain can be penetrated by water only if the water temperature is higher than the temperature of the core to be penetrated. Due to this fact, the surrounding water temperature must be raised sufficiently by hydration heat before the cement is fully hydrated and passes the colloidal state. If now the mixing water is heated up enough, the penetration of cement grains by water is considerably accelerated. At temperatures above 90°F, flush setting may already occur. Due to this fact, and for reasons discussed before, the softening period, at least 4 hr at lower temperature (70°F), is very important.

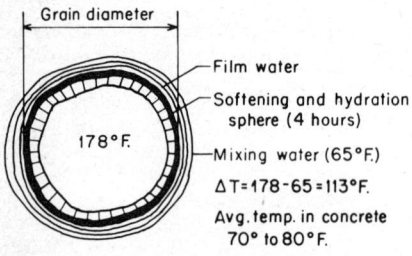

Fig. 208

When the temperature rises, the rate of evaporation increases; to avoid this, the relative humidity must be 100 percent. In addition, the difference in thermal expansion, which is dependent on the heat conductivity and specific heat of the concrete constituents, increases, resulting in internal stresses in the concrete. It is possible to control this only by reducing the rate of temperature rise and allowing time for the cement paste to develop tensile stress to resist the differential expansion of the constituents. Experience and tests show that the rate of temperature rise should be gradual—not more than 20°F/hr up to approximately 140°F and after that no more than 40°F/hr. Up to about 180°F (core temperature 178°F), it is not difficult to maintain the 100 percent relative humidity. Above this temperature, however, it is extremely difficult, even in closely controlled plant operations, to avoid drying out of concrete because of a drop in relative humidity. Due to this, it is not recommended that steam curing be used above 180°F. By curing for about 5 hr at 180°F, approximately 4,000 psi strength is commonly obtained. By raising the temperature for 1 hr up to 200°F, the strength can be increased about 10 percent. Very often the risk of dropping relative humidity is taken for two reasons: (1) to obtain lighter and more uniform color for the concrete and (2) because, for some reason, the minimum required 4,000 psi is not reached at 12-hr curing (4 + 3 + 5). The disadvantage of the 1-hr curing at 200°F, besides the possible crack formation, is the increased temperature difference between concrete and surrounding atmosphere ($\Delta T = 200 - T_A$).

Steam curing with forms in place is not effective since it is actually only heating and drying out, as is the case with hot oil circulating in pipes arranged close to the forms. Steam curing, in the real sense, also requires moisture; thus the concrete surface must be exposed to the saturated steam. The 4-hr softening and hardening time is usually enough to allow the forms to be opened and the concrete exposed to life steam. However, as the steam leaving the nozzles is relatively hot (up to 300°F), its direct hit of the surfaces must be avoided. For architectural concrete, care must be taken that water used for steam is free of iron; otherwise, the surface will be stained by rust.

The heated concrete should not be exposed to atmospheric conditions without adequate transmission time if temperature cracks are to be avoided. The adequate transmission period and rate of cooling are about 40 to 60°F/hr. The period of cooling can be shortened if warm water is used. Using cold water immediately after exposing the surface results in cracks. Further, it is advisable to use moist curing for at least a short period of time in order to slow down too rapid drying out. If this is not done, the strength development in concrete practically stops after steam curing; the 7-day strength may be somewhat lower than immediately after steaming and also less than the 28-day strength if cured only in moist air. This phenomenon is explained by rapid evaporation after the concrete is exposed to the atmospheric temperature, resulting in unbalance of temperature and moisture in concrete. To establish equilibrium requires considerable time, but if adequate moisture is available, the "thermal shock" is rapidly overcome.

39. SURFACE TREATMENT

A variety of textures and colors and improved durability of concrete surfaces can be obtained by the use of different methods, mixes, and surface treatments. Basically, there are two methods for obtaining a desired surface texture: form finish and exposed-aggregate finish.

With form finish, the surface texture is the imprint of the form surface, which remains after the forms are stripped. The surface is smooth when plastic-coated plywood, fiber glass, or steel is used for form skin and is rough when relatively narrow board or plywood covered with rubber having the desired texture is used.

The exposed-aggregate method requires the removal of mortar from the surface up to a desired depth in order to expose the aggregate. This can be done by sandblasting, application of muriatic acid, use of a retarder, or bushhammering.

The quality and texture obtained by these methods depend on the uniformity and denseness of the surfaces, the size of coarse aggregate, and the depth of mortar removed. If the size of coarse aggregate used is more than ½ in., there is a possibility of loosening and disruption of large particles, especially at corners, from applied forces or impact and from differential volume change due to temperature and moisture variation. Further, such disruption of surface is promoted by the close embedment of coarse particles and by variation in density and strength of the mortar. The softer areas in the mortar will be penetrated considerably deeper than the harder ones by the exposing process, resulting in disuniform texture and reduced bond for larger particles. Due to this, the exposed-aggregate surface is less durable than the form-finished surface. However, experience shows that adequate durability is obtainable for exposed surfaces when the following requirements are met: leakproof forms, excellent gradation of finer natural sand, coarse aggregate (preferably quartz, marble, or granite), maximum size not more than 5/8 in., packing density not more than 45 percent of surface area. Form vibrators, if used, should be of low frequency, up to 3,000 rpm (see Sec. 38).

The most durable surface is obtained by sandblasting. The degree of exposure, uniformity, and texture is easily controlled. Sharp corners can be secured by rubber taping or by covering with water-soluble mastic. The only disadvantage of sandblasting is the relative dullness of the finished surface because the fine sharp blasting sand hits the aggregate and damages the crystal surfaces. The finished surface loses the concrete character and looks very much like a plastered surface, especially when white cement and marble aggregate are used.

In terms of appearance, exposed-aggregate (quartz, granite) surfaces obtained by acid exposer are far superior to sandblasted surfaces. However, the depth of acid penetration into the surface is rather difficult to control because of the timing and variation in density of concrete. Commonly, the surface is wetted thoroughly before acid is used to control, at least to some degree, the depth of cement paste removed. After the paste is destroyed, it is removed by brushing and a thorough water wash. To neutralize the traces of acid, the final wash water should contain a small amount of lime.

The use of retarders for exposing aggregate is practically limited to precast panel manufacturing. A thin plastic layer of water-soluble retarder is spread on the mold. After the panel is poured and the concrete hardened enough to allow stripping, the retarder and the destroyed cement paste in contact with the retarder are washed away by water. The surface quality is almost the same as with acid process.

The color and shade of surface are obtained by the choice of coarse aggregate, sand, and cement. Quartz, marble, and granite aggregates are available in considerable variety of color and shade: green, pink, yellow, white, black, etc. Natural sand is commonly available in yellow, white, and reddish. Cements on the market are white, light green, and dark brownish. If the desired color cannot be obtained by a combination of aggregates, sand, and cement, a limited amount of inorganic color pigment can be used to affect the color of the cement paste.

It is essential to secure the durability and permanent appearance of all exposed architectural concrete surfaces. Regardless of the fact that high-quality concrete has both of these attributes, under field conditions uncertain human factors, defects in materials, and unfavorable curing conditions may result in defects. To counteract the undesirable results of such defects, special treatment of finished architectural concrete surfaces is possible.

In accordance with practical experience, the best results are obtained by treating the finished surfaces with aqueous solution of sodium methyl siliconate. The silicon solution is thoroughly brushed onto the concrete. Filling the voids and cracks, it does not function as a continuous surface film but instead reacts with soluble calcium compounds present at the surface layer and converts them to insoluble calcium silicates. The duration of the effect of this treatment depends on the amount of free lime available for transfer of silicon dioxide to calcium silicate, the degree of concentration, and the depth of penetration. The treatment reduces water absorption mainly because of the water repellence of sodium methyl siliconate and the resulting more closed surface. Discoloration due to rainwater efflorescence and accumulation of dust is also considerably reduced.

To obtain effective results, the surface treated must be dry, well cured and hardened, and free of form oil and dirt. Commonly, two layers are required; the first coat is more diluted to obtain deeper penetration, and the second coat is more concentrated to close the voids and cracks. If insufficient free lime is present to form insoluble calcium silicate, the treatment does not have a lasting effect. This deficiency can be counteracted by adding a small amount of calcium hydrate to the mix.

Where, for economical reasons, two-layer concrete is used, the architectural, exposed aggregate layer should be at least 2 in. thick and the coverage for reinforcing should be 1½ in. The w/c ratio for the exposed layer, using white cement, should be selected slightly higher than for regular backup concrete. In this way, and by using proper internal vibration (Fig. 207), the movement of water carrying cement is toward the backup concrete and staining of the white paste is avoided. If form vibration were used, the movement of water would be reversed, resulting in severe staining of the exposed layer. Furthermore, the white cement is sensitive to iron oxide. If curing water or steam is not free of iron, practically unremovable rust staining will occur. No pigments should be used in combination with white cement.

40. PHYSICAL PROPERTIES OF CONCRETE

Compressive, Tensile, and Shear Strength

The obtainable qualities of concrete were thoroughly discussed under "Mix Design," Sec. 36. The structural and architectural design determines the required quality of concrete to be used. Commonly, for contemporary structures the minimum 28-day compressive strength is, for superstructures, $f'_c = 5,000$ psi and can be as high as $f'_c = 7,000$ psi or even higher. The 3- or 7-day strength depends on cement type and curing condition; as an average, it can be assumed to be 0.50 f'_c to 0.65 f'_c and 0.65 f'_c to 0.75 f'_c, respectively. The maximum values apply for high-early cement.

The flexural tensile strength as an average expressed in terms of compressive strength is $f'_T = \frac{1}{10} f'_c$ to $\frac{1}{8} f'_c$, and direct tensile strength is about half of flexural tensile strength. The difference between the two tensile strengths is caused mainly by two factors: The direct tensile strength is difficult to determine because even a small eccentricity and disuniformity involved in mix or testing results in bending stresses in specimens. The strain in concrete increases more rapidly than stress, resulting in deviation from the Navier or linear stress distribution, so that the computed flexural fiber stress is only an apparent stress. As the nonlinear stress distribution exists in almost all structural concrete elements subjected to bending, flexural tensile strength controls crack formation in structures and as such is considered in design or evaluation of the tensile quality of concrete.

The shear strength of concrete, like tensile strength, has two features: flexural and direct shear strength. In accordance with Mohr's theory, the flexural shear strength of concrete is

$$V_f = \tfrac{1}{2} \sqrt{f'_c \times f'_T} \tag{556a}$$

Also, it is a combination of compressive and flexural tensile strength. The direct shear strength, unlike the direct tensile strength, is twice as high as the flexural shear strength. Thus,

$$V_d = \sqrt{f'_c \times f'_T} \tag{556b}$$

The bonding strength between concrete and reinforcing bars or strands is an apparent strength; it is a combination of adhesion, friction, and direct shear strength of concrete. For plain bars or wires, the adhesion and friction control the bond, which may be as high as 280 psi. For deformed bars and strands, the bonding strength may approach the direct shear strength of concrete [Eq. (556b)].

Deformations

The total deformation of concrete may consist of four different kind of strains: elastic strain and permanent initial set, shrinkage, swell, and plastic flow or creep. All these deformations have a different effect on internal forces as well as on stress distribution over any given cross section of an element or member and, therefore, will be discussed separately in detail in the following.

Elastic strain and permanent initial set:

The modulus of elasticity is a function of the strength, age, and degree of stress or strain existing at the time the increments of stress and strain are assumed to be measured. The influence of age appears in the strength of concrete, which increases with time and, therefore, will not be considered separately.

The elastic limit for concrete is not definite. There appears, however, to be a limit to the stress which can be repeated indefinitely without continuing to add to the deformation. The limit is about one-third of the ultimate strength, and the stress-strain curve is, until this limit is reached, nearly a straight line (Fig. 209). Beyond this limit, the stress-strain curve bends because the modulus of elasticity is dependent also on the stress or corresponding strain. It can be computed by the Roš equation

$$E_\sigma = \frac{d\sigma}{d\epsilon} = \frac{2f'_c}{(2a-1)\epsilon_b^2}(a\epsilon_b - \epsilon) \tag{557a}$$

where a is a material constant, ϵ is the strain corresponding to the stress σ, and ϵ_b is the strain at the stress $\sigma = f'_c$.

For $\sigma = 0 : \epsilon = 0$,

$$E_\sigma \to E_0 = \frac{f'_c}{\epsilon_b} \frac{2a}{2a-1} \tag{557b}$$

Substituting for f'_c in Eq. (557a) the expression from Eq. (557b), we obtain

$$E_\sigma = \frac{E_0}{a\epsilon_b}(a\epsilon_b - \epsilon) \tag{557c}$$

For $\sigma = f'_c : \epsilon = \epsilon_b$,

$$E_\sigma \to E_b = E_0 \frac{a-1}{a} \tag{557d}$$

The equation for the stress-strain curve is obtained by integration of Eq. (557a):

$$\sigma = \frac{f'_c \epsilon}{(2a-1)\epsilon_b^2}(2a\epsilon_b - \epsilon)$$

$$= \frac{E_0 \epsilon}{2a\epsilon_b}(2a\epsilon_b - \epsilon) \tag{558}$$

and is represented in Fig. 209.

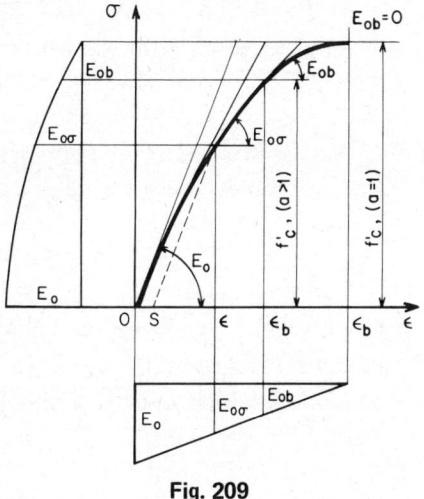

Fig. 209

According to Eq. (558), the stress-strain curve is a parabola. For $a = 1$, the $E_b = 0$; and for $a > 1$, E_b has a definite final value. For purposes of computing, it is determined at $\sigma = 0.85 f'_c$. The breaking strain ϵ_b depends to a large degree on the mix. It is smallest, about $\epsilon_b = 1.6‰$ (‰ = 1/1,000), for igneous-rock coarse aggregate with the lowest fine-coarse aggregate ratio (0.35) and is highest, $\epsilon_b = 3.0‰$, for finer and lightweight aggregate. The lateral strain, expressed by Poisson's ratio (ν), is almost proportional to the breaking strain. It may vary, respectively, from 0.16 to 0.24. The higher strain values measured (ϵ_{UL}), as given above, are considered caused by internal crack formation (shear cracks), which, according to microscopic studies, starts when the compressive stress reaches about 85 percent of ultimate strength. Due to this, the breaking strain ϵ_b and modulus of elasticity E_b are determined at 0.85 f'_c. The strain values can most easily be determined by direct measurement of test cylinders. The a values may vary from 1.0 to 1.5, depending mainly on the physical characteristics of the stones used for coarse aggregate and on the fine-coarse aggregate ratio. When the initial modulus of elasticity E_0 and breaking strain ϵ_b are known, the a value is computed by Eqs. (557). The E_0 values can be determined with sufficient accuracy by Eq. (64) or (65):

$$E_0 = 8.15 \times 10^6 \frac{f'_c}{2{,}300 + f'_c} \quad \text{regular concrete}$$

$$E_0 = 3.65 \times 10^6 \frac{f'_c}{1{,}500 + f'_c} \quad \text{lightweight concrete}$$

As can be seen from Fig. 209, the $E_{0\sigma}$ values, as a function of stress σ and determined by Eq. (557b), follow a curve and are directly proportional to the strain ϵ.

During unloading, the stress-strain curve of a concrete specimen follows an almost straight line parallel to the tangent at the origin of the initial stress-strain curve (Fig. 209) and meets the ϵ axis at point s. The distance $0s$ is the permanent initial set or strain. It is approximately 10 to 15 percent of the total elastic strain and is a result of uneven internal stresses in hardened concrete caused mainly by shrinkage. During the first loading, up to approximately $0.4\, f'_c$, these internal stresses are balanced, which explains the fact that in unloading, concrete obeys the initial modulus E_0 and not $E_{0\sigma}$. In this instance, the behavior of concrete is similar to that of steel, as will be discussed later.

The modulus of elasticity for tension is commonly assumed to be equal to that for compression. The shear modulus is approximately

$$G = \frac{E_0}{2(1 + \nu)} \sim 0.45 E_0 \tag{559}$$

where ν is Poisson's ratio. It characterizes the distortion of a section due to shear. For quality concrete, it is about 0.16.

The allowable stress in concrete under normal design load depends on the factor of safety required, which, on the other hand, depends on the type of structure and the loading condition. For average design-load condition, the factor of safety is 2.0, and for temporary loading, 1.5 can be considered reasonable.

Thermal Properties of Concrete

The coefficient of linear expansion (α_T) is the change in length per unit length per one degree of temperature. For high-quality concrete, the average value of $\alpha_T \simeq 6 \times 10^{-6}$ per 1°F.

The heat conductivity of concrete may vary considerably. As an average value for regular concrete, it is about 12.6 BTU (hr)(ft^2)(°F) and is approximately half of that for structural lightweight concrete. (1 BTU $\cong 0.252$ Kcal.)

41. SHRINKAGE AND SWELL

The origin of shrinkage and swell lies in the capillarity of concrete. It was first and most thoroughly explained and discussed by Freyssinet.* The capillary phenomenon and forces in action are illustrated schematically in Fig. 210. The solid volume of hardened concrete, as discussed before, varies from 0.73 to 0.93; thus the volume of voids and capillaries is 27 to 7 percent of the total volume of concrete. The larger capillaries and voids are filled with air (V_A), and the smaller with water (V_W). In accordance with thermodynamics, the normal tension p_n in capillary tubes, caused by the surface tension of water during evaporation and balanced by concrete, is

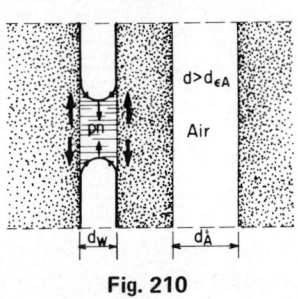

Fig. 210

$$p_n = \frac{2\gamma}{d_W} \tag{560}$$

where $\gamma = 8$ mg/mm is the capillary constant and d_W is the diameter of a capillary tube filled with water. The normal tension or pressure p_n, as expressed in terms of relative humidity ϵ_A, is

$$p_n = 1{,}300 \ln \frac{1}{\epsilon_A} \tag{561}$$

From Eqs. (560) and (561) can be computed the limiting value for the diameter d_{ϵ_A} of a capillary tube, down to the size in which complete evaporation of the water takes place under a given relative humidity. Thus,

$$d_{\epsilon_A} = \frac{2\gamma}{1{,}300 \ln(1/\epsilon_A)} \, 10^{-4} \text{ mm} \tag{562}$$

---

*E. Freyssinet, "Une révolution dans les techniques du béton," Leon Eytrolles, Paris, 1936.

Substituting $d_{\epsilon A}$ from Eq. (562) for dw in Eq. (560), the corresponding value of normal tension in these tubes is

$$p_{\epsilon A} = \frac{2\gamma}{d_{\epsilon A}} \text{ mg/mm}^2 \tag{563}$$

The limiting values for $d_{\epsilon A}$ and $p_{\epsilon A}$ under various relative humidities are given in Table 5. (The diameter of the water molecule is 0.26×10^{-6} mm.)

Table 5

ϵ_A	$d_{\epsilon A}$, mm	$p_{\epsilon A}$, psi	Climate
1.00	—	0	Saturated atmosphere
0.90	11.60×10^{-6}	2,000	Sea, coastal areas
0.80	5.50×10^{-6}	4,160	
0.20	0.76×10^{-6}	30,000	Dry continental

As can be seen from the values given in the table, the capillary forces in concrete in a dry continental climate can be very high and may cause correspondingly large plastic deformations and cracks when concrete is not properly cured during the hardening period and when the modulus of elasticity and shear are still relatively low. It is also interesting to note that the tensile capacity of water in a thin thread, approximately three molecules in diameter, is almost as high as that of the yield strength of steel.

The volume change due to shrinkage can be treated mathematically. Making use of the Eqs. (562) and (563) and applying Hooke's law, the unit average shrinkage ϵ_{sr} is

$$\epsilon_{sr} = \frac{P}{AE_{sr}} = \frac{p_{\epsilon A}(v_w A)}{A\psi_{sr} E_0} = \frac{v_w p_{\epsilon A}}{\psi_{sr} E_0} \tag{564}$$

where $\psi_{sr} = E_{sr}/E_0$, $E_{sr} = E_0 f(t)$, and A is the cross-sectional area of the element. The magnitude of ϵ_{sr} for a given concrete is a function of the rate of hardening and evaporation (curing). It may vary relatively widely, depending on the value of the modulus of elasticity at the time when capillary forces $p_{\epsilon A}$ start to develop. For proper steam curing followed by efficient moist curing, it may be as high as 0.9, and for poor curing, it may be as low as 0.5 or even lower.

As the drying out of concrete is an inevitable process, the value of $p_{\epsilon A}$ cannot be influenced. Thus the shrinkage can be reduced only by delaying the evaporation by curing until the strength of the concrete has developed sufficiently to resist the capillary forces. Further, it is interesting to note that the fineness of cement, in the case of proper curing, does not influence the shrinkage notably. This phenomenon is explained by the more rapid hardening and more complete hydration of high-early cement compensating the higher void volume filled with water (v_W) [see Eq. (564)].

The relative-humidity gradient in concrete rises sharply from the surface exposed to atmosphere toward the inside, and as tests indicate, it can be 100 percent in a relatively short distance from the exposed surface for several months. The capillary forces are a function of relative humidity [Eq. (561)]; therefore, considerable differences in magnitude of shrinkage during drying out exist inside a unit having even moderate dimensions. Since the shrinkage of paste is resisted by aggregate and reinforcing steel, as well as by the not dried-out interior, considerable internal stresses (tension outside and compression inside) are developed in concrete, which are beyond the tensile and bond-strength capacity of the paste if premature drying out of concrete is allowed to occur. The influence of aggregate upon the magnitude of shrinkage and internal stresses depends on the physical characteristics of the stone and the amount and size of particles present. Limestone, lightweight, and other porous aggregates are more compressible than granite or traprock aggregates and offer less resistance to the shrinkage of paste, resulting in higher shrinkage values, increased tensile strength, and reduced crack formation.

The shrinkage measured in laboratory specimens as a function of time is illustrated in Fig. 211 for cement paste and for plain and lightweight concrete moisture- and steam-cured. Since there is rarely a relative humidity equilibrium between the exposed surface and the surrounding atmosphere, the surface takes in moisture; the concrete expands or swells and, giving up moisture, shrinks. Therefore, it is practically impossible to determine the exact magnitude of shrinkage in structures. For reinforced prestressed high-quality concrete with granite and quartz aggregate, properly air-cured, the average shrinkage after prestressing is

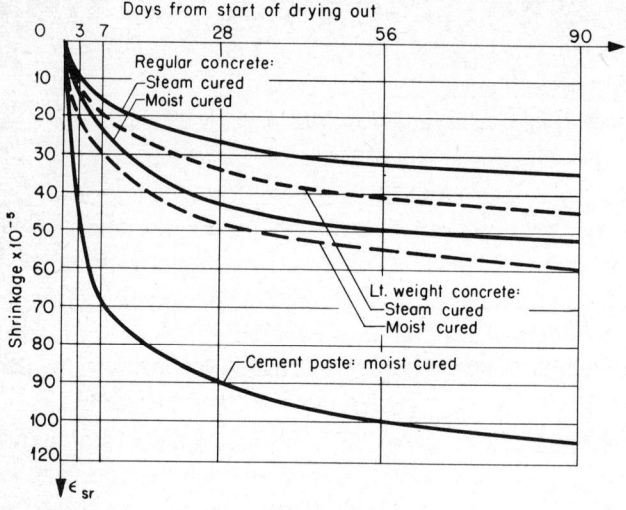

Fig. 211

applied can be assumed to be approximately 25×10^{-5} of unit length and even less when posttensioning is applied at a later date (Fig. 211). For lightweight and limestone concrete, the shrinkage is somewhat higher: up to 35×10^{-5}. For steam-cured concrete, the shrinkage is about 25 to 35 percent less. The time-shrinkage curves at approximately uniform relative humidity follow rather closely the exponential function

$$\epsilon_{srt} = \epsilon_{srn}(1 - e^{-t}) \tag{565}$$

where ϵ_{srt} is the time under consideration and ϵ_{srn} is the final value of shrinkage. In very dry, hot continental climates, shrinkage of medium-size structural elements can be considered finished in 6 months, but in moist sea climates and with larger elements, it may take several years.

It could be expected that when concrete is hardened and dried out, the capillary forces would diminish and then the concrete would expand again, at least to some degree. However, this is not the case; expansion takes place to a small extent only when water is absorbed and the capillaries are filled with water (swelling). This phenomenon is explained by increased intermolecular attraction in the course of drying out, resulting in a rigid capillary structure of hardened paste.

42. PLASTIC FLOW

Plastic flow is the physical characteristic of hardened cement paste which enables it to undergo plastic deformations under sustained load for a long period of time. As was discussed in the previous section, the origin of shrinkage lies in the action of capillary forces and its magnitude depends primarily on the limiting size and amount of water-filled capillaries (v_W) at a given relative humidity ϵ_A. In the case of plastic flow, the magnitude of deformations is dependent on the loading stress and on the total volume of voids $(V_W + V_A)$. In the range of stresses commonly used in design, the plastic flow ϵ_{pl} exceeds considerably the elastic deformations ϵ_{el} and shrinkage ϵ_{sr}. It is roughly proportional to the sustained stress and proceeds at a rather rapid rate initially, gradually slowing down over a period of time $(t = n, n = 1, 2, \ldots, 5)$ until it reaches a final value. Like shrinkage, plastic flow as a function of time follows very closely the exponential function

$$\epsilon_{plt} = \epsilon_{pln}(1 - e^{-t}) \tag{566}$$

where ϵ_{pln} is the total unit plastic strain and t indicates the time under consideration (Fig. 212).

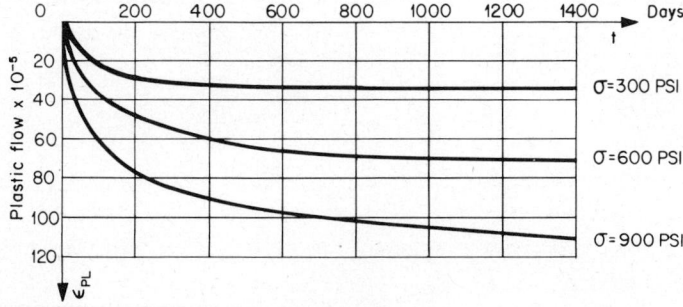

Fig. 212

Since both of these plastic deformation phenomena, shrinkage and plastic flow, have their origin in stresses—internal (capillary) and external (loading)—the factor affecting their magnitude is almost the same.

The mechanism of plastic flow is not yet fully explained. The most sensible explanation is that the sliding along crystal surfaces under sustained loading results in the flow of unhydrated amorphous material and water within the section, balancing the vapor pressure and equalizing the stress distribution. It is quite clear that such a mechanism always occurs when sustained stress (capillary and loading) exists; therefore, plastic flow proceeds also under shrinkage conditions. As there is a limiting value for deformations and time required for their equalization, it is reasonable to assume the total plastic deformation (shrinkage and plastic flow) for any concrete to be constant. On this ground, there is no possibility of dealing with shrinkage and plastic flow as independent and separate phenomena in practice (Sec. 9).

A small recovery of plastic flow after the removal of a sustained load is explained by the release of elastic strain in the aggregate, which forces some backflow of migrated water to the less stressed areas. The complete recovery of plastic flow is not possible because of the counteraction of increased intermolecular attraction due to decreased distances of constituents during sustained load.

For concrete in structure, the curves of plastic flow are not exactly continuous lines, like those gained by laboratory testing (Fig. 212), because they are influenced also by relative humidity and changes of atmospheric temperature, as is the case with shrinkage. However, the discontinuity of the curves has little influence on the final value of plastic flow (ϵ_{pln}) for a given condition.

The mathematical treatment of the influence of plastic flow on deformation, as well as their relationship, are thoroughly discussed in Part 1.

9
STEEL

43. REINFORCING STEEL

Steel used as deformed reinforcing bars is available in three ASTM grades: A-15, A-432, and A-431. The differences in physical properties of these grades (Table 6) are obtained by changing the carbon content and heat treatment.

The yield stress beyond which excessive elongation starts is measured at 0.2 percent offset. Minimum elongation at rupture is regarded as a measure of ductility.

In addition to deformed bars, welded wire fabrics complying with ASTM A-185 are also used as reinforcement for concrete. The wires up to ½ in. in the fabric are cold-drawn and have a minimum ultimate strength of 70,000 psi. The cold-drawn wires do not have a definite yield point (see Sec. 44). The intersecting longitudinal and transversal wires are fused together by a special automatic electric welding process.

400 Materials

Table 6

ASTM grade	Max. yield stress (f_y), psi	Min. ultimate tensile strength (f'_s), psi	Elongation at rupture (ϵ_{UL}) 8 in., %
A–15	40,000	70,000	12
A–432	60,000	90,000	12
A–431	75,000	100,000	5.5–7.5

Typical stress-strain curves of hot-rolled bars and cold-drawn wires are given in Fig. 213. As can be seen from the stress-strain curve of cold-drawn wires, the fabric tends to resist the stresses up to almost 85 percent of its ultimate strength. As the slippage of wires is prevented by welded cross wires spaced commonly not more than 12 in. apart, excessive crack formation is avoided almost throughout the wire strength range. The high ultimate strength of hot-rolled bars, regardless of deformed surface, cannot be fully utilized in reinforced concrete because of crack formation. In accordance with experience, the maximum acceptable crack width in concrete exposed to severe conditions is approximately 0.008 in. and is not more than 0.012 in. in protected conditions. Crack widths up to these limits, it has been proved, do not cause corrosion of steel. To secure durability, the allowable stresses

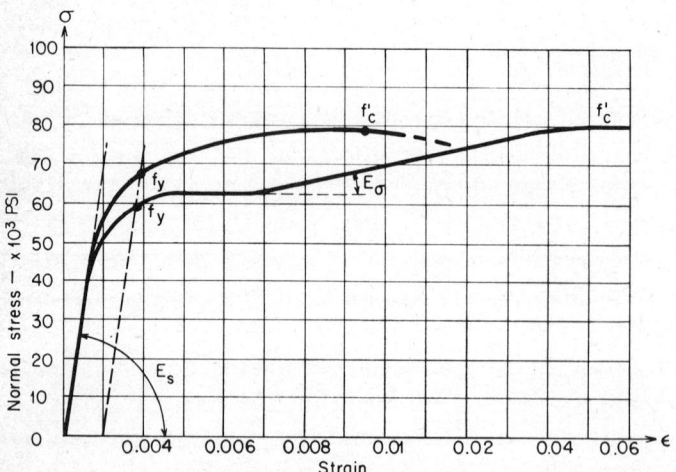

Fig. 213

should be limited to 20,000 psi and no more than 30,000 psi for any type of reinforcing steel. For waterfront works, they should even be limited to not more than 18,000 psi. The modulus of elasticity for hot-rolled reinforcing steel and for welded wire fabric is about 29×10^6 psi.

44. PRESTRESSING STEEL

Plain, relatively high-strength wires (ultimate strengths up to 350,000 psi) assembled into strands or cables are used for pretensioning or posttensioning. Also, high-strength bars up to 1-3/8 in. in diameter and with ultimate strengths up to 160,000 psi are available for post-tensioning.

The high tensile strength required in prestressed reinforcement is obtained by using the proper chemical composition in the manufacture of the steel, followed by certain shaping and treating operations. The chemical composition does not necessarily set the values of the properties to be obtained, but it definitely fixes the limits of the properties obtainable. The other factor, i.e., the shaping and treating operations, sets the particular values desired in the finished product. Strength and ductility are the two important characteristics of the steel. An increase in carbon content increases the strength and hardness of the steel considerably but reduces its ductility.

The melting point of iron is about 2800°F. When carbon is added to iron, it is found that iron holds carbon in solution up to 1.7 percent at approximately 2100°F (austenite) and only 0.007 percent in solid solution (ferrite) at normal atmospheric temperature. When not in solution, carbon exists in the ferrite matrix as carbide (Fe_3C). The carbon content in cold-drawn wires is up to 0.80 percent (or even higher). This is possible because of a special controlled cooling method for austenite. By slow cooling, the ferrite and carbide are rejected from the austenite matrix in alternate lamellae. As the cooling rate is increased, the spacing of the lamellae becomes smaller, and when the cooling rate is very rapid, the carbon does not have time to separate from the austenite or carbide, resulting in a highly stressed saturated structure of austenite (martensite). The carbide content present, its shape and distribution in the ferrite crystal matrix, determines the physical character of the steel.

The effect of carbon on the properties of steel can be modified to some extent by introducing various alloying elements. The addition of alloying elements does not change greatly the characteristic effect of carbon; it serves primarily to increase the dispersion of carbide in the ferrite, to change the properties of ferrite or carbide, and thus to contribute to the formation of a more uniform and finely grained austenite. Furthermore, the alloying elements, especially silicon (Si) and chromium (Cr), affect even the hardenability of steel in larger sections.

First, the steel is shaped from billets by hot-rolling above the critical temperature (austenite, $2100°F$) into rods. Considerable improvement of the physical properties of the steel is obtained by hot-rolling, which is especially beneficial in terms of ductility. The second step is the cold-drawing of the hot-rolled rods into wires, usually at atmospheric temperature. Steel in the hot-rolled state is a conglomerate of crystals which are small in size and can be seen only with special microscopes on a plane, finely polished, specially treated surface. Studies made of single crystals in recent years have shown that the physical properties of crystals depend greatly on the orientation of the axes of the crystals with respect to the direction of the tensile stresses to which they are subjected. In steel, the crystals are scattered at random; thus the crystallographic planes and axes of the individual crystals may be more or less favorably aligned relative to the direction of the stretching force applied. The plastic deformation of steel consists of slidings in certain directions along crystallographic planes. The magnitude of the sliding is dependent on shearing stresses acting on these planes; the stresses, in turn, are dependent on the angle of the planes with respect to the stretching force. The sliding is accompanied by an increase in the resistance to sliding, which represents the strain-hardening and indicates an increase in uniformity of the stress distribution along the entire crystallographic plane as a result of the stretching action. When steel is subjected to stretching beyond its initial yield point during cold work, there results a more favorable orientation of the individual crystals with respect to the stretching forces; thus a more uniform distribution of stress throughout the material is achieved. The effect of sliding will be further reduced when the crystals are small in size, as may be concluded from the foregoing reasoning. As the diameter of wire subjected to cold-stretching is increased, the resulting residual stresses increase, and therefore the ultimate strength decreases accordingly.

The ultimate strength of fine (up to 0.105 in. in diameter) cold-drawn wires available is about 350,000 psi. Wires from 0.105 to 0.283 in. in diameter, used for wire ropes and strands, are more ductile than the very fine wires. Ultimate strength for smaller-diameter wires is up to 280,000 psi and for larger diameters is about 250,000 psi. The yield stress is approximately 0.9 f'_s at 0.2 percent offset, and the elastic limit is about 0.75 f'_s. Elongation in 24 in. after rupture is 4 percent. Modulus of elasticity is approximately 29×10^6 psi.

For seven-wire strands, because of the twisting during manufacturing, the modulus of elasticity is slightly less than for the individual wires that compose the strand; it is approximately 28×10^6 psi. A typical stress-strain diagram for stress-released single wires and seven-wire strands is given in Fig. 214.

High-strength steel bars in diameters up to 1-3/8 in. and even larger are used mainly for posttensioning. These bars are manufactured from carbon-alloy hot-rolled steel by cold-stretching. The cold-work process is carefully executed to obtain the desired yield and ultimate strength.

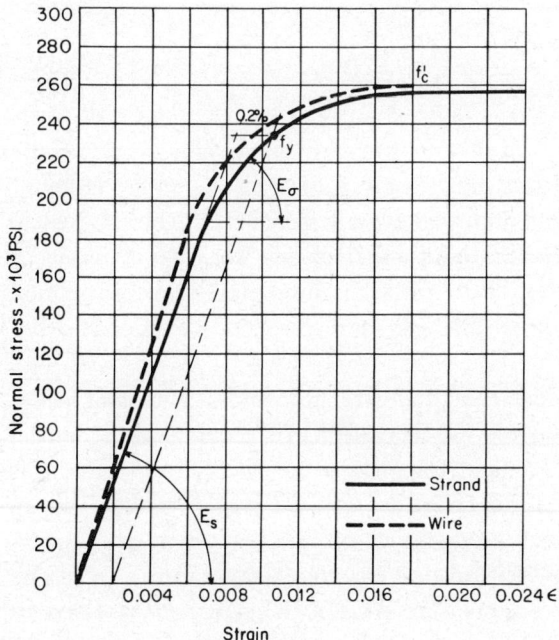

Fig. 214

Then the bars are stress-released in a gas furnace to secure adequate ductility and uniform stress-strain characteristics.

A typical stress-strain diagram for high-strength stress-released bars is given in Fig. 215. The physical characteristics are: minimum ultimate strength, 160,000 psi; yield stress at 0.2 percent offset, 140,000 psi; minimum elongation in 20 diameters after rupture, 4 percent; minimum reduction of area after rupture, 20 percent; modulus of elasticity, 30 × 10^6 psi.

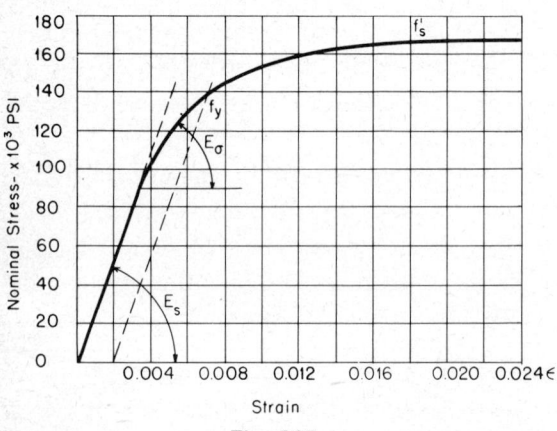

Fig. 215

The moduli of elasticity given are true only up to about the elastic limit (~ 0.65 f'_s); beyond this limit, the ratio is $E_\sigma = d\sigma/d\epsilon$ and Hooke's law does not apply. The maximum allowable initial stress (before losses) for seven-wire strand cables and bars is 0.70 f'_s and, for a short period of time, up to 0.75 f'_s. If the 0.80 f'_s limit is used, technical creep may occur; i.e., the crystals will be damaged, and so-called "stress corrosion" may occur.

Tests and experience show that high-strength cold-drawn steel under sustained load loses stress due to relaxation. In general, if the initial stress does not exceed 0.7 f'_s and for a short period 0.75 f'_s, the loss is about 2 percent of initial stress for stress-released strands or bars and up to 4 percent for non-stress-released strands or bars.

All high-carbon steels are sensitive to elevated temperatures, as can be concluded from the above discussion. At 800°F, in strands stressed to 0.65 to 0.70 f'_s, the loss in ultimate tensile strength is approximately 35 percent of the original strength and in yield stress is about 50 percent.

The losses at elevated temperatures can be substantially reduced when the carbon content in steel is slightly reduced and the silicon-chrome ratio increased. Such a small change in chemical composition does not alter notably the other physical characteristics of high-carbon steel. This fact must be kept in mind when considering concrete coverage to protect the steel against fire hazard.

In general, steel containing less than 0.15 percent carbon can be welded without special precautions. Steel bars having a carbon content of more than 0.3 to 0.4 should not be welded. Below the 0.3 carbon level, steel can be welded by electric arc when preheated and slowly cooled. Commonly, the strands and bars are flame-cut. The melting heat and following rapid cooling produce a hard and brittle martensite. The damage to the steel depends on the diameter size. Experience shows that for strands and smaller bars, a 1-in. minimum from the wedge end to the cut face is required to avoid anchorage damage; for larger-diameter bars, 1½ in. is required.

45. CORROSION OF STEEL

Basically, the corrosion of steel can be a purely chemical or electrochemical process. In most cases, the corrosion process starts with a chemical reaction or oxidation and then proceeds by electrochemical action. However, electrochemical corrosion is also possible without prior steel oxidation.

The oxidation of steel in concrete can develop in the presence of moist air (relative humidity above 50), in oxygen-rich water, and in alkali-poor (pH < 9) aqueous solutions. In alkali-rich (pH > 9) aqueous solutions, steel does not corrode because the hydration products of cement alkalinize and form with oxygen a thin protective iron or iron calcium oxide layer at the steel surface which is unsoluble in the alkaline medium. If the concrete coverage is insufficient, porous, or cracked, moist air rich in carbon dioxide (CO_2) has access to the concrete and carbonization occurs [$Ca(OH)_2 + CO_2 \rightarrow CaCO_3 + H_2O$] or the soluble calcium hydrate [$Ca(OH)_2$] is bled out and the pH value decreases, resulting in disruption of the protective layer. The protective layer can also be destroyed in an alkaline medium with chlorides ($NaCl, MgCl_2, CaCl_2$) by forming soluble iron chloride instead of iron hydroxide or iron oxide. After the protective coating is destroyed, rusting of steel proceeds. As the volume per unit weight of rust is about

4½ times greater than that of steel, pressure is exerted on the concrete, resulting first in cracking and finally in the popping off of the concrete coverage.

Electrochemical corrosion occurs when electric current is generated where dissimilar metals such as steel, aluminum, zinc, or copper are in contact in the presence of moisture or when stray currents find their way to the steel. In locations where the protective layer of steel is destroyed, oxidation takes place and electric cells are formed. The iron oxide acts as the cathode because of its higher potentiality in comparison with steel, and oxygen-free places function as the anode. At the anode, the steel is transformed into iron salt, releasing free electrons, which form OH ions at the cathode with oxygen and water. By such oxygen polarization and electron current, the steel surface at the anode is continuously carried away as long as moist air is available to condense and serve as the electrolyte. The speed of corrosion depends on the voltage of current and the electrical conductivity of the electrolyte. It is highest if some salts are present in the water (seawater).

When the steel is in contact with an electrolyte containing chlorides, nitrates, sulfates, etc., the free electrons from the anode form atomic hydrogen. In high concentrations, the atomic hydrogen (pH value < 9) penetrates into the steel surface through microcracks along loosened crystal boundaries and forms molecules associated with extremely high pressure. When the prestressing steel is stressed beyond the technical-creep limit ($0.8 f_s'$), some local loosening of crystal boundaries starts and the possibility of atomic hydrogen corrosion is seriously increased. This type of corrosion, loosely called *stress corrosion*, can cause a sudden break of the steel.

Stress corrosion is, for obvious reasons, higher for highly stressed small-size high-carbon steel wires than for large-diameter wires or bars. The danger of stress corrosion is greatest in poorly grouted small wire tendons when they are grouted with water-rich fly ash cement mix. Since free water cannot evaporate in metal tubing and the fly ash, if not properly tested, may contain chlorides, sulfates, or nitrates, the conditions for development of oxidation or electric cells are favorable. The danger is increased by the dissimilarity between the tubing metal and high-carbon steel. Also, for the same reasons, aluminum powder should not be used as an expansion agent to counteract shrinkage. The relatively small amount of such an active additive, commonly considered harmless, may interact with the cement ingredients and lead to a concentration of chlorides, sulfates, etc., and thus to a dangerous corrosion level.

To avoid corrosion of tendons, limestone powder instead of fly ash, fine sand, or pure cement should be used in the grouting mix. The excess water in the mix is absorbed by limestone powder, and free water is reduced or even entirely absent in hydrated grout. Besides, limestone is affined with cement, the fine limestone particles remaining in suspension under high grouting pressure, which is not the case with fine sand.

As a conclusion, the corrosion of steel of any kind does not develop in sound, dense concrete having a pH value in the normal range (11 to 13) and when chlorides, nitrates, sulfates, etc., are not present. The possibility of corrosion is greatly reduced if the temporary stress limit for high-carbon steel is kept below $0.75 f'_s$, if surface damage of steel during handling and prestressing operation is prevented, and if proper concrete coverage for the steel is provided.

10

BEARING NEOPRENE

46. PHYSICAL CHARACTERISTICS

Structural elements change their volume and length, deforming under loading, temperature change, shrinkage, and plastic flow. If such changes are restrained, secondary stresses and uneven stress distribution on bearings will develop which, if not properly considered in design, may cause serious damage to the structure. Also, to reduce dynamic effect upon a structural system, it is required that the elastic response of the system be as large as possible; i.e., the bearing at intersecting elements should have a high damping value, and the system as a whole should have a high ductility.

The most economical and practical way to counteract the inherent disadvantages of concrete structures is to provide neoprene bearings and separating layers between intersecting units or elements to control the elasticity and sound propagation and to avoid large stress concentrations.

The neoprene used for bearing purposes is a high-strength, quality product processed to have required resilience and durability. It is an oil-resistant synthetic rubber made by polymerizing chloroprene ($CH_2 - CClH - CH_2$), which is produced from acetylene and hydrogen chloride. The neoprene on the market commonly contains up to 20 percent of admixtures, fillers, and fibers. Depending on the composition and method of manufacturing, the physical characterisitcs of the finished product may vary within rather wide limits.

Basically, the raw neoprene under compression behaves like water, which means that it deforms under constant volume. Thus, to obtain the desired compressibility and resilience and to reduce lateral expansion, fillers and fibers are used. If a sufficient number of strong fiber sheets in several plies are embedded in plain neoprene, the bearing pad retains substantially its original length and width under compression and impact. Under sustained load, the neoprene creeps slightly (up to 3 percent).

The modulus of elasticity for compression, as can be seen from the stress-strain diagram (Fig. 216)*, is a nonlinear function of the

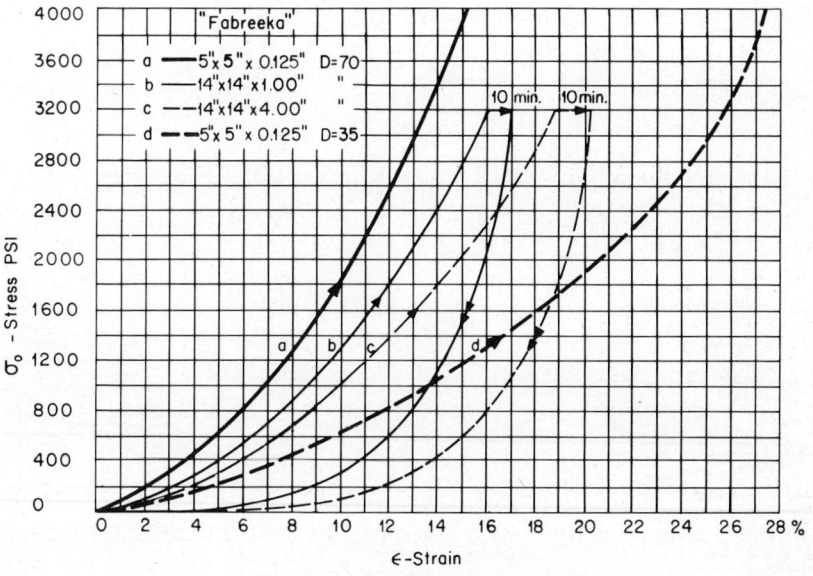

Fig. 216

*Tests carried out for Habitat construction, Montreal, 1966.

compressive stress σ_0 and thickness t. The initial set of neoprene is 1 to 3 percent, depending on σ_0 and grade. When the load is removed, the elastic recovery is not instantaneous but has a limiting time factor (up to 4 hr) for thick layers. Such elastic behavior is most advantageous with regard to slow irreversible movements.

Commonly, the relative compressibility (strain) of neoprene is estimated by the following formula:

$$\epsilon = \frac{\sigma_0}{4Gf^2 + 3\sigma_0} \qquad f = \frac{ab}{2t(a+b)} \qquad (567)$$

where f is the so-called *form factor*, depending on the thickness and size of the sheet, and σ is the shear modulus of neoprene. The formula is valid for a condition of even and restrained deformations of the contact surfaces.

The compressive strength of neoprene is relatively high; it varies from 15,000 to 35,000 psi, depending on the thickness of the pad and the fibers used. Therefore, the allowable compressive stress is controlled mainly by shear stress and strain. The maximum shear stress due to σ_0 develops at marginal regions of the pad. It is approximately

$$\tau_{max} = 1.5 \frac{\sigma_0}{f} = \frac{3t(a+b)\sigma_0}{ab} \qquad (568)$$

and the related maximum tensile force at the center of the pad is

$$T_{max} = 1.5 \sigma_0 t \qquad (569)$$

The shear stress τ_{max} is increased by the relative rotation α of the contact planes of the pad. Provided that α is small, so that no uplift occurs, the shear stress between the pad and concrete is

$$\tau_{\alpha, max} \simeq \frac{Ga^2}{t^2} \alpha \qquad (570)$$

When the pad is acted upon by a horizontal force H, the additional shear stress is

$$\tau_{H, max} = \frac{H}{ab} = \tan \gamma \, G, \quad \left(\tan \gamma = \frac{\Delta L}{t} \right) \qquad (571)$$

The maximum shear and tensile stresses are obtained by superposition. The tensile stress of plain neoprene is about 2,700 psi and the elongation is 325 percent of its original length. The shear strength is up to 1,200 psi. The elastic shear distortion of neoprene up to twice its layer thickness is almost directly proportional to the horizontal force H, and it is only slightly affected by the compressive stress σ_0. Beyond this limit, permanent distortion occurs. However, to secure durability, it is recommended that the $\tan \gamma$ value be limited to 0.70 (Fig. 217).

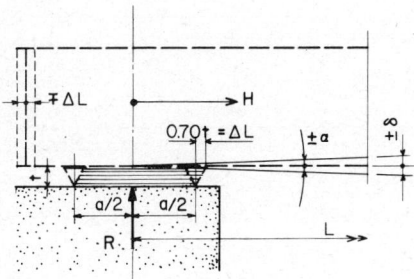

Fig. 217

The shear stresses must be balanced by the frictional stresses. Thus the frictional coefficient required to avoid slippage is

$$f_r = \frac{\Sigma \tau_{max}}{\sigma_0} \tag{572}$$

The frictional coefficient is not a constant value but a function of compressive stress σ_0. It can be as high as 1.0 for relatively small σ_0 and decreases rapidly with the increase of σ. As an average value, it can be assumed to be 0.3.

The shear modulus of neoprene increases with age and temperature, but only slightly, down to $-30°F$. For still lower temperature exposure, the G value increases rather rapidly. For neoprene having shore durometer 70, the average initial shear modulus is approximately $G = 230$ psi and the final is about $G = 260$ psi. For softer grades, it drops approximately 35 psi per 10 durameters.

In order to control the shear stresses, vertical flexibility, shear distortion, and horizontal force, multilayer bearing pads, assembled of n neoprene layers (up to 1/2 in. thick), separated by stainless-steel plates (up to 1/8 in. thick), and bonded to unity, can be built up to the required total thickness. In this way, a considerable variation of vertical stiffness can be achieved without changing the shear characteristics (Fig. 218). Thus the shear stresses and rotation angle are decreased and the lateral movement is increased by n-times the individual layer. Furthermore, the hardness as well as the type of neoprene (plain or reinforced by fibers), pad size, and side relationship can be chosen so that even the most complex requirements can be satisfied.

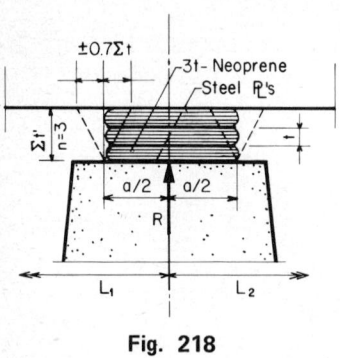

Fig. 218

The heat conductivity of neoprene is relatively small. Due to this, a neoprene bearing exposed only by sides does not lose its elasticity and strength even when exposed to extreme temperature for a considerable time. Experience also shows that the physical characteristics are not much affected by sustained high stress, dampness, or temperature change.

The neoprene is manufactured in sheets, with thicknesses of 1/16 to 1 in. and hardness grades (durometers) of 90, 70, and 30. Softer grades and smaller thicknesses are used principally for separations and to obtain uniform, watertight bearings.

To avoid failures, the particular neoprene to be used should be thoroughly tested and its physical properties determined prior to manufacturing the bearings.

Furthermore, the tensile-stress condition in concrete under the bearing pad is critical, especially when horizontal force and rotation are involved. To avoid edge and corner breakage and vertical cracking, the compressive stresses at the edge regions of the bearing pad must be reduced and adequate reinforcing provided to balance the tensile stresses. Also, as the compressibility of neoprene is far higher than that of concrete, any bonded steel bars through the neoprene pad will be highly overstressed. The control of these steel stresses is possible only by providing adequate nonbonded length for the steel bars or tendons above or below the neoprene pad.

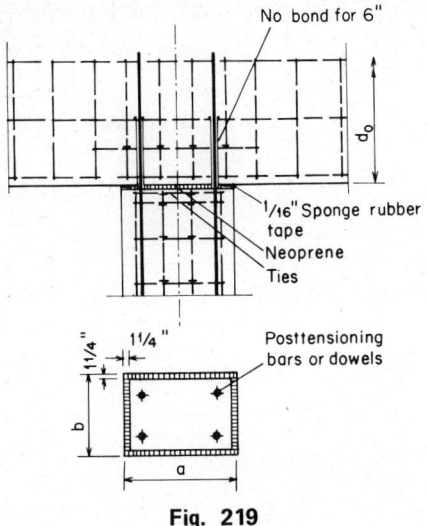

Fig. 219

According to experience, the best and safest results are obtained by the joint design, as illustrated in Fig. 219. In cases in which the tendons are only throughgoing, the conventional bars should be stopped about 1/4 in. below the neoprene.

PART 3

STRUCTURAL SYSTEMS AND METHODS OF CONSTRUCTION

The critical evaluation of a structural system is based mainly upon the end-means relationship. The principles of design, the carrying actions, and the means for realization at the designer's disposal have been thoroughly analyzed and discussed in Parts 1 and 2. The application of this knowledge to obtain rational and meaningful end-means relations in a design and construction is the subject of the following chapters.

Numerous structural systems can be combined or assembled from the same structural elements: slabs, beams, girders, columns, and shells. How these elements are combined and their relationships determine the meaning and efficiency of the system. The character of the elements themselves, on the other hand—their materials and shapes—determines the economy and variety of their application. The shapes of reinforced or prestressed elements and the structural systems combined from the basic elements most often follow the shapes and systems of steel or wooden structures, i.e., are practically steel substitutes. This is especially true when the structural systems are assembled from precast concrete elements. Concrete has its own physical characteristics, requiring special shapes, construction methods, connections, etc., and offering opportunities for structural systems not obtainable with steel or wood. The mechanical installations, which control in a large degree any contemporary design, can be easily incorporated into the overall structural system. Also, mass-production methods can be used for plant and site precasting, as well as for poured-in-place systems, more easily than with any other available materials at present.

The selection of a structural system is not arbitrary. The proper structural system is determined by the nature of the architecture and the function of the building. It must be kept in mind that a structure is not the end itself but is a servant to the function of the building. Due to this fact, the proper structural system grows out of the function and not the other way around, as is often the case, when a system selection is based on easy, simple design considerations or even on the availability of ready-made formulas for a design.

To simplify the treatment of problems involved with design and methods of construction, the structural elements and systems for various purposes will be discussed in the following identical or related main groups:

1. Basic structural elements
2. Structural systems
   One-way systems
   Two-way systems
   Space systems
3. Prefabricated houses
4. High-rises
5. Bridges
6. Miscellaneous structures
   Water towers, tanks, silos, poles, etc.

11

BASIC STRUCTURAL ELEMENTS

Basic structural elements are various types of slabs or planks, slab-beams, beams, girders, columns, and wall panels. These structural elements can be used independently or in combination to establish a structural system. In most cases, these elements are required in relatively large quantities; therefore, the method used for construction must be based on mass-production principles. Selection of the method of construction to be used, that is, prefabrication or poured in place, depends mainly on local conditions, equipment and time available for construction, and economy. In accordance with experience, prefabrication is more economical if a prefabrication plant is available nearby and the units can be transported to the site without difficulty. When the number and size of elements warrant it and the required space or area is available, prefabrication on the site may offer considerable economy.

47. PREFABRICATED SLABS AND PLANKS

In this category belong rib slabs and hollow-core slabs or planks. Rib slabs, as illustrated in Fig. 220, can be manufactured in the rib-upward position (Fig. 220a) by extrusion or slip-form method and turned into the reverse position (Fig. 220b). The extrusion device, as illustrated in Fig. 221, is simple and inexpensive. The concrete is fed from a ready-mix truck by belt conveyors into a hopper and is forced by a screw conveyor into a compression chamber and mold. The compression in the chamber is resisted by friction between the concrete and a special rubber-covered bed and forces the device in the direction of least resistance. To regulate the pressure and speed of movement, the device is equipped with a pressure-synchronized motor drive. The sides of the mold are provided with rubber gaskets and anchored against uplift by devices sliding along the guiding rails. The friction between the mold and concrete can be reduced by a high-frequency small-amplitude vibrator attached to the mold.

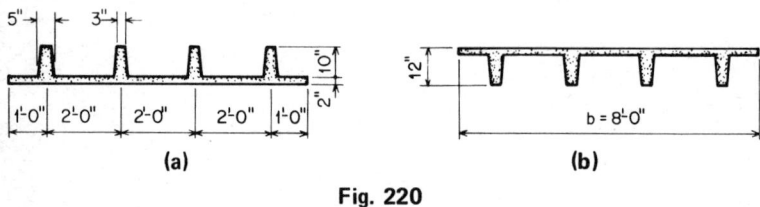

Fig. 220

The concrete, depending on the length l_m of the mold, is consolidated by relatively high pressure in the chamber and by vibration. After the concrete leaves the mold, it holds its proper shape mainly by capillary forces. The pretensioned strands and guiding rails hold the device in proper position. Welded wire fabric in sheets, provided with plastic spacers, is fed into the mold from the front of the extrusion device. The position of the pretensioned strand is determined by the position of the slab to be used. After the concrete has developed the required strength, the rib slabs are cut into desired lengths.

The economical length of the prestressing bed is up to approximately 600 ft, and the width of the rib slab is 8 ft. Since the mold part is attached to the extrusion device, the depth, width, and spacing of the ribs can be changed as required. The minimum depth is 4 in., and the

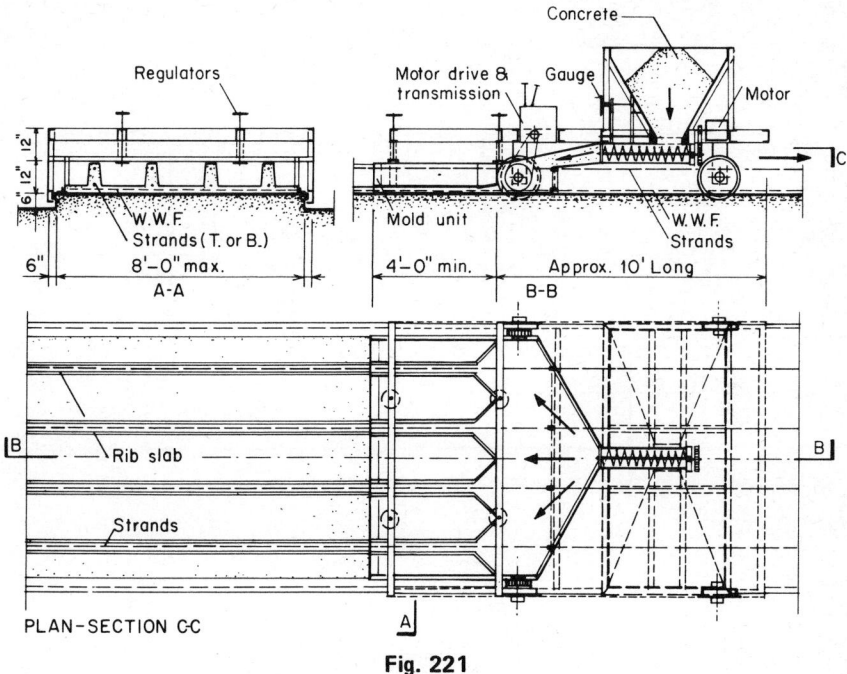

Fig. 221

maximum is up to 12 in. Commonly the span-depth ratio for simple spans can be as high as 35 and considerably higher for continuous slabs.

The economy of the rib slab with exposed ribs (Fig. 222) can be realized when one considers the relatively large lever arm (e_0) for prestressing strands. For floors with smooth ceiling (Fig. 223), the structural efficiency of the rib slab is lower because of the relatively small lever arm for strands under dead-load condition. The spacing between the ribs can be used for piping, wiring, and ducts for heating and cooling. The leftover space can be filled with lightweight materials for soundproofing and to serve as form for topping. For live-load conditions, the topping acts as a compressive member.

Also, rib slabs can be used as forms and tension members for bridge decks or for any other heavy-duty floors. If the space between the ribs is left empty, a chicken-wire net can be stretched over the ribs to serve as form for topping.

Numerous other types of lightweight and heavyweight slabs, as illustrated in Fig. 224, are available. Such rib slabs are commonly 2 ft

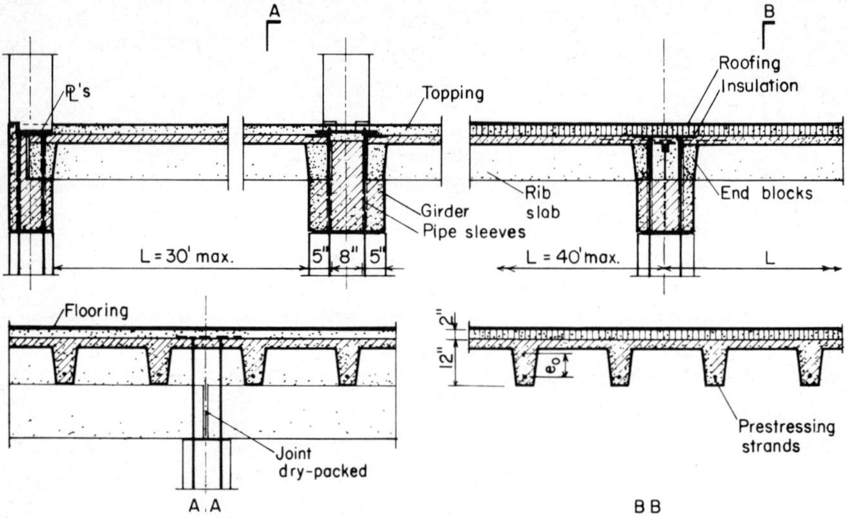

Fig. 222

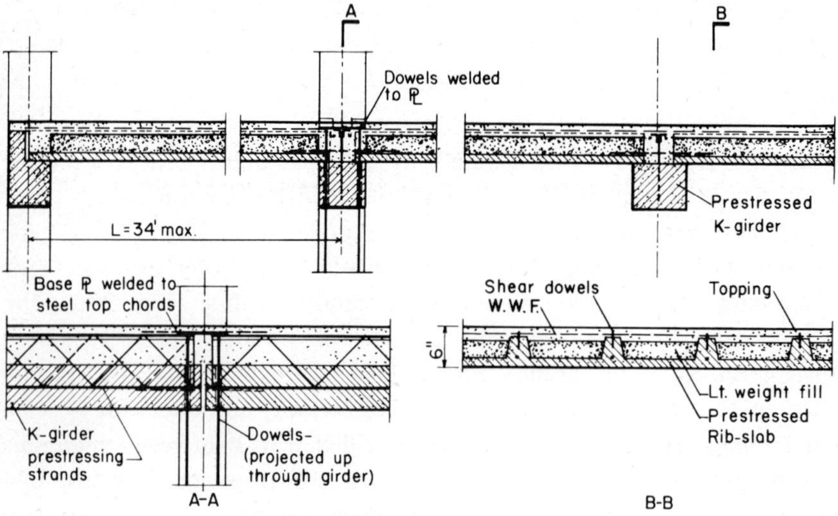

Fig. 223

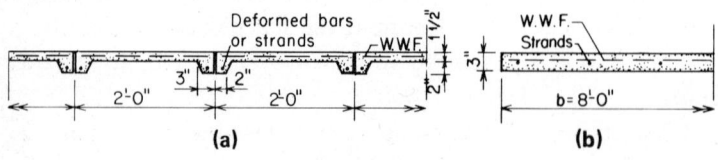

Fig. 224

wide and 3½ in. deep and have spans up to 9 ft. By application of prestressing, the span can be increased up to 12 ft. The simple planks (Fig. 224*b*), having uniform depth, usually 2 to 3 in. and widths up to 8 ft, are mostly used as roof planks, forms for floors, or bridge decks. The rib slabs and planks are most conveniently connected with the main supporting elements by epoxy glue.

Hollow-core prestressed planks, called *spancrete* (Fig. 225), are also manufactured by the extrusion method. Structurally, the spancrete is not as economical as the rib slab with exposed ribs. The extrusion equipment is quite expensive, but considering that this method is meant for large-scale manufacturing, it is, on the whole, rather economical. The extrusion device pours the slabs on top of each other up to several layers in the full length of the bed. The time lapse between the pours of layers is commonly 24 hr; in this time, the concrete of the last-poured layer is sufficiently hardened to carry the next layer. Before the next layer is poured, the surface is covered by a bond-breaking agent and then the prestressing strand is tensioned between the end supports of the bed. The end supports commonly are rather heavy because they have to resist the total prestressing force of several layers. When the concrete of the final layer is sufficiently hardened to resist the prestressing force by bond, the planks are cut from layer to layer into desired lengths. The spancrete pouring device can pour the individual planks up to three layers of different-quality concrete. The width of the slabs or planks is 40 in., and the span-depth ratio is up to 35. The minimum depth is 4 in., and the maximum is up to 10 in.

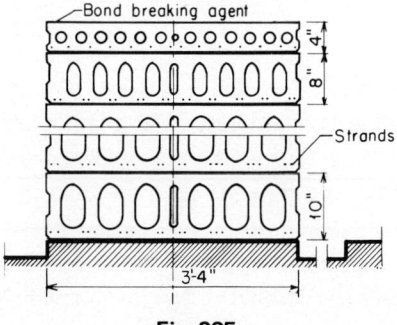

Fig. 225

There are other hollow-core prestressed planks (Fig. 226) on the market, with widths up to 8 ft and depths up to 10 in. The cores are created by thin round metal tubing. This type of hollow-core plank is manufactured in a flat, universal-type prestressing bed. The span-depth ratio is up to 45.

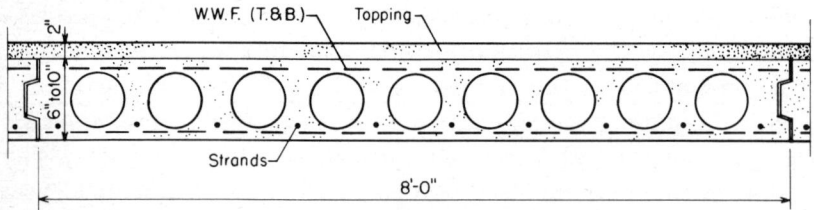

Fig. 226

Qualitatively, these planks are better and safer than spancrete planks. The cores in spancrete very often collapse when the mix is too plastic or too dry, and the ribs between cores will have poor quality mainly because of the frictional cracks, which can cause—and actually have caused—shear failure. Furthermore, the use of welding plates and transversal reinforcing is practically impossible. On the other hand, the planks manufactured in flat beds may have hair cracks caused by the metal tubing, due to the resistance it offers to shrinkage, unless coated with so-called "soft" asphalt paint. However, such cracks are not always evidenced, especially when the concrete is properly cured. Flat-bed planks are more expensive to manufacture, but considering the relatively high initial investment for spancrete and its relatively narrow width, the two types are practically equal in economy, especially when moderate quantities are required.

The hollow-core plank illustrated in Fig. 227, called *flexicore*, is only partly prestressed to balance the shrinkage of concrete. The planks are cast in individual forms. Round openings in the planks are obtained by rubber tubes filled with compressed air. The tubes are tied to conventional reinforcing and prestressing strands to avoid displacement and uplift. The prestressing strands are tensioned against form ends. The concrete is filled into the forms around the rubber tubes by a special overhead hopper or simply by a belt conveyor from a ready-mix truck. The planks are cured by low-pressure steam in a special curing chamber. When the concrete has developed the required handling strength,

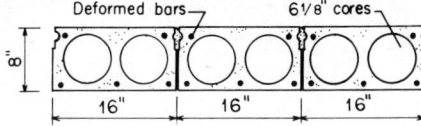

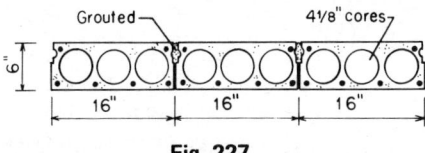

Fig. 227

prestressing is applied and the pressure is released from the rubber tubes so that they can be easily pulled out. Commonly, flexicore planks are 16 in. wide and 6 and 8 in. deep. They have a span-depth ratio up to 30, depending on the amount of reinforcing and intended use.

Hollow-core slabs are very often used also in poured-in-place work. The cores are provided by sonotubing or corrugated thin metal tubing placed between the bottom and top reinforcing and fastened firmly to avoid uplift and displacement during placement of concrete and vibration.

To obtain longer spans and control deflections, prestressing has been successfully used also in poured-in-place hollow-core floors. Prestressing is most easily applied by greased and wrapped ½-in. strands anchored individually with strand grips. The saving in weight due to cores, in comparison with solid slabs, is usually approximately 45 percent.

All hollow-core planks used for floors require a reinforced topping to avoid or balance differential sagging due to elastic and plastic deformations.

48. SLAB BEAMS

Slab beams are structural carrying elements in which the bridging element is an integral part of the beam, acting simultaneously as a slab in the transversal direction and as a compressive member for the beam in the longitudinal or span direction. In this category of basic precast structural elements belong the double T, single T, and channels (Fig. 228).

As can be seen in Fig. 228, the center of gravity of all these sections is very close to the top fiber of the slab; thus the element has a relatively large lever arm for conventional reinforcing or prestressing tendons.

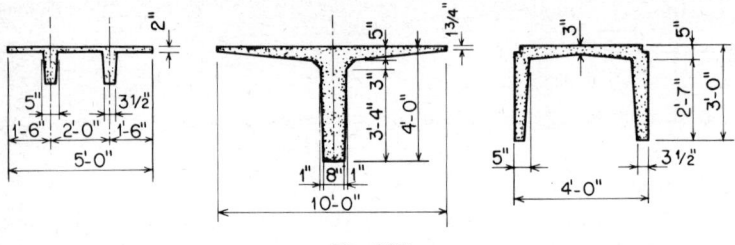

Fig. 228

Relatively large compressive and small tensile areas express most clearly the special characteristics of the concrete and make the beam-slab elements most economical. Due to this fact, relatively long spans are economically obtainable with such elements.

The deficiency of these sections is the relatively large and unstable camber if prestressing is used. Under prestressed conditions, the concrete in the bottom part of the beams is highly stressed, resulting in a large upward deflection of the element. Further, due to plastic flow, the camber increases considerably in the course of time [Eq. (67)]. However, this deficiency can be reduced and even entirely eliminated by either of two methods: (1) using two-layer concrete or (2) using two-stage prestressing.

The two-layer concrete manufacturing is rather simple and very economical. The webs of the section are poured with high-strength concrete, and the slab part with concrete having a relatively high fineness modulus and moderate strength. The strength relationship is selected so that under prevailing sustained loading, the element does not deflect [Eqs. (128) and (129)] when prestressing is applied. The high-strength concrete has a relatively small strain in comparison with the slab concrete; thus, depending on the strain ratio, the high-strength concrete (beam part) resists almost entirely the prestressing force ($t = n$), and as the web area is far smaller than the slab area, the amount of prestressing force is considerably reduced. However, the minimum required prestressing is controlled by the factor of safety. The pouring of two layers can be carried out with one operation. The finishing of the slab surface is easy because of the fineness of the mix. The concrete of these elements is cured by providing hot-oil heating to the steel molds, which causes drying out of the concrete, especially in the beam part (see Sec. 38). Due to this, the modulus of elasticity of the webs is considerably lower (for the same-strength concrete) than that of the slab because of the more favorable curing conditions. In two-layer operation, the steam from the more heated concrete of the webs cannot escape because of the high water-steam saturation in the slab. Due to this, the modulus of elasticity in the beam part is not much affected by so-called "premature" drying out, which is always the case in one-layer operation.

In two-stage prestressing, the member is prestressed in the first stage by pretensioning method just enough to balance the handling and also, partially, the dead-load stresses. The second stage of prestressing is applied by nonbonded posttensioning method at the latest possible time. This method is adaptable only for single T's because of the larger dimensions available for posttensioning tendons (Fig. 228).

Commonly, double T's are manufactured only 4 and 5 ft wide and 12 to 18 in. deep and single T's are up to 10 ft wide and 20 to 48 in. deep. The width of the channels is up to 4 ft (in exceptional cases, even wider), and the depth is up to 3 ft.

The variation in depth of the T's and channels is accomplished by the use of pilot lines in the standard casting mold or, most simply, by filling the grease-coated bottom of the form with lean concrete, which can be easily removed. The change of width is done by using movable side forms for slabs. The poured-in-place topping, usually 2 to 3 in. thick, can be used structurally for live-load conditions.

Very often, the channel members are used far apart and the gap between them is bridged over by lightweight concrete or by some type of pressed-fiber planks. The channels can most easily serve for the accommodation of ducts, lighting, and piping. The gap can also be used for the installation of skylights (Fig. 229).

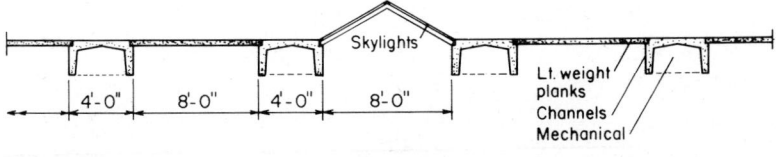

Fig. 229

In the case of poured-in-place work, all these sections (Fig. 228) are classified as T beams or slab-beam structural elements. In theoretical analysis, as well as in structural behavior in finished structures, there is no difference between T beams and precast sections except in the slab width participating in the carrying action of the beam and in construction methods. The width of the slab in precast elements is commonly controlled by handling and transport and also to some degree by torsion caused by unsymmetric loading.

In poured-in-place T beams, the spacing of the beams is mainly controlled by slab thickness and loading. The width of slab participation is commonly controlled by code; however, for greater accuracy in determining stresses and deflections, it can be computed by Dischinger formulas* [Eq. (573)] based on shell theory.

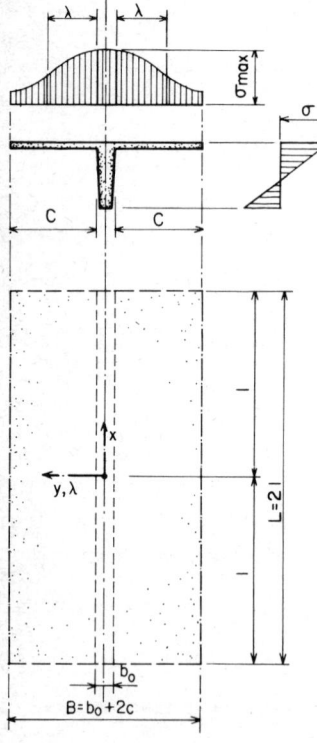

Fig. 230

For uniform symmetric loading w:

$$M_x = M_0 \cos \frac{n\pi x}{L} = M_0 \cos \alpha x$$

$$\alpha = \frac{n\pi}{L}$$

$$M_0 = \frac{wL^2}{8} = \text{max moment}$$

L = span
B = spacing of T beams

$$\lambda = \frac{2m}{\alpha} \frac{\sinh \alpha 2c + \alpha 2c}{(3m + 1) \cosh \alpha 2c + \frac{m+1}{2}(\alpha 2c)^2 + (5m - 1)} \quad (573)$$

$m \sim 9$ = Poisson's number
$n = 1$ = first harmony

*F. Dischinger, Massivbau, "Taschenbuch fuer Bauingenieure," Springer-Verlag OHG, Berlin, 1949.

$c =$	0.05	0.10	0.15	0.20	0.25	$\cdot L$
$\lambda =$	0.0491	0.0934	0.1295	0.1566	0.1752	$\cdot L$
$c =$	0.30	0.35	0.40	0.45	0.50	$\cdot L$
$\lambda =$	0.1873	0.1945	0.1985	0.2005	0.2014	$\cdot L$

$c = \infty \quad \lambda = 0.2046L$

The λ-values in the table are computed for midspans and T beams ($x = 0$). When the beams and slabs at the support ($x = a$) are rigidly connected by transversal beams, the moment of inertia I is constant and Eq. (573) gives rather accurate results, as proved by measurements of actual stresses and deflections under field conditions. If there is no transversal beam at $x = l$, the equation gives $\lambda = 0$, which means that the moment of inertia varies parabolically and, when this is not considered, the measured deflections show slightly higher values. However, the midspan stresses are not affected by the transversal beam.

Equation (573) applies also for uniformly loaded precast sections. In this category of carrying elements actually belong also the folded plates, as illustrated in Fig. 231.

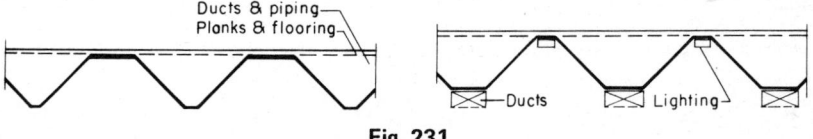

Fig. 231

Channels and folded plates are very suitable and economical for floors of laboratories and hospitals, where a considerable amount of space for ducts, piping, and wiring is required. The only deficiency is that the space available for this purpose is one-directional. However, for folded plates this deficiency can be eliminated by raising the floor over the section, as illustrated in Fig. 232.

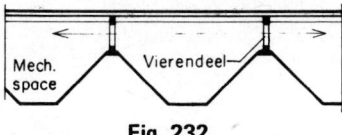

Fig. 232

428 Structural Systems and Methods of Construction

In cases in which a limited number of ducts and piping in the transversal direction are unavoidable, openings in the sides of the folded plates can be provided (Fig. 233).

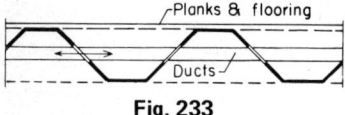

Fig. 233

Folded plates can be manufactured in a universal prestressing bed by installing coffers for the bottom form, into which the folded plates are poured most conveniently and economically by use of the slip-form method. To increase the lever arm (e_{pr}) for prestressing, the horizontal section of the folded plates should be heavier and the bottom one in tension lighter. The minimum depth of the sides is 3 in. For bridging, lightweight planks or form slabs (Fig. 224) are recommended because they can be easily removed when new piping or ducts have to be installed and major maintenance work performed.

The span-depth ratio for elements in this category is up to 35 and is approximately 20 percent less under vibrating live load.

49. PRECAST BEAMS AND GIRDERS

Precast and prestressed beams and girders are one-directional primary carrying elements to support the transversal bridging elements. As such, they are actually, in varying degrees, steel substitutes and do not express the basic concrete characteristics, except the box girder. The most common sections of beams and girders are illustrated in Fig. 234.

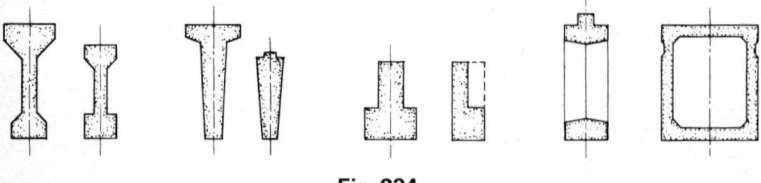

Fig. 234

As primary elements, beams and girders usually carry full dead load. Under live-load conditions, they may be designed as composite sections. In this case, the bridging element must be shear-connected and must act partially as a compressive section of the beam or girder. The long spans usually are relatively heavy and offer handling and some transport difficulties, especially in highly populated locations. However, girders with spans up to 100 ft and depths up to 5 ft have been used. For longer spans, the beams or girders must be designed in sections and post-tensioned to unity at the site. The top flanges may vary in width, depending on intended use. For supporting the bridge elements, the width of the seat must be 5 in. minimum. The minimum width for webs is 6 in. Very often, vertical ribs are used to strengthen the web, as is the case with steel-plate girders, which are not required unless relatively thinner webs are used.

The prestressing force required for long spans subjected to heavy loading is rather high and can be obtained usually by a combination of pretensioning and posttensioning. The amount of pretensioning is designed to balance the handling stresses, and the posttensioning is for the remaining stresses. The application of prestressing in two and even three stages is required to keep the stresses due to the loading and prestressing counteracting each other within desired limits. Initial temporary stresses up to $0.6f'_c$ and final stresses up to $0.5f'_c$ of the ultimate concrete strength are considered reasonable.

The economical span-depth ratio of beams and girders is far less than that of planks or slab beams; commonly it is 15 to 25, depending on the design load, type of prestressing, and sections. Seldom are the simple beams illustrated in Fig. 234 poured in place. However, if required in particular cases, it can be done without difficulty.

A different type of beam or girder is illustrated in Fig. 235. Called a *K system,* it is a composite prestressed beam whose top chord and web members are of steel and bottom chord is of prestressed concrete. The prestressing is applied for beams by pretensioning method and for girders

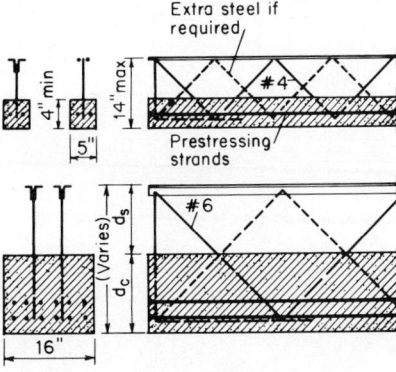

Fig. 235

by pretensioning, posttensioning, or a combination of both methods. Because of prestressing, the bottom chord of the composite section is in compression in its entire depth and the steel top chord is in tension or in relatively small compression, depending on the magnitude of prestressing and the location of the tendons in the section. The top chord is prevented from buckling by web members, the ends of which are fixed in the concrete bottom chord. The stresses in the beam or girder in the prestressing state are the reverse of the loading stresses; therefore, the relatively light beams are self-supporting under construction load for standard spans (span-depth ratio up to 25) with a reasonable margin of safety. However, for heavy bridge members and for exceptionally long spans, a light jack support at midspan may be required to control the deflections during the placing of the floor.

The bridge members resting on the concrete bottom chord of the beams can be lightweight machine-made concrete blocks, precast slabs, etc. After the bridging elements are placed, the topping and steel part of the beam is poured into a monolithic carrying system. Because of the open-web steel top part of the beam, it can be used as a simple or continuous span. For continuous spans, the steel top chords of the adjacent spans can be connected by welding, by bolting, or by special devices at intermediate supports. If the tensile strength of the top chord is inadequate to balance the tensile force, conventional reinforcing can easily be added (Fig. 236). Making the system continuous for all conditions of dead and live load not only results in more favorable stress conditions, with consequent economy, but also reduces deflections which extend the

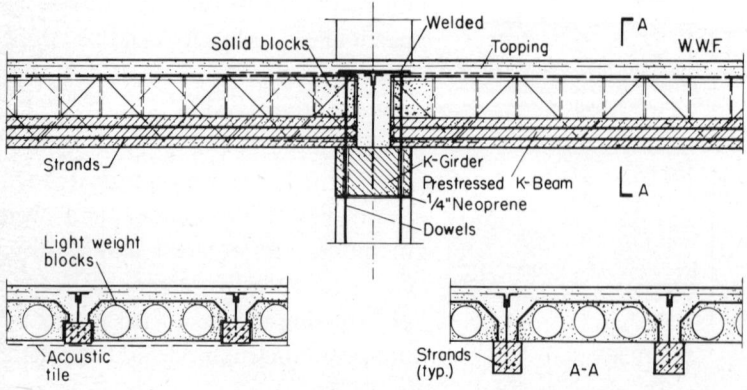

Fig. 236

range of span. K-system girders designed for heavy loading and long spans very often require a double steel top part which is transversally connected to avoid lateral buckling.

Experience indicates that the K system is very stable. This is explained by the steel top chord, which largely counteracts the influence of plastic flow and shrinkage. The relatively small deflection, even under vibrating loads, can be visualized by comparing the stress-modulus of elasticity ratio of the two materials:

$$\frac{f_c}{E_c} = \frac{2{,}000}{5.0 \times 10^6} = \frac{1}{2{,}500} \sim \text{concrete}$$

$$\frac{f_s}{E_s} = \frac{20{,}000}{30.0 \times 10^6} = \frac{1}{1{,}500} \sim \text{steel}$$

Thus, the f/E ratio is almost twice as high for steel as for concrete, and so also are the related deflections.

K-system beams are manufactured in a special prestressing bed approximately 500 ft long. First, the strands are pretensioned and the bottom chord is poured around the strands in a spring-type steel mold. After the bottom chord is poured, the prefabricated steel top part is vibrated into concrete. The concrete is air-cured for about 4 hr, and then the sides of the mold are released and opened by spring action, exposing the sides of the bottom chord for live-steam curing for about 12 hr. As the bottom chord is in compression and the steel top can carry temporary tension up to 30,000 psi, the 500-ft-long casting can be lifted out of the mold and then cut into the span lengths desired.

The width of the beams is 5 in., and the total depth 8, 10, 12, and 14 in. The depth of the concrete bottom chord may vary from 4 to 8 in. The span-depth ratio for simple spans is up to 35. The girders commonly will have a width of 16 in. and a depth as desired. The span-depth ratio for simple spans is up to 20.

50. COLUMNS

Poured-in-place columns are the most expensive structural elements, mainly because of the formwork involved and the pouring difficulties with long columns to avoid segregation. For these reasons precast columns are very often used, even when other elements are poured in place. A typical precast column is illustrated in Fig. 237. For convenience of formwork and to reduce their thickness, the base plates are made to match the column size. To allow fastening of the column, the corners of the bottom end are recessed. The concrete loss due to the recesses is balanced by extra reinforcing welded to the base plate. The size of the recesses must be chosen adequate to allow fastening of anchor bolt nuts. At the top end, the reinforcing must be projected as required. For multistory columns, the ends of projected reinforcing bars must be threaded because they serve as anchor bolts for the upper columns. The beam or girder ends supported by the column must be provided with pipe sleeves for passing the projecting top reinforcing bars. In cases in which the upper-column dimensions are different from those of the lower ones, the projecting anchors usually are extra bars located to match the upper-floor-column base plate.

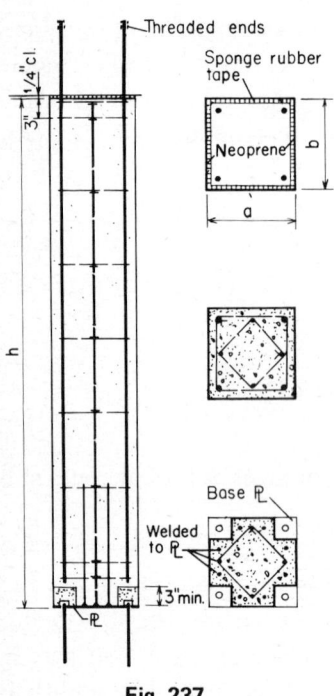

Fig. 237

Prestressed columns are seldom used, in spite of their economy and technical superiority, because any column is already under compressive stress and the prestressing is assumed to decrease the strength of the column by the amount of the prestressing force. From a commonsense point of view this cannot be denied, but experience and theoretical analyses prove that the column strength is not reduced by prestressing but is even increased. This fact is rather simply explained by the following analysis.

Let us consider the column section illustrated in Fig. 238 [see Eqs. (5) and (6)]:

$$P = 240\,k$$
$$M = 50\,k'$$
sustained loading

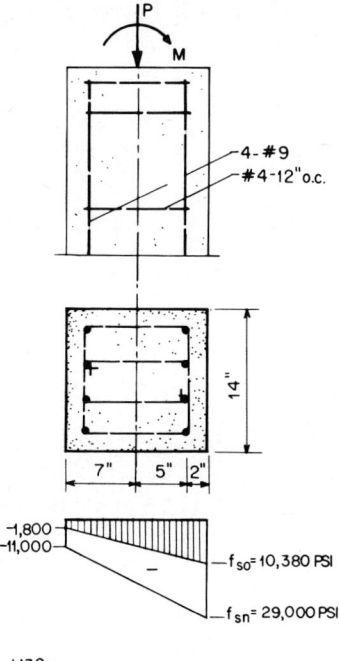

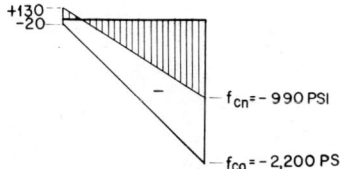

Fig. 238

$$f'_c = 6,000 \text{ psi}$$
$$f_{sy} = 40,000 \text{ psi}$$

$$E_c = 5.0 \times 10^6 \text{ psi}$$
$$E_s = 30.0 \times 10^6 \text{ psi} \qquad n = \frac{E_s}{E_c} = 6$$

$$A_s = 8 - \#9 = 8.0 \text{ in.}^2$$
$$A_c = 14^2 - 8.0 = 188 \text{ in.}^2 \qquad p = \frac{A_s}{A_c} = 0.0425$$

$$I_s = 8.0 \times 5^2 = 200 \text{ in.}^4$$
$$I_c = \frac{14 \times 14^3}{12} - 200 \simeq 3,000 \text{ in.}^4$$

434 Structural Systems and Methods of Construction

$$\frac{S_{ss}}{S_{cs}} = np = 6 \times 0.0425 = 0.255$$

$$\frac{S_{sF}}{S_{cF}} = n\frac{I_s}{I_c} = 6\,\frac{200}{3{,}000} \simeq 0.400 \qquad \begin{array}{l} F = \text{flexural} \\ s = \text{strain} \end{array}$$

$$\alpha_s = \frac{S_{ss}}{S_{ss} + S_{cs}} = \frac{0.255}{1 + 0.255} \simeq 0.203$$

$$\alpha_F = \frac{S_{sF}}{S_{sF} + S_{cF}} = \frac{0.400}{1 + 0.400} \simeq 0.285$$

Load distribution and stresses in steel and concrete for $t = 0$ are

$$P_{s0} = P\alpha_s = 240 \times 0.203 = 48.7\,k$$
$$P_{c0} = P\alpha_c = 240 \times (1 - 0.203) = 191.3\,k$$

$$M_{s0} = M\alpha_F = 50 \times 0.285 \simeq 14.3\,k'$$
$$M_{c0} = M(1 - \alpha_F) = 50(1 - 0.285) \simeq 35.7\,k'$$

$$\Sigma f_{s0} = -\frac{P_{s0}}{A_s} \pm \frac{M_{s0}}{I_s} z_s = -\frac{48.7 \times 1{,}000}{8.0} \pm \frac{14.3 \times 12{,}000}{200}\,5.0$$

$$= -6{,}090 \pm 4{,}290 \simeq \begin{array}{l} -1{,}800 \text{ psi} \\ -10{,}380 \text{ psi} \end{array}$$

$$\Sigma f_{c0} = -\frac{P_{c0}}{A_c} \pm \frac{M_{c0}}{I_c} z_c = -\frac{191.3 \times 1{,}000}{188} \pm \frac{35.7 \times 12{,}000}{3{,}000}\,7.0$$

$$= -1{,}020 \pm 1{,}000 = \begin{array}{l} -20 \text{ psi} \\ -2{,}020 \text{ psi} \end{array}$$

Stresses in steel and concrete for time shrinkage and plastic flow finished [see Eqs. (73) to (77)] for $t = n$ are

Shrinkage: $\quad \epsilon_{sr} = 30 \times 10^{-5}$

Plastic flow: $\quad \varphi_n = 2.0 \quad \left(\varphi_n = \dfrac{\epsilon_{pl}}{\epsilon_{el}}\right)$

$$\varphi'_s = \frac{1 - \alpha_s}{\alpha_s}(1 - e^{-\alpha_s \varphi_n}) \simeq 1.31$$

$$\varphi'_F = \frac{1 - \alpha_F}{\alpha_F}(1 - e^{-\alpha_F \varphi_n}) \simeq 1.10$$

$$f_{c0} = \frac{P_{c0}}{A_c} = E_c \frac{P_{c0}}{S_{cs}} \left[1 + \frac{\epsilon_{sr}}{\varphi_n} \frac{E_c}{f_{c0}} \right] = 1.735 \quad f_{c0} = -1,020 \text{ psi}$$

$$\begin{aligned}
\Sigma f_{sn} &= f_{s0}\left[1 + \varphi_s'\left(1 + \frac{\epsilon_{sr}}{\varphi_n}\frac{E_c}{f_{c0}}\right)\right] \pm f_{sF}(1 + \varphi_F') \\
&= -6,090(1 + 1.31 \times 1.735) \pm 4,290(1 + 1.10) \\
&\simeq -20,000 \pm 9,000 \simeq \begin{array}{l} -11,000 \text{ psi} \\ -29,000 \text{ psi} \end{array}
\end{aligned}$$

$$\begin{aligned}
\Sigma f_{cn} &= f_{c0}\left[1 - \frac{S_{ss}}{S_{cs}} \varphi_s'\left(1 + \frac{\epsilon_{sr}}{\varphi_n}\frac{E_c}{f_{c0}}\right)\right] \pm f_{cF}\left(1 - \frac{S_{sF}}{S_{cF}} \varphi_F'\right) \\
&= -1,020(1 - 0.255 \times 1.31 \times 1.735) \pm 1,000(1 - 0.400 \times 1.10) \\
&= -430 \pm 560 = \begin{array}{l} +130 \text{ psi} \\ -990 \text{ psi} \end{array}
\end{aligned}$$

The stress diagrams of steel and concrete for times $t = 0$ and $t = n$ are illustrated in Fig. 238.

Prestressed Columns

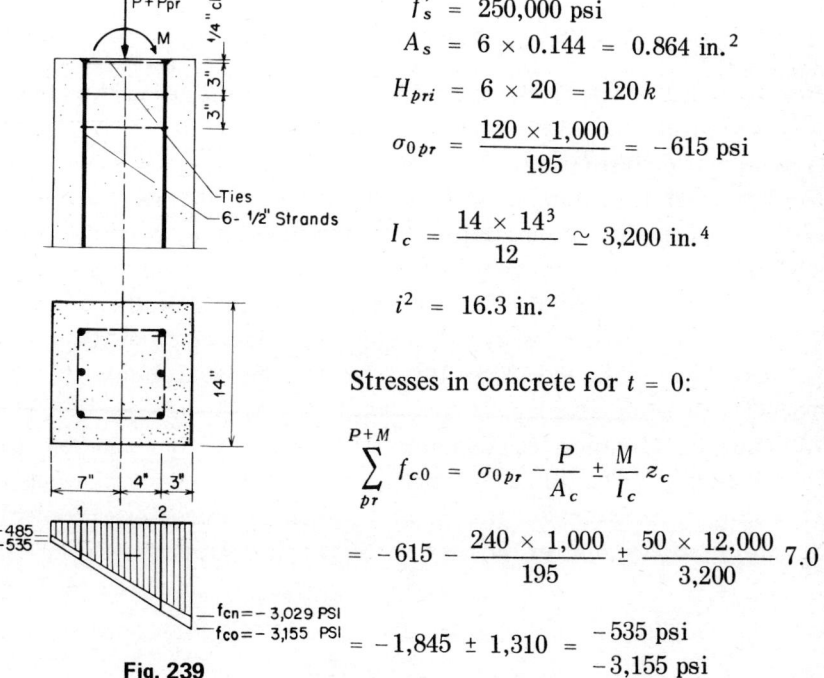

Fig. 239

$f_s' = 250,000$ psi
$A_s = 6 \times 0.144 = 0.864$ in.^{2}
$H_{pri} = 6 \times 20 = 120\,k$

$$\sigma_{0pr} = \frac{120 \times 1,000}{195} = -615 \text{ psi}$$

$$I_c = \frac{14 \times 14^3}{12} \simeq 3,200 \text{ in.}^4$$

$i^2 = 16.3$ in.^{2}

Stresses in concrete for $t = 0$:

$$\sum_{pr}^{P+M} f_{c0} = \sigma_{0pr} - \frac{P}{A_c} \pm \frac{M}{I_c} z_c$$

$$= -615 - \frac{240 \times 1,000}{195} \pm \frac{50 \times 12,000}{3,200} 7.0$$

$$= -1,845 \pm 1,310 = \begin{array}{l} -535 \text{ psi} \\ -3,155 \text{ psi} \end{array}$$

Stresses in steel and concrete for $t = n$:

$$\Delta\sigma_{2pr} \simeq \left(\frac{f_{c2}}{E_c} + \epsilon_{sr}\right) E_{pr} \quad f'_{c2} = \frac{3{,}155 - 535}{14} 11 + 535$$

$$\simeq -2{,}600 \text{ psi}$$

$$= \left(\frac{2{,}600}{5.0 \times 10^6} + \frac{30}{10^5}\right) 30 \times 10^6 = 24{,}600 \text{ psi}$$

$$\Delta\sigma_{1pr} = 15{,}050 \text{ psi} \quad f_{c1} = 1{,}085 \text{ psi}$$

$$\sum_{t=0}^{t=n} \Delta P_{pr} = 3 \times 0.144 \,(15.050 + 24.600) \simeq 17.20\,k$$

$$\Delta\sigma_0 = \frac{17{,}200}{195} = +88 \text{ psi}$$

$$\eta_{pr} = \frac{15.050 \times 3 + 24.600 \times 11}{39.7} \simeq 8 \text{ in.}$$

$$e_0 = 8 - 7 = 1 \text{ in.} \quad \psi_B = 1.43; \quad \psi_T = 0.57$$

$$\Delta\sigma_{pr,\max} = \Delta\sigma_0 \psi_B = 88 \times 1.43 = +126 \text{ psi} \quad \Delta\sigma_{pr,\min} = +50 \text{ psi}$$

$$\sum_{pr}^{P+M} f_{cn} = -3{,}155 + 126 = -3{,}029 \text{ psi}$$
$$= -535 + 50 = -485 \text{ psi}$$

If the moment is one-directional only, the stress distribution (Fig. 239) can be improved and maximum stress in some degree reduced by the use of unsymmetric prestressing.

Consider that the strength of a column is exhausted when stresses in steel reach the yield point, tensile stresses develop in concrete, and compressive stresses approach to about 85 percent of the ultimate strength. Using this criterion as a basis for evaluation of the carrying capacity of the columns analyzed, it can easily be seen from the stress diagrams (Figs. 238 and 239) that the prestressed column in this particular case is superior.

In reinforced columns considerable stress transfer from concrete to steel takes place in the course of time ($t = 0, t = n$) due to shrinkage and plastic flow of concrete. For the time $t = 0$, the concrete stresses control the factor of safety, but for the time $t = n$, the steel stresses control.

$$\text{F.S.}_c = \frac{0.85 \times 6{,}000}{2{,}020} \simeq 2.5 \quad t = 0$$

$$\text{F.S.}_s = \frac{40{,}000}{29{,}000} \simeq 1.38 \quad t = n$$

The use of high-carbon steel with a higher yield stress does not greatly improve the strength because of the development of tensile stresses in the concrete. In accordance with plastic theory, the increase of steel in a column beyond a certain percentage may result in stress transfer from concrete to steel to such an extent that the entire load is carried by the steel and even tensile stresses may develop in the concrete. This phenomenon explains the often noticed horizontal cracking of heavily reinforced columns. Such cracking is further promoted in columns when large-diameter bars are used because of the resistance they offer to transversal shrinkage. After the cracks have developed, the compressive stresses in concrete increase rather rapidly from this state on and the column commonly fails by the simultaneous buckling of the highly stressed longitudinal bars and the crashing of the concrete.

In prestressed columns, the change in stresses due to shrinkage and plastic flow is rather small and favorable. The factor of safety is controlled, as expected, by stresses in the concrete:

$$\text{F.S.}_c = \frac{0.85 \times 6{,}000}{3{,}155} \simeq 1.6 \quad t = 0$$

$$\text{F.S.}_c = \frac{0.85 \times 6{,}000}{3{,}029} \simeq 1.68 \quad t = n$$

For both times, the factor of safety is higher than for reinforced columns for $t = n$. Furthermore, tensile stresses do not develop even under the ultimate-strength range of the column (see also Fig. 58). The higher ultimate-strength capacity of prestressed columns has been proved by actual large-size tests. Numerous tests also prove that ties do not contribute to the ultimate strength of a column, except to prevent buckling of the longitudinal bars. As the prestressed strands or bars are in tension and remain so up to the collapse of the column, there is no requirement for ties in a prestressed column, except two or three at each end of the column to avoid splitting when prestressing is applied.

Posttensioned columns for prefabricated structures are often required for lateral stability. To provide ductility and a certain degree of elasticity and to reduce the magnitude of dynamic forces (wind, seismic forces, etc.) and temperature and shrinkage stresses, the posttensioning bars should be bonded not in their entire length but only enough to control the lateral movement of the structure. This is simply done by asphalt-coating some section of the bars (usually at floor levels) to allow their independent elongation. The metal tubing for posttensioning bars should be of the lightest possible type to avoid shrinkage cracks along the tubing. Base plates are structurally not required for posttensioned columns. However, a relatively light base plate at the bottom is required for erection purposes.

The percast and posttensioned columns are cast in individual molds, and prestressed columns are formed in a prestressing bed in the horizontal position, with one side hand-finished. Large-size columns or columns with complicated cross sections very often have to be cast in vertical position (Fig. 240).

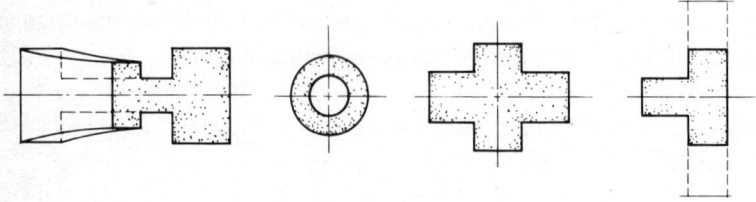

Fig. 240

The external columns and marginal beams can also be combined into a unit element (Fig. 241). The beam as well as the column part can be straight, curved, or tapered, depending on architectural requirements.

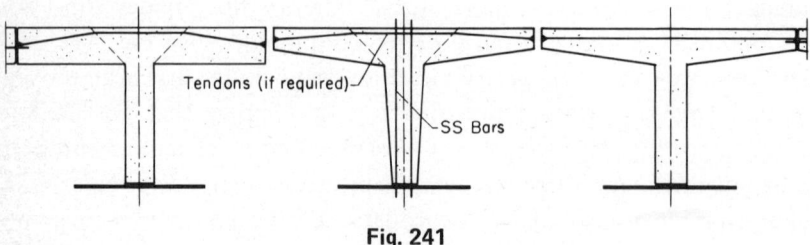

Fig. 241

As the height of such elements is seldom more than 10 ft, the beam element can be rather long without causing transport difficulties. However, as the beam acts statically as a cantilever, for longer lengths posttensioning is required if the element is to be subjected to relatively heavy loading from floor and walls. The posttensioning layout is illustrated in Fig. 241. To avoid shear and bending cracks due to posttensioning, the tendons should be bent downward at the ends, especially when large posttensioning forces are used. The joint between the beam ends can be straight and connected by dowels only to provide stability and balance the end deflections of the adjacent beams. This type of connection is possible because the shear for uniform loading is zero and for uneven live load is relatively small.

51. WALL PANELS

Two basic types of prefabricated wall panels are mainly used: curtain walls (Fig. 242) and vertical-load-carrying wall panels (Fig. 243). The second type of wall panel has two functions: to provide enclosure and to support the floor members. Also, since the panels have a greater depth, the ducts for heating and cooling can be easily accommodated between the ribs of the panels. Due to these features, the two-function wall panels offer more flexibility and economy than simple curtain wall panels. In order to keep the panels light and crack-free, moment connections between the panel and floor beams or slabs should be avoided. Therefore, the horizontal load (wind, seismic forces, etc.) should be carried by stair-elevator towers or shear walls.

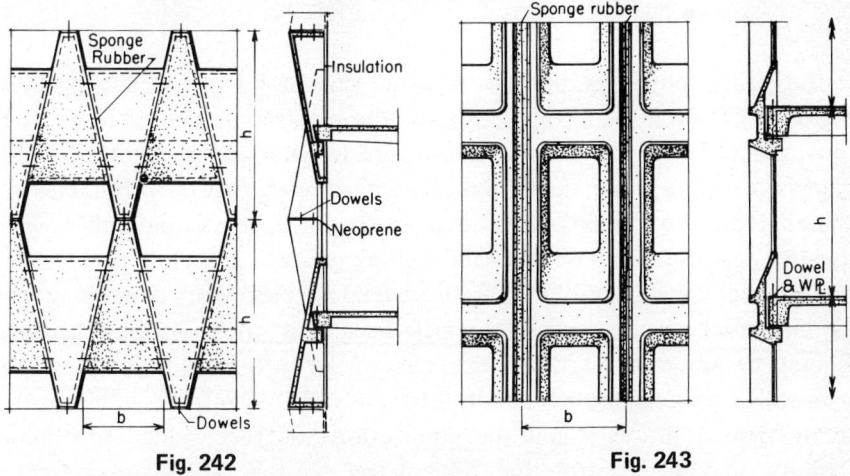

Fig. 242   Fig. 243

However, for considerable floor heights and walls for halls, etc., the load-bearing wall panels have to be moment-connected with the floor or roof members to secure the lateral stability of the structure.

The proper design of such a panel type is illustrated in Fig. 244. The moment loading requires increased panel ends to carry the moment and to keep the stresses below the allowable limits. At the midheight of the panel, the moment is almost zero, or relatively small. Thus the panel is loaded primarily by centric load, and the design should express it.

As the reinforcing does not prevent cracking but can only reduce it, all wall panels should be designed so that for existing loading conditions the tensile stresses remain below $1/10 f'_c$.

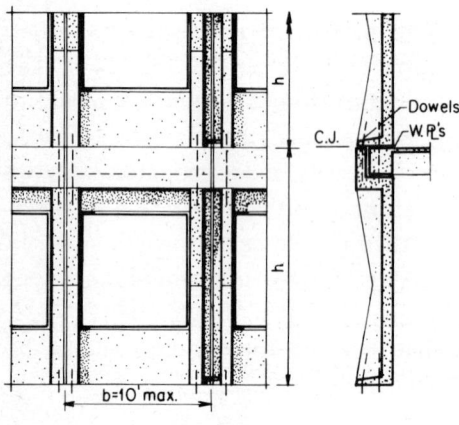

Fig. 244

For utility buildings, precast slabs and slab beams can be successfully used also for wall panels. In this case, the prestressing should be centric or form-true to avoid warping due to prestressing and plastic flow. Wall panels for more sophisticated buildings usually are cast on vibrotables or shock tables to obtain dense, high-quality surfaces. Commonly, high-quality aggregate (exposed by sandblasting or acid) and white cement are used. The panels can be one or several stories high, depending on architectural requirements. The width, because of transportation, usually should be less than 10 ft. The thickness of the curtain-wall-type panels is controlled principally by handling stresses. The reinforcing should be prefabricated in cages and the connections of bars welded to obtain rigidity and to control the dimensions so that adequate concrete coverage for steel is secured.

The fastening of panels in place should be designed so that concrete tensile strength ($1/10 f'_c$) will not be exceeded. Also, any type of stress concentration should be avoided by providing neoprene seating at the center of gravity of the panel, etc. The joints are best made watertight by sponge-rubber gaskets glued to both faces of the joint; no tongue-and-groove joints, which usually make the erection difficult and may cause stress concentration, are in this case required.

For walls having a different type of facing (stone), the shear stresses between the two layers can be computed as in Sec. 50 ("Columns") for load distribution between steel and concrete for times $t = 0$ and $t = n$. If such analysis is not carried out, there is a possibility that the two different materials will separate because of high shear stresses. In accordance with experience, such separation has occurred in high piers having stone facing and also with concrete walls with brick facing. If the shear stresses are relatively high, the joints of the facing stones can be designed to balance the deformation differences.

12

STRUCTURAL SYSTEMS

Any structural system is bound to the laws of nature, and structural shapes are bound to the physical characteristics of the materials the structure is made of. Due to this fact, the choice of a structural system for a certain building is not quite arbitrary. It depends on the layout of the plan or the function of the building and also in a large degree on conditions of the location of the building. Furthermore, experience shows that a structural system may become rather rapidly inadequate to carry the loads if it is in a developing area or that the spans may be too short if the functions of a building, especially of a utility or commercial building, have to be changed.

The carrying capacity of a structural system can be easily raised, but the spans cannot practically be increased without extensive and expensive reconstruction. To avoid this, it is highly recommended that in any contemporary design relatively long spans or large column-free spaces should be designed to start with.

Until prestressing was developed, long spans were rather expensive, mainly because, to avoid extensive crack formation and deflections, the depth-span ratio for reinforced concrete was limited to approximately 1/15 and stresses in reinforcing steel to 20,000 psi. This resulted in increased dimensions and dead weight of the structural elements. And since the moment is directly proportional to dead load and increases with second power of the length of span, the tensile force, which is proportional to the moment, increases also with second power of the span. To accommodate reinforcing steel requires considerable increase in the width of the beam, which also adds to the dead load. It is quite obvious that these facts set limits to the economically justified span lengths for conventional reinforced concrete. Such limitations are not present in prestressed concrete, as can be most simply illustrated by the following. The relationship between the external moments, stresses, and sectional dimensions is

$$\sum_{D}^{L} M = f_c S = f_c \frac{bd^2}{6} \qquad (574)$$

Thus,

$$d = \sqrt{\frac{6 \Sigma M}{f_c b}}$$

where f_c is the allowable stress for concrete, b the width, and d the effective depth of the sections. In the case of prestressed beams, the prestressing moment counteracts the external moment ($\sum_{D}^{L} M$) due to the loading and is $M_{pr} = H_{pr} e_0$, where e_0 is the eccentricity of the prestressing force H_{pr} in respect to the center of gravity of the section. If $M_{pr} = \sum_{D}^{L} M$, the beam is free of flexural stresses and subjected to uniform compressive center stress only and the deflection is zero. This type of prestressing is called *form-true prestressing*. In practice, the prestressing force is usually determined so that the compressive as well as tensile capacity of concrete in both fibers remains within allowable limits under construction and final design load conditions. This means that $\sum_{pr}^{D+L} f_c \leq 0.6 f'_c$ and $\sum_{pr}^{D+L} f_{cT} \simeq \frac{1}{10} f'_c$.

When the prestressing force or moment is determined for dead-load condition, the superimposed stress in the top fiber of the section is zero [Eqs. (56) and (57)], as illustrated in Fig. 245.

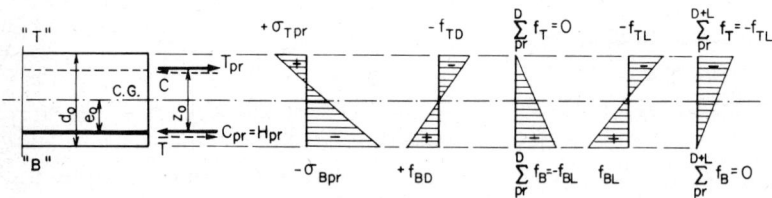

Fig. 245

As can be seen from these stress diagrams, the total compressive-strength capacity of concrete is available for carrying the live load. Thus the required depth of the beam is

$$d = \sqrt{\frac{6 M_L}{f_c b}} \tag{575}$$

Also, the total dead load is carried by prestressing tendons, and under entire design load condition the concrete acts simply as a fully homogeneous material. Thus the efficient use of material is increased by prestressing from 15 to 50 percent.

Besides being dependent on architectural and structural objectives and requirements, a structural system is also dependent on mechanical installations, which very often even control the entire design. For example, the mechanical installations involved in providing a variety of services for the proper functioning of a contemporary laboratory are extensive and complex. They must be controlled, maintained, and changed and new services must be installed without interfering with the laboratory work. In addition, almost any research laboratory must be kept noise- and vibration-free. This applies also for most other buildings. Some industrial buildings require considerable mechanical installations which are of different nature—less sensitive and usually exposed so that maintenance and new installations can be easily performed.

It is obvious that the structural systems for such complex buildings differ considerably from those for simple industrial buildings, such as garages and warehouses. Apartment and office buildings, hotels, schools, museums, theaters, air terminals, etc., again require quite different structural systems.

To secure structural soundness and economy, there are a number of principles to be followed for any prestressed long-span systems.

Precast and Prestressed Systems

1. To control stresses and stress concentration in precast units, the units should be separated from each other by bearing neoprene pads or sheets. The thickness of the neoprene depends on the extent of the movements practically tolerable.

2. Each member must have a seat, so that easy, quick, and safe erection is possible.

3. The principle of connection design should be that the connections will be weaker than the tensile strength of the concrete affected by the connections. This is especially true for wall panels and other relatively thin section connections.

4. In general, welded connections should be avoided because of discoloration of concrete and crack formation due to heat expansion. If they are unavoidable, the edges of welding plates should be covered with soft asphalt or rubber-based paints to allow free expansion.

5. Dowel–pipe-sleeve connections should be used where possible.

6. Tongue-and-grooves for alignment should be avoided where possible because they may cause stress concentration and make erection difficult.

Poured-in-place Prestressed Systems

Care should be taken that shortening and rotation of members during prestressing are possible. Members resisting these unavoidable displacements cause uncertainties in the prestress and may also cause undue stresses in resisting members. This can be avoided by introducing bearing neoprene plates and wrapping a short length (2 to 6 in.) of the vertical reinforcing passing the neoprene to eliminate bond and reduce the resistance of the reinforcing. Furthermore, it must be realized that multistory structural systems differ during construction from story to story; this must be considered in design, especially in long-span systems where the lower floor has to carry the total construction load of the floor under construction.

Various structural framing systems, multispan and multistory, complicated and relatively simple, for various purposes, and with related construction methods, will be discussed.

52. ONE-WAY STRUCTURAL SYSTEMS

Long spans are most easily obtainable with one-way prestressed systems; these systems dominate in contemporary design. The use of one-way systems is further promoted by the fact that mechanization is more easily adopted with both precasting and poured-in-place methods than with two-way systems or space structures, which in most cases have to be poured in place and which will be discussed separately. In addition, most architectural layouts demand one-way structural systems.

In general, the one-way systems can be classified into two main categories: narrow-grid and wide-grid systems.

Narrow-spaced Systems

In this particular structural system, the narrowly spaced floor beams, simply supported or continuous, control the structural design. The floor beams are supported directly by columns, walls, or short-span girders.

For stability, the connections between the beams and supporting elements must be rigid or elastically controlled, at least to the extent of satisfying the factor-of-safety requirements. As the building in its longitudinal direction, in comparison with the framing direction, is long, no longitudinal beams over columns are required for stability reasons, in spite of the relatively thin floor slab. A typical layout of a narrow-spaced prestressed system especially favored for garages or similar buildings is illustrated in Fig. 246.

Because longitudinal beams are not required, the cross-ventilation is good, and in spite of the long spans, the construction depth required is relatively small. All this results in the flexibility, simplicity, and economy of narrow-grid systems. If, for architectural or any other reasons, the narrow spacing of external columns is not desired, the outside beam ends must be supported by spandrel beams.

In order to eliminate different rigidities of floor beams and to avoid complicated column-beam connections, columns supporting the spandrel beams should not be lined up with floor beams. A recommended

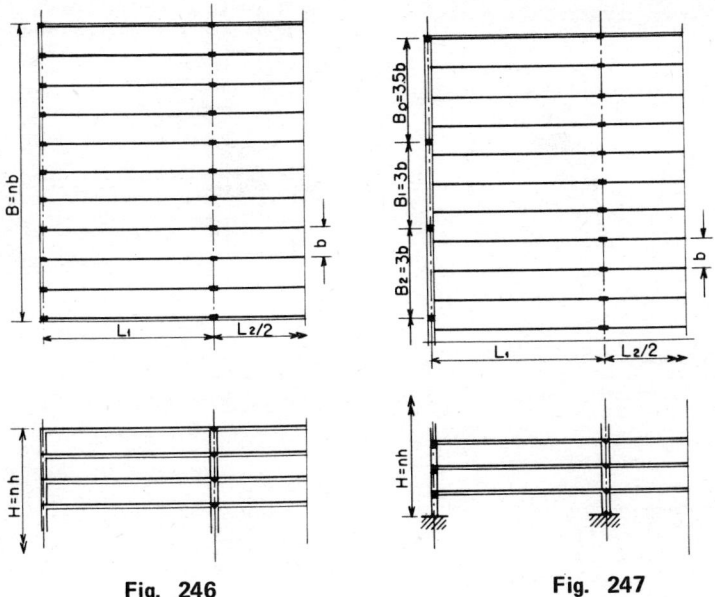

Fig. 246   Fig. 247

balanced framing layout for such a system is illustrated in Fig. 247. The floor beams, spandrel, and column connections must be such that rotation and lateral displacement of beam ends are possible to avoid high torsional stresses in spandrels and flexural stresses in columns, especially during prestressing.

For the same reasons, during construction the connections of interior columns and beams also must be designed so that slight rotation and lateral displacement are possible. The column connections at the foundation depend largely on soil and foundation type as well as on general layout. After construction is finished, the system will be tied together by posttensioning the columns for lateral stability. Structural details of such end supports and intermediate points are illustrated in Fig. 248.

The arrangement of greased and wrapped tendons for poured-in-place systems is illustrated in Fig. 249. Sharp curvature of tendons should be avoided to reduce frictional losses and tensile stresses along the curved portion of the tendons. It must be realized that the vertical load (w_{pr}) supported directly by curved tendons is

$$w_{pr} = \frac{T}{r} \tag{576}$$

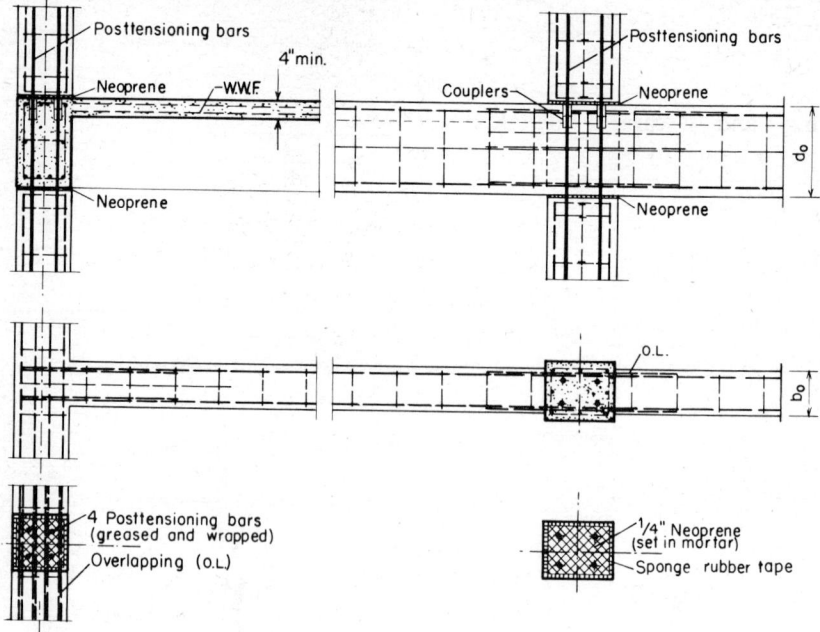

Fig. 248

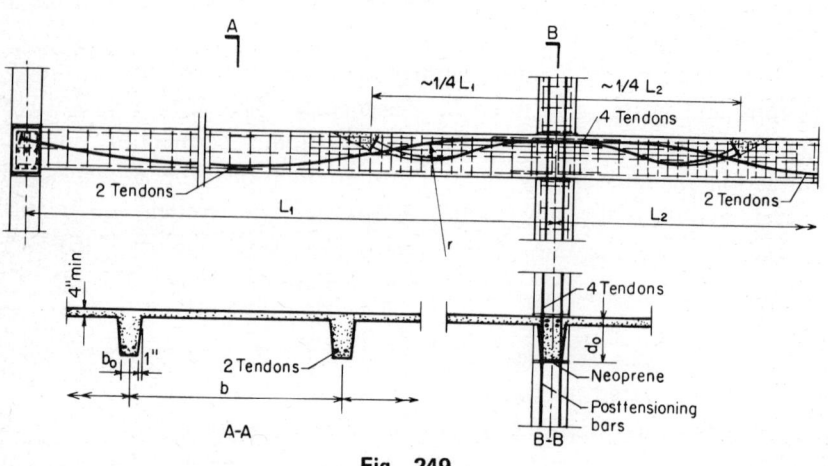

Fig. 249

where T is the tensile force in tendons and r is the radius of the curvature of tendons. The internal stresses in the surrounding concrete due to the w_{pr} loading together with frictional forces can be relatively high and, if care is not taken, will cause severe crack formations along the sharply curved portion of the tendons.

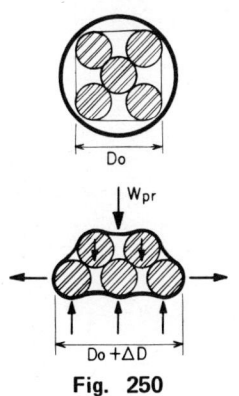

Fig. 250

The frictional forces are largely increased when the parallel-wire or strand tendon shape is allowed to be flattened at the curvature, as illustrated in Fig. 250.

Due to the wedge action caused by flattening of the tendons, the lateral forces can disperse the beams. This can easily be avoided by proper wire wrapping at sharp curvatures.

To make use of mass-production methods, conventional reinforcing is prefabricated in sections. Anchor plates for tendons are welded at proper locations to rebar. The complete reinforcing in its entire length is assembled from sections and tendons installed at the top level of forms. After the reinforcing cages are complete, they will be dropped into place, starting from one end of the beam. Then the welded wire fabric is placed and the floor poured. The layout of conventional reinforcing is shown in Fig. 251.

The easy and simple manufacturing and assembling of the reinforcing cages, carried out at floor level, including the placing of posttensioning, can be seen in Fig. 251. This all results in quick construction and economy.

By prefabricating this type of structure, a considerable amount of site work and time is saved. For this case, the structural systems have to be modified so that the beams can be manufactured and transported in sections. The location of the joints should be chosen slightly closer to intermediate columns than the inflection points of a continuous beam. The reason for this is to have the joints in the negative-moments region for superimposed dead and live loads. In this case, the top of the section at the joint location is in tension and the bottom is in compression (Fig. 252). The relatively small tensile force can be easily balanced by welding plates, and the compressive force by dry-packing the joints. The bottom joint will not open up because of the sustained compressive stresses. The

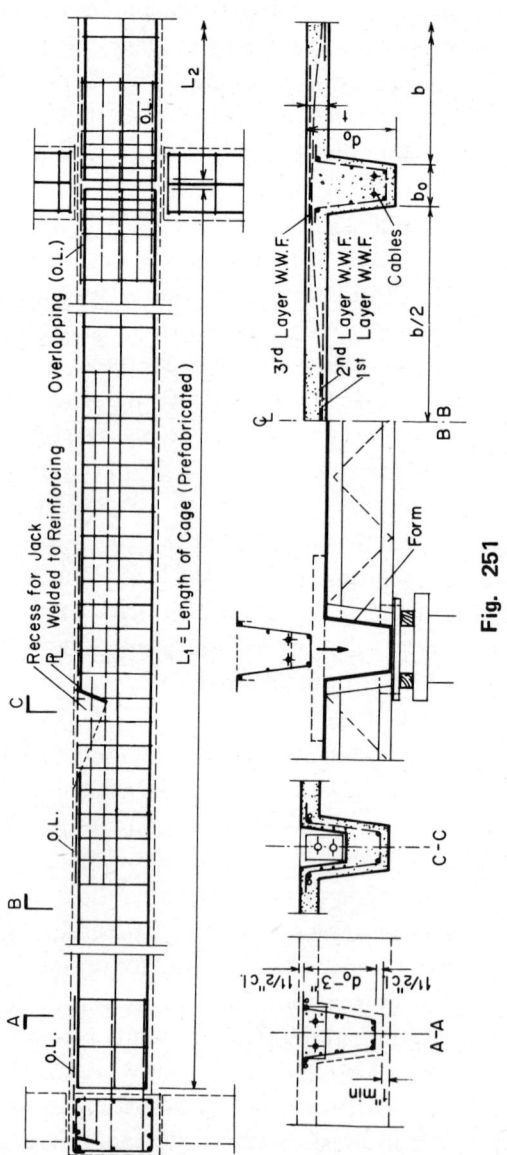

Fig. 251

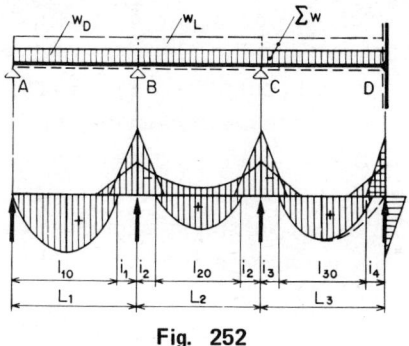

Fig. 252

location of inflection points from support can be easily computed or can be assumed to be roughly 0.20L.

The connections of the beam sections must be designed primarily for total shear force present at the joint. The top welding plates, required for safe erection, are commonly adequate to carry the tensile stresses due to superimposed loads ($w_{SD} + w_L$). The recommended joint types and reinforcing layout are illustrated in Fig. 253. The dowel-pipe-sleeve connection simplifies erection but can be used only for heavier sections because it interferes with anchorage of the single centric tendons commonly used for lighter sections.

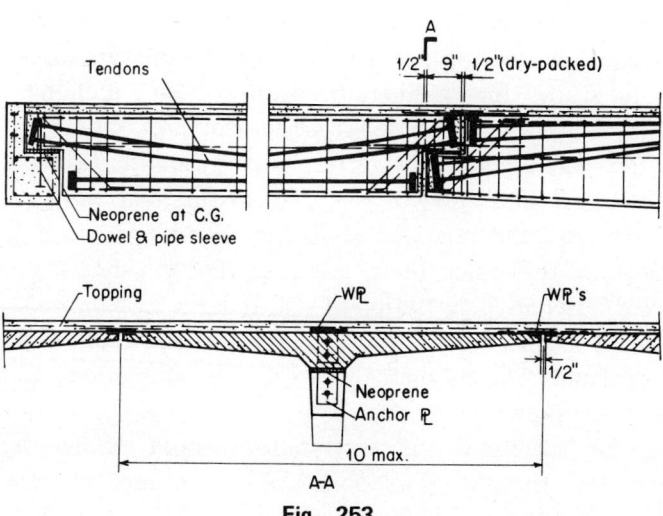

Fig. 253

Bearing neoprene interlay at seats is required to endure even seating and to allow rotation. In most cases, dry-packing of the sides of the beams at the spandrel, after the topping is poured, is structurally adequate. Dry-packing in the erection state may cause torsional cracking when the topping is poured and shrinkage occurs. For wide-spaced columns, girders similar to spandrel beams are required at intermediate column lines. In this particular case, the beams are simply supported at both ends. The recommended beam-girder connection is detailed in Fig. 254.

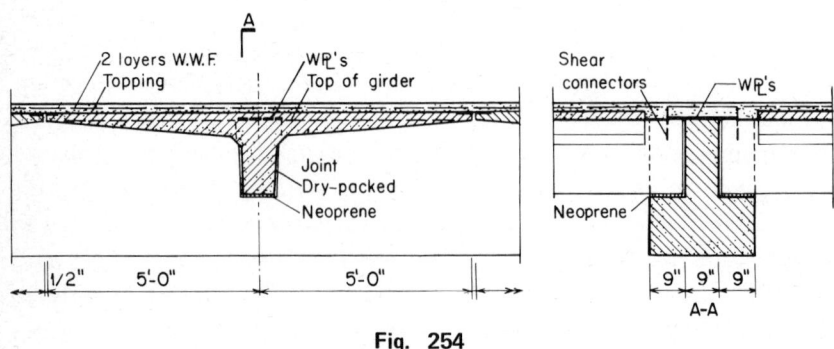

Fig. 254

For architectural reasons, the spandrel-beam joints should be straight. Thus the shear force has to be transmitted by steel, as indicated in Fig. 255. Due to such joint location and connections, the finished precast structure behaves like a poured-in-place continuous structural system.

Expansion joints in the longitudinal direction for long buildings, required for precast as well as for poured-in-place structures, should not be located more than 150 ft apart. The recommended joint location and type are similar to those illustrated in Fig. 255, except that the steel dowel should be greased and wrapped to reduce bond and to allow horizontal movement of the beams. In the slab area, the joint should also be doweled to avoid differential deflection of slab ends.

Single prestressed T beams are the most appropriate and economical for narrow-grid systems. The recommended layouts of sections and prestressing tendons are shown in Fig. 256.

The cantilever, for structural and economical reasons, should be tapered, mainly so that the same T-section type can be used and the prestressing force reduced to within manageable limits. The rather high compressive force at the bottom can be counteracted by conventional

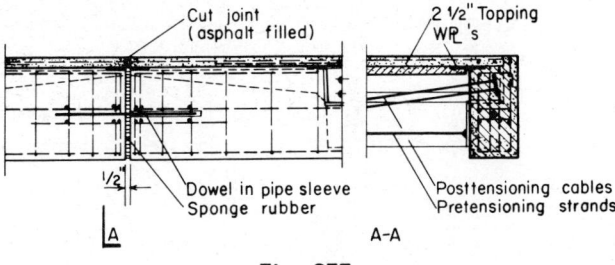

Fig. 255

compressive reinforcing. The drop-ins can be posttensioned or pretensioned, or both simultaneously, whichever method is more practical and economical.

The intermediate columns, tapered or straight, must be fork-type (Fig. 256) because of the narrow width of the beam. The bottoms of the columns should be provided with base plates for easy erection. The posttensioning of columns and the grouting operation can be carried out in one operation from roof level. However, for relatively high structures, especially those located in high wind and seismic areas, intermediate posttensioning is required for safety.

To reduce the dynamic forces (wind and earthquake) upon a structure, any degree of ductility can be obtained by providing nonbonded lengths at each side of the column-beam joints for posttensioning bars and by varying the thickness of neoprene.

Prefabricated structures require poured-in-place 2- to 3-in. topping reinforced with welded wire fabric. A narrow (approximately 18 in. wide) second layer is recommended over joints. The joints can be sealed best by sponge-rubber tape. The surface of the broom-finished T's should be thoroughly wetted and cement-dusted before topping is poured to ensure adequate bonding. Dry cement avoids the so-called *water interface* between the old and new concrete, which may cause considerable bond loss. It also increases, because of its fineness, the contact area of the surface. When wet mix is placed on hardened concrete, a considerable reduction of contact area is caused by surface tension and capillary forces.

Folded-plate Systems

The folded plates, single-span or multispan, supported individually by columns or by girders (Figs. 257 and 258) belong to the narrow-grid-system category.

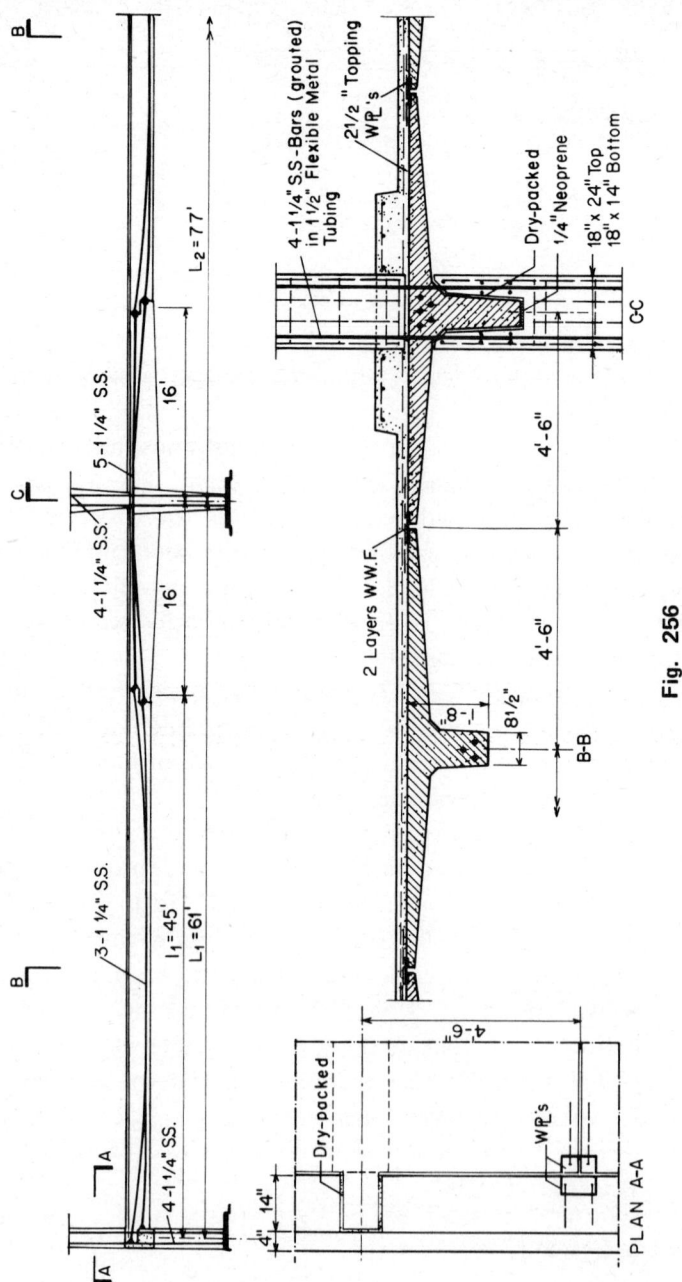

Fig. 256

The arrangement of the multispan folded plates is higher in the center span than for side spans. The overlapping at intermediate girders leaves openings and spaces which can be used for windows, louvers, or fans for cross-ventilation. Such a system can be successfully used for shopping centers and also for industrial buildings. The folded plates can have a considerable span, and the entire structure is easily prefabricated. The cross section and structural details are illustrated in Fig. 257.

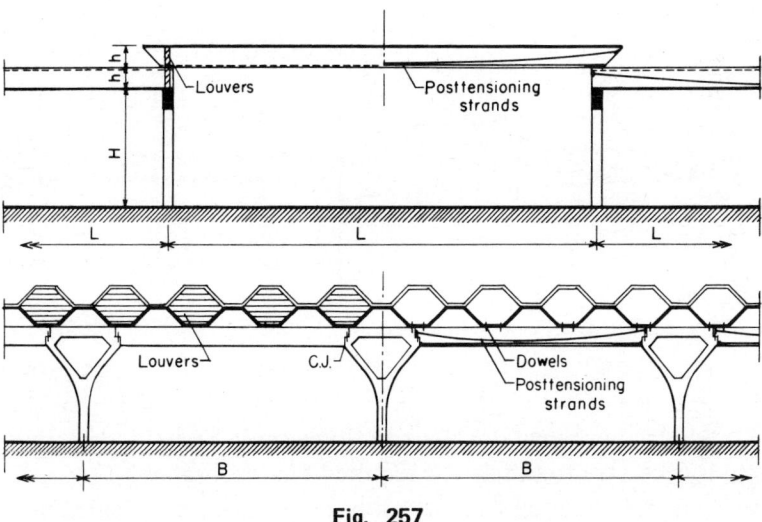

Fig. 257

The folded plates are most successfully used for cantilevered hangers. In order to reduce the moments for long cantilevers (more than 200 ft), the width of the plates is designed variable (Fig. 258) so that the carrying capacity of the sections follows the moment diagram. At the cantilevered end the moment is zero, and at this point the folded plates flatten practically to a relatively thin slab. At the far end, where the posttensioning tendons are anchored, the sectional dimensions are controlled by the size, number, and arrangement of anchor plates. Also, to balance the uplift and to secure a reasonable factor of safety, vertical posttensioning cannot be avoided.

As such long cantilevers are mainly loaded by dead load, greased and wrapped tendons can be used. When a symmetric arrangement of cantilevers is possible, the moments at the center supports are balanced and no uplift exists—except in hurricane areas, where a combination of

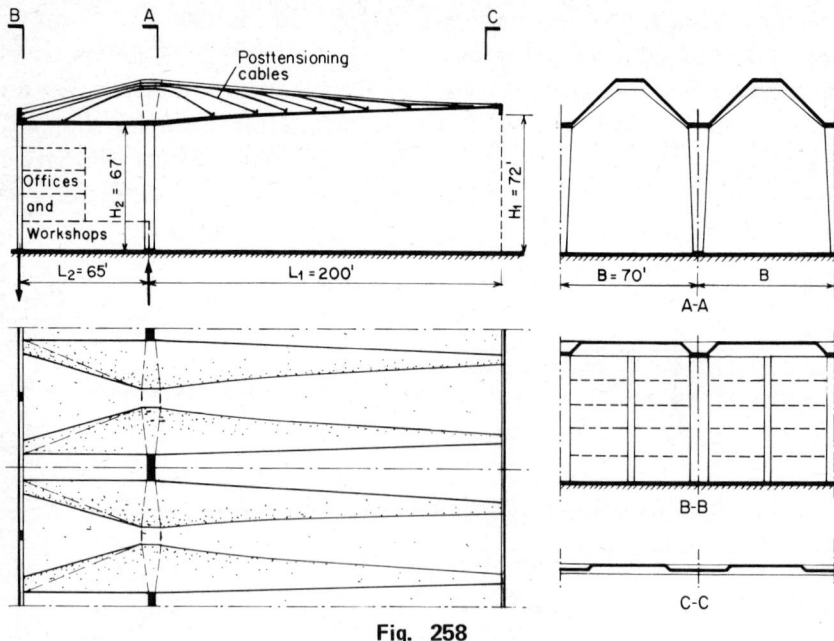

Fig. 258

pressure and vacuum may create conditions in which vertical posttensioning is necessary to obtain the required factor of safety.

For cantilevered stadium-tribune roofs (Fig. 259), the folded-plate system should not obstruct the visibility. To satisfy this condition, wide spacing of the columns is required and transversal framing, if any, should be arranged above the roof. The straight horizontal crest results in simple posttensioning but creates drainage problems. However, since the

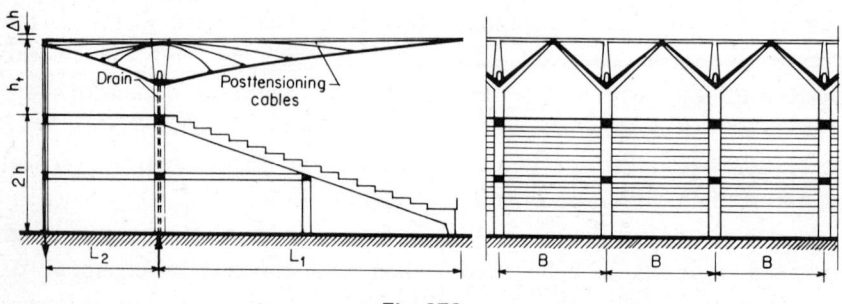

Fig. 259

columns, mainly for stability reasons, are relatively heavy, drainage for the roof can be arranged without structural difficulties through the columns. The rather deep roof slopes of this design and the crack-free concrete due to prestressing make the roofing unnecessary. Also, the arrangement of expansion joints at the crest is simple in comparison with reversed-type folded plates (Fig. 258).

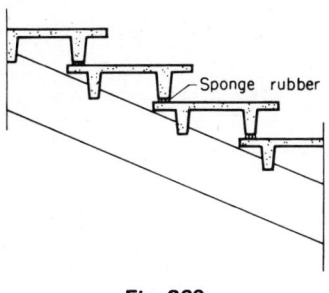

Fig. 260

For seating of the tribunes, precast prestressed double T's (Fig. 260) are most economical. The watertightness is achieved by sponge-rubber joints and by tying the T's vertically.

Relatively large column-free areas can be economically overroofed by cantilevered folded plates arranged face to face, as illustrated in Fig. 261.

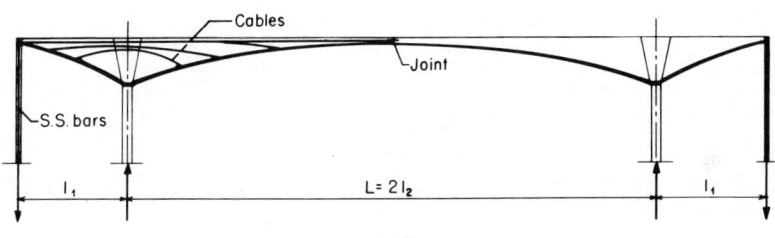

Fig. 261

Skylights can be placed in the longitudinal direction between the cantilevered ends and transversally at the crest by separating the adjacent plates horizontally.

Structurally, the straight-crest folded plates (Fig. 259) can be prefabricated (also in sections) most easily at the site, if economically justified.

Circular One-way Systems

For multistory circular buildings, two basic structural narrow-grid systems dominate (Figs. 262 and 263). The choice of system for a particular building depends mainly on the architectural design, the number of stories, and also the desired length of cantilever to open up the plaza level.

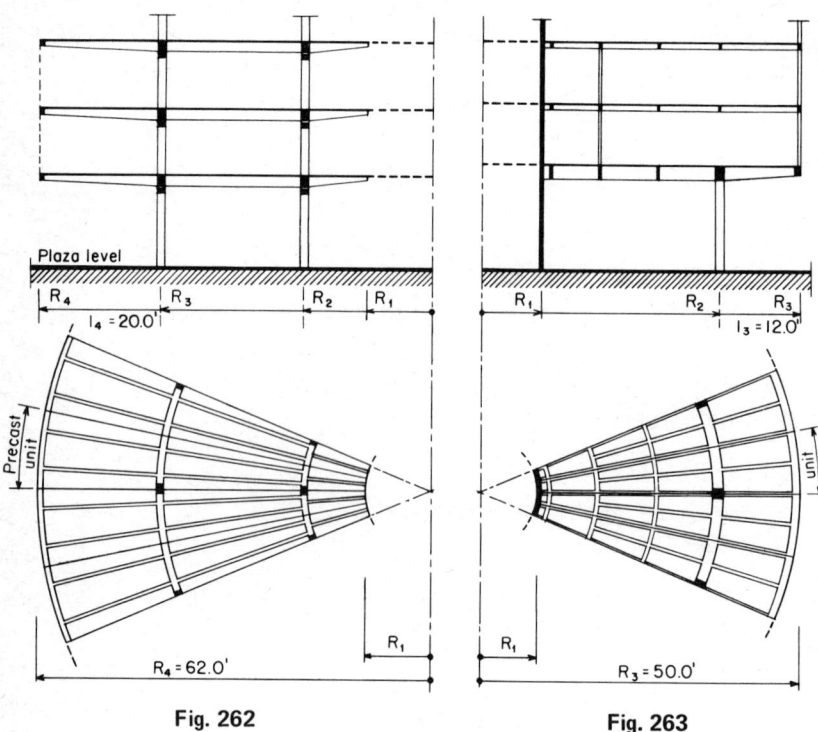

Fig. 262   Fig. 263

As can be seen, the system illustrated in Fig. 262 has a relatively large cantilever and has to have internal columns. The external walls are non-load-bearing curtain walls. The elevator and stair tower may or may not be located at the center of the structure. The height of the columns is from floor to floor to allow easy erection. The floor beams are of the narrow-rib trapezoidal type. The number of floors still economically possible can be up to 35.

The system illustrated in Fig. 263 has a short cantilever, no internal columns, and a poured-in-place elevator and stair tower at the center of the structure. The external wall panels, supported by cantilevered first-floor beams, are load-bearing and continuous up to the roof.

As the first-floor cantilevers support all the outside ends of the floor beams, the first-floor structure usually becomes rather heavy and highly prestressed. To keep these floor beams within acceptable size, the number of floors is limited to eight. Beyond this height, the system having internal columns (Fig. 262) becomes more economical. It is also easier to erect. The trapezoidal floor beams may or may not be prestressed, depending on the span.

There are a variety of structural solutions for one-story circular buildings. The one illustrated in Fig. 264 is a cantilevered structure. The beams are cantilevered from a center-core structure that accommodates services, stairs, etc. The roof area is practically level and can be used for a garden, observation deck, etc.

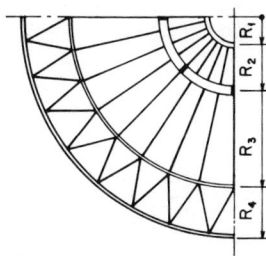
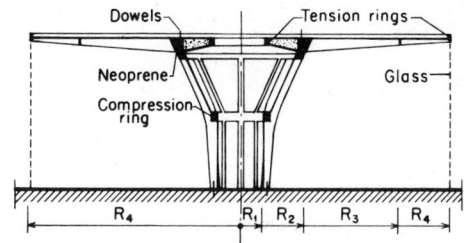

Fig. 264

The cantilevered trapezoidal beams are prestressed in radial direction, usually by posttensioning methods. The beam ends are tied to unity by circular prestressing applied by dead load, by overweight at the cantilevered ends, or by flat jacking.

The circular beam connecting the columns is heavily reinforced and dowel-connected with the panel diaphragms, forming a composite section. If the core structure is prefabricated, the connecting beam and compression ring should be channel-type, reinforced and filled with concrete after erection.

Combining rotation-symmetrically arranged folded plates—suspension system with multistory tower, acting as counterweight, considerably long cantilevered roofs for hangers, etc., are structurally and economically feasible (Fig. 265). The relatively wide space between the folded plates is roofed by a gunited concaved shell (Fig. 265, Sec. B).

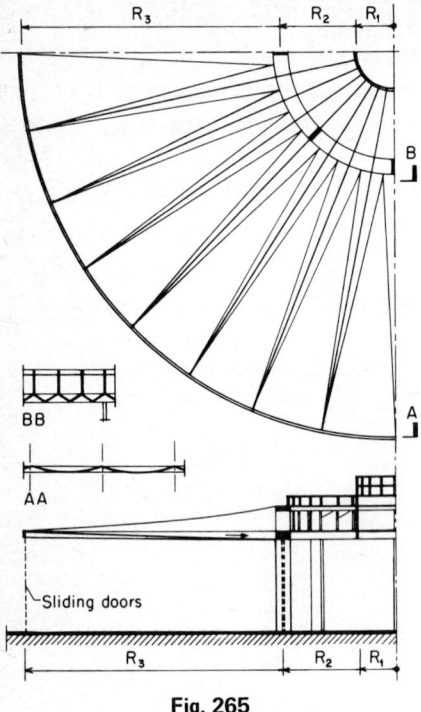

Fig. 265

Wide-Spaced Systems

This type of structural systems is primarily controlled by girders and is widely used for prefabricated industrial, commercial, and public buildings. The spacing of girders supporting the floor beams or bridging elements depends on the construction height, the soil conditions, and the type of bridging system used (double T's, single T's, or channels). The girders may run in the transversal, longitudinal, or radial direction. The connections with columns can be rigid, pin-type, or elastically controlled. The bridging elements are commonly simply supported at girders. Very often for relatively long and wide buildings, especially high-rises, where planks are used for bridging elements, extra beams at intermediate column lines in the direction of the planks are required for stabilization of the structure.

The girders carrying heavy loads require relatively large depth, up to 1/15 of the span. The prestressed beams or girders can be any of the types illustrated in Fig. 234 or can be K girders (Fig. 235). For

Structural Systems   461

economical reasons, the wide-grid systems are seldom poured in place. This is understandable because the continuous floor slabs control the spacing of the floor beams and limit the system down to a narrow-grid range. To increase the economy, a secondary beam-slab system between the girders is required. Also, the system in this case is composed of beam slabs, girders, and columns.

As stated before, the fewer the number of elements a structural system is composed of, the higher the economy. In this particular case, four elements are required—one more element than in the case of prefabrication.

Some typical prestressed precast systems in this category are illustrated in Figs. 222 and 266. The topping is commonly 2 to 3 in. thick and reinforced with welded wire fabric. The cavities between the bridging elements and girders are filled with poured-in-place concrete, so that the system becomes monolithic. Thus the girders for live-load condition can be considered as composite sections. However, the planks will remain as simply supported. If K-system joists and girders are used, the continuity of girders and joists is obtainable even for full design load.

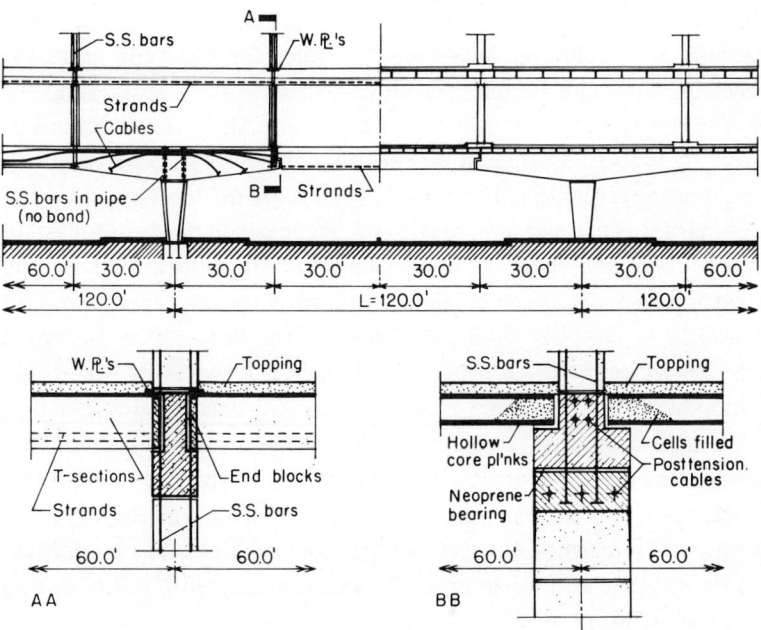

Fig. 266

In Fig. 266, the heavy lower-level posttensioned girders and columns are poured in place, except the drop-ins, which are prefabricated so as not to interrupt traffic on the underpassing highway during construction. The higher levels are prefabricated. To reduce the own weight, sonotubes can be used at areas of relatively small positive moments of the heavy girders.

For roofs, and halls, the posttensioned girders can be curved and can have relatively long spans (Fig. 267).

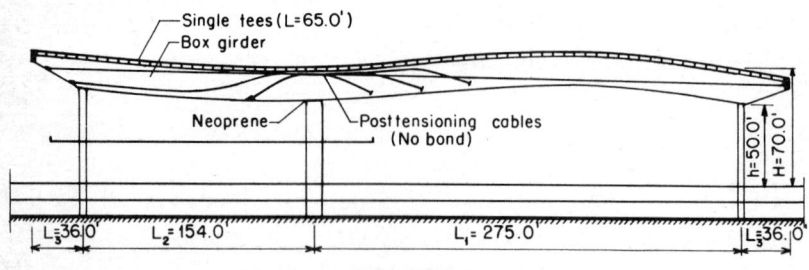

Fig. 267

For architectural reasons, solid box girders are preferred to trusses. Such girders usually carry considerable loads and require a large posttensioning force. The arrangement of cables for box-type girders is simple because they can be exposed during placing and prestressing for most of the length of the girder (Fig. 268). Also, the mechanical installations are easy to arrange in box sections, whereas in other kinds of sections such arrangements may create considerable difficulties.

The box girder illustrated in Fig. 268 is designed to be precast in sections. The size and weight of the sections are determined by the hoisting capacity of the crane available. For erection, two trestles (*a* and *b*) are required to support the heavier half of the box. After the joints have been dry-packed, the sections are connected by welding plates and dowels, and then posttensioning is applied for full dead load of the box girder. The lighter half is erected on the finished half, fastened, and posttensioned for its dead load. The two halves of the box are tied to unity by transversal posttensioning at the top and by welding plates at the bottom. The remaining posttensioning is applied in steps to control the concrete stresses and to keep them within acceptable limits during gradually added loading.

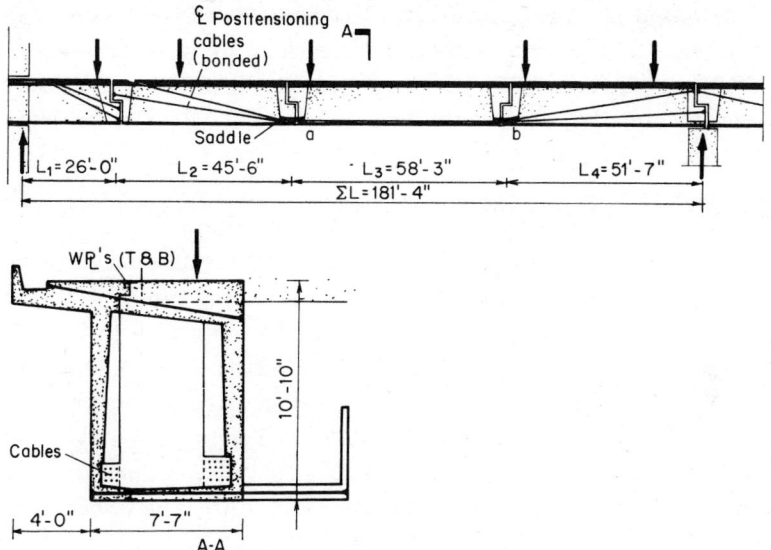

Fig. 268

In the category of one-directional wide-grid systems belong also the K systems. The K joists supported by K girders are narrow-spaced, up to 10 ft, depending on the type of secondary bridging elements. When machine-made lightweight concrete blocks are used, the spacing is limited to approximately 2 ft 8 in. The depth-span ratio for simple span joists is up to 1/25 and for continuous spans up to 1/35. The K girders may be of considerable span, but the depth-span ratio is not more than 1/20. Typical K-system floors are illustrated in Figs. 223 and 236.

Because of the steel top chords of the K girders, which can be welded, even the most complicated beam connections are possible, as illustrated in Fig. 269. As the K joists and girders are relatively light, no heavy equipment for erection is required.

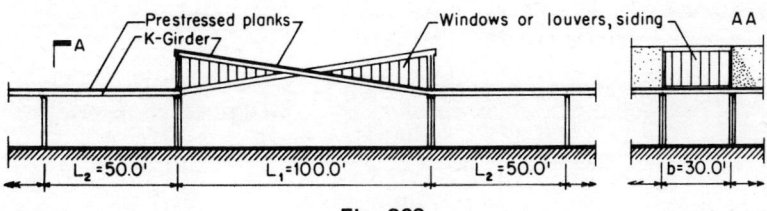

Fig. 269

To allow temperature and shrinkage expansion of relatively long spans (Figs. 267 and 268), neoprene or teflon bearings are most acceptable. Roller bearings, which are very often used, are not justified aesthetically or required structurally.

53. VIERENDEEL SYSTEMS

For relatively long spans and for heavy mechanical installations, trussed girders are appropriate because large ducts and piping can be arranged and incorporated in both ways. The spacing of the trusses, up to 50 ft or even more, depends on the type of bridging elements. The elements can be double T's, single T's, channels, or folded plates. If simple planks are used, the spacing is limited up to approximately 35 ft.

For buildings such as laboratories, in which mechanical installations are extensive, a double floor system is required (Fig. 270). The upper floor rests on the top chord of the truss, and the lower floor on the bottom chord. Thus the space between the floors is used entirely for mechanical installations. Maintenance and new installations can be carried out at any time without interfering with the regular activities in the building. Also, services can be given downward and upward in any location in the building simply by drilling openings through the slab or core part of the bridging elements. When the installations can be exposed, the lower floor

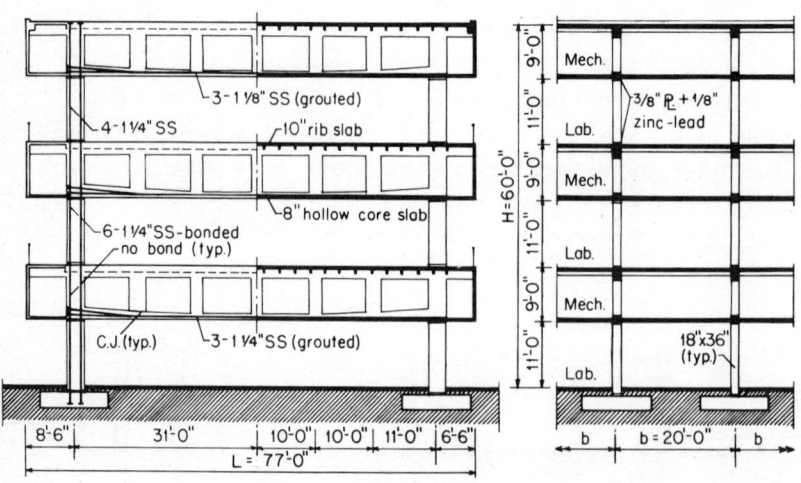

Fig. 270

is not required and the equipment, piping, ducts, etc., can be suspended from the floor.

The systems illustrated in Figs. 271 and 272 have relatively long spans supported by service and stair-elevator towers. The large negative moments of the trusses at the fixed ends must be balanced by posttensioning bars in the outside wall anchored in the foundation. The inside wall is heavily stressed by external loading and prestressing force, balancing the negative moment, but regardless of this, in earthquake areas nontensioned or partially stressed bars should be provided for safety. Also, large-diameter dowels are required to resist the base shear and to avoid rupture of the high-carbon posttensioning bars. The bonding length of the bars is controlled by ductility requirements.

The trusses in Fig. 271 are designed to be prefabricated in three sections of approximately equal weight. The bridging elements can be poured in place or prefabricated, whichever method is more economical. To reduce dead load and provide sound damping, styrofoam between the two slabs has proved extremely practical, adequate, and economical.

The truss-folded plate system illustrated in Fig. 272 can be poured in place or prefabricated. The folded plates are overbridged by lightweight rib-slab forms, reinforced for full design load and with 2½ to 3 in. topping bonded to the form. This type of bridging is structurally sound and economical. However, when sound damping is required, sandwich-type planks (Fig. 271) are more practical and economical.

The construction of expansion joints at the centerline of trusses is structurally acceptable. However, for expansion joints dowels should be provided to avoid differential settlement of the cantilevered top and bottom chords of the truss. The joint is best made watertight by sponge-rubber filler glued to concrete.

Long-span continuous trusses (Fig. 273) are mostly used for cold-storage and other warehouses. To make maximum use of the floor area for storage, the mechanical installations and handling systems are hung from the bridging elements between trusses, which are spaced 50 ft apart. The two-directional air-distribution system, which is required for cold-storage warehouses, can be easily and properly arranged in a trussed system suspended also from bridging elements. Therefore, the bridging elements must be designed for relatively high equipment, duct, and storage-handling loads, in addition to the normal dead and live loads. Prestressed double T's are most suitable structurally and economically

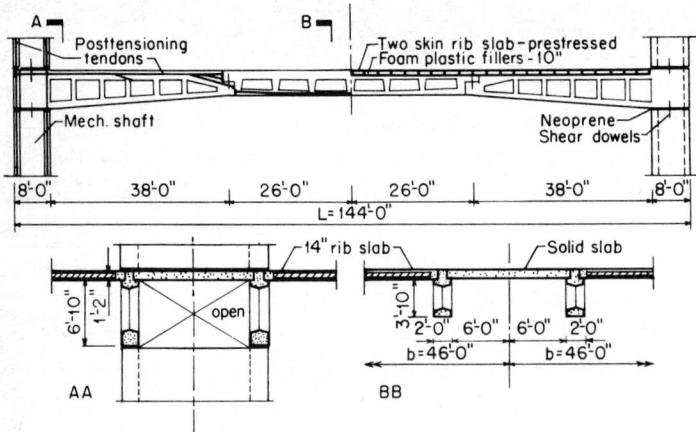

Fig. 271

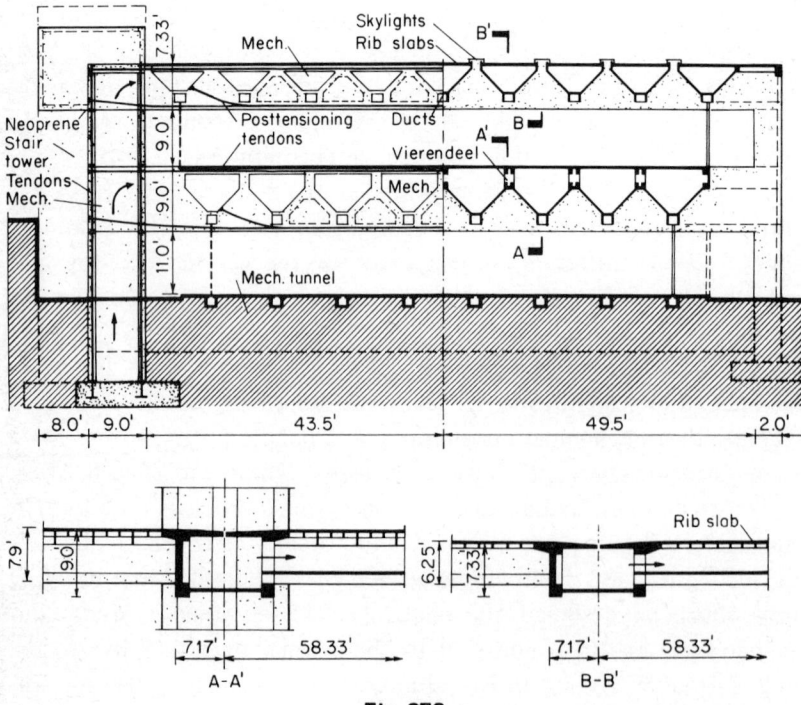

Fig. 272

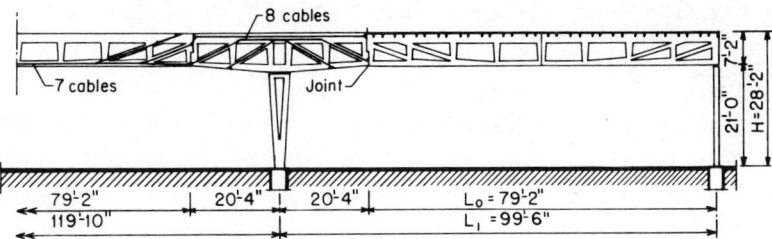

Fig. 273

for this purpose. However, considering that the prestressing force is determined for total load and the relatively high live load occurs only temporarily, the plastic flow is considerable and results in intolerable camber in a course of time. This deficiency of T's can be largely counteracted by using the two-layer concrete method described in Sec. 49. As the T's are relatively rigid, and provided that they are properly seated and connected to trusses, no additional transversal beams are required for lateral stability of the structural system.

To eliminate expansion joints for the trusses, a two-point support at intermediate columns is recommended (Fig. 274). The dowels or prestressing bars connecting the columns and trusses are not bonded, in order to allow controlled lateral movement of the columns to balance the change of length mainly due to shrinkage, temperature change, and plastic flow. The external columns for multispan systems must be simple pendulum types (Fig. 274). For single spans, two-point support, at least

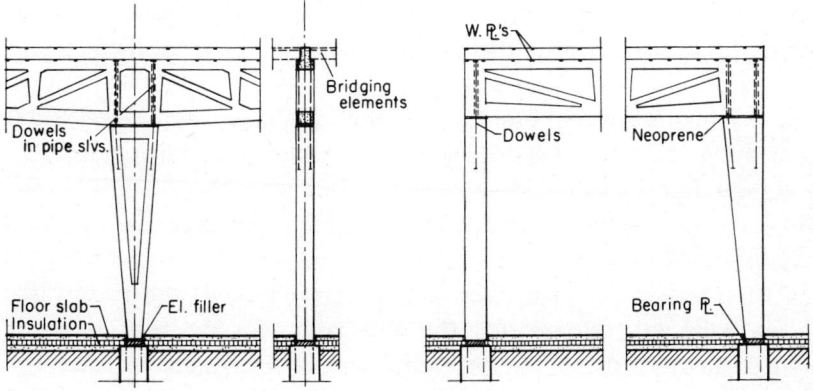

Fig. 274

for one end of the truss, is required for stability. However, to reduce the bending moment at midspan, two-point support at both ends of the truss is recommended, especially where high dynamic forces are expected. By elastic lateral displacement, the ductility of the structural system can be controlled as desired by thickness of neoprene bearings and unbonded length of column tendons.

The depth of truss as well as the dimensions of truss members can be easily determined from the roughly estimated external moment. For example, let us assume that the external moment at span is $\sum_{D}^{L} M_m = 3{,}562 \ k'$ and that a top chord is required by practical considerations, as indicated in Fig. 275:

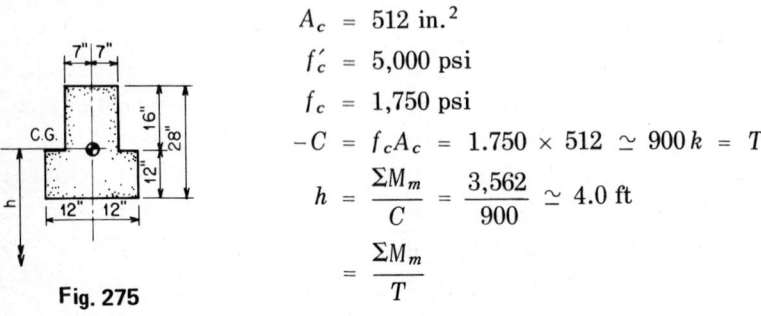

$$A_c = 512 \text{ in.}^2$$
$$f'_c = 5{,}000 \text{ psi}$$
$$f_c = 1{,}750 \text{ psi}$$
$$-C = f_c A_c = 1.750 \times 512 \simeq 900 \ k = T$$
$$h = \frac{\Sigma M_m}{C} = \frac{3{,}562}{900} \simeq 4.0 \text{ ft}$$
$$= \frac{\Sigma M_m}{T}$$

Fig. 275

If this depth is not acceptable, the concrete area or allowable stress must be changed. In a prestressed truss, the concrete in tensile chord and web members has secondary significance only; thus their dimensions should be kept to minimum. The flexural stresses in members are also secondary; however, in the final stress analysis of the truss they should not be overlooked.

In a Vierendeel truss, besides the normal forces N_T, N_B, flexural moments M_{TB} are present and in most cases control the design more than the normal forces (see Sec. 21). Due to this fact, the dimensions for chord and web members as well as the depth h_k of the girder are obtained by trial-and-error methods. First, the dimensions of all members and the depth of truss are assumed; then stresses are checked, and so on. But here also, the rule applies that the dimensions for the tension chord should be kept to the minimum. Since the verticals must balance the algebraic sum of the chord moments at the joint, their dimensions toward the supports, where shear is high, should be chosen slightly higher than the chord dimensions.

To start with, let us assume that the Vierendeel at midspan, where shear is insignificant, acts very much as a true truss; therefore, the dimensions of members and depth of the Vierendeel are determined as for a truss. The second extreme is the end bay, where shear is maximum. The bays between the two extremes change approximately in relation to the shear they carry. For illustration, the bay 1–2 in Fig. 276 is roughly analyzed:

First trial (midspan):

$A_T = 521$ in.^{2} $A_B = 24 \times 8 = 192$ in.^{2}
$I_T = 32,920$ in.^{4} $I_B = 1,024$ in.^{4}

$A_{B12} = 24 \times 14 = 336$ in.^{2}
$I_{B12} = 5,488$ in.^{2}

$A_{B12} = 24 \times 8 = 192$ in.^{2} $I_{B21} = 1,024$

$A_{VT1} = 30 \times 30 = 900$ in.^{2}
$I_{VT1} = 67,500$ in.^{4}
$A_{VB1} = 24 \times 12 = 288$ in.^{2}
$I_{VB1} = 3,456$ in.^{4}

$\sum_B^T I = 2 \times 32,920 + 5,488 + 1,024 = 72,352$ in.^{4}

$k_{T2} = \dfrac{32,920}{72,352} \simeq 0.455$

$k_{B1} = \dfrac{5,488}{72,352} \simeq 0.076$ $\sum k_2 = 1.0$

$k_{B2} = \dfrac{1,024}{72,352} \simeq 0.014$

$M_{T12} = k_{T2} \Delta M_2$
$\qquad = 0.455 \times 1,108 \simeq 505\,k'$
$M_{T21} = M_{T12}$
$M_{B12} = 0.076 \times 1,108 = 84\,k';$
$M_{T10} \simeq 600\,k'$
$M_{B21} = 0.014 \times 1,108 = 15\,k';$
$M_{B12} = 16\,k'$

$$M_{VT1} = +(M_{T10} + M_{T12}) = 1{,}105\,k' \qquad M_{VB1} = +(M_{B10} + M_{B12}) \simeq 100\,k'$$

$$N_{T2} = -\frac{M_{20} - (M_{T21} + M_{B21})}{h_2} = -\frac{2{,}680 - (505 + 15)}{4.0} \simeq -530\,k$$
$$= -N_{B2}$$

$$= -\frac{M_{10} + (M_{T12} + M_{B12})}{h_2} = -530\,k$$

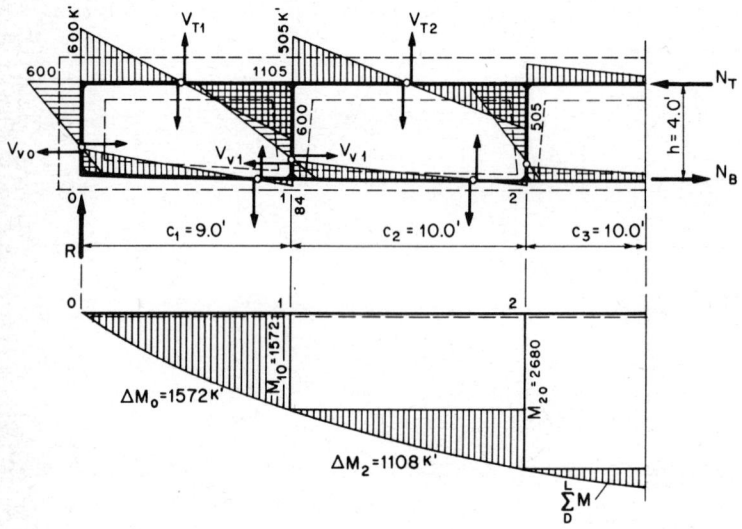

Fig. 276

From the moments, the inflection points and shear can be determined. If the resulting stresses and reinforcing required are within acceptable limits, the assumed dimensions and depth are sufficiently accurate for more rigorous analysis of the Vierendeel. If not, a second trial must be made.

The almost straight posttensioning does not affect appreciably the shear forces and moments in Vierendeels. However, in true trusses where the tendons are also in diagonals, the external shear and moments are in large degree carried by suspension action. The change of bottom-chord depth at joints is mainly required to provide space for posttensioning anchor plates (6 in. minimum).

The curvature of tendons at joints is rather sharp; to avoid overstressing and undue frictional forces, saddles must be provided (Fig. 277).

For practical and economical reasons, the multistory one-way systems are designed so that during construction the load of a higher floor under construction is carried fully by the finished floor below. Usually the construction load is higher than the live load the floors are subjected to, but considering that the construction load is only temporary, higher stresses can be used for this condition. The construction loading, therefore, is simultaneously also a test load for each floor except the roof.

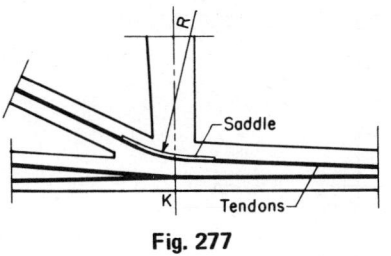

Fig. 277

The trusses are normally poured in horizontal position in sections. The size and weight are controlled by handling equipment.

54. FRAME AND ARCHED SYSTEMS

Frame and arched systems are used when relatively long spans are required or for architectural reasons. To obtain economy, the spacing is chosen as wide as the available bridging elements allow. As one-story structures carry mostly only dead and roof load, slenderness of the prestressed bridging elements can be chosen up to 1/45 for simple spans and even higher for continuous spans.

The construction of forms and scaffolding for high frames and especially for arches is expensive and time-requiring; therefore, prefabrication is generally used. Manageable-size sections are manufactured in horizontal position in a nearby plant or on the site to eliminate transportation problems. Transportation is especially difficult and expensive for long-span arches. Frame systems are most easily prefabricated; therefore, these systems dominate.

A relatively long-span two-hinged prefabricated frame system is illustrated in Fig. 278. Structurally, the carrying action of this frame is accomplished by suspension and prestressing. The cable force is resisted by suspended beam. By proper choice of beam cross-section and location of prestressing force induced into the beam, the direct and flexural stresses as well as the crown elevation can be controlled within acceptable limits. As the beam is subjected for the most part to relatively high compressive stresses, it can be manufactured in sections without providing complicated connections of the sections.

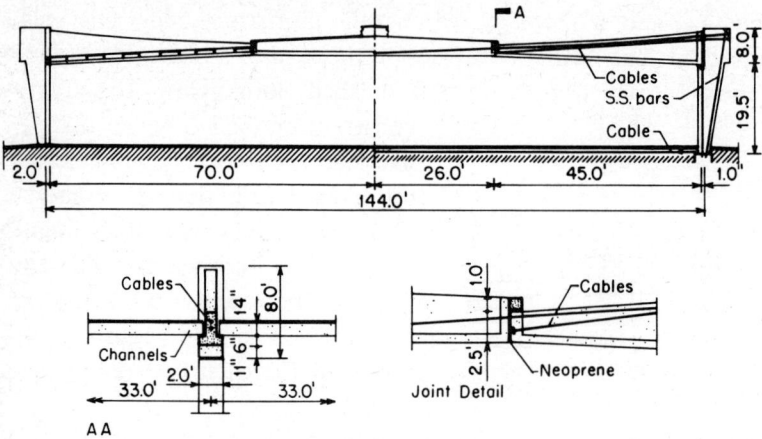

Fig. 278

The bridging elements at the center part of the beam are supported at the top of the section and at the end parts on the bottom chord. Thus windows can be arranged to close the gap at the changeover of the bridging. These locations are simultaneously used to provide lateral stability for the frames. The erection of the frames is rather simple and requires only a limited amount of scaffolding. To control the deflections and stresses during erection, at least a two-stage application of posttensioning is required.

For factory buildings where overhead services are required, a channel is appropriate as the frame beam (Fig. 279). The space required for

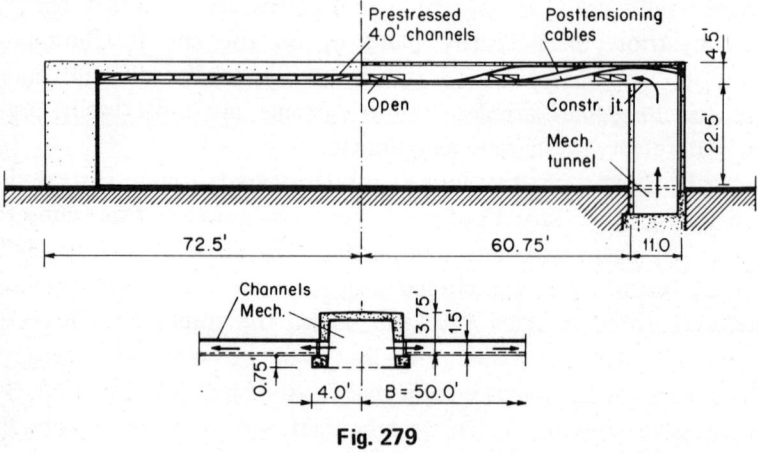

Fig. 279

Structural Systems   473

mechanical installations controls the size of the channel and the method of construction—prefabrication or poured in place. Prefabricated channels between frames allow lateral distribution of services.

For cases in which double columns are architecturally desirable or relatively heavyweight mechanical equipment and services are arranged above the roof in penthouses and heavy crane rails are suspended from the roof, a framing system to satisfy such requirements is given in Fig. 280. The fork-type poured-in-place cantilevers and columns are commonly more economical than prefabricated ones. However, to simplify construction and posttensioning of the cantilevers and to reduce secondary stresses, the drop-ins and decking should be prefabricated.

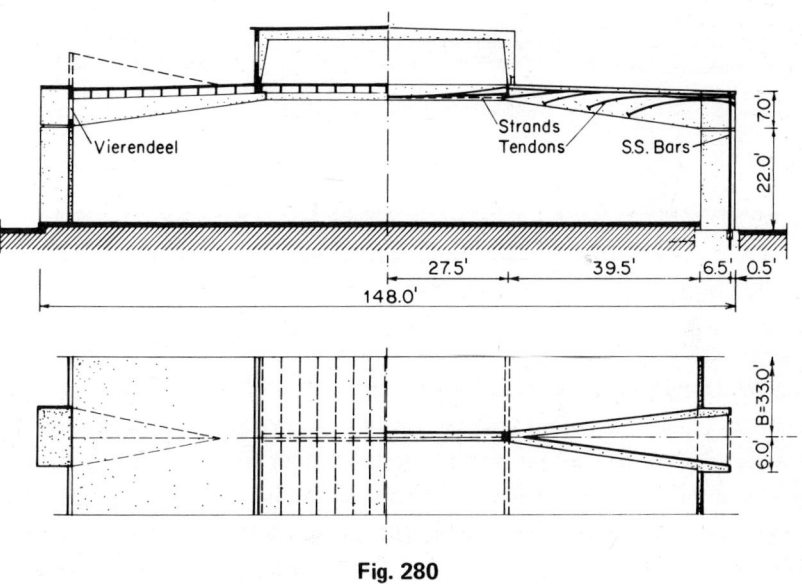

Fig. 280

When the spans of the frame are short but heavily loaded vertically as well as horizontally, the most proper shape of frame is as illustrated in Fig. 281. This type of heavy frame is poured in place and may or may not require prestressing of the beam, depending on the location and degree of loading.

V-shaped verticals, double or single, are structurally and aesthetically most acceptable. Columns with uniform width, regardless of how closely spaced, and also V columns having openings of any shape visually appear unsafe and therefore are not recommended, especially when the superstructure is high and the columns relatively tall.

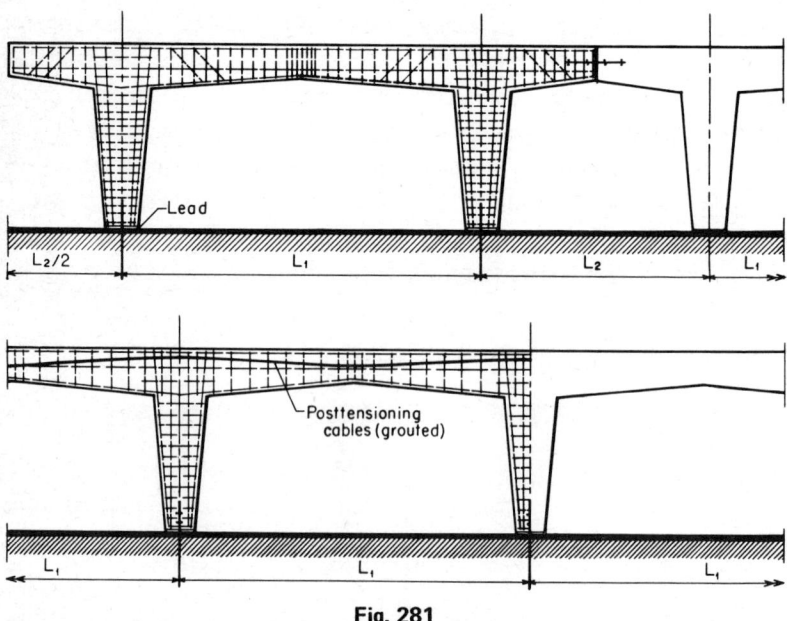

Fig. 281

The north-light frame-shell system (Fig. 282) to roof a health-center swimming pool is designed to be entirely prefabricated. The beam of the frame is a box to allow installation of special radiation-treatment equipment and lamps inside the box to avoid breaking glass, etc., dropping into the pool. The shells for bridging the frames are favored mostly because of their property to reflect uniformly the light projected from beams and windows and also because of their easy manufacturing and light weight.

The approximately 114-ft-long beams, window frames, and shells are prefabricated in three sections. Thus two pipe trestles are required for assembling the sections.

The relatively high precast columns are elastically fixed at the foundations and are moment-connected with the box beam by posttensioning bars.

The tendons in the box beam are exposed up to the columns for easier placement. For the column part, they are placed in flexible tubing. After

Structural Systems

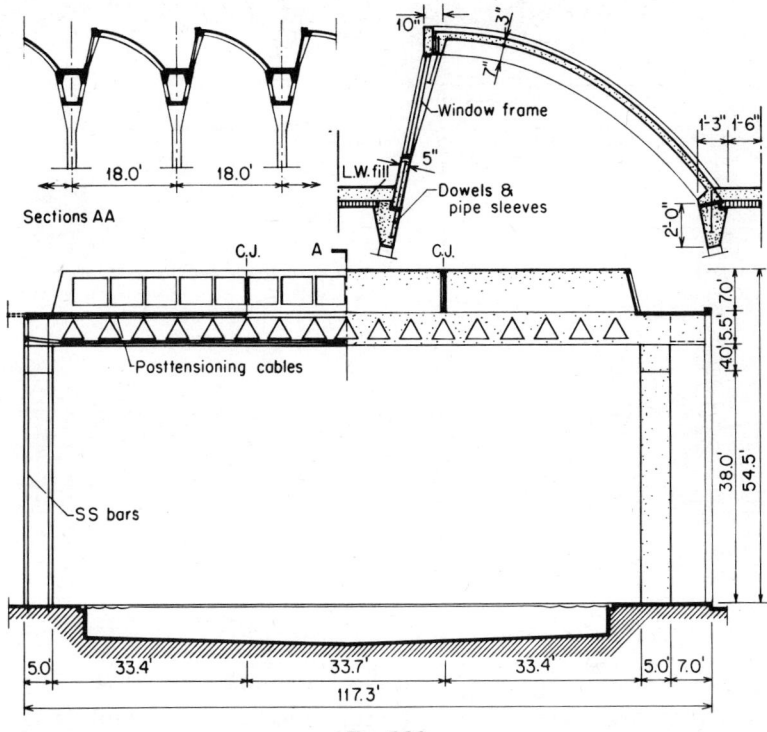

Fig. 282

prestressing is applied, they are simply poured in. The upper tendons, to balance the negative moments, are greased and wrapped in their total length. Greasing and wrapping are not recommended for the lower cables, mainly because to satisfy the factor-of-safety requirements too much conventional reinforcing is required, for which sufficient place is not available.

The box beam has an open top for stripping the inside mold, to allow easy placing and pouring in of tendons, and for installation of services. The top is closed by prefabricated planks covered with lightweight concrete to provide slope for drainage.

A different type of Vierendeel arched frame for a north-light structure is illustrated in Fig. 283. The bridging planks are supported at one end on the upper chord and at the other end on the lower chord of the arches. The arched frames are designed as prefabs to be manufactured in sections at the site.

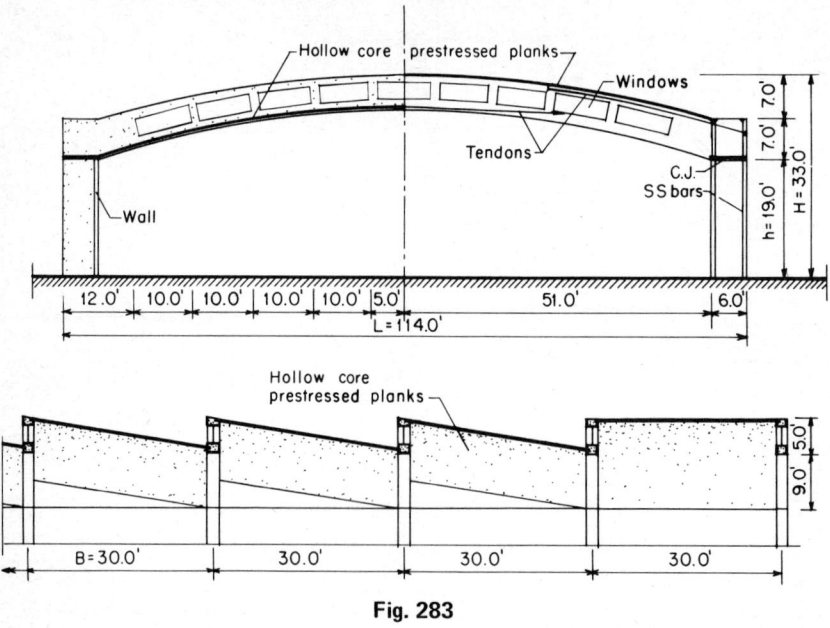

Fig. 283

For auditoriums, concert halls, etc., circular buildings are very often used. As a principle, the roofing structure should correspond to the seating arrangement. The cantilevered frame (Fig. 284) best satisfies this principle.

The 12 cantilevers and columns in this particular case are designed as one piece to be fabricated on the site. The deck slab, light dome, and ring beams are poured in place.

For spans longer than approximately 150 ft, prefabricated arches are generally more economical than frames. A three-hinged prefabricated arch is illustrated in Fig. 285. To obtain some natural light, the windows are arranged at fourth point. The concentrated loads caused by this window arrangement have the great advantage of reducing crown settlement due to plastic flow, so characteristic of all relatively flat three-hinged arches [Eqs. (144) and (275)]. Such a window arrangement offers also an opportunity to provide proper lateral stiffness to the overall structural system. When soil conditions do not warrant an arch, because of high horizontal thrust, a tied arch must be used. This applies also for long-span frames.

For structural clarity, the abutments should be visible. Thus the tie rods must be buried underground below the floor connecting the

Structural Systems 477

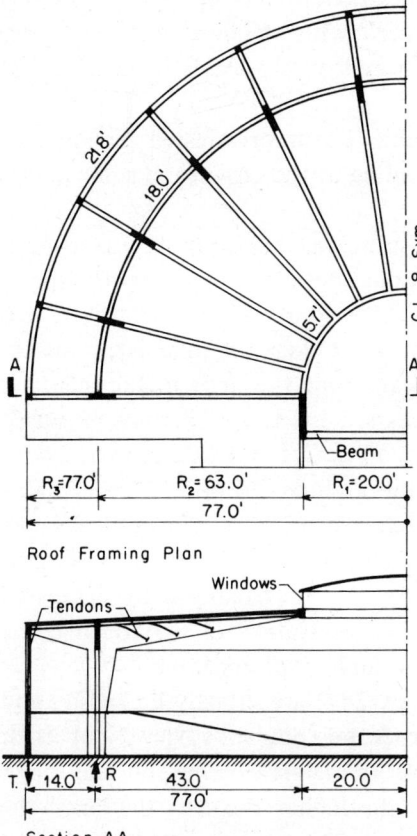

Fig. 284

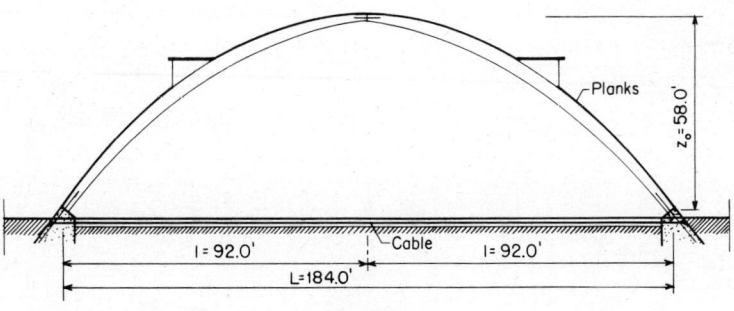

Fig. 285

abutments to avoid their lateral displacement. Structurally, the neoprene or teflon lamella-type hinges, reinforced with stainless steel sheets, are adequate. The thrust is resisted by a bundle of steel rods at the center of gravity of the spring of the arch.

The bridging elements can be double T's or prestressed planks. The erection of the arches requires scaffolding at the crown and some guides for stability during erection.

As can be noted in Fig. 285, this particular arch curvature is unpleasing, depressive, and not functional because it is flat at the spring and to increase the useful floor area leads to a high rise. However, it is economical because the curvature is rather close to funicular shape for full dead and half live load; this means that the arch under this load condition is almost entirely in compression and thus the concrete is used most efficiently. From an aesthetic and also a functional point of view, an elliptic or cycloid centerline of the arch would be more acceptable (Fig. 286), but for both of these curvatures the uneconomical beam action ($M = \Delta z H$) is heavily present and practically controls the design. The inefficient use of concrete has to be balanced by increased dimensions, resulting in heaviness. By application of prestressing, the economy of elliptic and cycloid arches can be increased.

For short spans, high-rise arches (Fig. 287) are visually impressive and are economical from a purely structural point of view, but their construction is difficult and expensive. In high-rise arches, normal forces mostly dominate the design because the higher the rise, the larger the compression and the smaller the bending. The moment at crown is zero

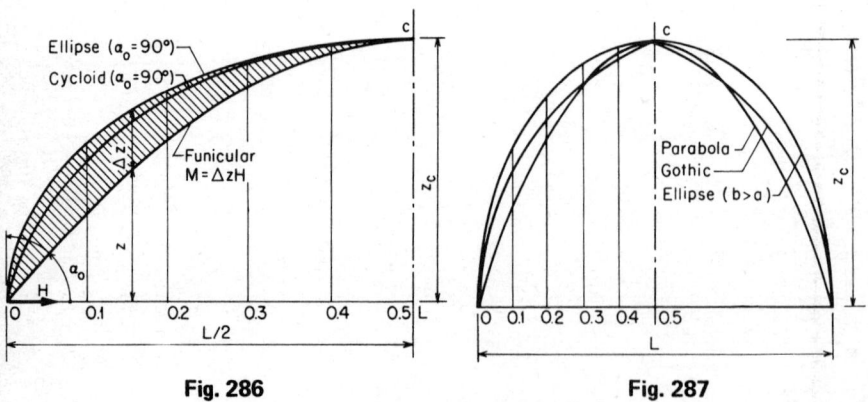

Fig. 286   Fig. 287

or rather small. Thus structurally the proper shape of these arches is similar: the dimensions are smallest at the crown and increase slightly toward the spring in accordance with the increasing normal force and moment.

The so-called *corrugated* or *folded-plate* arches (Fig. 288) are aesthetically impressive and from a purely structural point of view have their advantages, such as that no bridging elements are required and rather long spans are possible.

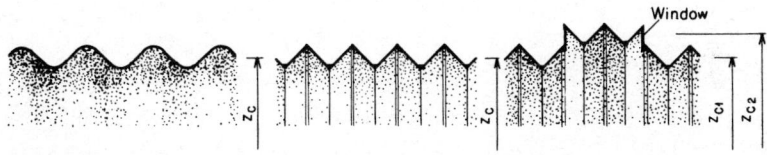

Fig. 288

Folded plates with different rises and spans, arranged as shown in Fig. 289, are architecturally flexible and attractive. The area left between the adjacent arches can be used for lighting, for ventilation, and even for mechanical installations in the spaces between the overlapping plates.

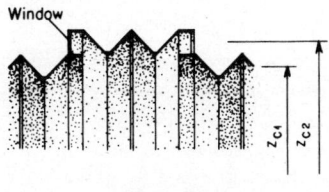

Fig. 289

The construction of this type of arch is difficult and expensive, especially the corrugated ones. Structurally, precasting in sections is possible provided it is economically warranted and adequate space at the site for manufacturing, storage, and erection is available.

55. TWO-WAY STRUCTURAL SYSTEMS

Two-way carrying systems are structurally more economical than one-way systems because the floor slab is supported on all sides and transmits only one-fourth, instead of one-half, of its load to the floor beams, resulting in decreased dimensions for slab and beams. The depth of the slab is usually controlled by stresses and the amount of reinforcing steel that is acceptable. The deflection of the slab is seldom a problem for slenderness (L_{min}/d_0) less than 35 for simply supported slabs and 45 for both-ways continuous slabs, but the slab thickness (d_0) should be no less than 3½ in. The effectiveness of two-way action decreases very rapidly with the increase of the spans' ratio and beyond the ratio $L_x/L_y \geq 2$, it is practically zero.

The two-way systems are most appropriate for approximately square, towerlike buildings (Fig. 290 and 291) or for column layouts with almost equal spacing in both directions (Figs. 292 and 293). For column spacings larger than about 30 ft, the solid slab becomes too heavy and uneconomical. To reduce the slab thickness to within economical limits, a secondary grid, arranged within the main grid, is required. The secondary grid can be wide or narrow (Fig. 291), depending on the overall layout, the mechanical system, and architectural considerations. If the beams interfere with the proper function of the building or are not desirable for any other reason, they can be eliminated; the system then becomes a flat slab or, for longer spans, a waffle slab (Fig. 294). If a smooth ceiling is desired, the two-skin narrow-grid system is appropriate (Fig. 295). In buildings in which mechanical installations are extensive, the beams can be designed as Vierendeels to allow arrangement of installations within the construction depth of the structural system.

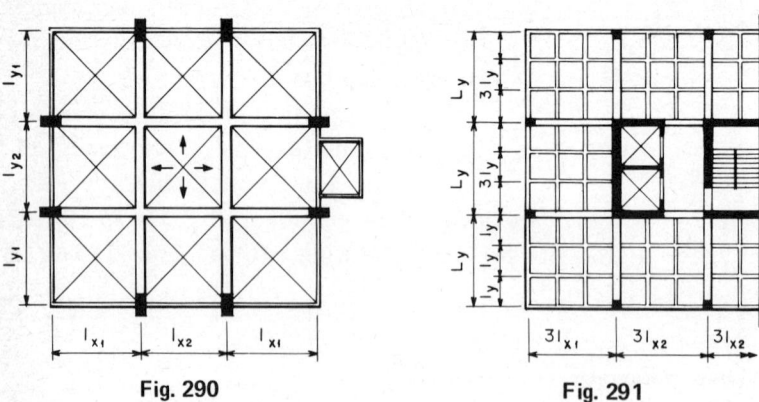

Fig. 290　　　　　　　　　　　Fig. 291

The prestressing of two-way structural systems that are poured in place and with modified spans, especially systems such as that illustrated in Fig. 293, should be avoided mainly because the numerous rigid connections between the columns and beams prevent effective prestressing and cause relatively large moments in the columns. For longer girder spans with secondary grid systems, prestressing of girders is

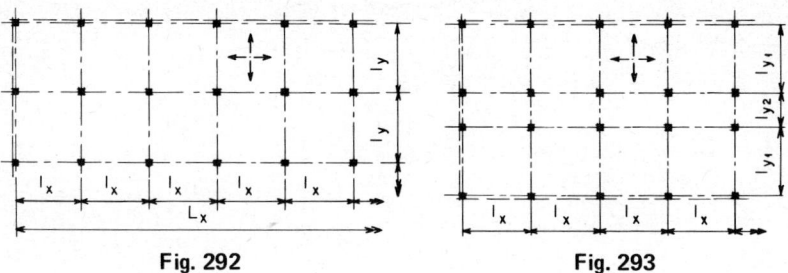

Fig. 292   Fig. 293

unavoidable for structural as well as economical reasons, but the column-girder connections must be designed flexible to allow horizontal shortening due to prestressing (see Fig. 248).

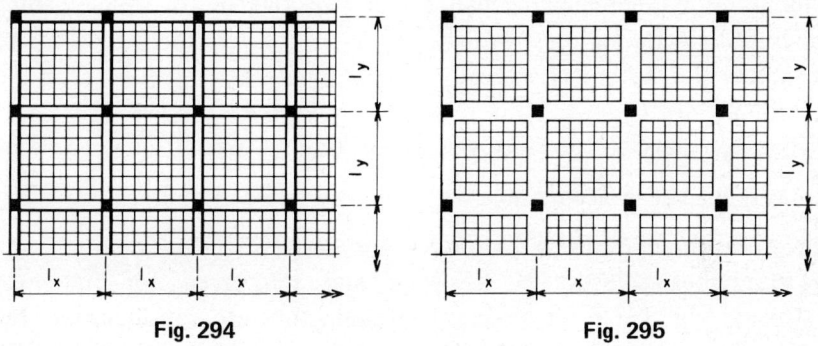

Fig. 294   Fig. 295

Statically, the waffle and two-skin narrow-grid systems with widened solid column strips can be forced by prestressing to act as two-way slabs. The proper prestressing method in this particular case is form-true prestressing. The prestressing force is determined commonly for dead plus one-half live load $(w_D + \frac{1}{2} w_L)$. Thus the suspension force [Eqs. (186) and (187)] is

$$T_{pr} = \frac{\left(w_D + \frac{1}{2} w_L\right) L^2}{8 z_0}$$

and

$$w_{pr} = \frac{8 T_{pr} z_0}{L^2} = w_D + \frac{1}{2} w_L$$

It is determined for the longest span(L_{max}). As the loading $w_D + \frac{1}{2}w_L$ and suspension force T_{pr} are constant, the difference in spans will be balanced by rise z_0 of the cable. Thus,

$$z_0 = \frac{\left(w_D + \frac{1}{2}w_L\right) L_n^2}{8T_{pr}}$$

Since the cross-sectional area is rather large, the influence of normal stresses, caused by the reaction to suspension force T_{pr}, is negligible.

For this loading condition, the deflection of column strips is zero; that is, the $w_D + \frac{1}{2}w_L$ load is transmitted to columns directly by cable (suspension action) and no flexural stresses are developed in beams and columns. For the remaining half of the live load or for other loading conditions (including prestressing), as well as for determining the factor of safety, the system is statically indeterminate and must be analyzed accordingly.

The secondary-grid beams act as continuous beams on unyielding supports loaded in accordance with two-way slab design $\sum_D^L w = w_x + w_y$. Therefore, the grid beams seldom require prestressing. In flat-slab design, the grid beams must be designed and reinforced for full load ($\Sigma w = w_D + w_L$) in both directions. Taking this into consideration, the qualitative superiority and economy of this system, in comparison with the flat-slab system, can be easily realized.

The voids in two-skin systems can be most conveniently provided by styrofoam during the pouring operation. First, the bottom slab is poured; then the styrofoam is set, welded wire fabric is placed, and the pouring is finished. The time interval between the two pours depends on the type of cement used, the climatic conditions, etc., but should be not more than 2 hr, so that the two layers of concrete can still be vibrated thoroughly to avoid cold joints. The styrofoam voids, besides reducing the dead load, provide excellent heat as well as noise insulation.

In the layout of the structural systems it should be realized that the moments in end spans are larger than in intermediate spans and also that the corner columns are far less loaded than other columns. For architectural reasons, the depth of spandrel must be uniform and the dimension of facade columns equal. Since the corner columns are exposed from two sides, they appear visually too heavy in comparison with other columns. The remedy is to eliminate the corner columns by cantilevering the spandrel so that the end spandrel and column adjacent to the corner and the intermediate facade spandrels and columns will be loaded equally. The difference in the two column arrangements is illustrated in Fig. 296.

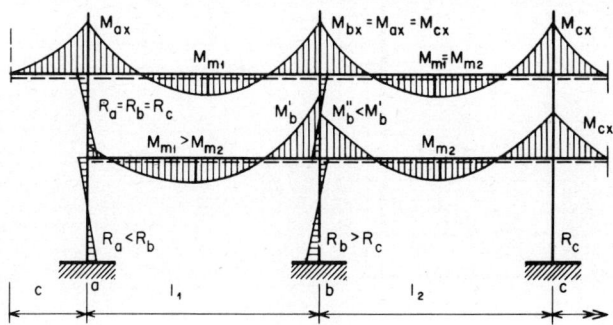

Fig. 296

In general, the two-way systems are too complicated for prefabrication. However, for multistory tower types and one-span (Fig. 297) structures having a large number of repetitions, the beams, girders, and columns can be most easily prefabricated, if economically justified. In this case, the beams or girders in one direction are prefabricated as one piece in full length and prestressed by either pretensioning or posttensioning methods. The beams or girders in the other direction must be manufactured in sections, erected on the seats at the full-length beams, and posttensioned to unity. To save time and avoid stress concentration, the joints between the intersecting members should have neoprene gaskets designed to transfer the posttensioning force. After posttensioning, the tendons and joints are pressure-grouted in one operation. The forms for poured-in-place slabs are supported by beams, so that no scaffolding and shoring is required for construction.

Structural layout and details of a long-span industrial building subjected to heavy loading and having extensive mechanical installations to be arranged above the bottom of grid beams and at the roof so as not to interfere with the two-directional conveyor system hanging from the beams are given in Fig. 297. The main-grid Vierendeel girders have a span of 90 ft and cantilever 22.5 ft over the corner columns at both ends.

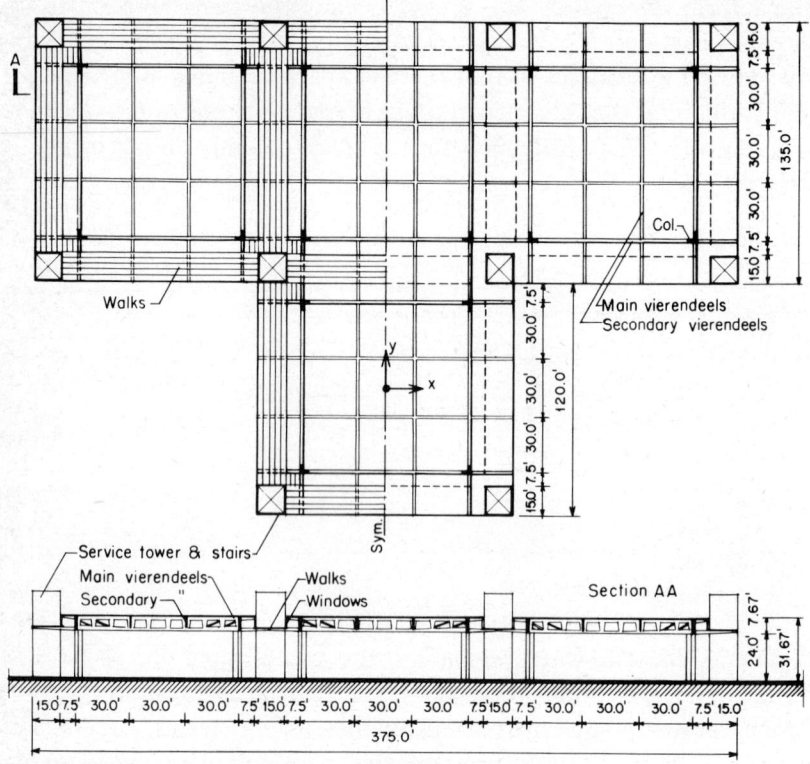

Fig. 297

They support secondary-grid Vierendeels spaced 30 ft center to center in each way. The Vierendeels, the decking of the cantilevered walks between the service towers and columns are designed to be prefabricated. The waffle-type roof slab is poured in place.

The main prestressed Vierendeels and the secondary ones (except the cantilevered ends) in the x direction are manufactured in full length at the site and in the y direction are erected in sections on the full-length Vierendeel and posttensioned into unity. The forms for waffle slab are supported by the grid. The decking of walks consists of hollow-core prestressed planks supported by cantilevers.

The Vierendeel girders and structural details of supports and connections are given in Fig. 298. The heavily loaded main Vierendeels of regular type are not able, in this particular case, to carry the high shear and moments in the end bays. Due to this, and to make use of suspension action, diagonals are introduced in the two end bays. For secondary Vierendeels, diagonals could not be used because they would have interfered with mechanical installations. To reduce the dead load and handling difficulties, the Vierendeels are designed as light as possible. However, to keep the stresses within acceptable limits, the waffle slab was shear-connected and used as a compression area for Vierendeels. The bearing neoprene between the columns and Vierendeels is required to ensure uniform load transference and to allow slight rotation and displacements due to prestressing, shrinkage, and temperature change. Before grouting, the dowels connecting the columns were asphalt-coated about 1.5 ft above the neoprene to provide sufficient elongation length.

The laboratory towers shown in Fig. 290 are also designed as a prefabricated system. However, the floor slabs in this particular case are poured in place (Fig. 299). The full-length Vierendeels are pretensioned

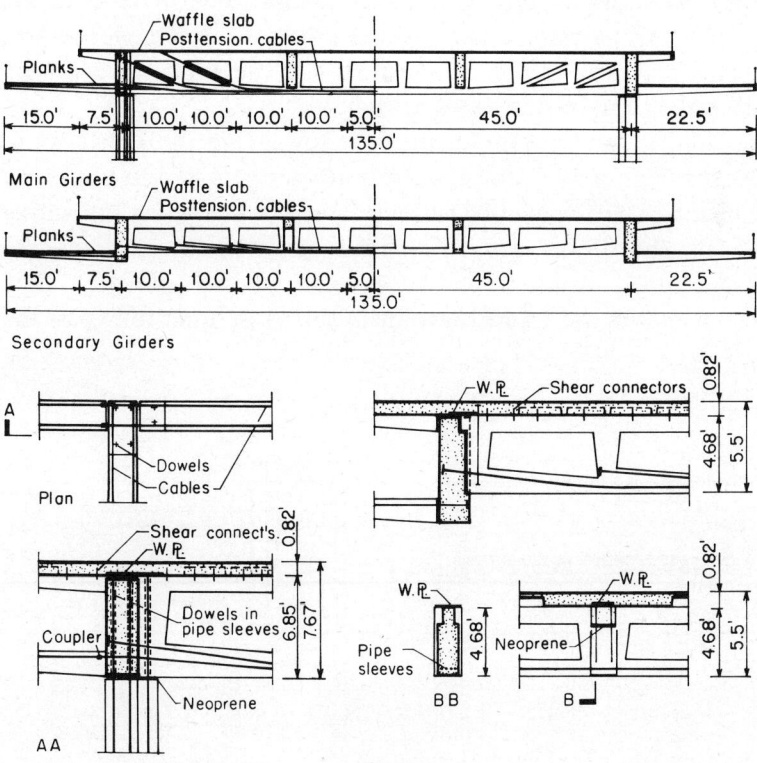

Fig. 298

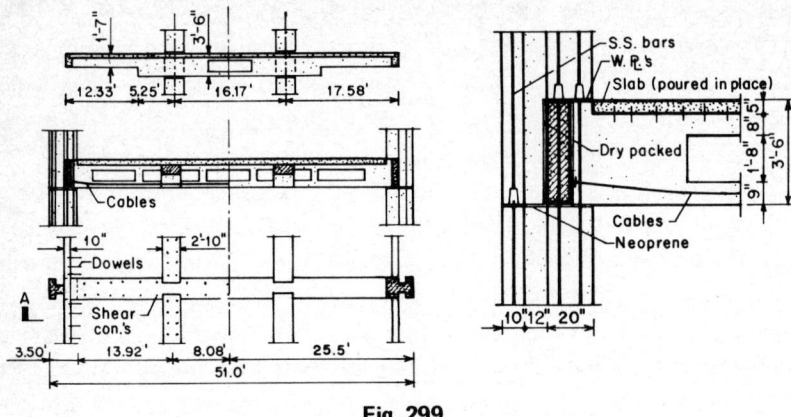

Fig. 299

with seven-wire strands. The Vierendeels manufactured in sections are posttensioned with stressteel bars. Spandrel beams are not prestressed. The 5-in. two-way slab is shear-connected with prefabricated members and serves as the compression area for live load.

A different structural system for the same tower (Fig. 290) is shown in Fig. 300. The main girders and spandrels are designed to serve as ducts both for heating and for cooling systems. Some of the piping and wiring is buried in the floor slab. The main girders and columns are prefabricated; the spandrels and floor slab poured in place. The prestressed spandrels are framed into main girders in order to bypass the

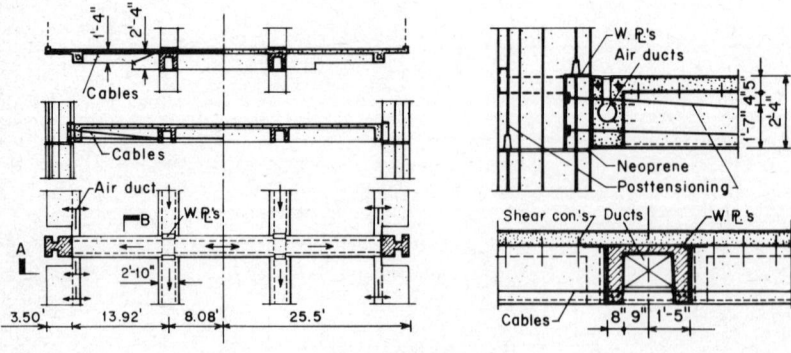

Fig. 300

posttensioning bars of the columns and to provide more space for ducts. Due to this spandrel arrangement, the columns are to a high degree eccentrically loaded and must be posttensioned beyond the amount required for the first layout (Fig. 299). The floor slab is cantilevered over spandrels about 3 ft 4 in. and supports the curtain wall. By this design, the useful floor area is increased about 15 percent and the story height about 1.0 ft. The method of manufacturing and erection is similar to that described above for the industrial building.

Where the story height is rather small and the columns widely spaced, the beams, especially heavy ones, projecting from the ceiling are architecturally not pleasing and also interfere with proper location of partitions. The typical flat-slab system for long spans in these cases becomes too heavy, resulting in uneconomical live-to-dead-load ratio (w_L/w_D). Besides this the noise propagation through a solid slab is considerable. Reducing noise transmission to within acceptable limits requires a rigid insulation layer at the top of the slab below the flooring, which is rather expensive and also increases the dead load. Due to this, the most acceptable and economical floor system in such cases is the two-skin narrow-grid structural system (Fig. 301). The minimum slab thickness (top and bottom) is 3 in., and the ribs should be chosen wide enough so that stirrups are not required. However, if the width becomes more than about 6 in., the use of stirrups at the end portion is more economical than the use of wider ribs. The 3-in. top-slab thickness reinforced with welded wire fabric is structurally adequate for 3-ft rib spacing.

If the widened column strips are not symmetric in respect to the column line or if the loading from adjacent bays is unequal $(R_L, M_L \neq R_R, M_R)$, the beam will be subjected to torsional moments. These torsional moments can be reduced by proper location of posttensioning cables and by the use of unequal prestressing force in cables. However, the location of cables should be chosen with a 45° angle $(a + 2d_o)$ at the column in order to guide the cable reaction directly to the columns without undue stresses. By the use of unbonded cables, the stresses in cables do not increase appreciably under live and ultimate load conditions because the cables are not an integral part of the beam action but act independently within the beam as a suspension system which is far less rigid than the beam. Due to this, conventional reinforcing is required to satisfy the factor of safety.

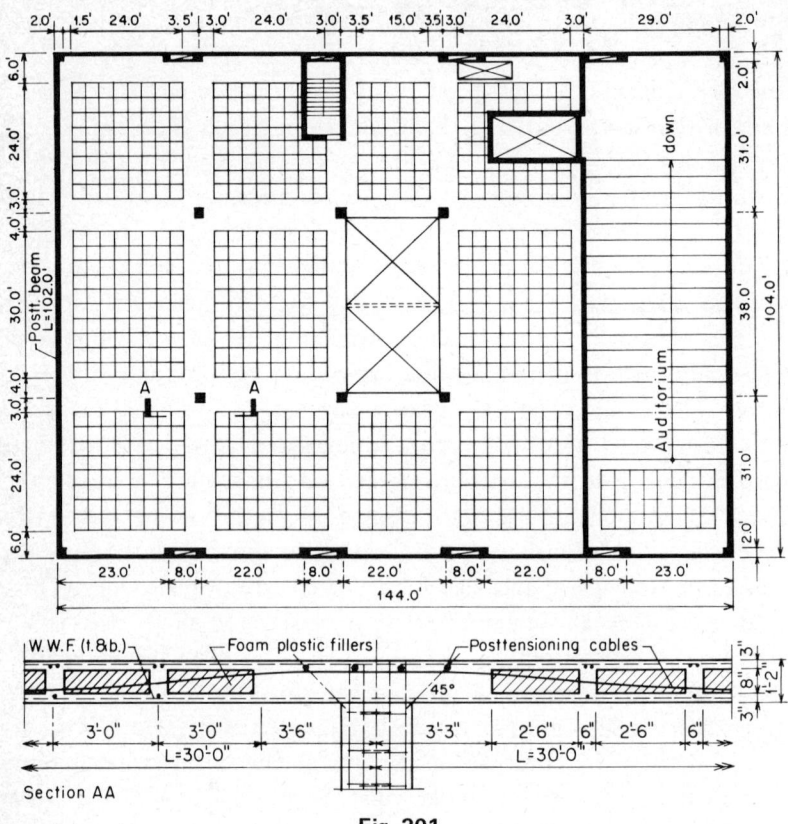

Fig. 301

It is not necessary that the grid beams intersect at a 90° angle; there may even be cases in which the angles for all beam intersections vary, as illustrated in Fig. 302. When the angle is too sharp, the beams or Vierendeels manufactured in sections may slip during posttensioning. To avoid this, the contact area must be thoroughly roughened and slightly recessed and welding plate must be designed adequate to resist the component in joint direction of the prestressing force.

For architectural reasons, it is also sometimes required to design and prefabricate a two-way structural system which is not straight, as indicated in Fig. 303. In this particular case, the whole roof structure is prefabricated. To achieve equal loading for the girders, the hollow-core prestressed spancrete planks are arranged in chessboard fashion. The

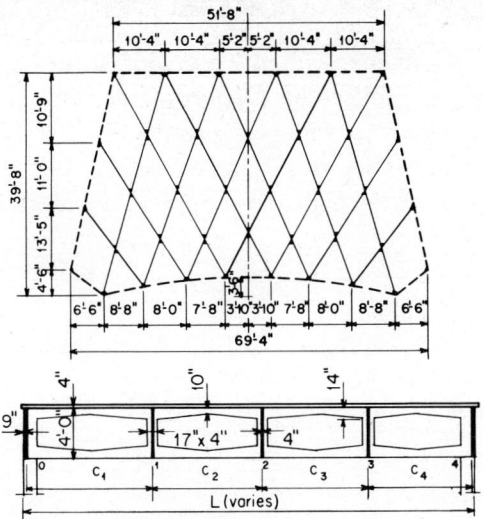

Fig. 302

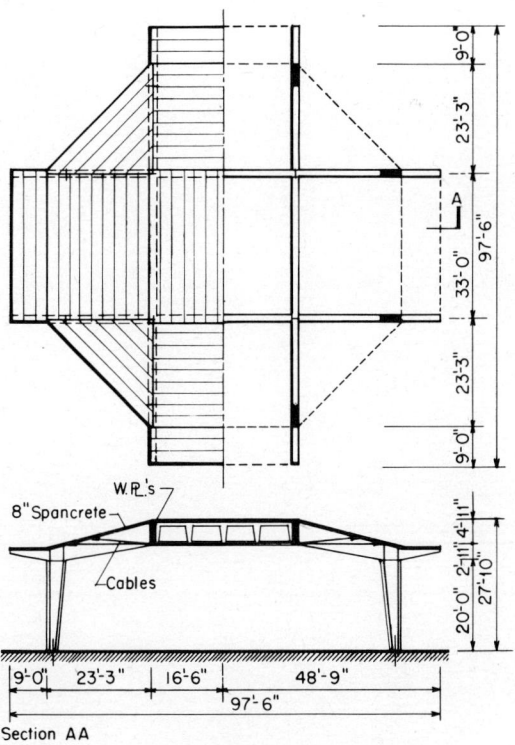

Section AA

Fig. 303

expected torsion moments, caused by one-sided loading of the girders, can be in some degree reduced by supporting the planks at the outermost ends on neoprene pads as close as possible to the centerline of the girders. In cases in which this is not possible, the torsional stresses developed can be relatively high and should not be overlooked.

13

SPACE SYSTEMS

56. PRISMATIC SHELLS

The three-dimensional nature of space systems limits their use for one-story buildings, such as garages, hangars, halls, arenas, gymnasiums, industrial buildings, silos, and water tanks. The three-directional carrying action is statically the most efficient. Due to this and to the relatively light loads structures of this type are subjected to, large column-free areas can be overroofed economically. However, the construction of one-way and especially two-way curvatures is difficult and expensive. Precasting, to reduce the forming costs, is possible only in exceptional cases. To overcome the forming difficulties for some complicated space systems, guniting in connection with partial precasting can be used. Precast sections, such as ridges, valleys, and stability rings, are utilized to establish a stable skeleton for the system. The curved surfaces between the skeleton parts are formed by closely spaced steel grids welded to the dowels projecting from the precast sections. The grid is commonly

492   Structural Systems and Methods of Construction

covered with two layers of chicken wire and then gunited in spiral layers, starting from the bottom of the system. To reduce the loss of gunite, it has proved successful first to cover the chicken wire with a thin layer of gunite from the inside and then, after some hardening and troweling of this layer, to complete the guniting from the outside.

The space structure illustrated in Fig. 304 has a two-way prestressed column capital upon which the shell rests. The upper edge of the shell is

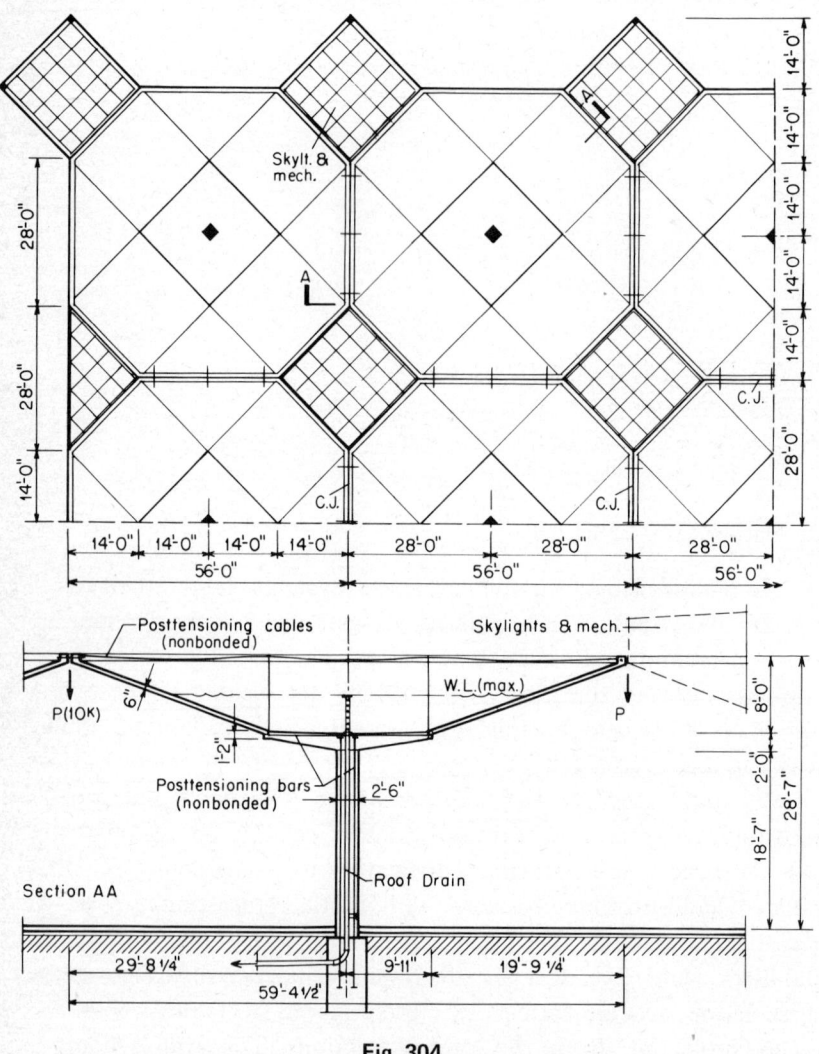

Fig. 304

also prestressed. The space left between the shells is used for skylights and the installation of mechanicals. Drainage of the large shell area is most acceptably accomplished through the columns. For stability, adjacent shell edges are connected with each other by bolts. The columns, capital slabs, and shells can be easily precast in sections and tied into unity by posttensioning. If the whole system is poured in place, the reinforcing, including prestressing rods, for capital slabs and shell sections can be prefabricated at the ground level and lifted into place by crane. After the reinforcing sections of the shell are completed and the tendons placed for the upper edge ring, the shell can be poured, starting from the capital slab. Considering the large number of units almost always involved, the repetition of forms, scaffolding, and prefabrication of reinforcing, this system is rather economical.

A similar but pyramidal type of space structure is illustrated in Fig. 305. The long marginal beams as well as the sides commonly have to be posttensioned to avoid undue deflections. The economy of this system

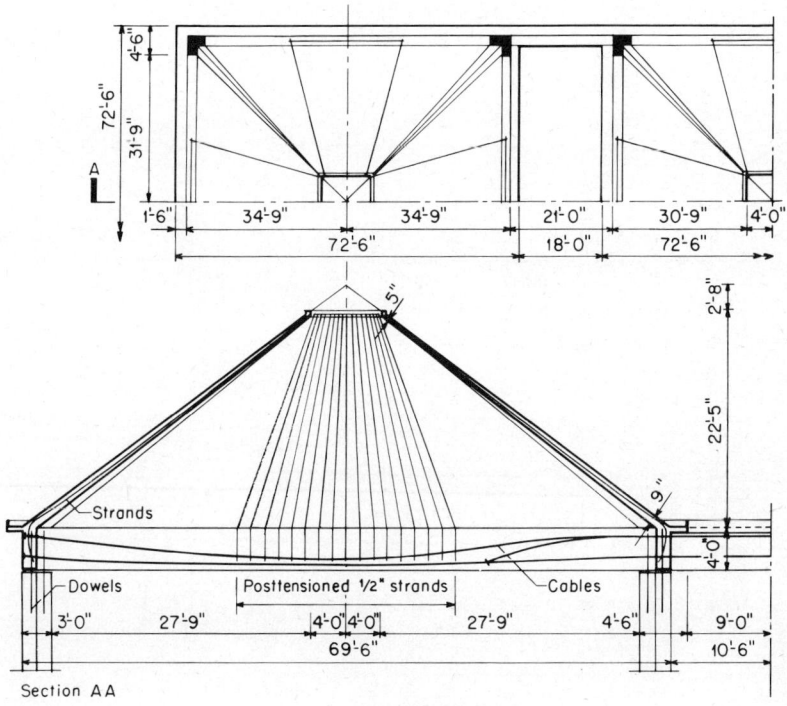

Fig. 305

494 Structural Systems and Methods of Construction

can be improved by using slightly curved triangular sides for the dome. In this case, the sides act as a cylindrical shell, transmitting part of the load to the ridges (see "Polygonal Domes," Sec. 24).

For moderate spans and when cranes with high lifting capacity are available, the pyramidal domes can be fabricated as a unit at the ground level and lifted into place (Fig. 306). The grooves between the units can be used for ducts and piping or for posttensioning. The tops of the domes can be cut for skylights, ventilation, etc. To obtain adequate seating and moment connection for four domes supported by one column, the most preferable column shape is the cross type (Fig. 306) tapering toward foundation. The columns are subjected to two axial bending moments acting at the top and only normal force at the bottom. Since the beam action is the most uneconomical carrying action, the large cross-sectional area at the top is structurally required to compensate for the inefficient use of materials and to keep the stresses within acceptable limits. At the bottom, the material is uniformly stressed and the concrete can be used to its maximum allowable compressive strength. Due to this, the bottom cross-sectional area required, in comparison with the top, is rather small. A tapered column thus pronounces most strongly

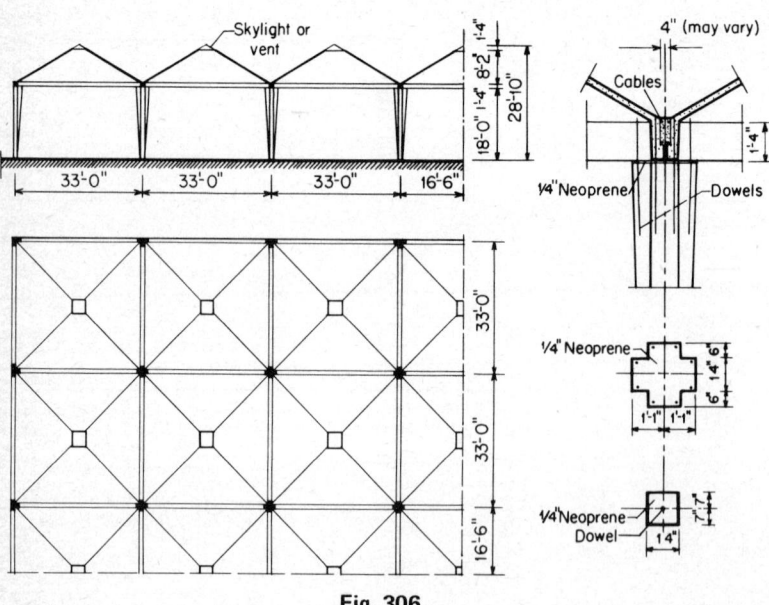

Fig. 306

the structural behavior and efficiency of material by which it balances the forces acting upon it. This is why such columns appear visually more safe than uniform-cross-section or reversely tapered columns.

The two-way folded long-span space structure illustrated in Fig. 307 can be used to overroof economically a relatively large area. The corner regions can be used for light wells or can be simply bridged over by two-way slabs, after the posttensioning is applied, for free edge of the folded plates. The crossing folded plates must have a moderate rise for drainage and to induce compression in the concrete to control shrinkage. Regardless of the flatness of the system, the structure is rather rigid and behaves almost exactly in accordance with theory.

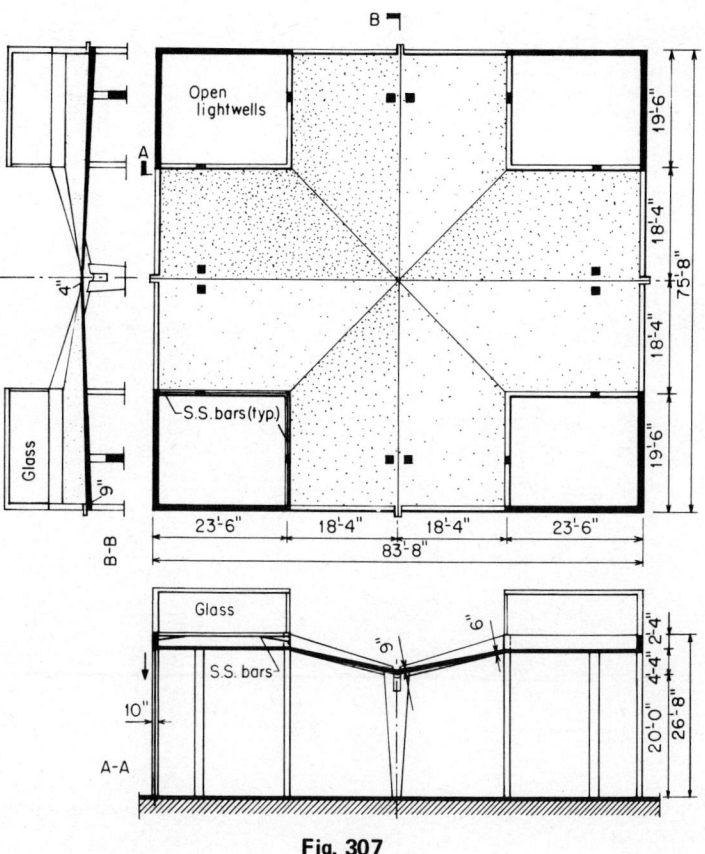

Fig. 307

57. DOMES, LONG-SPAN ARCHED SYSTEMS

There are numerous possibilities for arranging and combining basic folded-plate units, straight or curved, to obtain structural systems for various purposes. A folded-plate structure consisting of flat surfaces and corresponding in its structural behavior to a ribbed dome is illustrated in Fig. 308. The individual units lean at the crown against a central compression ring and at the outside ends against a circular tension ring supported by columns or walls. Thus the units act as two-hinged frames.

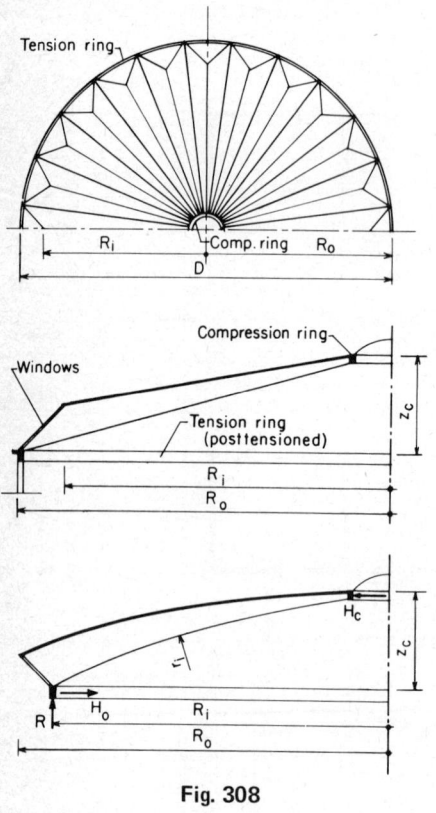

Fig. 308

Instead of straight units arranged in a circle different units and arrangements can be used to develop interesting and structurally sound domes, halls, etc. A triangular and rectangular base-and-crown arrangement is illustrated in Figs. 309 and 310. At the crown, where the units lean against each other, it is structurally preferable that the crown line $c - c'$ be kept at the same elevation. However, to lower the crown from c to c' does not offer insurmountable difficulties. As the thrust can be considerable for longer span, prestressed ties at abutments are usually required. The curved units have to be poured in place, and the formwork is expensive. The straight units, because of the large number of repetitions, can be prefabricated, if economically justified. The construction starts from the crown and works symmetrically and continuously outward, so that the thrusts are balanced in the highest degree possible. In this way, after the concrete of a section has developed the required strength, the forms and scaffolding can be stripped and reused. If narrowly spaced welded wire fabric is used as reinforcing, slopes up to 45° can be poured without outside forms, but to speed up the pouring operation, sliding outside forms can be successfully used.

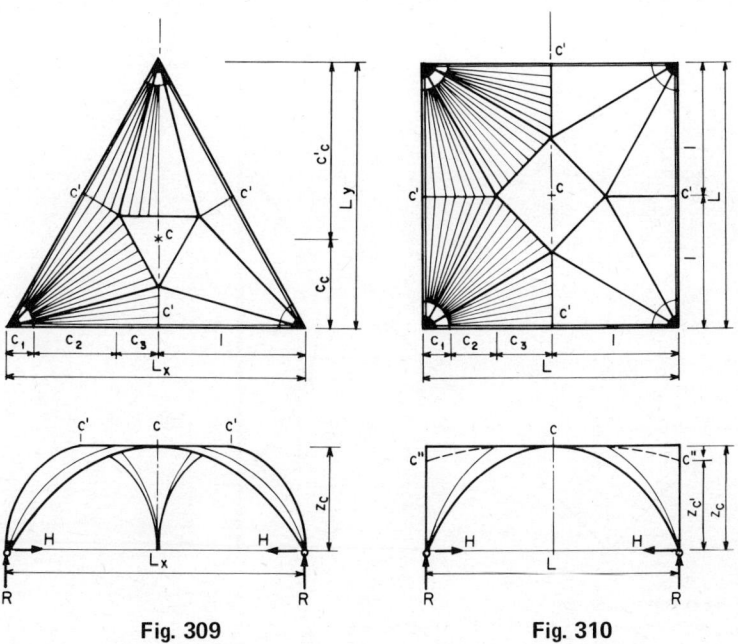

Fig. 309 Fig. 310

The elevations of adjacent sections can be adjusted by jacking the tie rods of the sections to be raised or lowered. When all sections have been finished and final adjustments, if required, have been accomplished, the joints are rigidly closed.

498 Structural Systems and Methods of Construction

Instead of folded plates as basic units, shell sections can be combined to obtain rational and attractive space structures.

Statically, the curved systems, regardless of the type of cross section—folded plates or shells—carry as two-hinged arches. Depending on the curvature, the beam action involved can vary with considerable limits. It is the least for a funicular centerline.

In space systems such as that shown in Fig. 311, in which the cantilevered shells do not have a common boundary at junction because of the difference in rise required to provide space for crescent-shaped windows, arch and shell actions accomplish the carrying capacity of the system. The arch action controls the design in the direction of the shell generating line between the abutments, and the shell action principally controls the cantilevers. Structurally, the free boundary of cantilevers commonly does not require any strengthening. For long arch spans, the thrust, normal force, and bending moments are rather high, and therefore, stability becomes a serious problem. Buckling can be

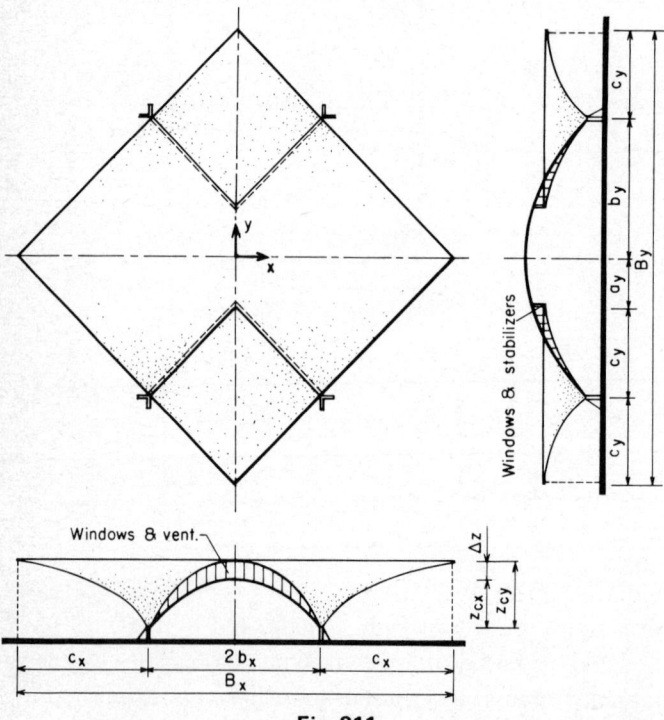

Fig. 311

counteracted by increased moment of inertia and reduction of thrust by avoiding span-rise ratios beyond about 6. As the cross section involved in arch action is widest at the crown and narrowest at the abutment, the most proper cross section for long spans is the two-skin shell section with varying slab thickness—smallest at the crown and largest at the abutment. In addition to high moment of inertia, the voids between the slabs offer adequate insulation for the roof. The common abutments, acted upon from the intersecting shells, are subjected to diagonal thrust. The proper shape, therefore, will be the tied-fork type, where not directly resting on the rock. Enormous areas can be overroofed by this type of space structure. The longest span (720 ft) at present is the Exposition Palace in Paris (1958).*

Structurally, a quite different space system for overroofing large areas economically is the hyperbolic paraboloid shell (Fig. 312). The carrying

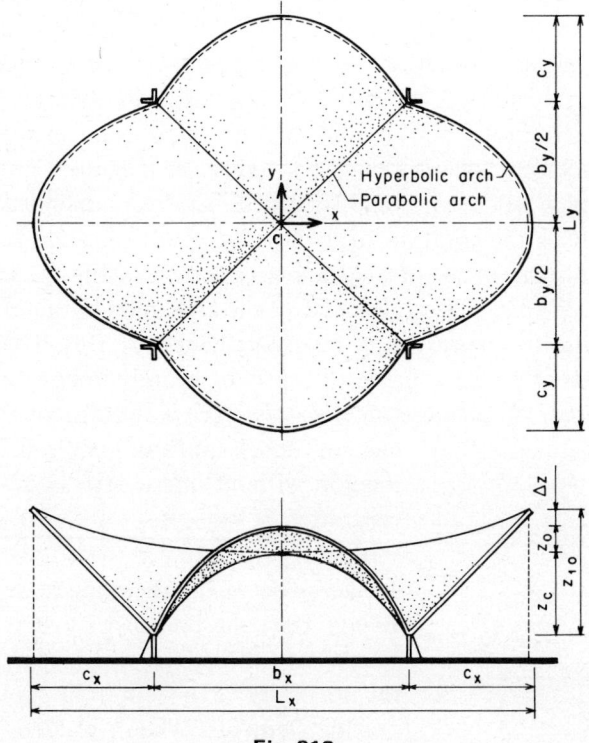

Fig. 312

*Nicholas Esquillan, Enterprises Boussiron, Paris.

action of this shell type is accomplished by arch and suspension action. The skin of the shell can be relatively thin, even for long spans, mainly because the buckling of the arch action is avoided by directly related suspension action. Due to the high flexibility of the shell, the bending moments involved are of only secondary order.

Diagonal arches are formed at intersections from the resulting valleys and are parabolic. The outside boundaries over squares follow the asymptotes of a hyperbola and are straight, forming a triangle. For rounded corners of the square, the boundary becomes a hyperbolic arch. Because of the reduced shell area at the crown, the diagonal arches act practically as three-hinged arches acted upon upward by suspension action at the crown and downward by arch action at the abutment regions. The two-hinged boundary arches are acted upon all the way downward by suspension as well as arch action. The shear along the arches commonly controls the thickness of the shell.

The common hinges and abutments must be designed to resist the reactions and thrusts of the arches for all conceivable loading during construction and in finished condition. As the reactions, thrusts, and rotations even for equal arches can differ to a considerable degree, especially during construction, the design of the supports is very demanding. If the rock lies deep, the thrust must be balanced by prestressing cables, which adds to the difficulties. The recommended hinge design, which allows the rotation of arches and is not sensitive to the normal expected displacement of abutments, is shown in Fig. 313. To keep the thickness of steel plates attached to the arch and abutment within acceptable limits, the plates must be strengthened by ribs. The elastic layer between these plates consists of layers of bearing neoprene separated from each other by stainless-steel sheets. Such a sandwich pad ensures almost uniform load transmission without undue stress concentration.

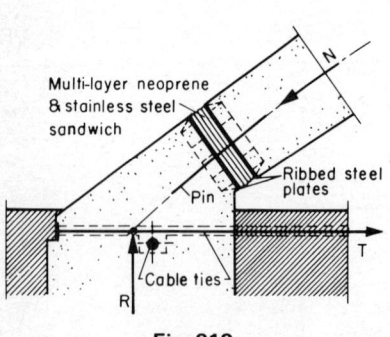

Fig. 313

In polygonal domes, the curved surfaces between the ridges or valleys transmit the load by shell action into the ridges, which carry the loads by arch action to supporting elements. When the curvature is a parabola or is funicular, the curved surfaces act as arches and do not transmit the loads into ridges. Therefore, the lower boundary of such surfaces must be supported steadily.

Possibilities for such shell-unit arrangements related to plan are illustrated in Fig. 314.

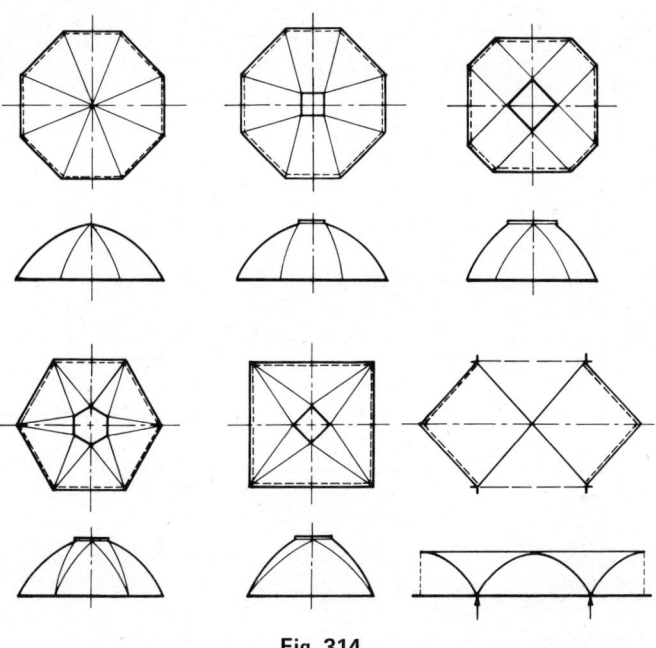

Fig. 314

There are other possibilities for combining shell units to meet any conceivable functional and architectural requirement. However, in each case, it is essential that the form of a combined shell is not capricious but follows a certain geometric law. Only if the form itself is consistent can an arbitrarily chosen linear grid be integrated into a spatial whole. It must be realized that in combined shells the shell units are not stable in themselves; the shell system achieves its structural strength only by interaction of components with each other or with other structural elements. Units with junctions which have single curvature and provide straight-line contact are the most easy to combine. If this is not the case, additional structural members are required for balancing the forces at unit boundaries to obtain stability.

502 Structural Systems and Methods of Construction

Some combined shell systems and shells with complicated curvatures can be most easily and economically constructed by combining precast sections with steel grid and guniting method (Fig. 315). This building is composed of four different shells. The lowest and intermediate ones are formed of cylindrical shell sections transmitting the loads, including the

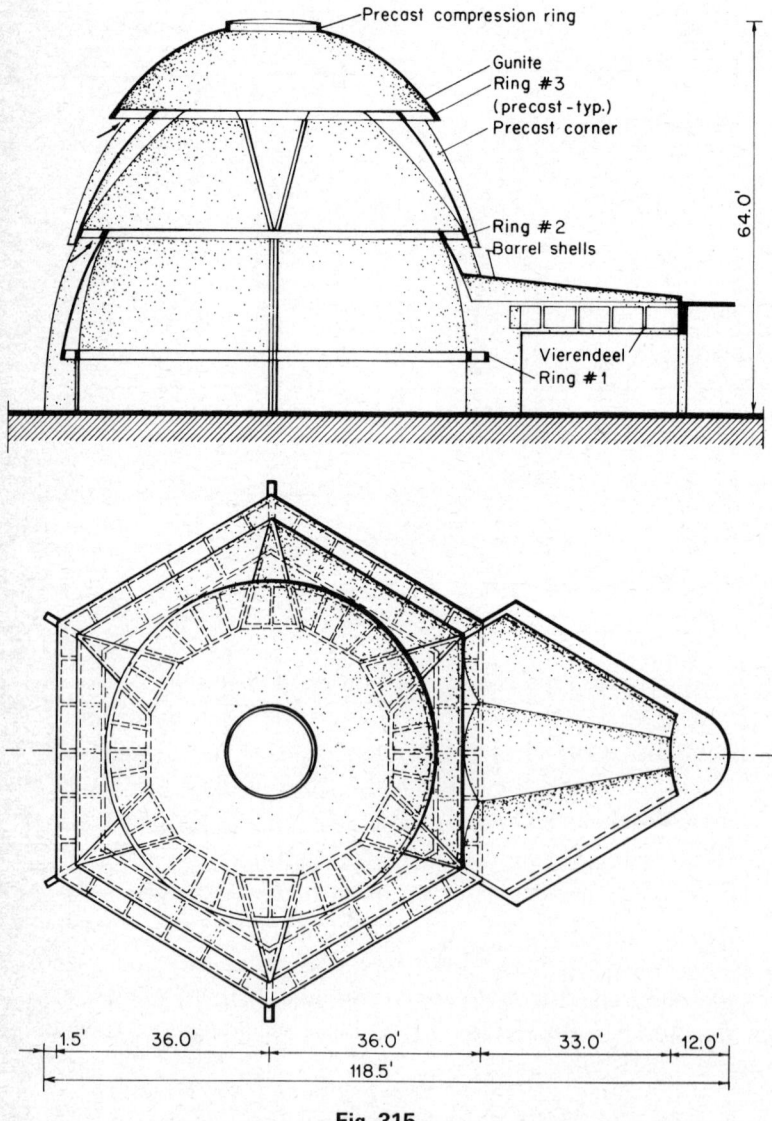

Fig. 315

partial dead load of stability rings, into corners by shell carrying action. The highest shell is a dome with a skylight at the crown. The entrance building is overroofed by barrel shells with linearly varying width and rise. At the horizontal junctions of the shells, Vierendeel-type stability rings are introduced for purposes of ventilation. This is a church intended for a tropical area in which air conditioning is not required and natural ventilation is vital and adequate.* The design foresees a combined construction method of precasting and guniting. The stability rings and corners are designed as precast units, the proper shape of the structure being established after the precast units are erected and connected by dowels and welding. The curved surfaces between the precast units are established by reinforcing grid welded to dowels projecting from the precast units (Fig. 316). To make the use of gunite possible, the grid is covered with two layers of closely spaced wire fabric.

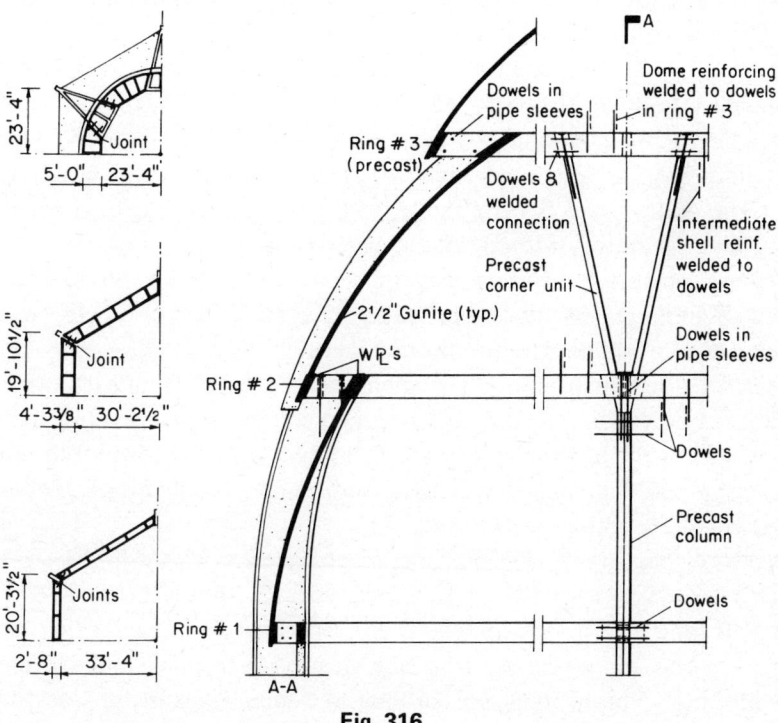

Fig. 316

*Architect: Henry Klumb; Del Carmen Church, Puerto Rico.

Gunite is applied in concentric layers, starting from the bottom stability ring of each individual shell. The barrel shells are designed to be poured in place.

The most difficult problem connected with these space systems is insulation. The rigid insulation at the outside surface requires a built-up roofing which considerably reduces the dramatic effect of this type of structure. The built-up roofing can be avoided by the use of lightweight concrete for shells and hypalon coating of the outside surfaces of concrete for moistureproofing and waterproofing. In tropical countries, where evaporation is very rapid, silicon treatment of surfaces is entirely adequate.

58. CURVILINEAR CYLINDRICAL SHELLS

Overroofing large areas can be most economical if the number of shells required is large, so that construction—poured-in-place or precast—can be accomplished with a limited number of molds.

A series of shells designed to be precast at the site for roofing a motel is illustrated in Fig. 317. The 38 shells involved can be precast in 1 month using a 24-hr pouring cycle and only two molds. In this particular case, concrete molds were used which could be pulled along the building, so that the crane could lift the shells from the mold and erect them in one handling operation, without changing position.

For lateral stability, the precast columns and shells were posttensioned to unity. Because of the short span, prestressing of shells was economically as well as structurally not justified. Furthermore, it would have delayed the construction. The shells are separated from each other by sponge rubber. The greased and wrapped column posttensioning bars are anchored at the bottom end of the columns and project from the top for receiving and fastening the shells. Columns are provided with anchor plates and bolted to the foundation.

A series of long-span curvilinear cylindrical shells with cycloidal cross section was used to overroof an area of about 47,200 sq ft (Fig. 318).* The 104-ft-long and 23-ft-wide shells are separated from each other by channel-type beams, which serve as edge beams for the shell and provide space for ducts for heating and cooling systems. The bottoms of the channels are closed with metal soffits, which are equipped with inlets

*Architect: Louis I. Kahn; Kimbell Art Museum, Texas.

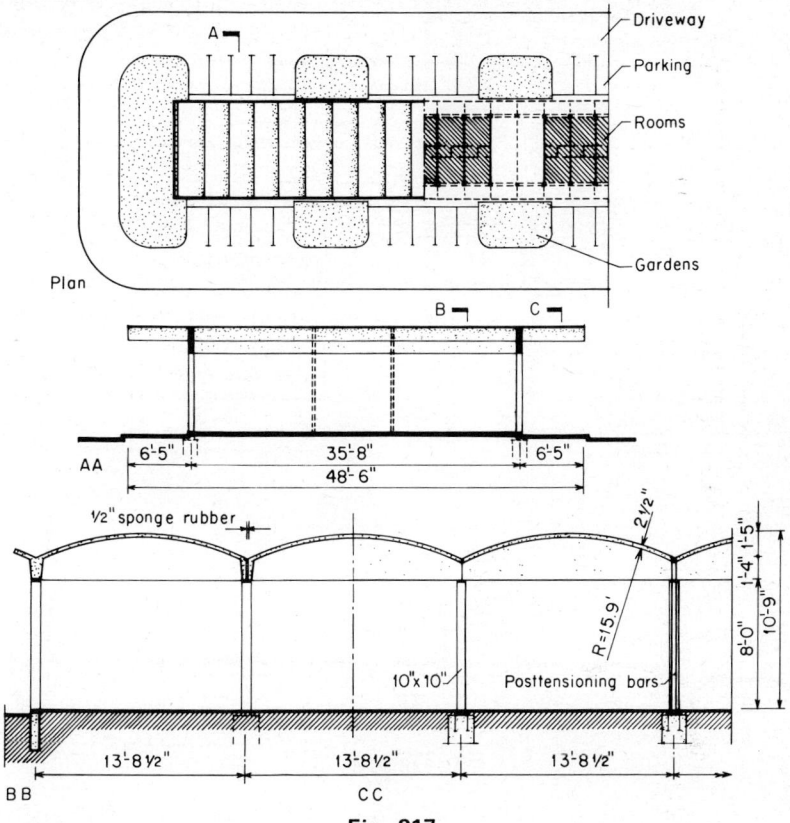

Fig. 317

and outlets. The ducts are supplied from the lower floor through the spaces between columns.

The shell system is designed to be poured in place. To allow proper posttensioning, the shells are separated from the columns by 1/4-in. neoprene joints, and for the same reason, the columns have a 1/8-in.-thick neoprene interlay at the floor slab. Since the building is in a tornado area, the neoprene joints are also required to increase the ductility of the system to reduce the dynamic forces upon the building. The end diaphragms are two-hinged arches dowel-connected with columns (Fig. 319).

Because of the skylight in the shell, the beams along the skylight are subjected to normal force N_x and to high torsional moments caused by the bending moments in the shell and the resistance of the struts spaced

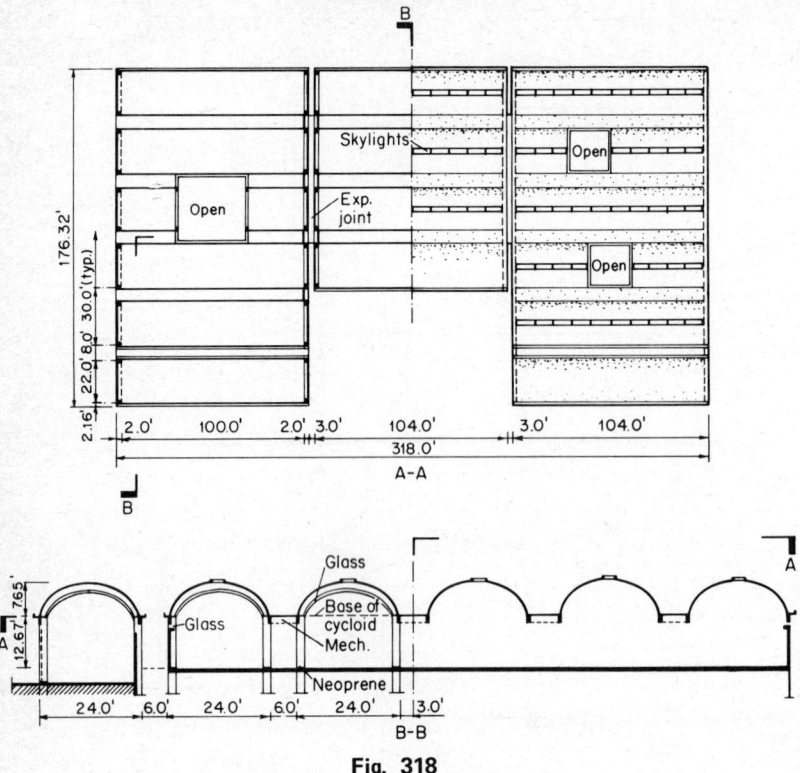

Fig. 318

10 ft center to center. The struts thus have to balance the bending moments of the shell and to resist the normal force N_α.

The concrete walls are separated from the end arches as well as from the edge beams by a 6- to 9-in. narrow gap to express clearly the shell action. The depth of the marginal channels also has been kept as small as possible to avoid the impression that these beams carry the shell, and not vice versa (which is actually the case, regardless of the size of the beams).

The posttensioning layout is shown in Fig. 320. The cables are greased and wrapped and follow in vertical projection a parabolic line. The shells are reinforced with two layers of welded wire fabric strengthened, where required, with conventional reinforcing. The cables are determined for full dead and half live load. To satisfy the required factor of safety, the edge beams are reinforced adequately to balance the half live load and to increase the ultimate moment of the cycloids.

As designed, the columns, walls, and marginal channels can be constructed separately prior to the cycloids. With such a work sequence,

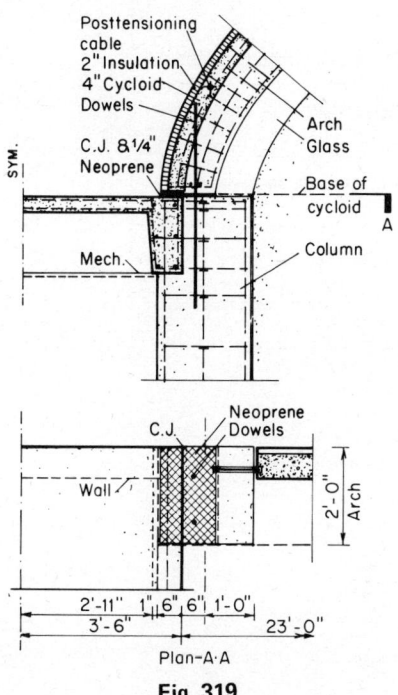

Fig. 319

a working platform for construction of cycloids is available. Due to this arrangement, posttensioning cables are placed entirely in the shell. The dowels projecting from channel beams are designed to carry the relatively heavy shear and bending moments, even when cold joint occurs.

The forms for cycloids are designed in sections to simplify stripping and handling for reuse. The skin of the forms consists of three layers of plywood so that they can be bent more easily to match the cycloid curvature. The skin is armor-coated to avoid damage to the plywood during stripping and handling. For outside forms, where the curvature is rather sharp, slip forms can be successfully used. Since narrow-spaced welded wire fabric is used, there is no difficulty in keeping the concrete from sliding down.

As the dimensions of the 16 cycloids are the same, only two sets of molds are actually required for pouring the shells. The two cycloids at the center will be formed at the same time, and after these shells are posttensioned, the forms can be stripped and reassembled for construction of the adjacent shells.

The shells are insulated with 2-in.-thick rigid insulation. The roofing over the insulation is lead-covered copper fastened to wooden strips arranged between the insulation mats and connected to the shell by screws and inserts. The drainage of the roof is arranged through the columns.

Since the cycloids are relatively light, they can be easily prefabricated. However, in this case, the marginal channel beams must be designed with

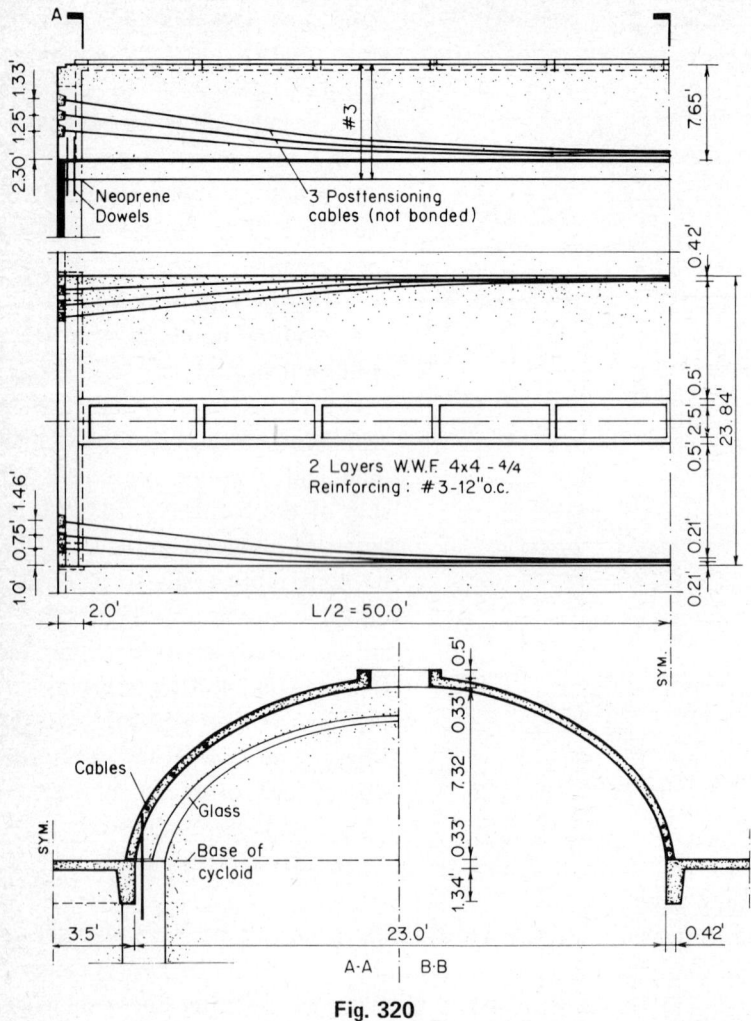

Fig. 320

a hinge at the midspan of the slab. Also, in order to reduce the weight, a hinge at the midspan of the struts is possible so that the cycloids can be manufactured and erected in halves. But prefabrication can be considered only if there is enough space at the site for storage and a high-capacity crane is available.

A similar structure can be obtained using folded plates or curvilinear shells instead of cycloids. The formwork is less expensive, but this saving would be more than balanced by increased reinforcing expenditures.

14

SUSPENSION SYSTEMS

59. CIRCULAR ONE- AND TWO-WAY SYSTEMS

For buildings such as halls, arenas, stadiums, and coliseums, relatively large column-free areas are required. Circular one- and two-way catenary roofs are the most appropriate and economical for such purposes. Basic types of long-span suspension structures are illustrated in Figs. 321 and 322.

The most common and earliest application of a catenary roof is the rotationally symmetric circular one (Fig. 321). The carrying action of a circular suspension structure is statically a closed system composed mainly of suspension and arch action. As both of these carrying actions involve mainly one stress type—tension or compression—they are the most efficient, which explains the economy of suspension systems. The radially directed cable tension is counteracted by a circular center tension ring and the outer compression ring supported by vertical columns. Under dead-load condition, the tension of each cable is equal and approximately constant. At the center ring the cables are closely

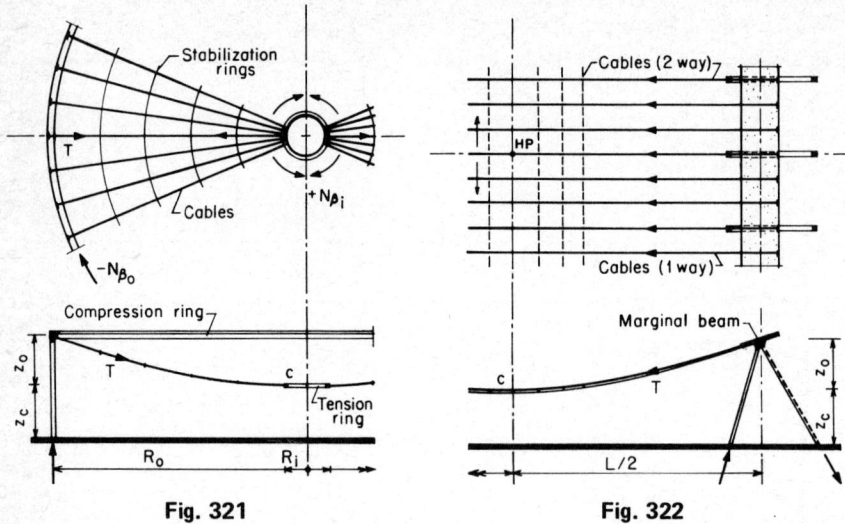

Fig. 321   Fig. 322

spaced, and therefore, the ring is practically free of moment and tensile stresses are almost uniformly distributed over the cross-sectional area of the ring. But regardless of this, the tensile stresses are enormously high and can be counteracted only by prestressing. For an unsymmetrically located center ring or nonuniform loading (construction and live loads), the cable forces are not equal, and this results in the development of moments and uneven stress distribution in the ring. At the outer ring, where the cables are more widely spaced, flexural moments in the cable direction are present under all loading conditions and can be rather high for an unsymmetrically located center ring. In the vertical direction, the ring girder must carry, besides its own weight, all the vertical components of the cable forces, equaling the total vertical load of the roof structure. Depending on the spacing of columns supporting the ring, the vertical and torsional moments can vary with wide limits. As the vertical loading is carried by beam action, the most uneconomical of the carrying actions, for long circular spans the vertical depth of the beam is rather large and posttensioning is commonly required to obtain economy and keep the stresses in the beam within acceptable limits.

For one- or two-way suspension systems, the cable layout is octagonal (Fig. 322). Statically, these are not closed systems, and the carrying action is accomplished by suspension and beam action. The cable tension is counteracted and balanced by a marginal beam-column system. In cases in which the columns can be spaced at the same line with cables, relatively small marginal beams are required, mostly for stabilization of the structural system as a whole. The cable forces are directly transmitted by fork-type or cantilevered columns to the substructure or foundation. For wider column spacing, the beam is subjected to relatively large vertical, horizontal, and torsional moments. To carry these moments requires a rather heavy prestressed column-beam structure. Depending on the function and architectural design of the building, the column-beam system at each end can be different. The basic cable-supporting systems for straight or slightly curved beams are illustrated in Figs. 323 and 324.

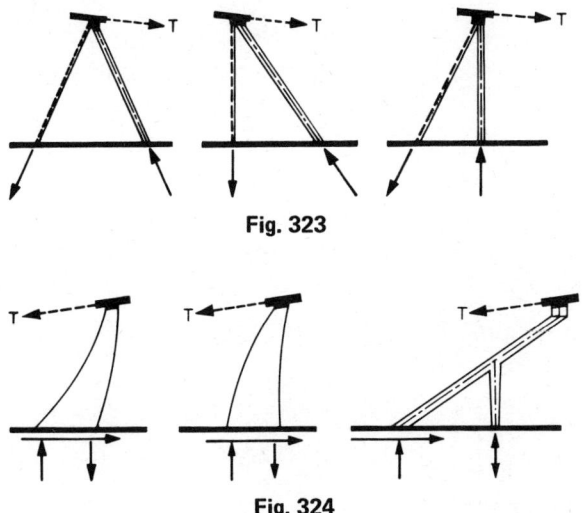

Fig. 323

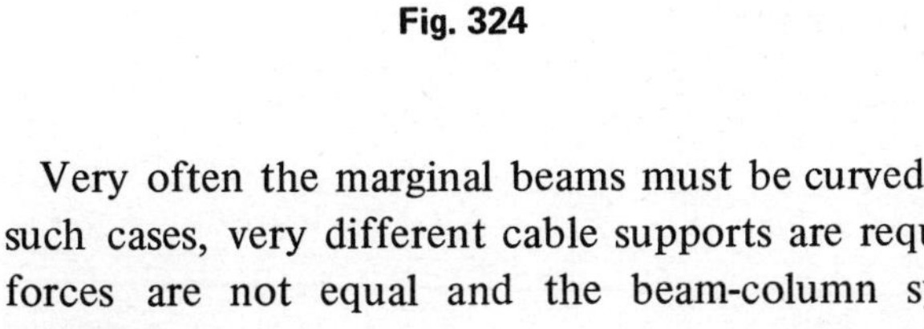

Fig. 324

Very often the marginal beams must be curved to match the plan. For such cases, very different cable supports are required because the cable forces are not equal and the beam-column system becomes rather complex.

The recommended prestressed marginal beam for wider column spacing is the box type because its weight is less than that of a solid beam and besides it has a high torsion-resistance capacity, which all adds up to increased economy. However, in order to reduce torsion, the cable forces must be applied as close as possible to the center of gravity of the section, which creates some difficulties with box sections. In special cases, cantilevering the marginal beams can be very helpful in overcoming these difficulties. Some of the possibilities for arranging the curved marginal beams, which act as arches in the cable direction and, therefore, have to be supported accordingly, will be discussed in the following.

The system illustrated in Fig. 325 has two arches which intersect at their ends. The arch thrust is partly counteracted by transversal cables, which do not participate in the carrying action and induce extra load on the suspension cable but are required as stabilizers. The rest of the thrust is balanced by the foundation. The space beneath the crossing of the arches provides a definite and appropriate entrance to the building.

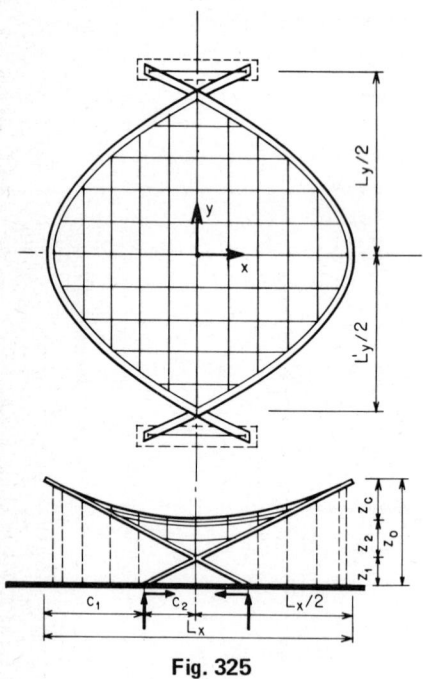

Fig. 325

For symmetric loading, the cable forces and the slanted arches form a closed, stable structural system. It should be realized that the stability of the system exists only within very narrow limits because it depends in large degree on the interaction of the highly elastic, flexible, two-way curved stressed cable grid and the comparatively far more rigid arches, which are independently not stable enough to secure equilibrium. Due to this fact, the equilibrium and stability of the system are lost for unsymmetric loading (live load, wind). To ensure stability for all conceivable loading conditions, the arches must be supported or tied down by cables along their sides. This requirement downgrades considerably the elegance and daring basic concept* of this system. The supports can be abandoned only if the arches are designed to be independently stable to carry unsymmetric loading. An example of such a solution is shown in Fig. 326. The fixed arches have a box cross section and are free-standing from abutment to abutment. The arches must be designed with sufficient rigidity and stability to resist possible unsymmetric loading. When the entrances are designed in locations other than at the ends, the arches can have a combined abutment.

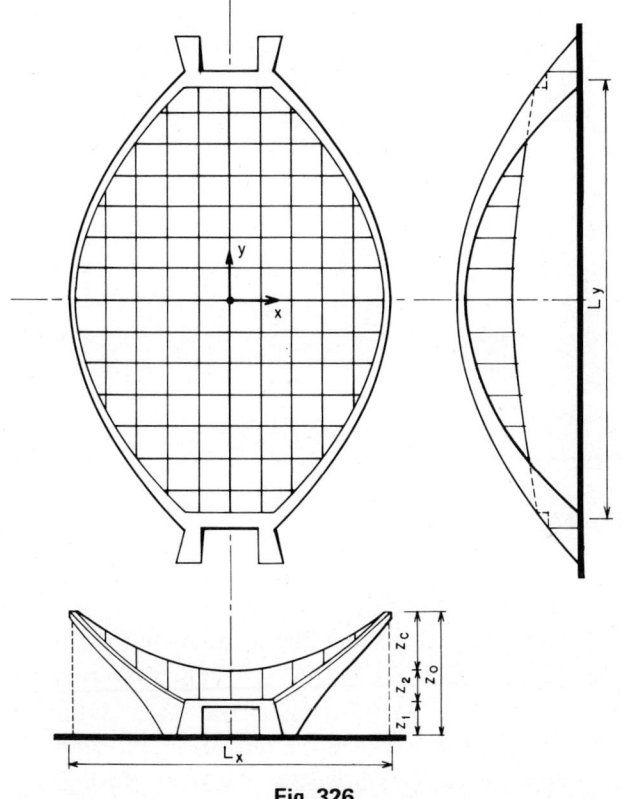

Fig. 326

*Severud, Elstad, and Krueger; Engineers: The Arena, Raleigh, North Carolina.

Instead of using fixed-end arches, marginal arches with hinges at the ends and column supports close to the abutments (Fig. 327) could also satisfy the stability requirements of the system.

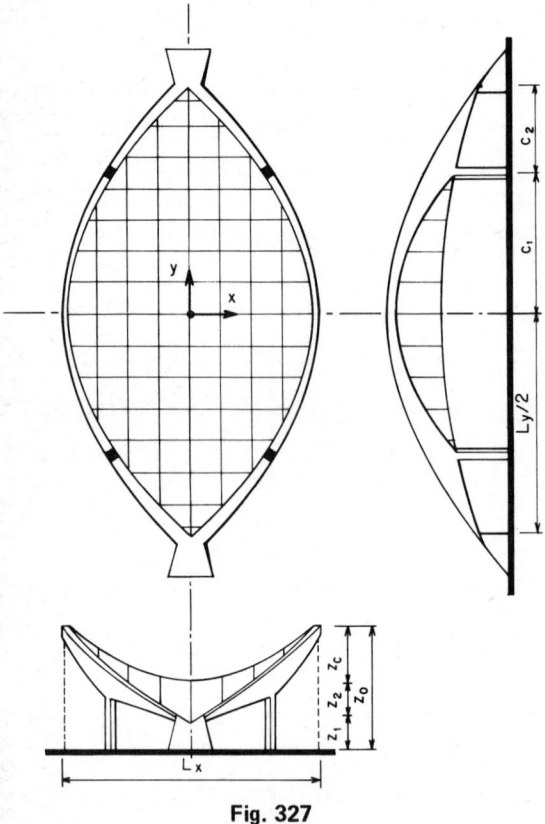

Fig. 327

For an elliptical-base plan, as illustrated in Fig. 328, the marginal curved beam is subjected to compression and high bending moments. To keep the elliptical ring beam within acceptable size limits, it must be buttressed. However, by the use of buttresses the arch action is downgraded, and if the relationship between the buttresses and arches is not well balanced, the system becomes statically unclear, confusing, and complex. The statical analyses are considerably simplified and the economy increased if the marginal ring can be composed of two elliptical

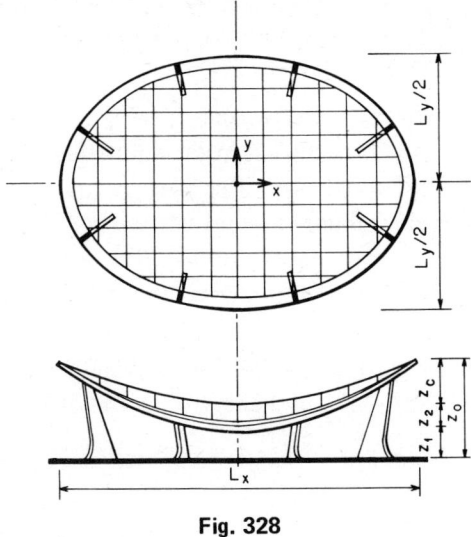

Fig. 328

arches, as shown in Fig. 329. The buttresses in this structural layout can be abandoned, but if used, they must be designed first to provide stability for unsymmetric loading and secondly to reduce the bending moments in arches.

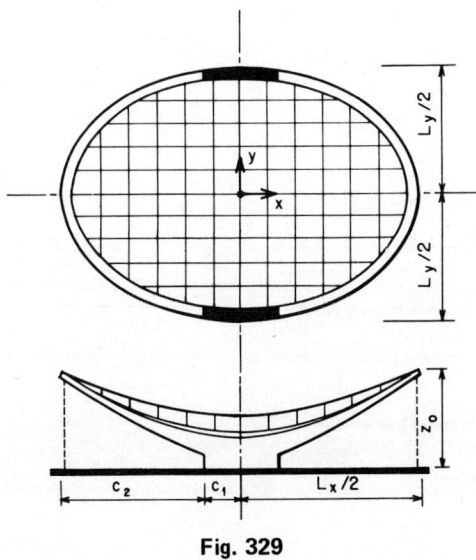

Fig. 329

60. RECTANGULAR LONG-SPAN ONE- AND TWO-WAY SYSTEMS

Long-span cable-suspended roofs over a rectangular base plan with columns in corners only are also possible. Such a roof is illustrated in Fig. 330. Prestressed cantilevers radiating from the four corners are designed to flatten the roof sag and to dampen wind vibrations. These rigid cantilevers carry the loads of center ring and dome. The box-type peripheral beams resist the tensile forces from suspension cables and dead load. The teflon-coated prestressed cables, running curved along the beams, are arranged to carry partially the cable tensional forces to the corners, thus reducing the shear and bending moments in the direction of suspension cables.

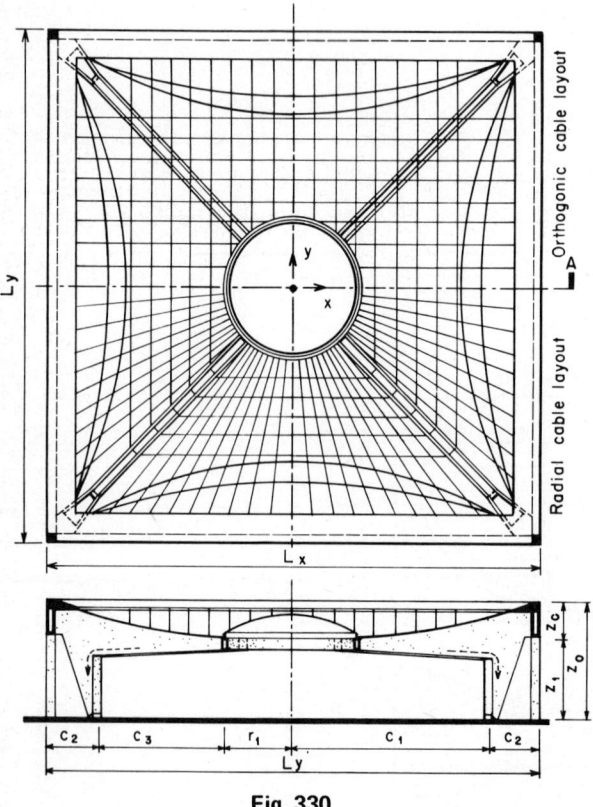

Fig. 330

The complex suspension-cable grid can be orthogonal or radial. The cables in a radial layout parallel to peripheral beams are supported by cantilevers. In an orthogonal layout, the cables intersecting the cantilevers can be designed so that in dead-load condition no vertical load from the cables is transmitted to cantilevers. Which of the cable layouts is preferred depends mainly on the type of decking: prefabrication or guniting. If guniting is used, the cables should be preloaded in the amount of dead load plus to balance the shrinkage and avoid cracking. The intersecting cables, especially the prestressed ones, must be clamped to avoid sliding and cracking.

The corner columns should be hinged so that displacement and rotation due to posttensioning, temperature change, and plastic deformations can take place without crack formation. The pendulum-type corner columns should be provided with nonbonded cables for control of center-ring elevation and to avoid possible tension in the columns. Tension and bending in the center ring are counteracted by prestressing applied by flat jacks. Drainage of the sagged roof is arranged through the cantilevers.

The system is adaptable also to triangular and trapezoidal base plans. Spans up to 500 ft are well within the comparative economical limits.

Suspension systems with rectangular base plan can be successfully used for roofs of large multispan utility buildings. The suspension system illustrated in Fig. 331 is supported by a continuous girder with a channel

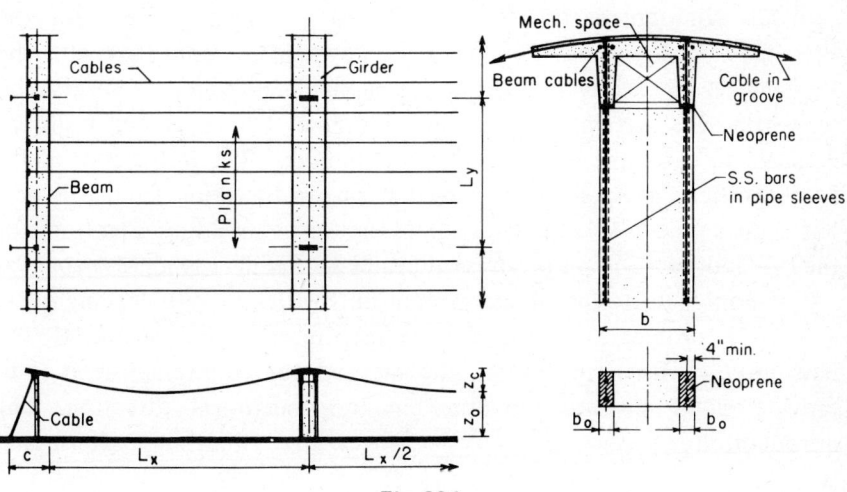

Fig. 331

cross section to accommodate services. In order to obtain acceptable curvature for the suspension cables, the slab of the channel is cantilevered and slightly curved. The columns are separated from the girders and foundation by neoprene bearings and are connected to an elastically controlled semirigid structural system with nonbonded post-tensioning bars.

A combined long-span rigid frame and suspension system* with rectangular base is illustrated in Fig. 332. The shape and structural system of the building were determined by the site and soil conditions. The bedrock at the location of the proposed building is overlaid with unstable strata of considerable depth. Therefore, the foundation for the structure had to be carried down to bedrock to avoid differential settlement. As such deep foundations are rather expensive, the number of open caissons had to be reduced to the minimum possible. Furthermore, the change of length due to temperature and shrinkage for a 460-ft-long building can be expected to be up to 3 in. In addition, the need to have the least number of obstacles at the open plaza level strongly favored the two-support solution for the structure.

As the building accommodates an amphitheater, a large multipurpose hall, and a roof garden, the long spans resulting from the two-support system have to carry considerable load. To reduce the span and counteract the positive midspan moments, pendulum-type walls supporting the huge frames were set back 40 ft from the ends. The suspension system carrying the theater is anchored into the verticals of the frame and acts simultaneously as an additional prestressing force for the Vierendeel box girders. The suspension system is connected with the Vierendeels by relatively light struts acting mainly as stabilizers and supporting the curtain walls for the wind load. The pendulum walls are designed to carry the entire load of the building. The inside ends of the frame verticals rest on a prestressed beam cantilevering from caissons, but these supports serve only as stabilizers for the building. The floor for the hall and the roof are prestressed beams with slabs top and bottom.

Commonly, fluttering of suspension roofs with concrete decking is not a serious problem, except for extraordinarily long spans. If the possibility of fluttering exists, the method for counteracting it is to provide slight upward curvature in the transversal direction. The introduction of reversed cables changes the vibration pattern and

---

*Architect: Louis I. Kahn; proposed to Venice Cultural Center.

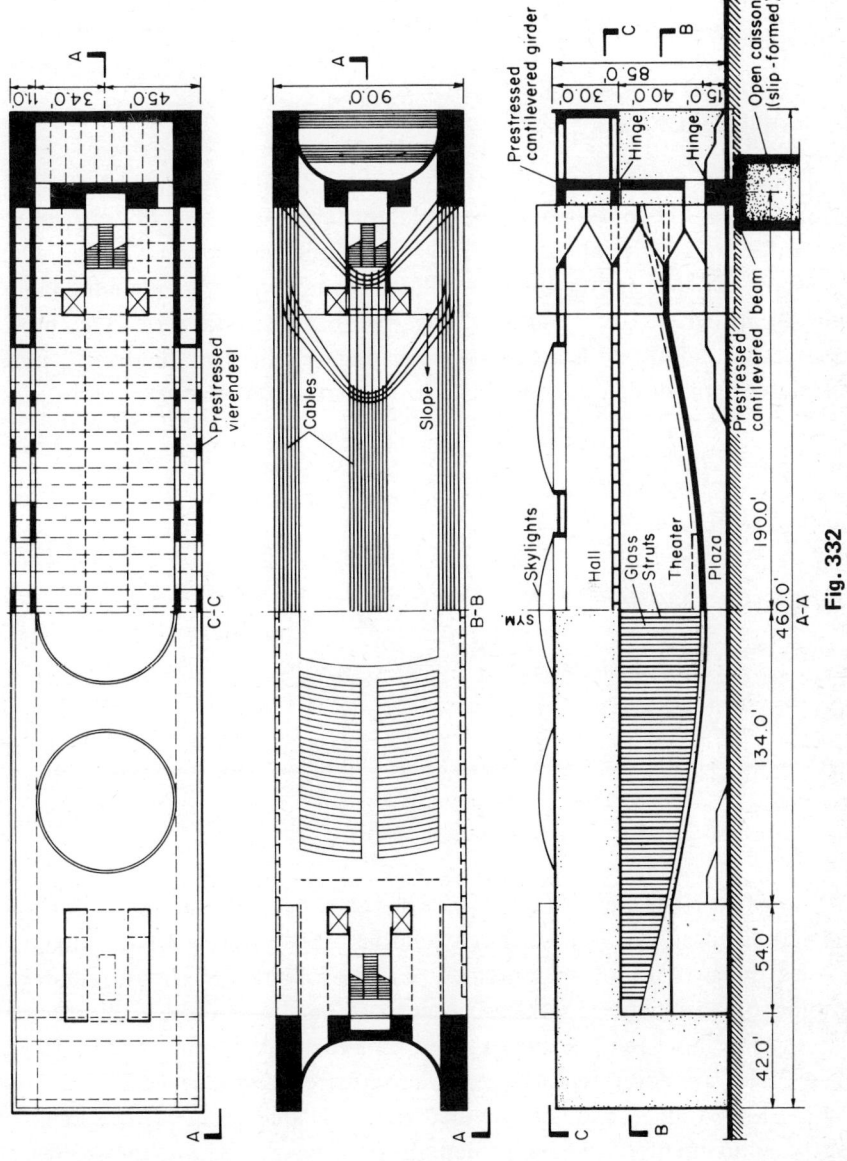

Fig. 332

increases the internal-damping capacity of the system. However, if the reversed cables are not symmetric, cable forces differ and moments increase in girders.

61. DRAINAGE, CABLE ANCHORAGES, AND DECKING

For one-way suspension systems, the use of the slightly upward reversed cables solves the problem of possible fluttering and also the otherwise difficult drainage problem. To solve the drainage problem for circular catenary roofs is rather difficult. Pumping rainwater is understandably not a proper solution, and it has not been satisfactory. The most acceptable solution is indicated in Fig. 333. For relatively long diameters, this design counteracts also fluttering of the roof.

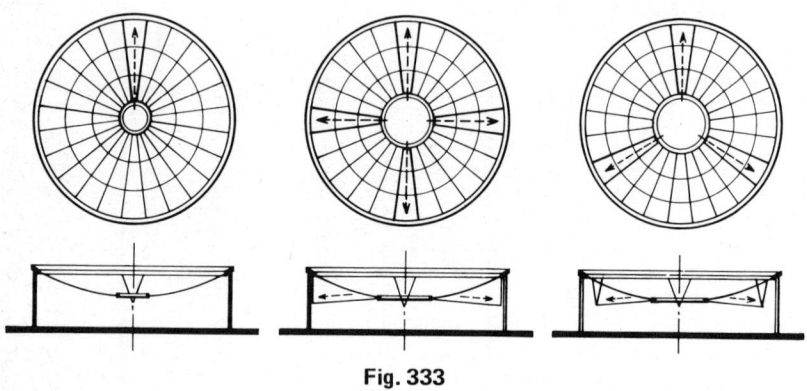

Fig. 333

The anchorage of cables in rings or marginal beams is similar to the end anchorage used for typical large tendons. Pipe sleeves for passing the cables through beams are provided at the roof side ends with neoprene tubing of sufficient length to allow soft bending of the cables when leaving the more rigid beams to avoid fatigue due to vibration (Fig. 334). The cables are composed of parallel seven-wire stress-relieved strands (up to 19 ½-in. strands) and are provided with threaded anchor fittings which allow adjustment of the cable length and ensure safe anchorage. Also, large-diameter (up to 3 in.) single cables (Roebling type) with special fittings and adjustment devices attached are available. The cable length is computed and stressed by dead loading and preloading. Greased and

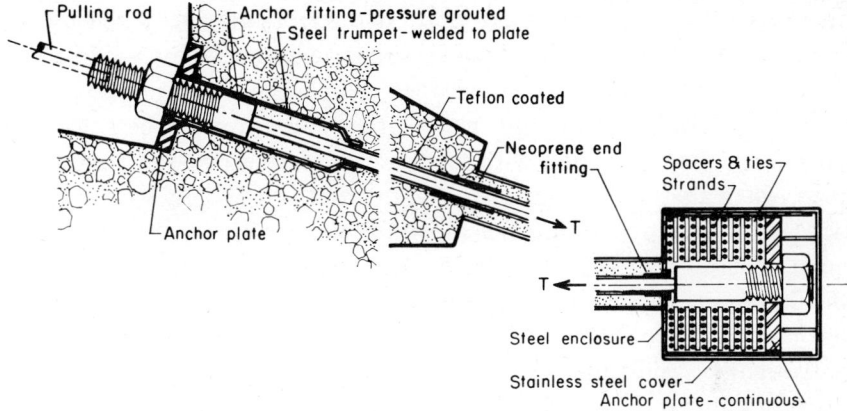

Fig. 334

wrapped cables stressed by jacking are seldom used, but it is possible, at least for adjustment of the deflections.

The center tension ring consists of teflon-covered strands encased by steel plates. The strands are wound in layers close to each other in the vertical direction and are separated in the horizontal direction by spacers, which also serve as ties, required to avoid lateral displacement (Fig. 334). The end anchor plate is designed to be continuous but is fabricated in sections to avoid overstressing during tensioning of the strands. This applies also for encasement plates. When the assembly is finished at ground level, it is erected on a temporary falsework at the designated position. Then the suspension cables are set, the falsework removed, the final adjustments made, and the center ring pressure-grouted in sections. The decking should be performed rotation-symmetrically, starting from the center ring. When the dead load is fully active, the trumpets are pressure-grouted and the steel encasing of the center ring, including anchor plates, is welded and the joints filled.

The posttensioning for marginal beams must be applied gradually to avoid overstressing. The use of greased and wrapped tendons is most convenient. As the dominant loading is dead load, deflection, usually associated with nonbonded posttensioning, is not a problem.

The economical bridging of the suspension cables is the most difficult problem, especially for circular and two-way systems. Because the cable curvature is not stable but is determined and controlled by loading, the shape of the bridging elements supported by the cables must match the cable pattern. The most acceptable method of bridging is the use of prefabricated form slabs or planks supported directly by cables. The cables running in the cavities between the planks are simply poured in (Fig. 335). Counteracting the change of cable curvature is best done by preloading the cables first with prefabricated, closely placed panels and then, when they are in place, loading the roof additionally with sand- or water-filled bags to balance the superimposed dead load, shrinkage, and approximately half live load. The loading should be symmetric to reduce flexural moments in ring and beams. The forms closing the cavities from the bottom are of thin steel plates covered with rubber and pulled by bolts tightly against the bottom of the panels. If the decking is poured in place or gunited, the cables must be preloaded for the entire dead load, shrinkage, and half live load. The decking is most economically poured or gunited on light concrete form planks. The spaces between upward ribs can be filled with plastic foam for insulation (Fig. 336); thus high-quality composite sandwich-type decking is obtained.

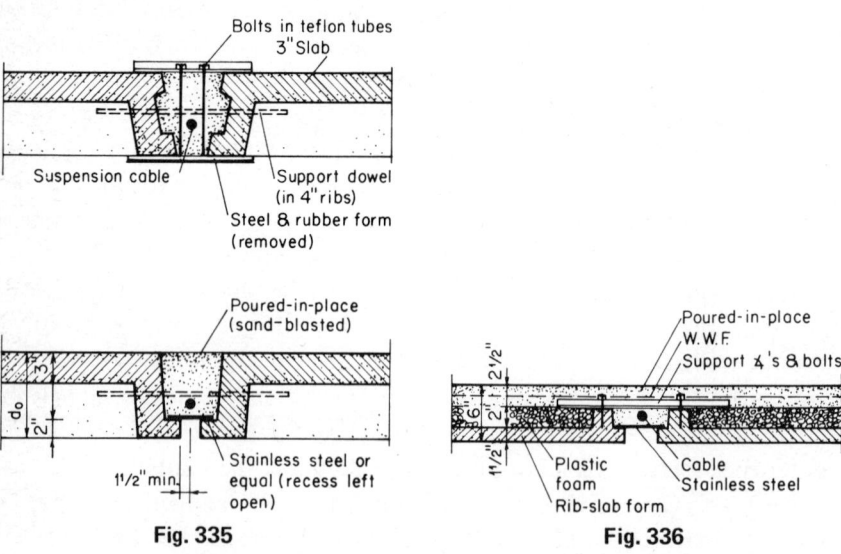

Fig. 335   Fig. 336

The bags are removed gradually as equal weight of load is added. When the roof is finished and the concrete hardened, the remaining bags are removed and the overstress in cables is released and transferred to concrete by bond. In this way, a crack-free roof is achieved.

If smooth-plane undersurface is required, guniting or plastering is usually necessary. For undersurfaces with pronounced cable patterns (Figs. 335 and 336), prefinished rib or form planks cast in molds to match the curvature of cables is commonly more attractive and economical than a smooth-plane undersurface. Very often, for sound damping the entire undersurface must be sprayed with special sound-absorbing material.

As the concrete decking is rigid for live load, the decking of a suspension system acts as a shell unless the curvatures are funicular or parabolic. Taking this into consideration, the transfer of live load to the marginal member can be quite different and should be checked in this aspect to avoid overstressing and cracking.

If thermal insulation is required, plastic foam or closed-surface rigid insulation must be used. For roofing, neoprene sheets bonded to concrete or insulation has proved most satisfactory.

15

PREFABRICATED HOUSING

Maximum results at minimum cost are obtained for prefabricated houses when the design and construction are based on the following principles:

1. Potentiality, safety, and permanence with least maintenance.
2. Easiest manufacturing and erection without complicated, unusual equipment and expensive relocation of manufacturing facilities.
3. Simple installation of the mechanicals and finishing of units.
4. Maximum possible completion of the units at ground level before erection.
5. The arrangement and connections of the components to structural unity must be flexible and accomplished with the fewest possible secondary structural systems.
6. The finished complex should be inviting, safe, homely in atmosphere, and architecturally flexible and acceptable.

There are basically two methods available for producing prefabricated houses: panel system and modular-box system.

62. PANEL SYSTEMS

Prefabricated panels are the basic units in erecting individual houses or house complexes one or many stories high. The panels are manufactured at off-site plants, transported to the site, and assembled to structural unity in place. All panels are load-bearing, connected with each other, and interlocked with room-size floor panels by grouted steel traps, dowels, loops, steel runs, or welding plates. The exterior panels are commonly prefinished sandwich types, with plastic-foam insulation. There is a variety of patented connection methods for panels and floor slabs. Panel sizes, types, and connections are practically the only difference in the load-bearing panel systems presently on the market. Commonly, the bathrooms are completely serviced, one-piece concrete units. The service wall of the units is at the kitchen side and accommodates also the services for the kitchen. The one-piece balconies can be fitted with exterior wall panels after they are in place.

The panels are fabricated in flat beds in the horizontal position. The manufacturing is simple and inexpensive. Any conceivable size and thickness of panel is easily obtainable, being controlled only by handling. The panels are commonly reinforced with two-layer welded wire fabric strengthened where required with bars. Windows, door frames, and services such as electric wiring in plastic conduits and piping, arc placed in the mold and poured in. Commonly the panels are fabricated in 24-hr cycles. The manufacturing operation usually takes place indoors, but in locations where climatic conditions are favorable, outdoor manufacturing can also be successfully used. The panels are cured by steam or moist air, depending on the type of plant and climatic conditions. After the panels are finished and dressed, they are transported by special trailers to the site. The panels are rather light, so that generally no heavy handling and erection equipment is required.

The first phase—production—of the panel system is simple, flexible, and economical, and the production quantity is controlled only by the erection and finishing capacity of the houses. The second phase—erection—is rather difficult, requiring time and skilled man-hours, especially if safety and quality of the work are regarded, because relatively large numbers of panels have to be placed, aligned and fastened. Besides this, a large linear footage of joints has to be sealed watertight, and any human error in setting the panels, such as misalignment (even though guiding dowels are used), delays the erection.

Also, improper dry-packing and fastening can cause safety risks, especially in multistory designs.

By its very nature, the vertically oriented panel system requires perfect alignment and dry-packing of the bearing walls to avoid stress concentration and undue flexural and torsional moments. The rather rigid vertical structural pattern of the system considerably reduces and limits structural as well as architectural flexibility, resulting in the monotony and regimentation associated with this type of prefabricated house (Fig. 337). The majority of attempts to improve the architecture have led only to decorations and reduced economy. The panel system can be made architecturally more acceptable but only by the use of rather expensive secondary structural systems.

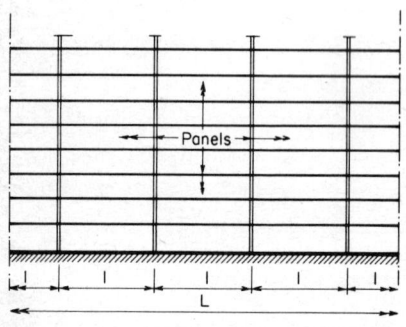 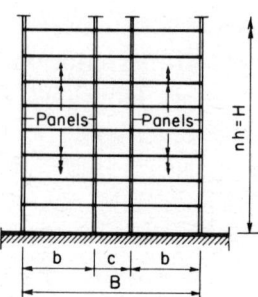

Fig. 337

The finishing work to be accomplished on the site is expensive and does not differ much from conventional construction systems, regardless of how perfectly and completely the panels are finished in the plant. As a result, the economical gains of the first phase of the operation are largely lost in the second phase, and in the best cases, the net economical gain, in comparison with conventional systems, is no higher than 10 percent.

As a conclusion, none of the contemporary requirements, as expressed in the six basic principles, is satisfied by the wall-bearing panel system to an acceptable degree. But in spite of this, these panel systems are widely used in Europe and also in many other countries. This is explained mainly by the fact that heavy handling, transport, and erection equipment is not available or is beyond the economical reach of these countries and also by the fact that highly skilled labor is plentifully available.

63. MODULAR-BOX SYSTEMS

A complete box is the basic unit in manufacturing individual houses or house complexes one or many stories high. The boxes can be fabricated in off-site plants or at the site, depending on transportation possibilities and box types used.

There are four basic types and methods of fabricating the boxes. These types in their final position are illustrated in Fig. 338.

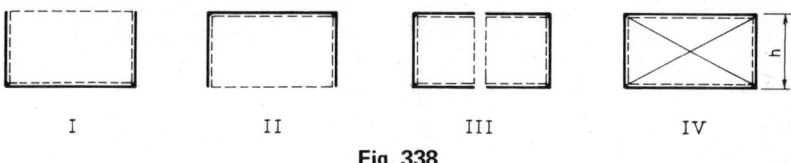

Fig. 338

In box type I, the bottom slab and sides are cast together and the top slab is manufactured in a separate operation. In box type II, the bottom slab is manufactured in the first operation and the sides and top in the second operation. Box type III is manufactured in two pieces, sides up, in the horizontal position and is turned to the vertical position and joined in midwidth. Box type IV has one or both ends open and, therefore, can be cast in one operation. The ends are used for doors, windows, and louvers or are closed by curtain walls. These four box types are manufactured practically in the same manner and using similar molds.

The inside molds are hinge-connected folding types, as illustrated in Figs. 339 and 340. For stripping the inside molds, the sides are shortened by pulling neoprene-hinged panels toward the center of the mold (Fig. 339) or by providing the mold with easily compressible interlays at the center in both directions so that the mold can be jacked to decreased size for stripping (Fig. 340).

By tapering the bottom of the box slightly, box type III can be lifted out without any change of mold dimensions. However, to avoid clamping it is necessary to loosen the box halves from the mold uniformly and simultaneously by synchronized hydraulic jacks (Fig. 341). For this operation, the mold is fastened to the bed floor and jacks, at least two per side, are built-in in the floor.

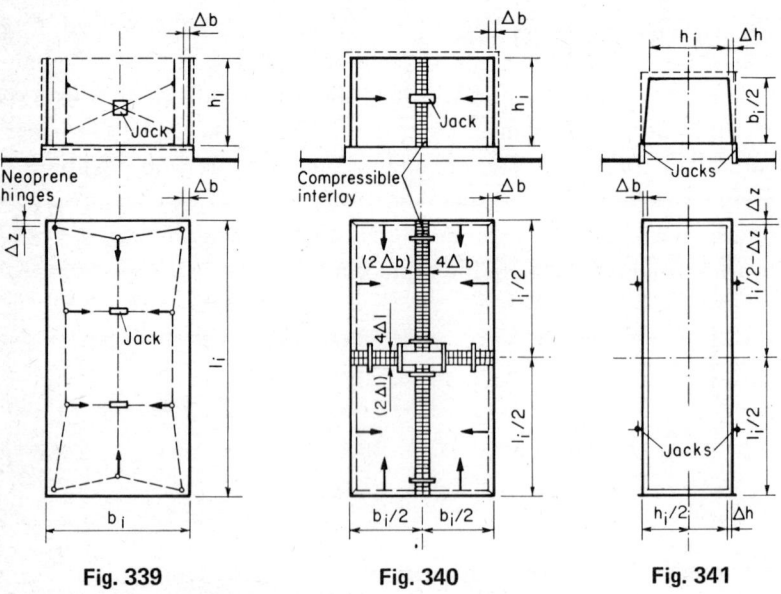

Fig. 339   Fig. 340   Fig. 341

The outside forms are mostly hydraulically tilt-up hinged types, as illustrated in Fig. 342. The operation starts for all boxes with the outside forms in the down position. For box types I and IV, the prefabricated bottom reinforcing is placed and the one-piece inside form is set. Reinforcing is prefabricated in wall-size mats and placed separately for each wall. The mats overlap at the corners and are connected by welding. The use of welded wire fabric, strengthened where required with bars, results in considerable time saving. For box types II and III, which have

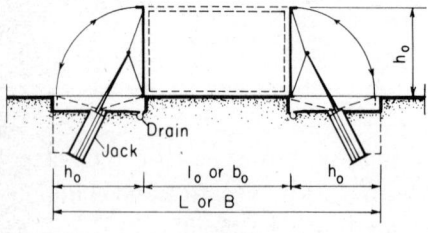

Fig. 342

an open bottom, the inside form is stationary, sliding on base; therefore, wall and slab reinforcing can be fabricated and placed as one piece. For proper concrete coverage, heavy-duty plastic spacers are used. The door and window frames are provided with sponge-rubber gaskets and attached to the inside form; welding plates, tubing, piping, etc., are fastened to the reinforcing. After the accessories are firmly in place, the side forms are tilted up hydraulically and closed. The joints are made tight by rubber gaskets attached to the edges of the forms. The bottom slab of box types I and IV is poured first, up to the widened edge of the inside form, which is also used as a guide for screeding. After the concrete is sufficiently set, the walls are poured from the top. The extended bottom of the inside form must be wide enough to resist the spreading of concrete from the walls. For consolidation, flat vibrators are used. After the bottom slab is hardened, setting of the inside form is also possible, but cold joints between the castings are a strong possibility. Concrete commonly is pumped to the mold.

Before steaming, the concrete is air-cured for about 4 hr. In that time the inside form of box type I can be stripped and removed for reuse. For other types this is not possible, mainly because the inside form carries the top slab. In about 1 hr of steam (from the inside), the outside forms are first loosened slightly and then opened for about 6 in. at the top, the top and corners are enclosed, and the concrete is exposed to live steam. The temperature of steam is raised approximately 20°F for the first hour and then gradually to about 40°F/hr up to 180°F. At this temperature, the 4,000-psi concrete strength required for safe handling is obtained in about 6 hr. Thus, per one set of mold a box can be fabricated in a 24-hr cycle.

The top slab is connected by slag screws or threaded-end dowels projecting from walls and nuts. The bottom slab of box type II and the halves of box type III are connected by special screw devices or by welding plates. To avoid dry-packing and to ensure watertightness, all joints are sealed with neoprene and sponge-rubber gaskets. The structural details are illustrated in Fig. 343.

All these box types can be practically completed and finished, including services, partitions, insulation paneling, windows, doors, floors, etc., at ground level, so that only piping and electrical circuit connections have to be performed in place.

Evaluating these box types from a structural and manufacturing point of view, box type III is the most acceptable for the following reasons:

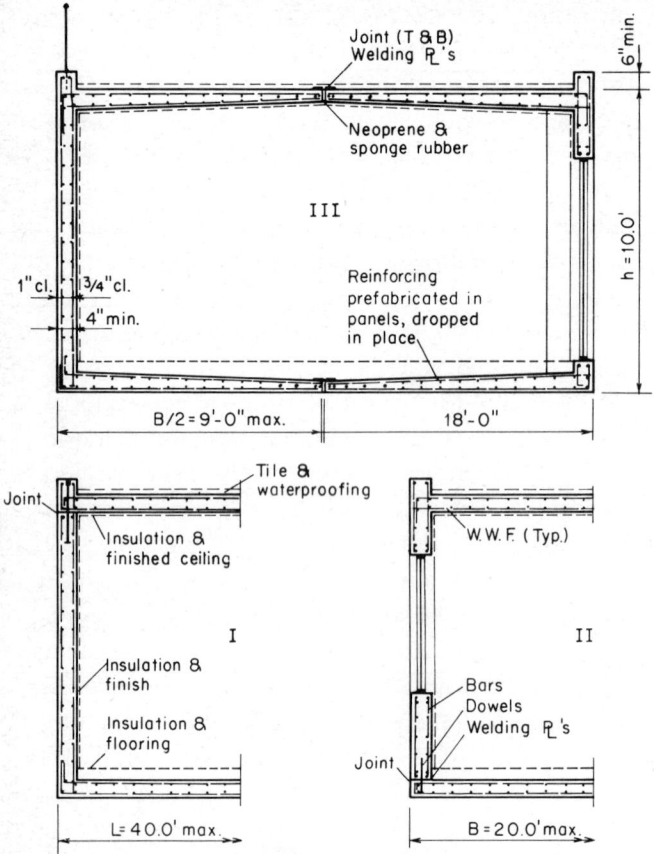

Fig. 343

1. The individual sidewall thicknesses can be variable, and buttresses can be provided where required without difficulties.
2. Manufacturing is simple, and fewer man-hours are required because of the stationary inside form. Window and door framing and other poured-in items are installed in horizontal position during manufacturing. Casting is simple.
3. The vertical joint at midbox is least subjected to stresses, regardless of the supporting and loading conditions.
4. Transport (off-site plant) and handling are less expensive and easier.

Box type II (Shelley system) is flexible and has manufacturing and economical features similar to those of box type III, but the location of the joint practically eliminates bottom-slab participation in the overall carrying action. This is particularly true for cantilevered boxes.

Box type I is difficult to manufacture and requires more man-hours, but the location of the joint is structurally more favorable and the use of inside form for at least two castings per cycle makes this box type competitive.

Box type IV is structurally the least acceptable, mainly because it lacks torsional rigidity, which reduces its potentiality considerably.

There are some rules which apply for all four box types:

1. Concrete partitions should not be used.
2. Reinforcing for slabs and walls, except end walls, should be in two layers. Outside concrete coverage should not be less than 1 in. Minimum wall thickness is 4 in. for two-layer and 3 in. for one-layer reinforcing.
3. No framing should be used; plain heavier walls are simpler to form, resulting in considerable economy.
4. Attachments of any kind should be on the inside form. This applies also for form vibrators. This requirement is necessary to secure the quality of exposed outside surfaces.
5. For safety in multistory complexes, the boxes must be tied vertically by posttensioning.
6. To avoid stress concentrations, balance stiffnesses of walls, and allow changes of length due to shrinkage, temperature change, plastic flow, and support settlements, the boxes at support and contact points must be separated from each other by thin (1/8 to 1/4 in.) neoprene or teflon interlays. Dry-packing should be used only if structurally unavoidable. All joints should be sealed by sponge rubber.
7. Form finish, cleaned and siliconed, should be preferred to any other type of surface treatment because of high durability and smoothness of exposed surfaces.

Taking these rules into account, the first phase—manufacturing—of the box systems is far more expensive than that of the panel systems. However, the reverse is true for the second phase—erection and finishing. As a conclusion, for mass-production the box system is more suitable structurally, productively, and economically than the panel system.

64. HOUSING COMPLEXES

Because of their rigid nature, the arrangement of boxes into a housing complex is an open vista for architectural imagination. Structurally the boxes can be staggered, stacked, and cantilevered horizontally and vertically into a self-sufficient stable carrying system. For example, an inverted-pyramid type of design, as illustrated in Fig. 344, can be economically constructed without difficulties.

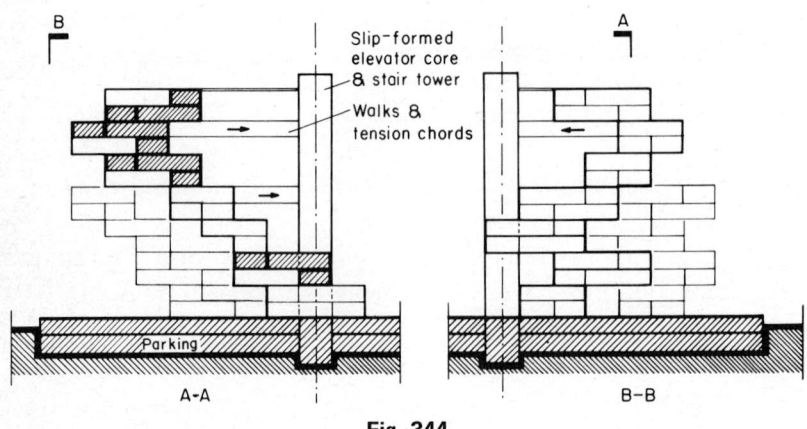

Fig. 344

The stair-elevator towers are slip-formed first, and a beam crane is equipped with counterweight or derrick mounted at the top for erection of boxes, light prefabricated walks, etc. The stair-elevator tower accommodates also vertical services, is connected directly with box clusters or overwalks, and serves as the stabilizer of the complex. The semibox-type prestressed walks act as horizontal ties counteracting the overturning and are used to balance the loading of box walls. They also carry the pipes, ducts, wiring, etc., from clusters to the tower. The economical height limit is approximately 20 stories. The box system with boxes separated from each other by neoprene interlays and tied to unity by partially nonbonded vertical posttensioning results in high ductility and reduced dynamic forces and noise propagation. Also, the multimass system has a relatively high shock and vibration dumping factor. Due to these characteristics, the box system is suitable to use for locations where buildings are subjected to high dynamic forces (earthquake, tornados, hurricanes, etc.), but the number of stories should be kept within safe limits, as determined by design characteristics.

For high-dynamic-force areas, panel systems are not appropriate because the rather weak connections of joints and the inherent rigidity of the panels (lack of ductility) do not meet the safety requirements associated with the increased dynamic forces acting upon the structure.

As can be concluded from the above, the box system satisfies almost all the six principles to an acceptable degree, except the requirement for availability of heavy, expensive handling equipment. Also, the box system has far larger areas exposed to climatic conditions than does the panel system. However, this applies only for unorthodox cluster-complex designs. The disadvantages of the box system in comparison with its advantages are practically negligible.

The potentiality of the box system in its architectural aspects and structural challenges is best demonstrated by considering some of the plan layouts proposed by various architects which have not been executed because of structural and economic problems. Extensive advancements in recent years have brought material technology and structural engineering to a level where the architect's dreams of such extraordinary projects can become reality. The base-plan layout illustrated in Fig. 345, with upper-level clusters staggered, cantilevered, or

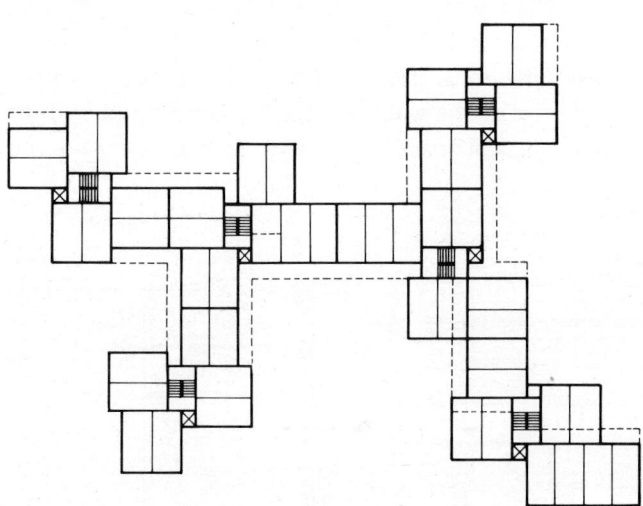

Fig. 345

slanted, can be executed by the use of slip-form and box systems rapidly and within economical limits.

A vertically oriented layout (Fig. 346) favors the Shelley high-rise system. The boxes are stacked and staggered in checkerboard pattern up to 20 stories high. By staggering the boxes, extra space is obtained, with floors, ceilings, and sidewalls provided by the surrounding boxes. Such an arrangement of boxes in a complex results in considerable economy.

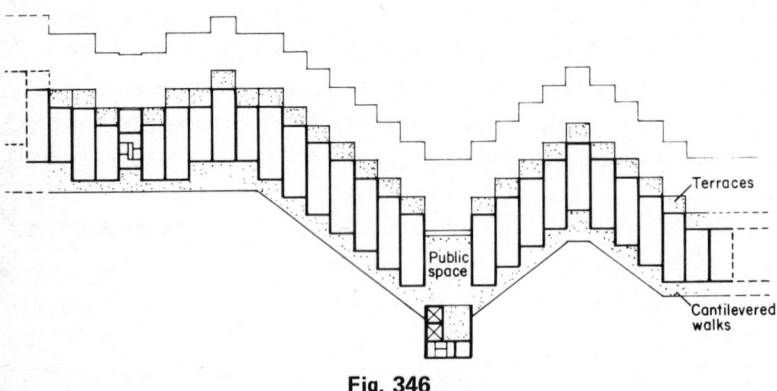

Fig. 346

The buttresses of overlapped boxes match vertically and carry the loads to the foundation. The contact areas of buttresses are separated by 1/4-in.-thick bearing neoprene, and the outside ones are tied to unity by partly bonded posttensioning rods. This posttensioning is also required to balance different rigidities of the walls and the tension caused by high wind or earthquake. The vertical space between walls is continuous and is used for utilities (Fig. 347).

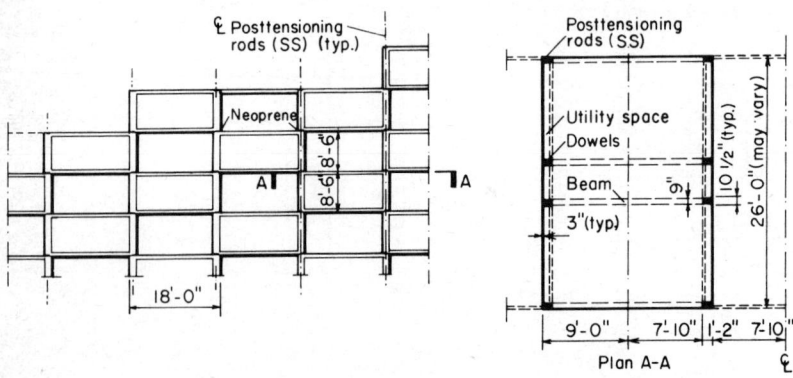

Fig. 347

The boxes can be of different lengths, stacked at angles and cantilevered to provide balconies for upper apartments. The stair-elevator towers are precast or slip-formed, depending on architectural design. To reduce the number of towers, they can be connected by light precast prestressed covered walks. The areas around the towers can be used for public places. By arranging the boxes at angles, the balconies have more privacy and protection. However, for this case the location of buttresses in the boxes is determined so that they match vertically. The heights of the house rows can be varied, and the roofs can be used for gardens.

A large box-system complex, the first of its kind to be constructed, is Habitat '67;* the base plan is given in Fig. 348. It involves stacking 354 precast boxes, each 38.5 ft long, 17.5 ft wide, and 10.0 ft high, weighing 70 to 80 tons, into a 12-stories-high composite structure of 158 housing units. Since the basic principles of design and structural details follow similar rules for all box systems, the project Habitat '67 will be discussed and illustrated in more detail.

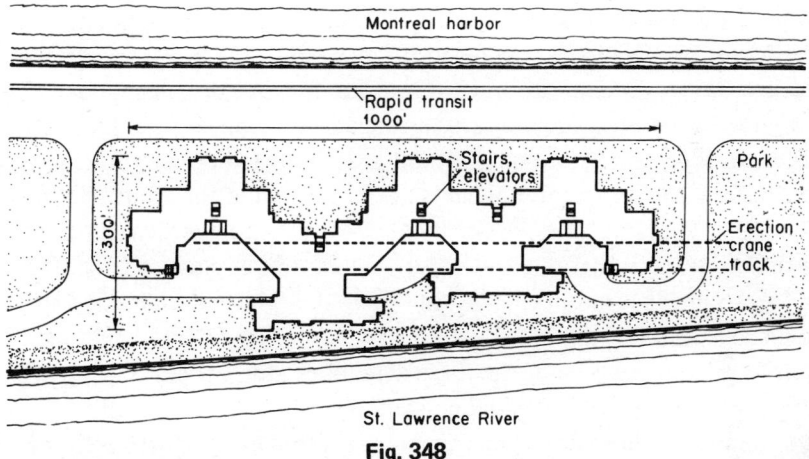

Fig. 348

The basic circulation system, completely separated from the service road and garages one level below plaza, connects all parts of the project through walks, bridges, and public plazas, with vertical circulation through three elevator and seven stair towers.

*Associated architects: Moshe Safdie & David, Barott, Boulva; EXPO, Montreal.

The overall structural system consists of three basic structural components: box clusters, street girders, and supporting elements (columns, elevator towers, and stair towers); the system is illustrated in Figs. 349 and 350. The components are joined into the carrying system by controlled elastic connections (neoprene interlays and posttensioning). Thus all supports are yielding except the supports at plaza level. Such connections are required to allow relative displacements and rotations of the components due to temperature change, shrinkage, posttensioning, and elastic-plastic deformations and to decrease noise propagation and increase ductility of the system (seismic area). The box clusters as well as street girders are inclined about 60° in respect to each other.

Each cluster consists of eight boxes—two boxes per level—arranged in an L-shaped pattern, with the halves of lower boxes cantilevered in order to provide each apartment with a garden on the roof of the boxes beneath. The cantilevered ends of the two boxes at the same level are connected flexibly by dowels and neoprene interlays. The center of gravity of the boxes themselves is slightly inside the four support points at the lower box. This was arranged by staggering the boxes of adjacent levels by 3.5 ft, mainly to provide space for utilities. The boxes at street-girder level (5 and 9) are arranged to connect the individual clusters vertically and horizontally and are supported directly at clusters beneath. The top and center clusters are supported at the front by street girders (A) and at the back by boxes at street-girder level (B) counteracting the uplift due to the 60° inclination of clusters (Fig. 351). The lower cluster is supported by the poured-in-place plaza level. As the clusters are simply supported (statically determined), the yielding of supports at the long-span street girders does not influence the load distribution within the individual clusters.

The standard dimensions, arrangement of boxes in cluster, connections, reinforcing, and posttensioning are illustrated in Fig. 352. As can be seen, all boxes have a frame-slab structure. The total thickness of framing is 12 in., including the 5-in. slab. The depth of the frames is 16 in. top and 18 in. bottom. The intermediate columns are 3.5 ft wide, and the rest are 12 in. At lower levels, the intermediate columns are strengthened by buttresses. All openings are within slab-area width, varying from 2.25 to 7.0 ft, height 7.0 ft. The top and bottom slabs are 5 to 6 in. thick.

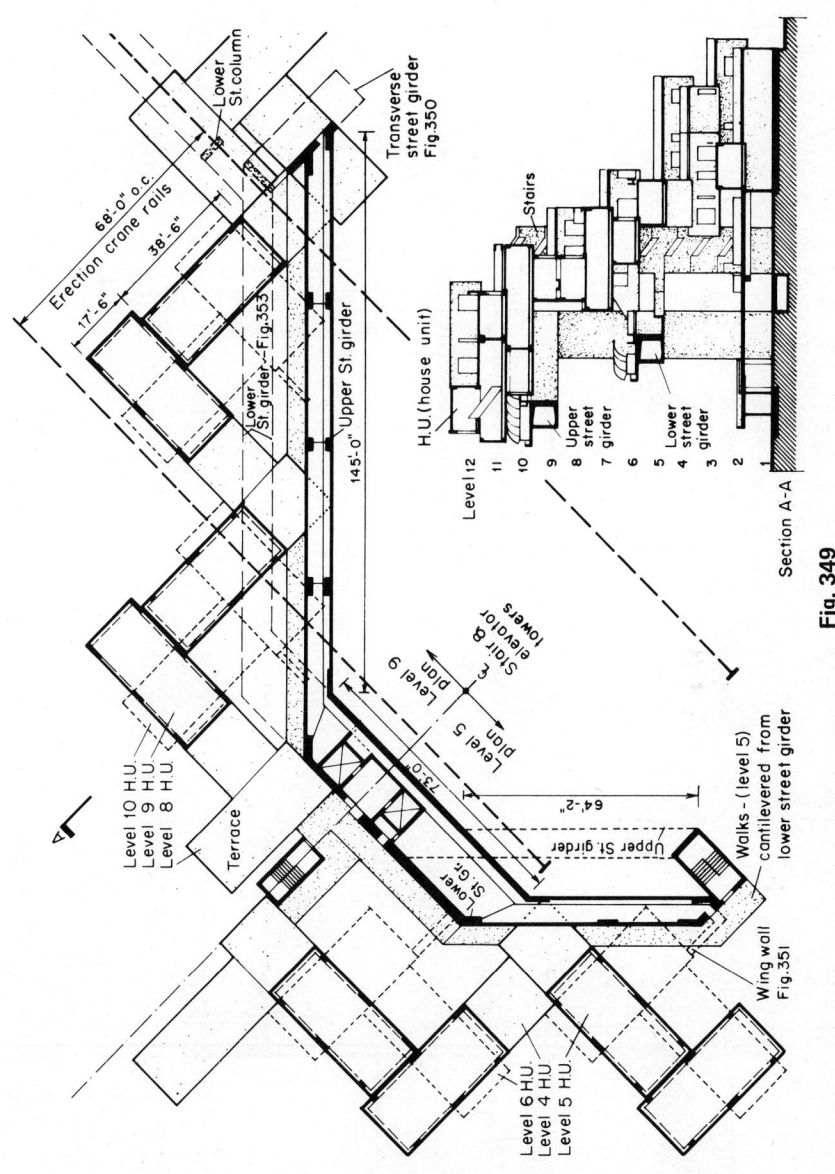

Fig. 349

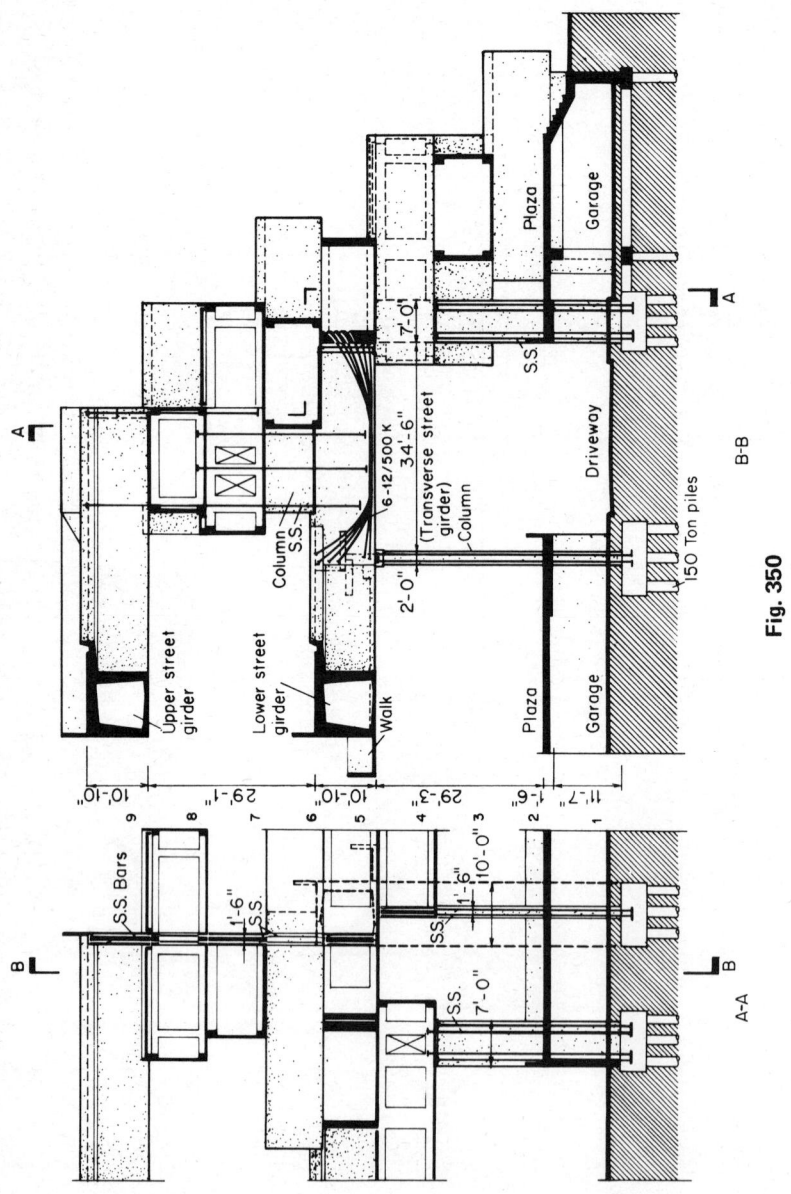

Fig. 350

Prefabricated Housing 539

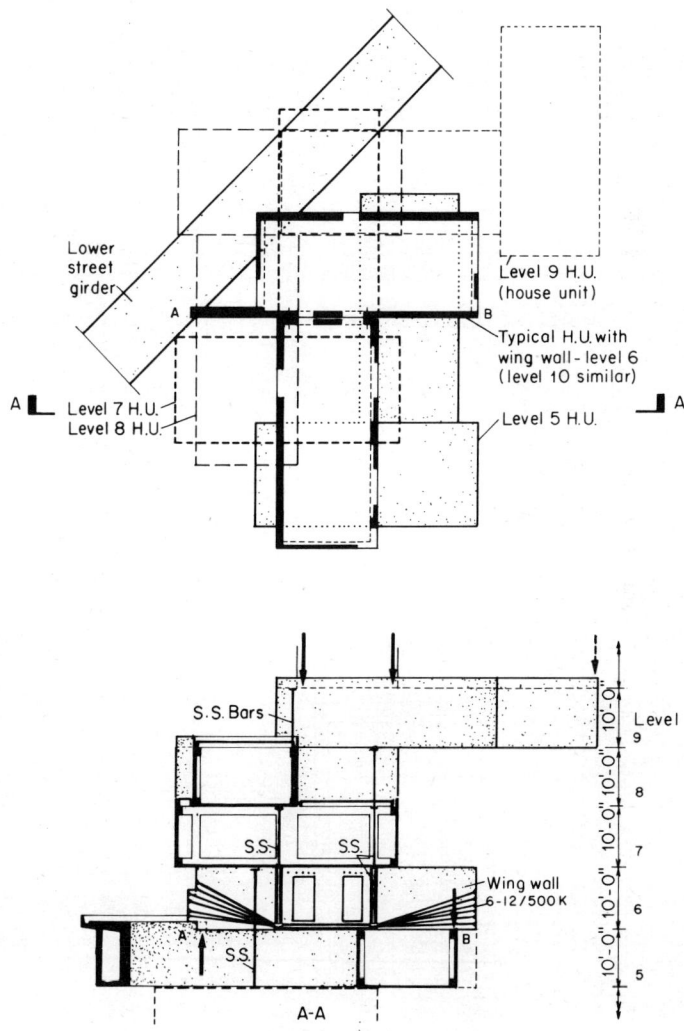

Fig. 351

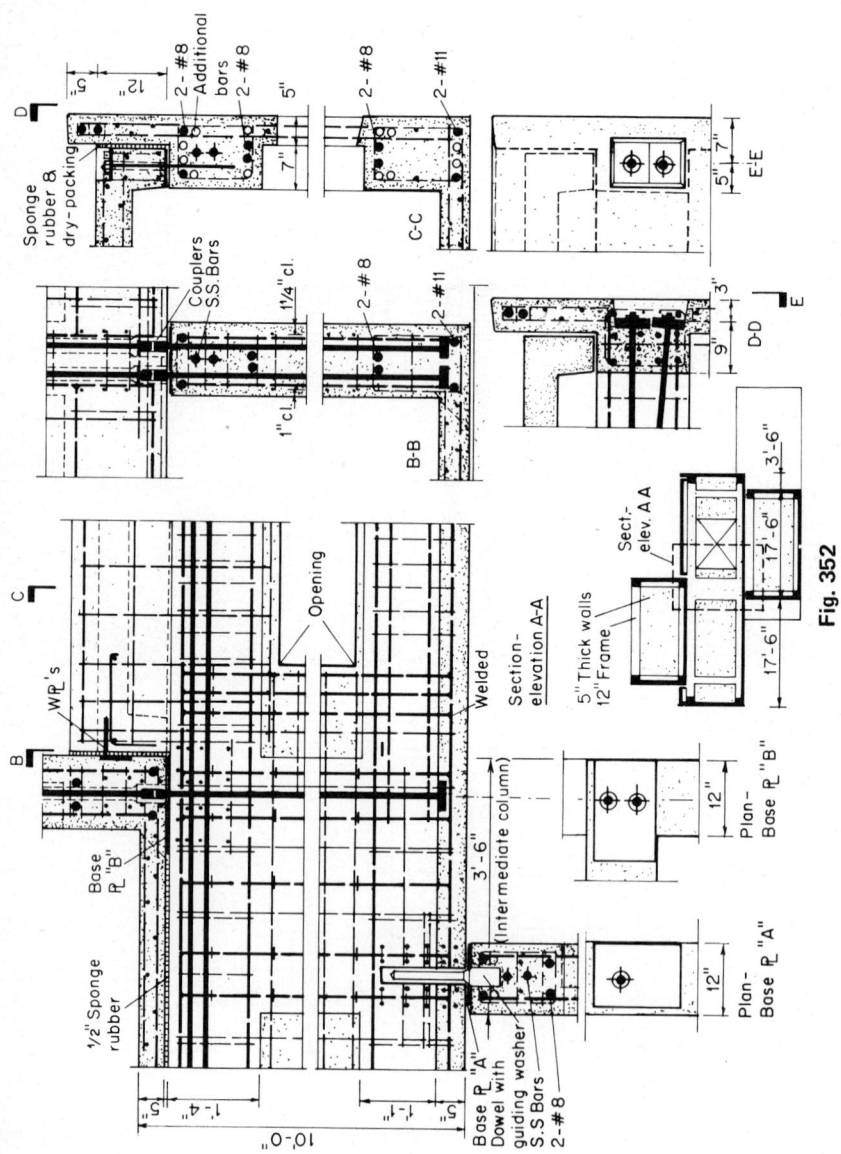

Fig. 352

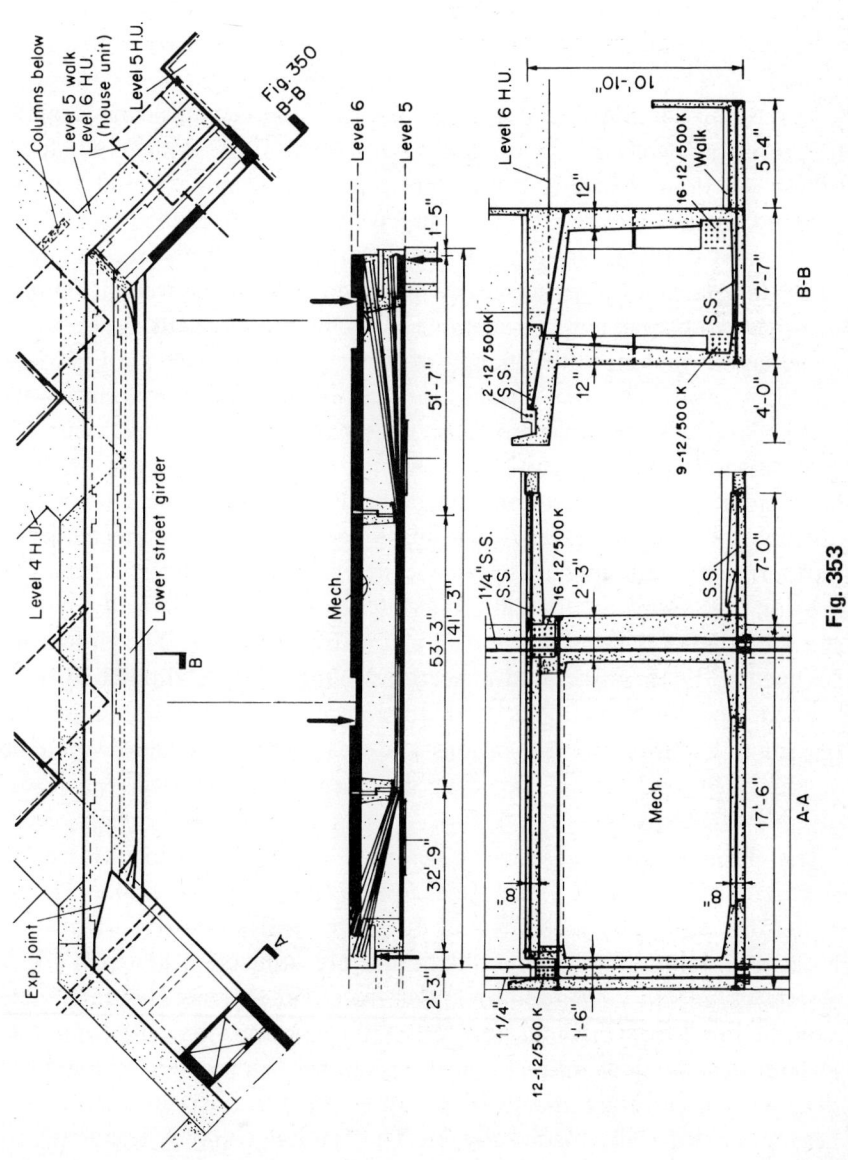

Fig. 353

Due to the cantilevering of boxes, the walls are unequally loaded and the box is subjected to relatively high torsion. Also, high shear across the openings is nearly constant because of the concentrated loads from upper boxes and can be carried only by embedding heavy rolled-steel sections over the openings.

The frame-slab box type is uneconomical and creates considerable structural as well as manufacturing difficulties. If typical windows instead of floor-to-ceiling types were used, the framing would not be required. If the thickness of individual walls were determined to suit the loading and the walls strengthened with buttresses only where required, computations and estimates clearly indicate that reduced weight, simple, easy manufacturing, and considerable economy would result.

The layout and structural details of street girders are given in Figs. 349 and 353. Statically the girders are simply supported by cantilevers passing through the elevator cores, extending out a maximum of approximately 34 ft, and by columns or escape-stair towers. The dimensions and cross sections are controlled by the overall layout of clusters and the minimum inside-space requirements for mechanicals. Furthermore, obtaining acceptable center of gravity and stiffness and keeping the stresses for all conceivable loading conditions within allowable limits determine the wall and slab thicknesses. The girders are designed to be manufactured in sections to meet the erection conditions and to control deflections and torsion due to posttensioning and one-sided loading. The dimensions and dead load of girders would be considerably reduced if no deep recesses were required for boxes supported by girders.

The elevator cores and stairs are designed in sections for manufacturing and handling reasons. The stair towers are split vertically at the centerline; each half includes half of the platforms and one flight. The horizontal joints follow the flights and are controlled at street-girder elevations by supporting requirements for street girders. The units were manufactured in horizontal position. After erection, the units were tied in both directions to unity by posttensioning. The location and amount of posttensioning are mainly determined by stress conditions of the heavily, eccentrically loaded towers. The Habitat complex was designed so that no scaffolding or even temporary supports were necessary for erection.

The erection was carried out by a 100-ton-capacity derrick mounted on a specially designed undercarriage to allow passage of travel lifts in order to feed the derrick from the front. The heavily loaded derrick track is on pile foundation. Its location is indicated in Figs. 348 and 349.

To ensure proper alignment during erection, the boxes are provided at the top with two dowels equipped with conical guiding washers and at the bottom with pipe sleeves. The safe, uniform seating and computed elevations at the receiving box were established at the supports by neoprene pads. Heavy torsion was kept within acceptable limits by setting the supports in different elevations and by posttensioning.

The soundness of a box system in structural context was tested by the construction of Habitat '67. However, economically it was not a success for reasons other than the concept.*

All prefabricated house complexes should have a plaza level and at least one level for parking beneath it. This is required to open up the street level and to have a solid base for the house clusters. This is the only rational way to keep the streets open for traffic.

*A. E. Komendant, Post-mortem on Habitat, *Progressive Architecture,* March 1968.

16
HIGH-RISES

65. MEDIUM-HEIGHT HIGH-RISES

Up to approximately 25 stories the most widely used structural system for reinforced-concrete high-rises is the flat slab. Structurally the flat slab is not an efficient system, but considering its easy construction, especially the simple forming, the flat slab is highly competitive with other structural systems. The most favored layout and column arrangement is illustrated in Fig. 354.

The ducts, piping, wiring, etc., are arranged below the slab between the intermediate columns and covered by hanging ceiling. The spandrel beams can be omitted, if desired, without structural sacrifice.

The vertical loading of the columns is rather high, and horizontal wind loading in buildings up to a 25-story height is only of secondary importance. Because of the high relative stiffness of the floor, especially of the short corridor span, in comparison with column stiffnesses, the internal columns are subjected to only small moments. The external columns, because of the unsymmetric loading, have to carry relatively large moments and also higher vertical loads, especially if heavy center walls are used. If the external columns are provided with semielastic hinges at the top and bottom, these columns will be almost moment-free. Thus the columns are subjected structurally in high degree to centric compressive stresses resulting in slender columns because of the almost 100 percent efficient use of concrete and reinforcing-steel strength. If the location and stiffness of staircases or shear walls are adequate to resist the wind, the moments of internal columns are also considerably reduced.

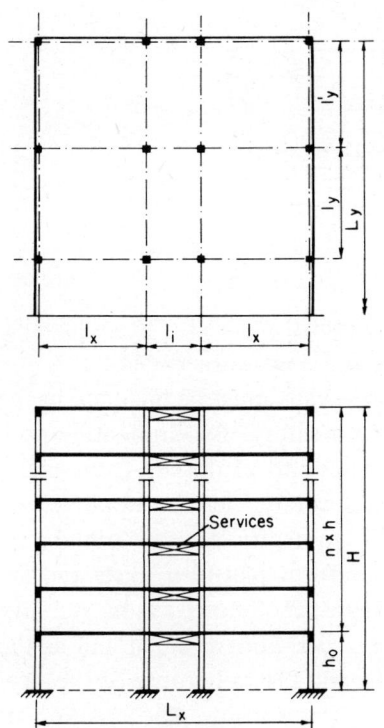

Fig. 354

Capitals for columns interfere with partitions and mechanicals and are architecturally not desirable, except for plaza and parking levels. High shearing stresses associated with highly loaded slender columns can be balanced by special shear reinforcing.

The rather large volume of concrete required for slabs can be most efficiently delivered by high-capacity concrete pumps. Heavy-type welded wire fabric reinforced with high-strength bars or even by the use of prestressing can be very rapidly installed. This all results in relatively short construction time.

In this story range is another conceivably competitive structural system: slip-formed bearing walls and prefabricated floors (Fig. 355). The external and corridor walls and stair-elevator towers are slip-formed in one operation up to the roof.

In favorable climatic conditions and with walls having no openings or only occasional ones, the speed of pours can be up to 1 ft/hr. For walls with extensive windows and doors, as is the case with high-rises, the

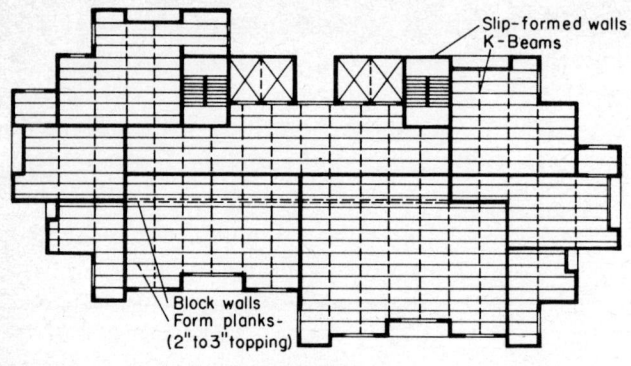

Fig. 355

speed is considerably reduced, mostly by interruptions for exact installation of framing for openings and grooves and dowels for floors.

For relatively high thin walls, inner crosswalls must be included in the slip-form operation to ensure stability and safety. By eliminating nonstructural walls and installing floor beams as part of the slip-form operation for stability, considerable economy is obtainable. For example, the relatively light K beams, which are lifted by electric winches running on rails fastened to lower flanges of the slip-form platform joists and are fixed in the boxed-out seats, are more than adequate to ensure stability of the walls. As the number of K beams per floor is small and as the beams are lightweight and easy to handle and fasten because of the steel top chord, there is adequate time for their erection. The erection of beams can be started after the slip form is about one floor higher.

The formwork and placement of reinforcing for poured-in-place floor slabs are time-requiring and expensive because the handling of materials is difficult between closed-in narrow spaces. Economy and speed of construction can be considerably increased by the use of reversed rib or form slabs. The slabs are about 4 ft wide and can be lifted through the openings in corridor bays and erected by means of electric winches. The construction of floors is independent of the slip-form operation. Both the slip-form operation and floor system are illustrated in Fig. 356.

If stone facing is desired for external walls, it can be done by feeding the stones provided with anchors into the slip form. For exactness of joints, closing, and quick setting, the stones must be provided with neoprene tape around the edges. The use of white cement for external walls is economically feasible because of the relatively small volume required. However, the work requires two separate concrete feeding

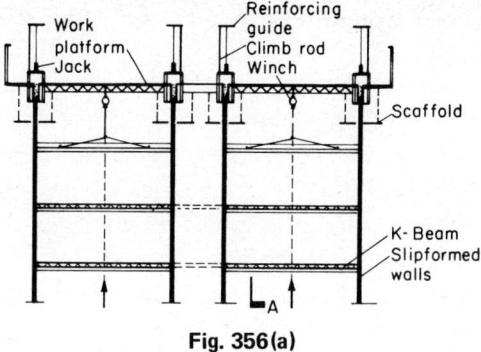

Fig. 356(a)

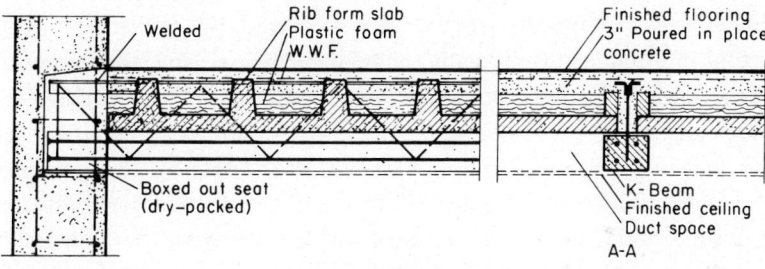

Fig. 356(b)

systems, which increases the cost considerably above the already high price of white cement, but regardless of that, it is more economical than stone facing.

The slip-form method requires a firm, even base on which to assemble the form and service platforms to start the operation. For this reason, the plaza level and parking floors are poured in place. Commonly, the time required for fabrication of the slip form is equal to that for about two floors of conventional-type construction. For materials handling, tower cranes installed on the slip-form platform are used. However, if concrete is pumped, only light electric winches are needed to handle the reinforcing and frames for doors and windows. The higher plasticity of pumped concrete in comparison with that of bucket-delivered concrete can be easily balanced by admixtures or by longer waiting time before placing the concrete in the slip form. By the use of this method, it is possible to decrease the construction time to half of that required for conventional construction.

66. HIGH-RISES—SUSPENDED SYSTEM

Generally, steel skeletons are considered more economical than reinforced-concrete structures. This is justified mainly by the higher dead load of concrete framing associated with larger dimensions of components, by longer construction time, and by more complex design and construction. However, taking into account the rapid development of concrete and high-carbon steel strand technology, as well as the improvements in construction methods in recent years, concrete high-rises can be competitive with the steel structures.

To reduce the dead load in areas seriously affecting the economy of conventional reinforced-concrete rigid-frame systems, the system itself must be abandoned. The construction time can be shortened considerably below that required for steel structures by adopting mass-production methods and radically changing the existing forming and casting methods. Furthermore, the mechanical part of the high-rises consumes close to 35 percent of the total construction cost, regardless of whether steel or reinforced concrete is used. In the case of steel framing, there is no possibility of reducing the price of mechanical installations, but in contemporary reinforced-concrete high-rise designs, there are great opportunities to accomplish this.

Taking under consideration all the above, a feasible, simple, safe, and economical structural system (Fig. 357) will be described which can challenge the steel high-rises up to 60 stories and even higher. In the layout of the carrying system, the conventional large number of rather closely spaced columns are replaced by a few relatively large *column stations*, which act simultaneously as supports for long-span girders between the stations and to accommodate mechanical ducts. The prestressed floor system is coffer type, having a minimum slab thickness of 3 in. and spacing of tapered ribs 5 ft o. c., suspended from the trussed girders at the outside edge and supported at the other end by the center service core. At higher levels, where the center core is reduced because of fewer elevators, columns and beams are required to support the floors. Along the outside perimeters of floors, air-supply ducts are poured in as an integral part of the marginal beams. These ducts are fed from the column stations. The exhaust ducts are cast in the walls of the center core. Vertical air-distribution ducts as well as lighting fixtures are incorporated into the open spaces of the coffer-type floor system (Fig.

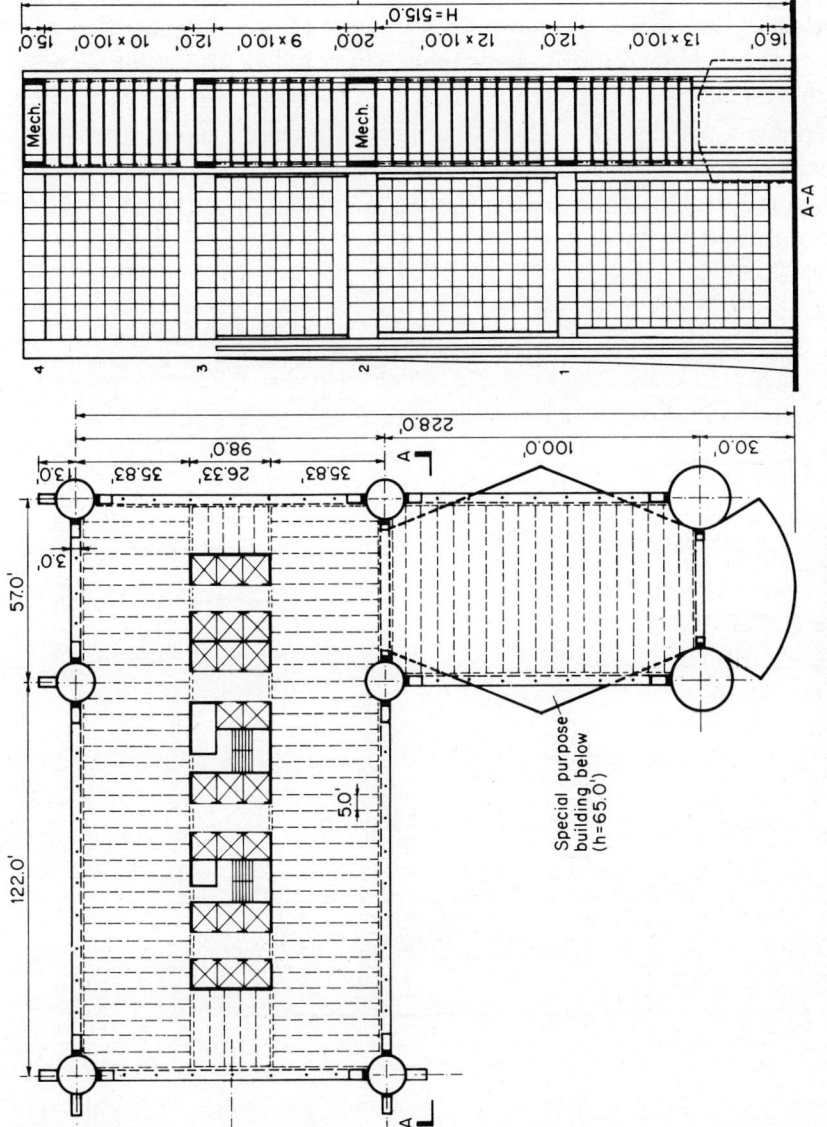

Fig. 357*

*Architect: Louis I. Kahn, Philadelphia.

358). The spaces can be closed by installing acoustic panels between the ribs or by a hang-ceiling attached directly to the ribs. The reinforcement of floor ribs is prefabricated and is complete with posttensioning tendons in place, teflon-coated and provided with plastic spacers so that they can be dropped quickly into the forms (Fig. 358). The floor slab is reinforced with welded wire fabric. The conventional reinforcing is determined so that no posttensioning is required under dead-load condition. This is necessary to allow independent operation of casting and posttensioning. The posttensioning is applied from the center-core end or from floor level.

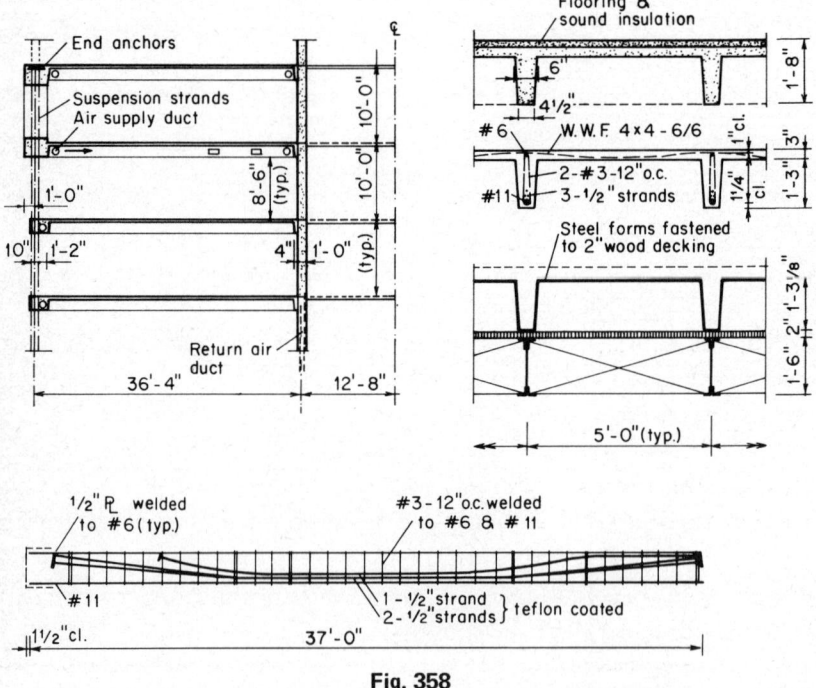

Fig. 358

Due to this design, the floor depth can be reduced to considerably below that of steel structures, which commonly require a hang-ceiling to hide the horizontal ducts arranged beneath the floor beams and girders.

The shape and size of the column stations are determined in large degree by structural and mechanical requirements and construction methods used. The dimension of the walls and fin-type buttresses increases downward in accordance with the increase of loading—vertical and horizontal. The trussed girders carrying up to 20 floors are composite steel-concrete posttensioned Vierendeel type. The verticals of the trusses are spaced at the location of suspension cables carrying the floors. They are erected and fastened on buttresses during the construction of column stations to ensure their lateral stability (Fig. 359).

The column stations and center core are slip-formed. The form depth is about 4 ft, and the average casting speed is approximately 15 ft per 24 hr and in discontinuous pouring about 8 ft per 8 hr in favorable climatic conditions. After the concrete leaves the form (in about 4 to 5 hr), the surface is still soft for treatment to obtain desired quality. During upward movement of the slip form, bent greased dowels are poured in and openings and recesses are obtained by feeding into the slip form removable or permanent boxes or framing at desired locations for doors, trusses, floor ribs, supports, stair steps, etc. When the slip-form operation has passed the trussed-girder level, the steel trusses, which are equipped with posttensioning cables, reinforcing, and anchor plates for suspension cables, are lifted by winches from the work platform of the slip form and are safely fastened by bolts or by welding on buttresses. For lateral stability against buckling, top chords of the steel trusses are connected by removable steel beams with the center core.

When the slip-form operation has reached the full height of column stations and the last trussed girders have been installed, the work platform is rearranged into three independent sections (I, II, and III) and the steel coffers are installed for casting the floors (Fig. 360). The work-platform sections are designed relatively rigid and heavier than structurally required. This is done to avoid undue deformations and to simplify the pullout of coffers by gravity. The floors are constructed in three operations: section I, reinforcing; section II, casting; section III, hardening (including lowering and cleaning). After the concrete of section III has developed required strength (approximately 3,000 psi), this section of the work platform is disconnected from the slip form and connected to a winch system arranged on and supported by the top floor. The section is lowered to the level beneath, and the cycle starts again in the same order.

The top chords and web members (diagonals and verticals) of the steel trusses are poured in simultaneously with the top floor, and the bottom chord is poured with the floor at its level. As the work platform has to

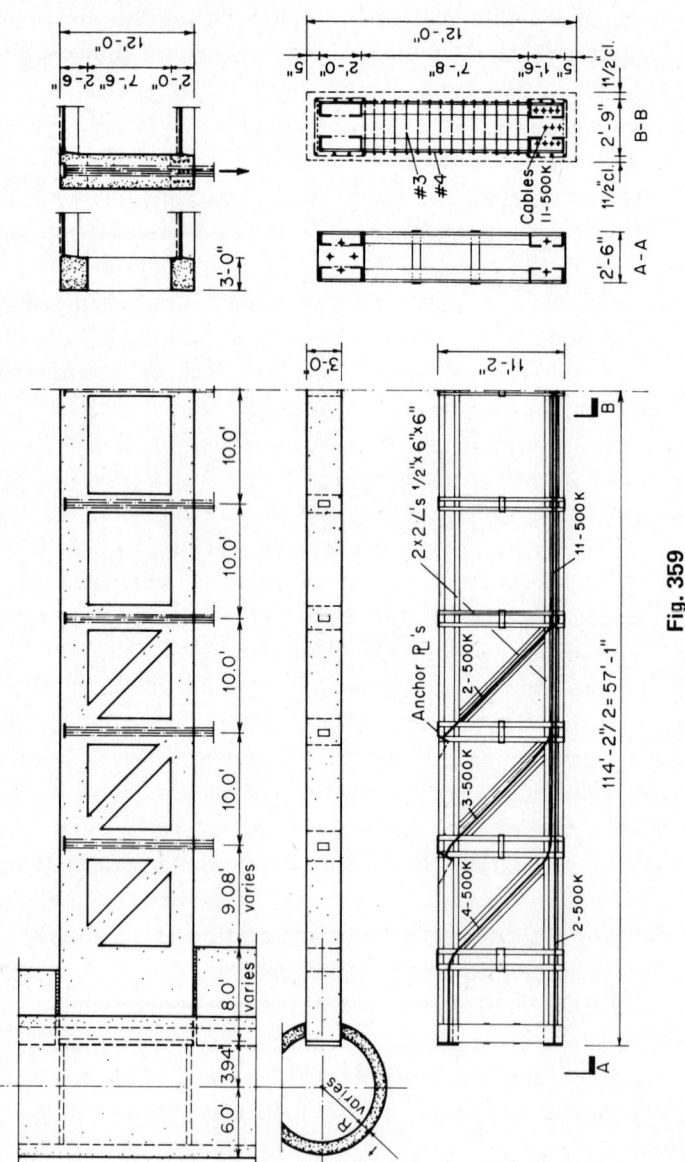

Fig. 359

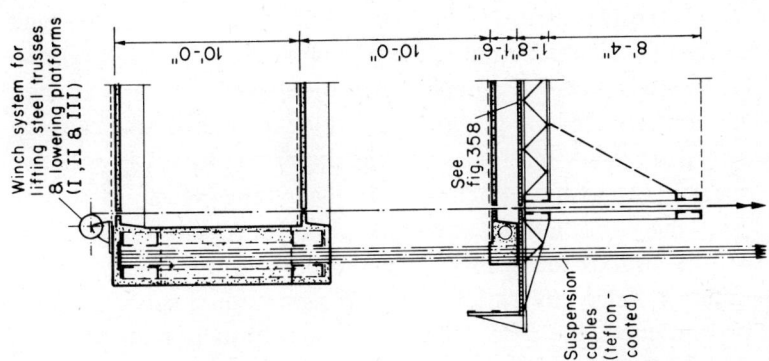

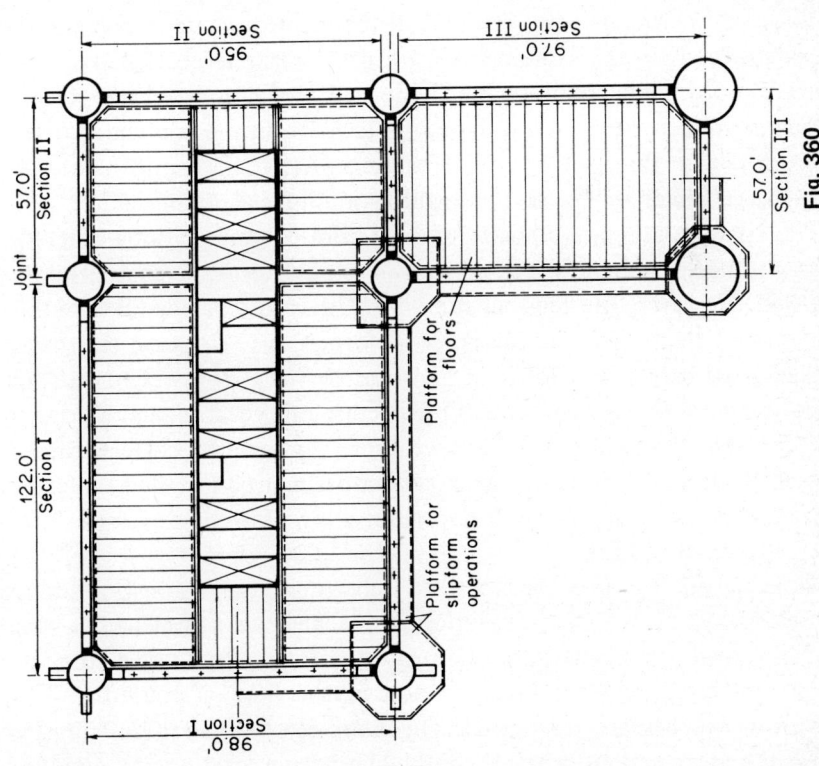

Fig. 360

bypass the steel trusses, hinged attachments are provided to support forms for marginal beams and walks.

The suspension cables, consisting of seven-wire strands, are prearranged and marked, and individual strands are provided with end anchorage so that each floor is anchored directly and independently to a trussed girder. The strands carrying a particular floor have two anchors, the lower one to carry work platforms and the upper one to be cast into the marginal beam. After this attachment is made, the winch cables are loosened and the total load of the work platform is transferred to the strands. As the strands must also carry the dead load of the concrete, they are almost fully pretensioned by construction load. After the concrete is hardened, the load of the work platform is transferred to lowering winches by disconnecting the lower strand anchors. When all floors down to the next trussed-girder top-chord level are poured, the winches are rearranged and installed at the next-lower truss level and the work proceeds in the same manner until the plaza level. If there is any deviation from the theoretically computed elongations of the strands, adjustments can be made by hydraulic jacks from the top chord of the truss for any floor in any one section. After the adjustment, the suspension cables are encased by window mullions and grouted in.

The trussed girders are posttensioned from column stations, after the concrete has developed the desired strength, by form-true posttensioning method. However, the tendons can be slightly posttensioned before they are poured in to allow checking of deformations. The steel trusses are designed to carry two floors in addition to their own weight to allow time for hardening of concrete. For speedup of hydration and hardening of concrete, especially in cold-weather works, the coffers can be provided with oil pipes and heat generators can be installed under the work platform. Such concrete treatment allows lowering of forms after about 10 hr of heating.

Time saving by this method, as well as by rapid construction, is increased by starting the installation of mechanicals and elevators when the slip form has reached only halfway up. The finishing of a story can be completed after the upper section of the floor is poured and the lowering winches are rearranged. The stairs can be poured and finished from the plaza and from the second-highest truss level up. The window and curtain walls are set from inside. The entire construction work can be performed without heavy equipment. The construction site is clean and dust-free because practically no formwork in the usual sense is required. The system is simple and flexible and allows the use of special construction trades (steel and concrete workers, carpenters, etc.) continuously. As computations and estimates indicate, it is highly competitive with steel-skeleton structures.

The multipurpose high-rise shown in Fig. 361* is designed to be constructed by the same method as that described above. It has a helicopter port and a swimming pool at the top part of the building. In this particular case, the column stations and center core are slip-formed up to the top of the column stations. From this level, the steel trusses are installed, and after the surface of the work platform for the two-skin floor system (Fig. 301) is prepared, the floor and top chord of trusses are poured. After the concrete is hardened, the work platform is disconnected from the slip form, connected to the winch system, and lowered to the bottom-chord-level floor. From this level on, the work proceeds simultaneously downward by suspension method and upward (special floors, helicopter port) by conventional method.

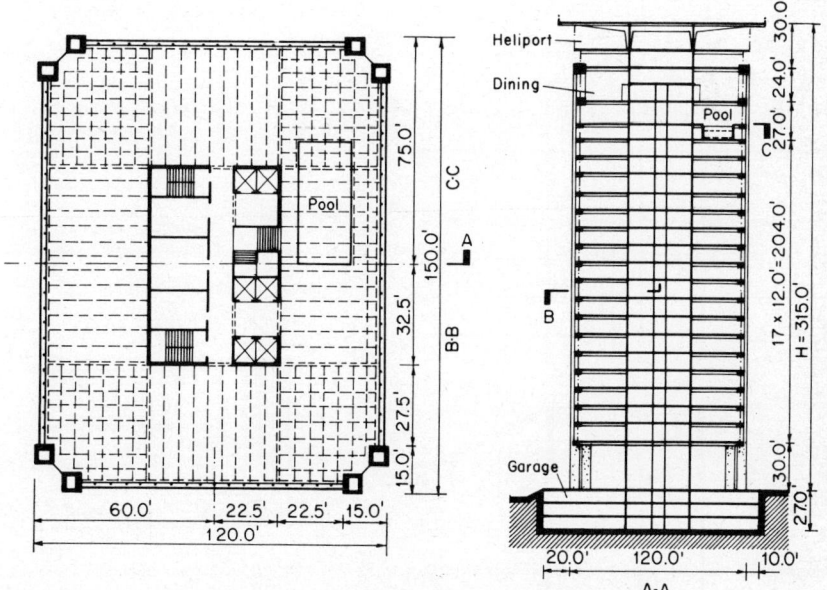

Fig. 361

*Architect: Louis I. Kahn, Philadelphia.

556 Structural Systems and Methods of Construction

If column stations are not desirable, the floors can be suspended by cantilevers from the center core (Fig. 362). The height of this high-rise system depends on the type and dimension of the center core carrying the total load of the building and also on the horizontal wind forces. In severe earthquake and hurricane areas, this system is not adaptable because of the difficulties in satisfying the safety requirements for lateral stability of the building. For this particular suspended system, prefabricated wide-grid floors are more appropriate and economical than

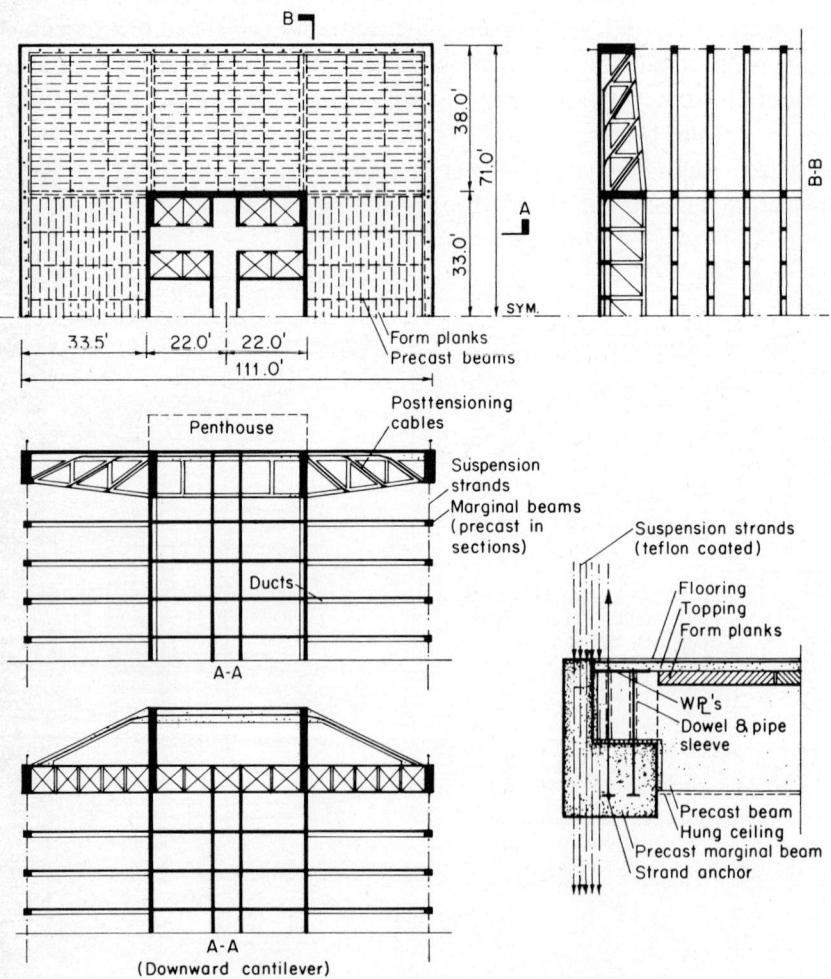

Fig. 362

narrow-grid floors poured on the slip-form platform. Starting from the top, first a stable grid consisting of marginal and floor beams for each floor is erected. Then the prefabricated form planks are set from the bottom upward. In this way, the electric winches for lifting and setting of planks can be arranged at higher floor beams. Also, no safety net is required, which is not the case when working from the top downward. The floors are tied into a unit whole by 2- to 3-in. topping.

There are two possibilities in designing the cantilevers: downward and upward tapering, as indicated in Fig. 362. Upward tapering is more simple and economical than downward tapering, which is usually clumsy and heavy. For both cases, a prestressed composite steel-concrete system is appropriate, mainly because steel trusses can be erected practically without scaffolding from the center-core slip-form platform with least time. This applies also for marginal beams supported at the ends of cantilevers and carrying the suspension cables.

The number of floors suspended from a trussed girder depends on many factors and on architectural design. An arrangement of intermediate-height girders is required for tall high-rises in order to reduce the number of strands in suspension cables and the dimensions of the girders. Structurally and in terms of construction, such a design creates almost no difficulties.

17

BRIDGES

Since the development of prestressing, prestressed bridges up to 1,000-ft spans have been challenging steel bridges aesthetically as well as economically. Almost all contemporary concrete bridges, except for relatively short spans, are prestressed; therefore, only prestressed bridges will be discussed in this chapter.

Bridges are a special field in structural as well as highway engineering. Since highway engineers lay out the alignment and establish the general features of a highway project, they should be familiar with the basic ideas in contemporary bridge design. In most cases the approach to a bridge design is purely functional and the aesthetic quality of the design is simply overlooked or even not recognized at all. The reason for such a prevailing attitude is primarily ignorance, based on the assumption that aesthetic considerations in a functional project such as a bridge are economically an unjustified luxury. The actual facts are: An aesthetically high-quality bridge has nothing to do with economy but has to do rather with the thorough knowledge and good taste of the designer. Experience shows that any up-to-date, proper, simple, and functional design, combined with efficient use of materials, is of itself aesthetically acceptable. Furthermore, experience also proves that simplicity and efficient use of materials, associated with appropriate construction method, always result in high economy for any structure, including bridges.

The general rules governing proper design and the factors contributing to the aesthetic qualities of structures were thoroughly discussed in Part 1. The application of these rules and factors to individual bridge types will be dealt with in detail in the following.

Bridges, in general, can be divided into two main categories: river and valley bridges and bridges for grade separations. The first bridge group commonly comprises large, long-span bridges, and the second group comprises relatively short- or moderate-span smaller bridges for overpasses, underpasses, and intersections of superhighways, railroad tracks, etc. Almost all bridges are exposed from every angle to the eye during their entire lifetime; therefore, the structural system of a bridge must be carefully studied also from this aspect, and not only the elevations as they look on the drafting table or in small-scale models or perspectives. A revised approach which recognizes and takes into account this important factor has resulted in contemporary bridge design. This new approach is based on the concept of unified design, structural clearness, completeness, and simplicity. The structural system must harmonize with its function and the nature of the site. The construction must be easy, quick, and economical. Great contemporary bridges, such as the ones designed by Maillart, Finsterwalder, and many others, are famous not only because of their exceptionally long spans but also, and rather more, for their unique, unified, high-standard designs, advanced construction methods, and overall economy.

The dead load of prestressed concrete bridges is higher than that of steel bridges. To make concrete bridges competitive, first of all the dead load must be reduced to the minimum. This can be done by the use of high-quality concrete and tendons. As a result, the slenderness of prestressed concrete bridges is commonly higher than that of steel bridges.

For uniform-depth continuous spans, the span-depth ratio can be as high as 25. For variable-depth arch beams and continuous beams, the ratio is up to 65 at midspan and 18 to 20 at supports. With higher slenderness, there may be vibration problems. To control the flexural and shear stresses at supports and to obtain greater lever arms for tendons, the wall and bottom-slab thicknesses are increased; at midspan, where the shear and moments are relatively small, the sectional dimensions can be considerably reduced.

All things considered, and even if the higher maintenance cost for steel bridges is overlooked (a common case!), prestressed concrete bridges have already surpassed the economy of steel bridges, especially in the medium- and longer-span ranges and in locations where structural steel is not readily available. This has been possible mainly because of the great liberty of design of prestressed concrete, the fast development in concrete and high-quality steel-cable technology, and advanced construction methods. Making use of these advantages, the design can be accomplished so that the entire dead load of the bridge can be carried directly and most economically by cables. On the other hand, the design of steel bridges is more restricted, and developments in structural systems and construction methods have been practically at a standstill for the past two or three decades.

67. BRIDGES FOR GRADE SEPARATION

Much attention has been paid to long-span bridges, but almost none to smaller ones, even though they numerically dominate and are more exposed to travelers than the long-span bridges. Due to this, bridge designs for grade separation will be discussed first and in more detail.

The bridge types for overpasses and underpasses can be simple or multispan, straight or skewed, and beam, frame, or slab, depending on the span, the construction depth available, and the crossing characteristics. Also, concrete arches for wide, deep cuts in mountain terrains or for underpasses through high embankments are structurally most appropriate. However, since any type of short-span arch is more expensive than other bridge systems, arch bridges are justified only to obtain change in the relatively monotonous picture of frequent beam bridges.

Possible structural bridge systems for grade separations are given in Fig. 363. The lengths of bridges for overpasses and underpasses may vary from 30 to 150 ft, depending on the number of lanes to be crossed, width of center strip, shoulders, clearance required, and restrictions for construction. Furthermore, to avoid monotony, not more than two bridges of the same type for overpasses should follow each other along the line. The design of all these bridge types is very demanding, mainly because they are small and exposed and, as such, can impress only with their aesthetic quality. The aesthetic quality does not depend as much on the structural system as on the balanced span-clearance ratio and the relationship of components: abutments, piers, beams, cantilevered slab, and railing.

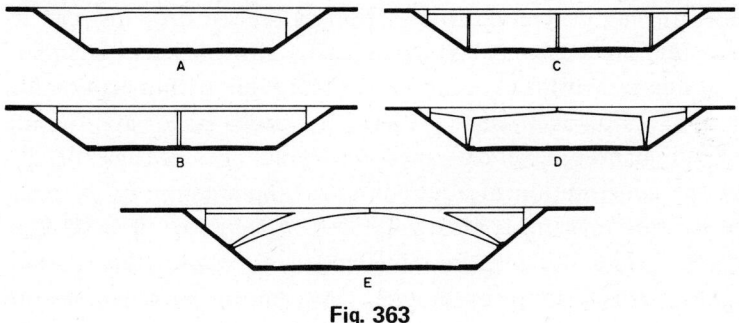

Fig. 363

An acceptable span-clearance ratio can be obtained by choice of the locations of piers and abutments. Too large a ratio results in depressive, heavy beams and an unbalanced pier-beam relationship. On the other hand, too small a ratio results in a bridge which looks more like a wall with openings overbridged with lintels. Concrete railings should be avoided, and the cantilevered slab should be kept thin, so that the slab part does not dominate over the beam. The horizontal lines along the entire length should be parallel and, if possible, continuous. Haunches and change in depth of beams violate this rule and, therefore, should be avoided. As the negative moments at intermediate piers are approximately twice the span moments, haunched beams are structurally justified. However, their disturbing effect should be reduced by mushroom-type increased depth or by the use of relatively flat haunches at piers (Fig. 366). It is also possible to balance the moments by jacking the beam or slab ends at abutments, thus creating positive moments at the bottoms of the two- or three-span continuous beams. The abutments carry comparatively less load than the piers; therefore, they should be kept small. The material is more efficiently used in piers than in beams—a fact which is emphasized by keeping the columns narrower in comparison with the depth of the beams. For constant-depth simple beams, a span-depth ratio up to about 25 is considered within economical limits. Beyond this ratio, the economy is reduced rather rapidly and vibration of the bridge becomes notable. Roller bearings in these span ranges are not justified; neoprene or teflon is completely adequate and far less expensive.

For flat terrain in which a bridge must be constructed over an existing highway without interrupting the traffic, the most appropriate bridge systems are type A and type B (Fig. 363). For type A, both abutments—with cantilevers and midsection as precast drop-ins—can be constructed simultaneously and independently. Provided that adequate construction depth is available, spans up to 150 ft are within economical limits (Fig. 365). For deep cuts, the type A bridge is not appropriate because of unbalanced span-clearance ratio and also because of difficulties in the construction of abutments and the erection of drop-ins.

The type B bridge system is most suitable for deep as well as shallow cuts. For wide center strips, two intermediate columns are usually required. For relatively long continuous spans, mushroom-type increase of depth at piers is structurally as well as aesthetically preferable to two simple spans.

The type C bridge system is easier to construct than the type B system for relatively deep cuts, and it is aesthetically and economically superior to two-span haunched beams (Fig. 363). However, for shallow cuts, the end spans of this system are too short in comparison with the intermediate spans, resulting in unbalanced relationships. The balance is considerably improved when the center pier is omitted. Structurally, in dead-load condition the system consists of two cantilevers (type D). By posttensioning, the system can be modified to continuous frame for live load. A slightly modified version of the type D bridge can be successfully used also for relatively deep cuts (Fig. 367).

The arch bridge (type E) illustrated in Fig. 363 is appropriate for deep cuts and long spans. The arches can be easily prefabricated and erected by pulling and lifting the crown end with winches from the abutment or lifting by crane at the middle strip and tying it back by cables. The superstructure can be prefabricated or poured in place; the forms and shoring are supported by the arches themselves.

Bridges for underpasses and through embankments are commonly of short spans and relatively wide. The structural bridge systems for these purposes are simple precast spans, poured-in-place frames, and short-span arches.

All these bridge types, except abutments and short-span arches, can be completely prefabricated or poured in place. Which construction method is used is determined by local conditions and economy. As the structural layout and details of these bridge types are principally the same, whether poured in place or prefabricated, only prefabricated types will be discussed.

The width of a two-lane bridge is 40 ft (two lanes, 15 ft each, and two sidewalks, 5 ft each); three lanes, 55 ft; and four lanes, 70 ft. Very often a two-way highway consisting of six lanes with about 10-ft dividing strip is carried by a single 110-ft-wide bridge. Commonly, two separate bridges are used to carry eight-lane superhighways. The cross section of the beams can be a channel, single T, box (square or triangular), or double T as form for poured-in-place slabs. Precast beams with poured-in-place slab, also commonly in use, are not economical and flexible enough for spans beyond 100 ft, as compared with channels or single T's. Prefabricated boxes and solid or hollow-core slabs are economical only up to 70-ft spans. The width of channels and single T's is commonly 8 to 10 ft, and the depth is up to 5 ft. For boxes and double T's, the width is about 4 ft; the height for boxes is 3.5 ft and for double T's is 18 in. The width of flanges for beams is about 2 ft, and the depth is up to 4 ft.

Thus, using the maximum span-depth ratio of 25, the maximum span for simply supported channels or single T's is 125 ft and for boxes and hollow-core slabs is about 80 ft. Poured-in-place boxes for grade-separation bridges are seldom used. To keep the weight within acceptable limits for handling, the thickness of webs and slabs of the cross section of the beams is reduced to the minimum possible. Thus, in a true sense, all precast beams serve in some degree as forms for the finished bridge cross section.

The channel cross section is structurally the most flexible, although it is more difficult to manufacture than single T's, mainly because it can be manufactured in sections, prestressed by pretensioning method for handling, and posttensioned most easily at the site for continuous spans.

As can be seen from Fig. 364, by arrangement of joints at third points the posttensioning force can be determined for the total moment in these points. The pretensioning strands provide balance for the maximum midspan moment. Thus manufacturing in sections does not increase the cost as far as prestressing is concerned. However, two sets of scaffolding are required for erection.

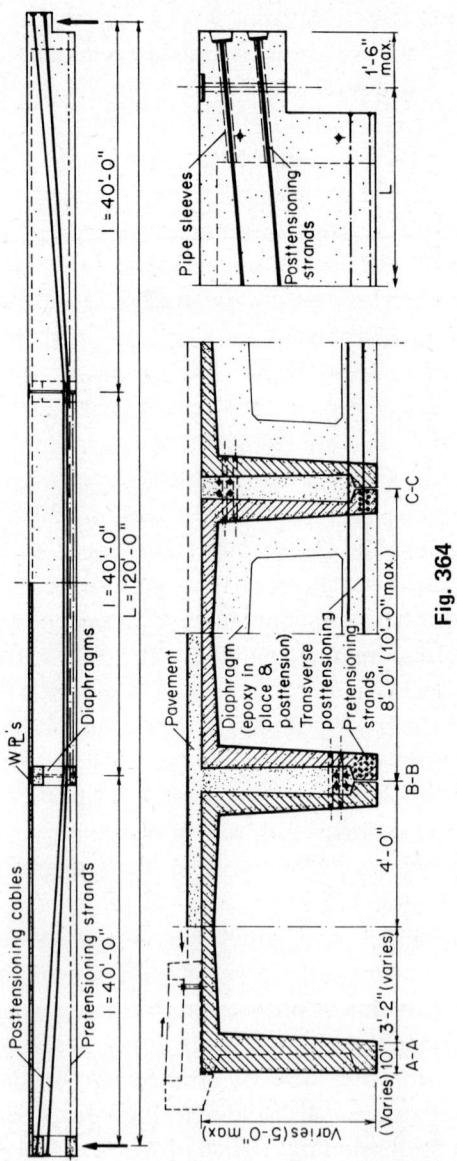

Fig. 364

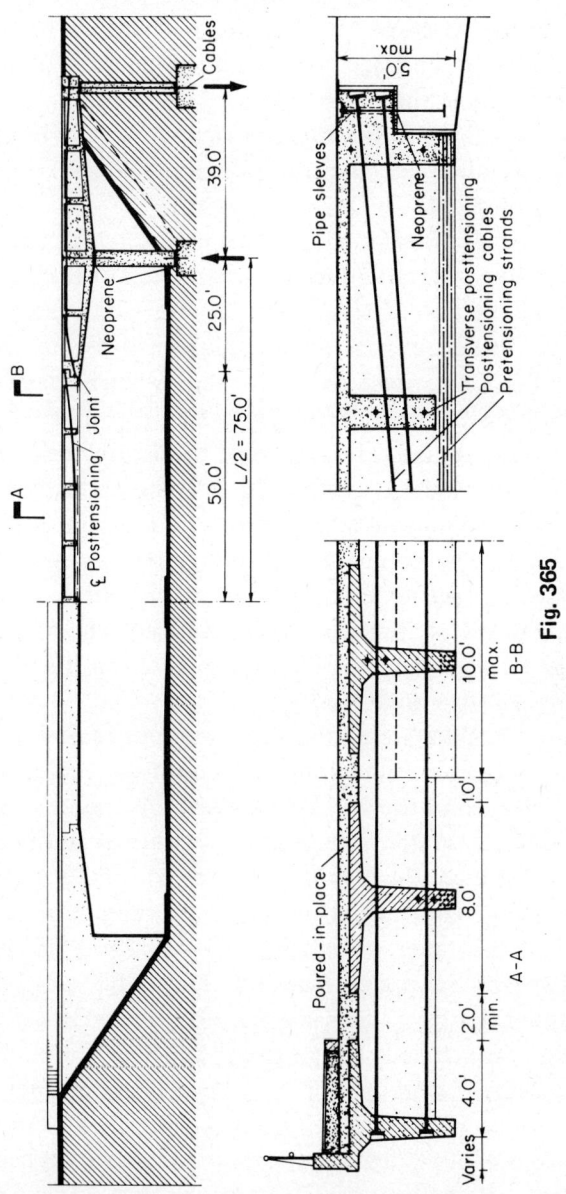

Fig. 365

Structurally the box section is the least economical, and it is also difficult to manufacture. No posttensioning is possible at the site. Furthermore, the bottom of the box must be provided with an adequate number of holes for drainage to avoid accumulation of water in case there is any cracking in the top slab.

A type A three-lane bridge over an eight-lane superhighway, having poured-in-place abutments and prefabricated channel drop-ins, is illustrated in Fig. 365.

The weight of a 100-ft-long channel and that of a single T are approximately equal—about 50 tons. The erection of the six drop-ins can be easily handled by two 45-ton typical construction cranes.

A three-lane bridge over a six-lane superhighway in flat terrain is illustrated in Fig. 366. Since the construction height has to be kept to the minimum possible to reduce the height of embankments, a slab bridge of type B is chosen. In order not to disturb the normal flow of traffic on the superhighway, the 18-in. double T's with increased slab thickness serve simultaneously as forms and as an integral part of the finished structure. The double T's are used in reversed position, with the slab prestressed to the extent required to prevent cracking until the posttensioning of the continuous cables is applied. The ribs should have a high amount of compressive reinforcing to keep the stresses under control during construction.

The continuous cables are teflon-coated and posttensioned gradually so that the stresses in concrete will remain within allowable limits under all conceivable load conditions. The cantilever is poured in place and posttensioned in the cantilever as well as the transversal direction.

In order to reduce the flexural stresses in cantilever ends due to horizontal loading, temperature change, and shrinkage, the connections—cantilever, pier and pier, foundation—are designed to be elastically controlled two-point supports (see Fig. 274). The end support of the slabs at abutments is, for the same purpose, also elastically controlled. The elastically controlled connections are especially important for structural systems having points with considerably reduced rigidity, causing stress concentration that results in fatigue and crack formation.

For overpasses of superhighways at deep cuts, bridge types C and D are the most acceptable and proper. As these systems differ only in respect to the length and number of spans between the outside shoulders of the highway, only bridge type D, as illustrated in Fig. 367, will be discussed.

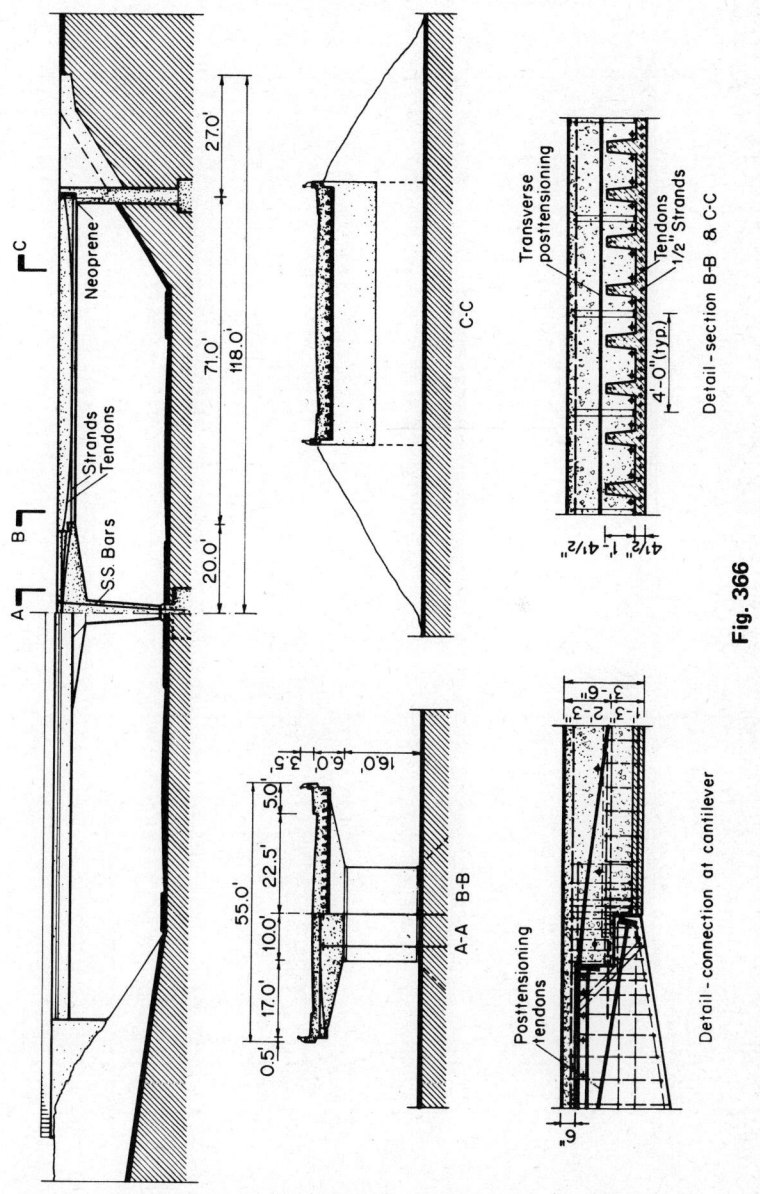

Fig. 366

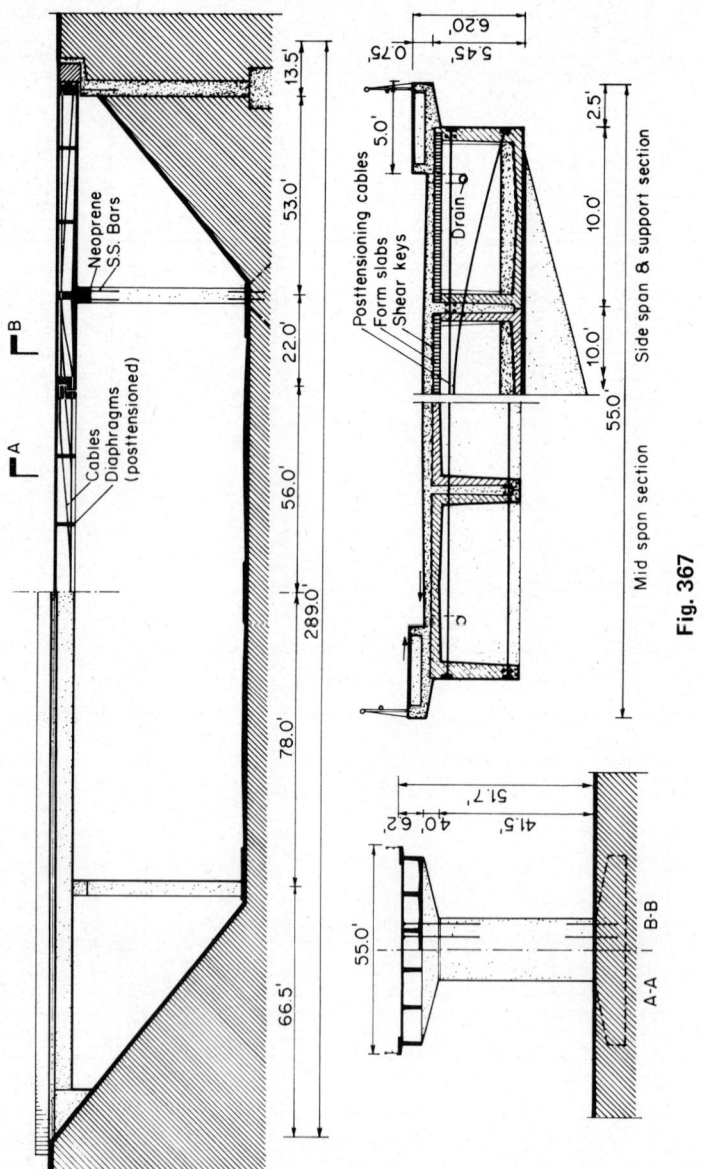

Fig. 367

The shape of piers is controlled by the depth of the cut and width of the bridge. In this particular case, to obtain a balanced relationship for the piers, their width has to be considerably narrower than that of the bridge. Double-column piers are not justified aesthetically and economically. The only acceptable solution is to use a constant width and thick pier with cantilevered top to provide proper support for the beams. However, the unified-design principle requires that the cantilever should not dominate the system but should act as a unifying element between the pier and the beams. This principle can be satisfied most acceptably by considering and designing the cantilever as a part of the superstructure, gathering the reactions and transmitting them into the pier.

The superstructure is composed of channel beams (Fig. 364). For given span-cantilever ratios, the side spans under design load condition are acted upon by negative moments in their entire length; therefore, the channels in cantilevered side spans are placed in reversed position. The slabs are strengthened as required by poured-in-place concrete. The prefabricated cantilevers over piers, having cutouts for channel webs, are partially prestressed for dead-load condition and tied elastically to piers by posttensioning rods projecting from the piers. To obtain required displacement and rotation for the superstructure, the cantilever is seated on bearing neoprene pads, and the projecting posttensioning rods are teflon-coated to provide adequate elongation length. The elasticity of the connection is controlled by the amount of posttensioning applied. The slabs of the channels also have matching cutouts at supports to allow erection and to provide proper interlocking with the support cantilever. The joints are thoroughly dry-packed before prestressing. After the transversal end beam is poured and cavities at the support filled, the cantilevers are posttensioned to unity with the channel beams. The posttensioning cables for dead load are anchored in the posttensioned transversal end beam, and the cables for live load are continuous from abutment to abutment. Prestressing is applied at abutment ends for both sets of cables gradually, in sequence with the erection and loading of main channel beams. The pavement slab over the reversed channel beams is poured on form slabs.

Using channel beams or single T's, simple span bridges up to 120 ft long are commonly more economical than type A or short multispan bridges. With a well-balanced relationship between abutments and beam depth, these bridges are more attractive than the short, multispan, simply supported bridges.

If the required construction depth is not available, the beam depth can be considerably reduced by using counterweights behind the front wall of abutments supporting the superstructure, as illustrated in Fig. 368. The amount of counterweight is controlled by compressive stresses in beams at the support section (normal case) and by tensile stresses at midspan. If the weight or length creates difficulties in transport and erection, the channels can be manufactured in sections (Fig. 364).

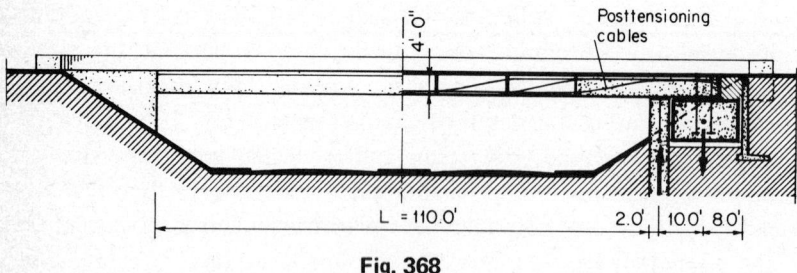

Fig. 368

68. BRIDGES FOR PEDESTRIAN CROSSINGS

Narrow long-span light concrete bridges are used for pedestrian crossings over highways, railroad tracks, or rivers. Since the live load is small, considerable single spans are structurally possible and well within economical limits. Most suitable for such crossings are arches. One of the most attractive crossings of this type is the Schiersteiner Bridge (Wiesbaden harbor),* illustrated in Fig. 369.

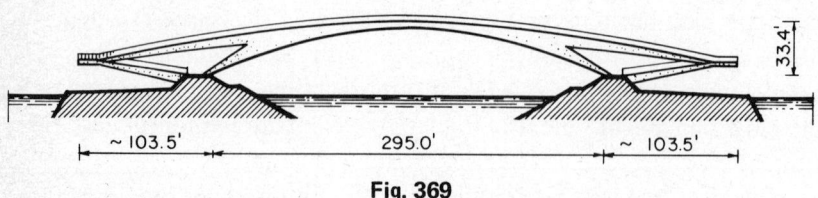

Fig. 369

This daring structure was constructed by the free-cantilever method over the channel (210 ft wide) simultaneously from both shores. To reduce the dead load of the arch and to balance the thrusts, high-strength lightweight concrete was used for the arch and standard concrete for outside-end cantilevers.

*Architect: Gerd Lohmer; engineering and construction: Dyckerhoff and Widmann.

Prestressed frame systems with slightly curved beams can also be successfully used for pedestrian crossings in flat terrain. The frames are simpler to construct and can be entirely prefabricated, if economically justified or required by local conditions. The aesthetic quality of these bridges is considerably improved when the depth of the beam is reduced to the minimum possible (Fig. 370a and b). This, however, is not possible if simple beams are used.

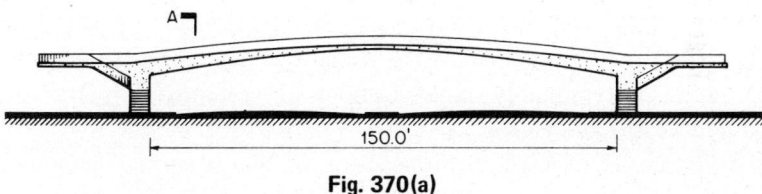

Fig. 370(a)

The lateral safety of such slender structures (approximately 10 ft wide) is most easily provided by combining the frame verticals and stair structure to a structural unity. Multispan beam systems for such bridges,

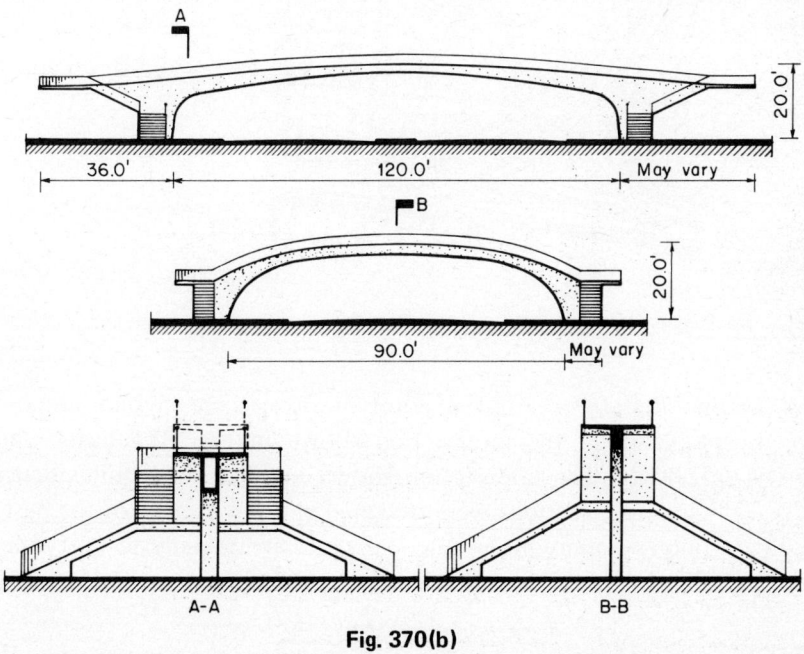

Fig. 370(b)

regardless of how well designed, commonly lack the grace of arches and arched frames.

69. BRIDGES FOR INTERSECTING SUPERHIGHWAYS

Bridges of this type are complicated to design, mainly because the crossings must follow the alignment curvature and because they occur at different elevations and levels. Due to this, the choice of column locations is highly restricted, especially if a small-radius alignment curvature has to be matched with straight segmental short spans, resulting in an unreasonably large number of columns or piers and distorted and obstructed views.

A typical example of such an intersection of two superhighways in a big-city suburb is illustrated in Figs. 371 and 372. The two- to four-lane bridge crossing occurs at two to four levels, with elevation differences of up to almost 100 ft and radii of 150 to 1,000 ft, forced into a relatively small area. The design of such intersections requires the utmost structural skill and understanding of aesthetic principles to obtain an acceptable, unified, harmonious solution.

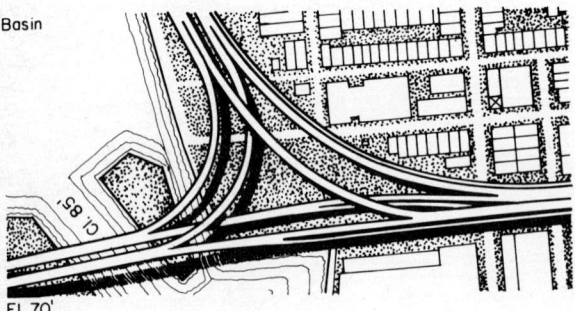

Fig. 371

Considering that from a traffic point of view, these layouts are the only acceptable ones, the intersection shown in Fig. 371 has to be entirely on bridges. The intersection shown in Fig. 372 requires seven bridges. If these intersections were designed in a way typical of the past, the first would be annoying aesthetically and structurally to everyone, but especially to those with trained or educated eyes. The second intersection, because the bridges are almost straight and the roadway is mostly on embankment, would not be much affected by the quality of bridge design. However, it cannot be denied that even in this case high-quality bridge design would add considerably to the aesthetic value of the intersection as a whole.

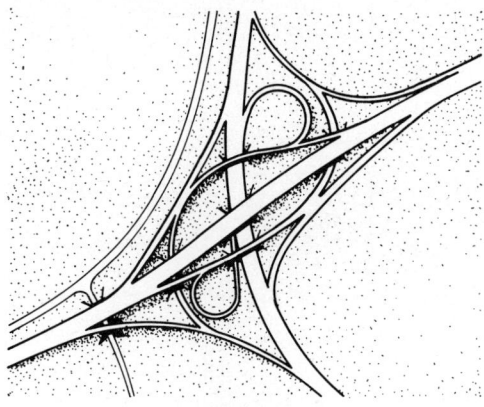

Fig. 372

The design of intersections based upon unified-design principles and making use of advanced technology would never be irritating, even for extremely sensitive eyes. In this context, "unified design" means: (1) The entire intersection is considered as a whole, and the bridges are treated as balanced, matching parts of it—not just as a number of individual bridges, unrelated in an aesthetic as well as structural sense—located on a small area for a common, purely functional purpose. (2) Individual bridges, short- or long-span, have affinity and continuity in the structural sense. (3) The construction material and method of construction used are the same throughout the intersection.

The most suitable structural systems for intersections are straight or curved long-span prestressed hollow-core slab or box-girder bridges supported by simple columns with mushroom-type or no capitals (Fig. 373).

To balance the torsional moments and eliminate shrinkage as well as temperature stresses, two-point elastically controlled connections between the superstructure and columns are the most appropriate (Fig. 274). Only in such a way can the sharp curvatures and relatively large

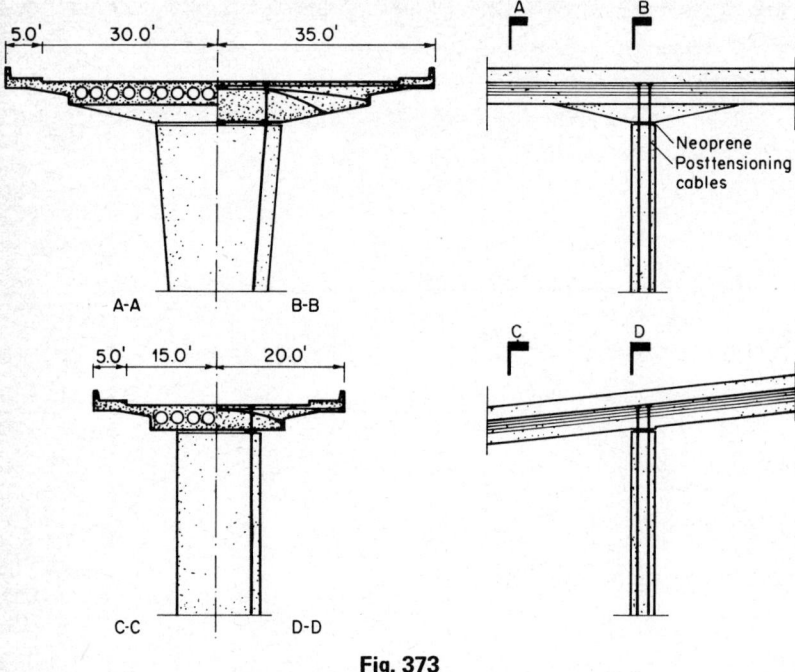

Fig. 373

slopes of the alignment be matched without difficulties. The long, curved, continuous spans require fewer supports and allow a more liberal choice of support locations.

Considering that concrete is more efficiently used in columns than in beams and that column width in a transversally prestressed cross section is controlled mainly by stability requirements, the one-column support is structurally fully justified, even for four-lane-wide bridges. The one-column support, besides permitting a reduced number of columns, allows simple solutions for branchings, as illustrated in Fig. 374.

For complex intersections, columns with constant thickness and width are aesthetically more acceptable than tapered ones, mainly because of the varying heights and widths of the columns. However, downward-tapered columns for isolated bridges (Fig. 373) that have moderate and almost constant column heights are superior to constant-width or trestle-type supports.

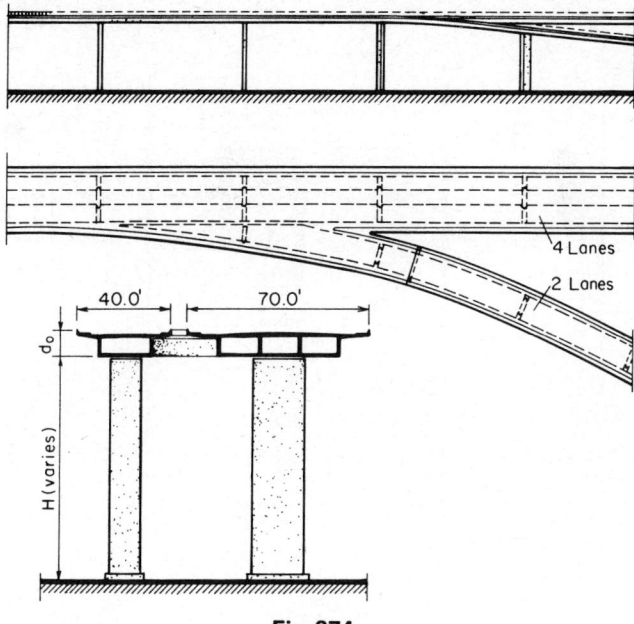

Fig. 374

70. MEDIUM-SPAN BRIDGES

Bridges in this category serve for crossing medium-size rivers, railroad tracks, and streets and also for elevated rapid-transit systems. Depending on the span and local conditions, these bridges can be prefabricated, poured-in-place, or even a combination of both.

The bridge type illustrated in Fig. 375 is simple to construct. The cantilevered sections at the piers are most economically poured in place. The span sections are prefabricated T's, boxes, or channels. Stability of the structural system is established by prestressing the construction joints in the end spans.

The arch-beam bridge types illustrated in Fig. 376 are commonly poured in place. To reduce the costs of forms and scaffolding, the proper cross section should consist of twin boxes constructed by the use of the same forms. The gap between the boxes is most conveniently overbridged with a simple slab poured by the use of form slabs.

The system illustrated in Fig. 376a is a three-hinged arch beam prestressed by cables. The cables run in loops from each cantilevered end along the top of the sidewalls at the level of the deck slab. They are

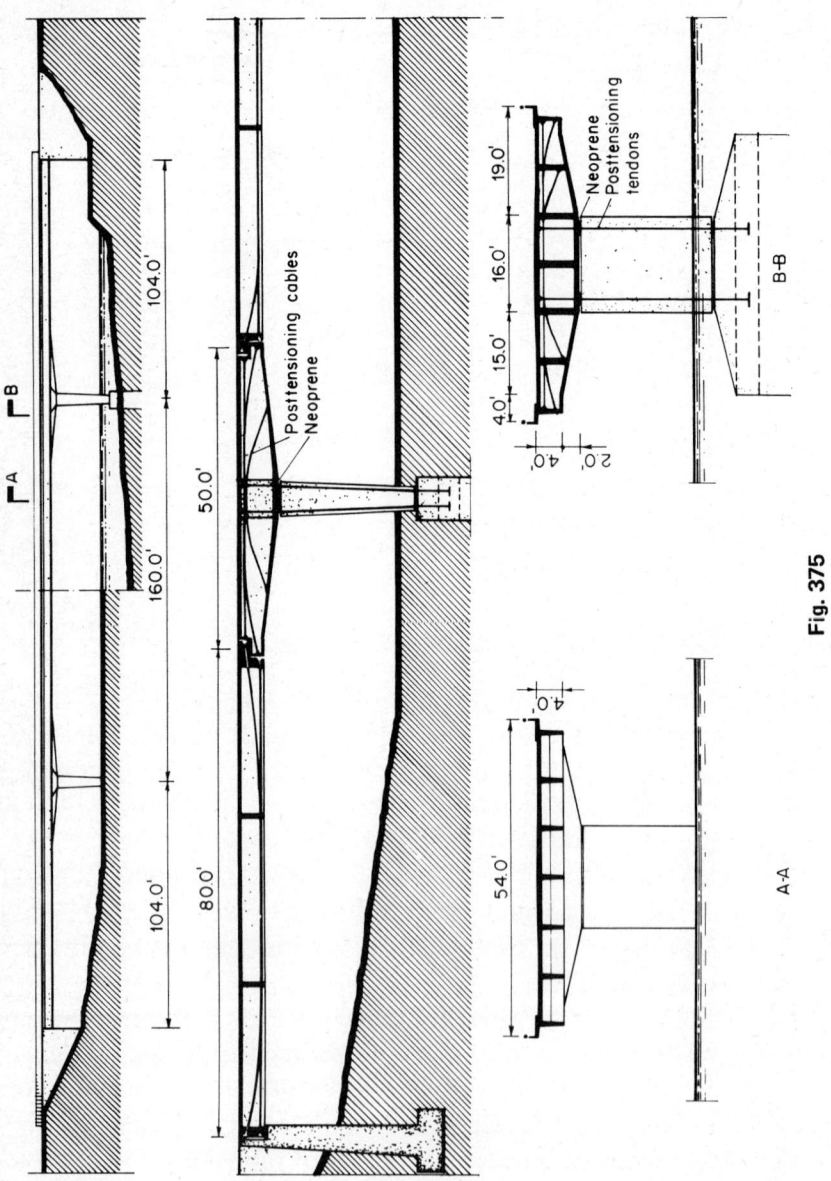

Fig. 375

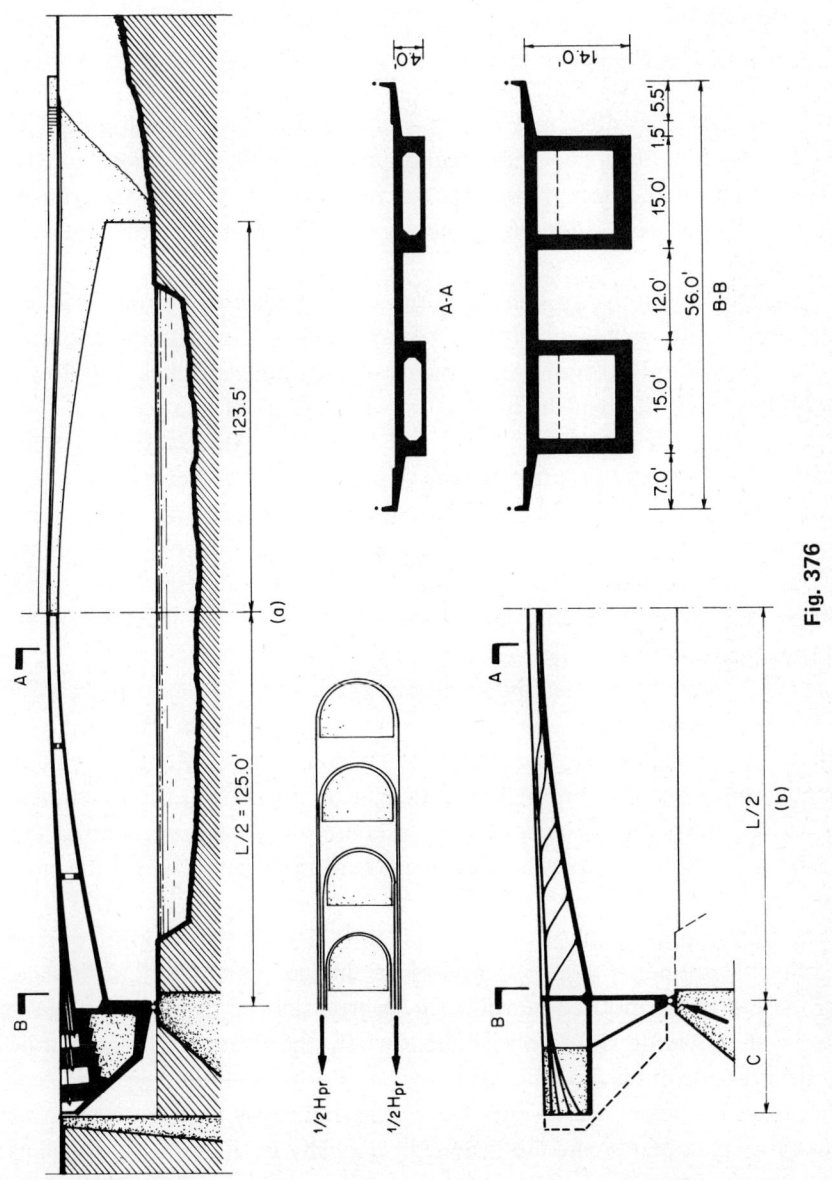

Fig. 376

tensioned by hydraulic jacks against the hardened concrete from the anchorage recesses. The concrete disks resisting the prestressing force are shear-reinforced and transmit the force into the deck slab and sidewalls. After the cables are tensioned, grooves and anchorage recesses are filled with concrete.

The bridge illustrated in Fig. 376b is a two-hinged frame arch. The cross section and prestressing technique are similar to those of the three-hinged arch beam. However, for relatively long spans, to control the shearing stresses the arrangement of cables shown in the figure is more appropriate.

Bridges for rapid-transit systems form a special class of railroad bridges. The span length commonly has to be variable because of the restrictions for pier location. Furthermore, in long sections the bridge has to follow a curvature, and its width has to be increased at station areas to accommodate the platforms, stairs, and escalators. Also, station sections have to be protected against climate. However, for aesthetic reasons, the construction depth should be kept uniform and constant. Considering the heavy loading, high speeds, and acceleration and deceleration forces involved, the design is a rather demanding one. In addition, since these bridges run through highly populated areas, noise and dust problems require the utmost attention.

A design which satisfies most closely these high demands is given in Fig. 377. As can be seen, the cross section is a box with relatively long cantilevers at both sides. The simple box widens gradually and becomes a two-chamber box at station areas. In station areas also, the cantilevers and piers have to be widened to accommodate the platforms and escalators. Depending on the design of stations, double piers are most often required.

The box girder is supported at each pier by double two-point neoprene bearings. Movements due to temperature change, plastic flow, shrinkage, and acceleration and deceleration forces are balanced by prestressing the piers and providing slip joints. In areas where the street traffic cannot be interrupted, drop-ins must be used.

To reduce noise, dust, and wear, the rails rest continuously over neoprene on concrete and are fastened elastically in the vertical as well as lateral direction. The ties (wooden or prestressed) in ballast bed, which are generally used, have far higher maintenance costs. Also, dust and cleaning problems for high-speed trains are considerable.

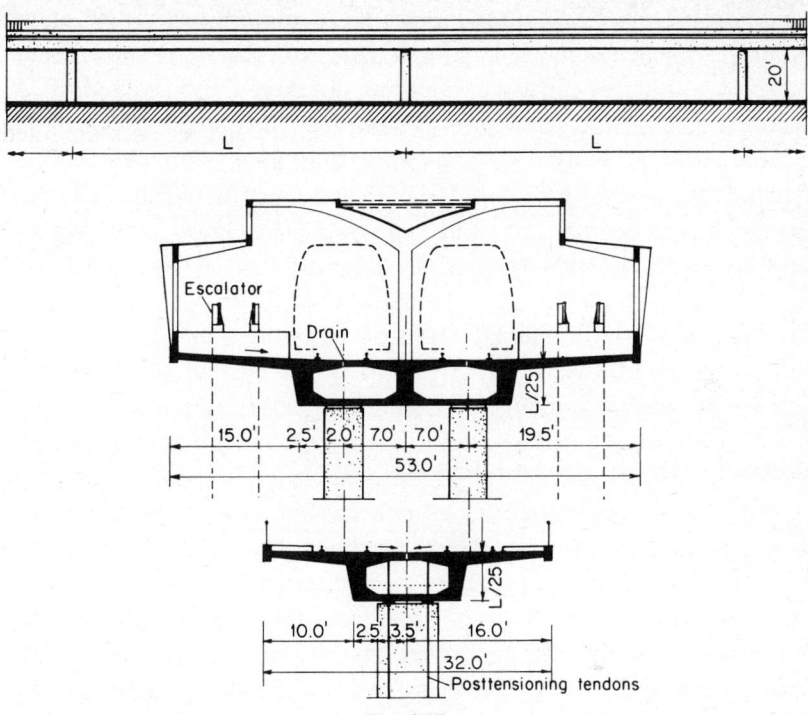

Fig. 377

71. LONG-SPAN RIVER AND VALLEY BRIDGES

Basically, the design principles for long-span bridges are the same as those for short-span bridges, discussed above. However, there is a considerable difference in construction methods. Since the cost in most cases is the decisive factor in the selection of a bridge type, prestressed concrete bridges designed and constructed in the traditional way cannot compete with steel bridges in the long-span range, mainly because of the enormous expenditures for stable, rigid scaffolding, the long construction time, and the limited spans, which require too many piers to be constructed in open water or in deep valleys, a procedure which is costly and also time-requiring.

Much of the credit for the development of long-span prestressed concrete bridges belongs to engineers, such as Finsterwalder and many others, who had the vision to realize that steel bridges in this span range can be challenged only by abandoning the traditional reinforced-concrete construction methods and developing and adapting entirely new, more economical construction techniques. One of the best-known new techniques is the free-cantilever method developed by Dyckerhoff & Widmann (Finsterwalder) and used to build the Nibelungen Bridge across the Rhine River at Worms (1952). Since that time, numerous bridges throughout the world have been designed and constructed in keen open competition successfully by this method. The latest and longest prestressed bridge of this type is the Bendorf Bridge, also across the Rhine (1964).

The layout of the western section of this bridge and the technical details are given in Fig. 378a. Structurally the system of the double box-girder (separated by continuous longitudinal joints) bridge is rather simple: a continuous beam with seven spans symmetrical about the shear hinge (Fig. 378b) at the center of the main span. The system is seven times statically indeterminate for all conceivable loading conditions but is only three times for symmetric loading. The hinge arrangement provides for longitudinal movement of the structure due to shrinkage and temperature change. The box girders are rigidly connected with the main piers and rest on bearings (roller bearings on a thin layer of teflon to allow transversal movement) of the other piers. To avoid uplift, the boxes at piers C and F are filled with gravel ballast and are prestressed to the piers. The shape of the box girders in elevation is determined in such a way that the tensile and compressive stresses increase approximately linearly from the end of the cantilevers (shear hinge) to the support. Such linear increase of stresses results in almost constant shear in the webs along the entire length of the cantilever. Thus the thickness of the webs and the amount of reinforcement can be designed as constant. The span-depth ratio of the main span at the supports is 20 and at midspan is about 48.5. The structure is prestressed with 1¼-in. high-strength bars in three ways: longitudinally and transversally over the cross section of the deck slab and diagonally in the webs. The bottom slab is prestressed only at the sections under the ballast. With this type of prestressing arrangement, the concrete behaves as homogeneous material and the entire cross section participates fully in the carrying action.

This structural system allows long free cantilevering without auxiliary supports because the structural behavior under dead-load condition during construction is practically the same as for the final state. Because of the shear hinge, the moments are zero at midspan and the depth of the box girders smallest there. Thus the dead load is minimum at

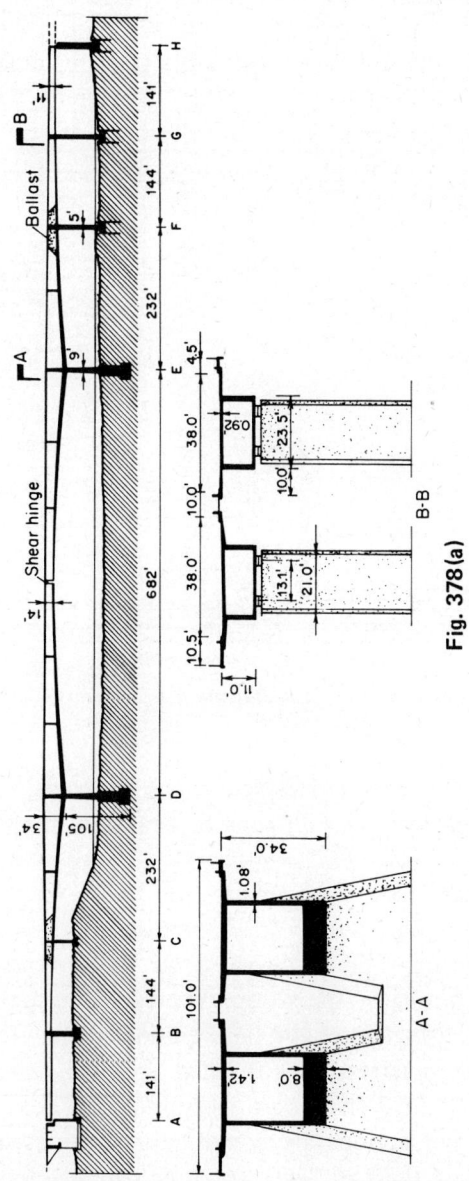

Fig. 378(a)

582 Structural Systems and Methods of Construction

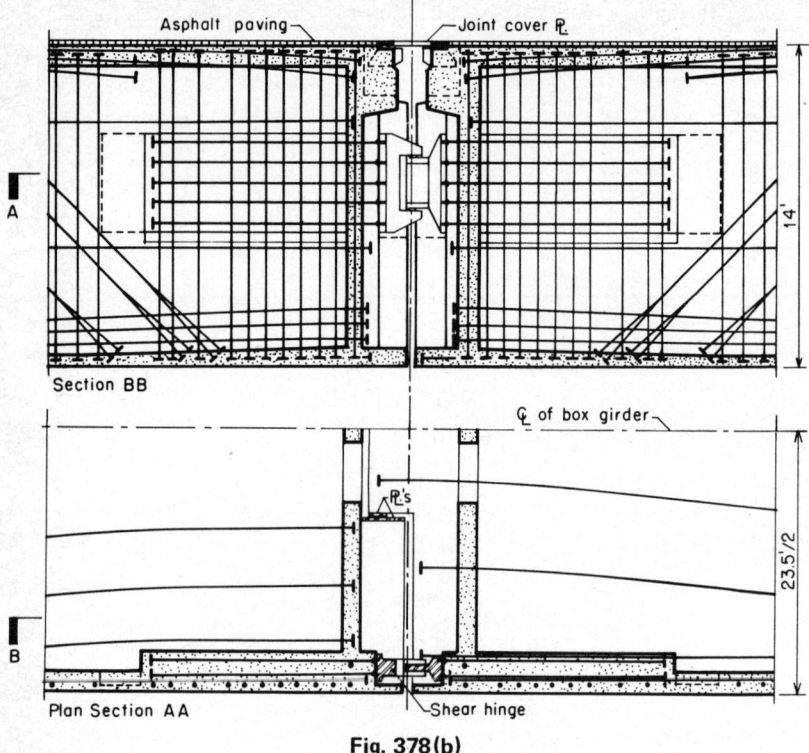

Fig. 378(b)

midspan and is concentrated closer to the supports, which are rigidly connected with girders. This all adds to structural soundness and higher economy.

The construction starts symmetrically from piers D and E and proceeds step by step, with 11.5-ft sections poured in place toward midspan until closing (shear hinge). After the concrete has developed about 4,000-psi strength, the section is prestressed against the previous one by means of high-strength bars ending at the front face of the section. When the prestressing is finished, so-called *cantilever trucks* are moved by winches 11.5 ft forward to support the next section. To start the operation, two sections are commonly constructed in the conventional way so that the cantilever truck can be erected and tied down properly. The truck has three or four levels, on which all preparations for reinforcing, prestressing, and casting are done. Construction of each section is strictly scheduled so that the balance of weights of the symmetrically executed cantilevers remains within safety limits at all times (Fig. 379).

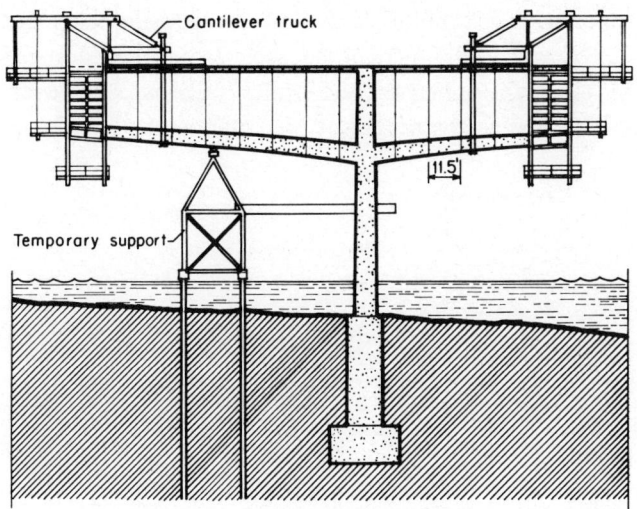

Fig. 379

Due to the step-by-step construction, the layout and arrangement of prestressing bars are closely related to the construction method and therefore deviate from the typical arrangement. For easy placing, the longitudinal bars are located in groups over the entire cross section of the deck slab. The bars are straight except at the ends, where they are bent and anchored in the webs slightly below the anchorage of the diagonal bars placed at 45° angle. A certain number of longitudinal prestressing bars are carried through unstressed at a length to balance the difference of the prestressing induced by the bars anchored regularly at the forefront of the sections and prestressed corresponding to the static requirements of the construction. All bars are grouted immediately after prestressing is applied. The transversal bars follow the transversal moments of the deck slab and are also grouted. The prestressing bars are all of the same size (1.26 in. in diameter) and are delivered in flexible metal tubing in unit lengths: longitudinal bars, about 3 × 11.5 = 34.5 ft long; transversal bars in full length, about 43.0 ft. The length of diagonal bars varies. The units are connected by threaded couplers to required length. The bars have ultimate strength of 149,000 psi and yield strength of about 113,500 psi. The allowable stress in the bars under design load is approximately 75 percent of the yield strength—thus about 85 ksi, or roughly 105 k per bar. Each box girder is prestressed at piers E and D by 560 bars. The moments and general layout of prestressing bars are illustrated in Figs. 380 and 381. As can be seen from the moment diagrams, there will be tension in concrete under full load; however, these stresses are below the tensile strength of the concrete used.

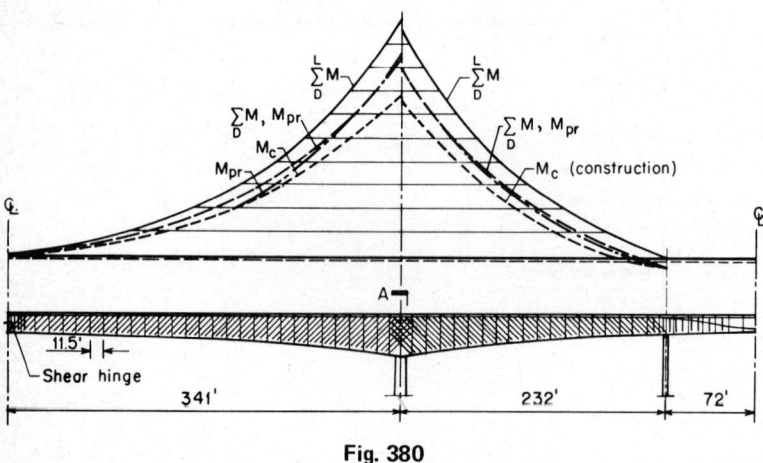

Fig. 380

The bottom slab is reinforced by conventional reinforcing in the longitudinal and transversal directions. The high amount of compressive reinforcing is used mainly to reduce the dead load, to provide longer lever arm for the prestressing bars, and to reduce the influence of plastic flow, shrinkage, and sensitivity of the permanent shear hinge.

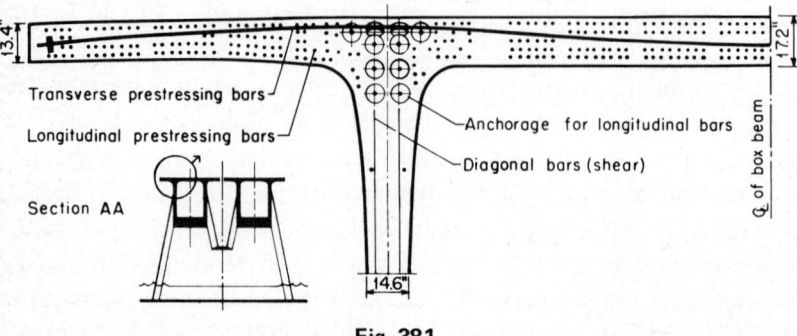

Fig. 381

Regardless of strict planning and construction and two-way symmetry, it is unavoidable that the stresses of a multispan system during construction cannot be kept below the final stresses without using artificial support settlement by hydraulic jacks or auxiliary support.

The free-cantilever method lends itself successfully also for the construction of long-span, constant-depth, continuous-beam bridges, such as the Viesville twin bridge in Belgium (Fig. 382).*

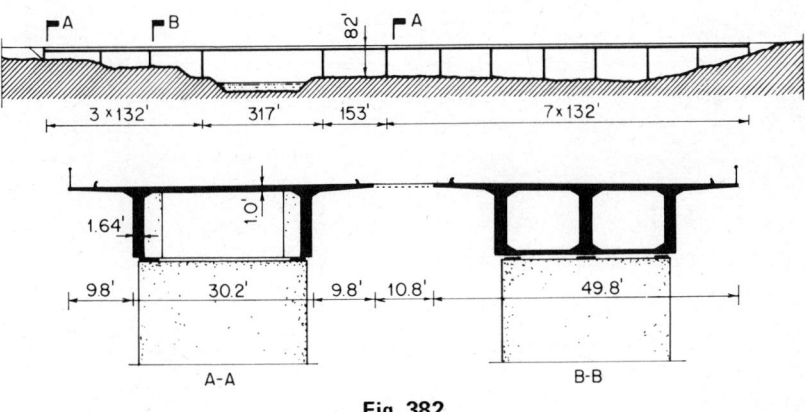

Fig. 382

The bridge was built in 1966 and has 12 spans. The main span, over the Brussels-Charleroi Canal, is about 317 ft long. The overall length is approximately 1,800 ft, and the height is about 82 ft above the high-water level. The main and two adjacent spans have a two-compartment box cross section, and the other spans have double-webbed T beams. The span-depth ratio of the main span is about 27. The bridge was constructed by four cantilever trucks with the aid of pylons and suspenders moving in one direction only. In the middle of the main span, the two cantilever arms are rigidly connected.

A series of frame bridges across deep valleys and densely populated areas was constructed by this method in Italy. Among them is the Bisagno Bridge in Genoa (Fig. 383),† with main spans of about 380 ft.

*Design: Arnold Bagon, Bureau d'Etudes, Brussels. Contractors: S.C.E. Eugema, Brussels.

†Design: Silvano Zorzi, Studio Tecnico, Milan. Contractors: Impresa Moviter, S.p.A., Milan.

The tall piers (up to 330 ft) of these bridges were constructed economically by slip-form technique. The cantilever arms are connected by shear hinge, except in the Nervi bridge, where drop-ins were used to close the gap between cantilever arms.

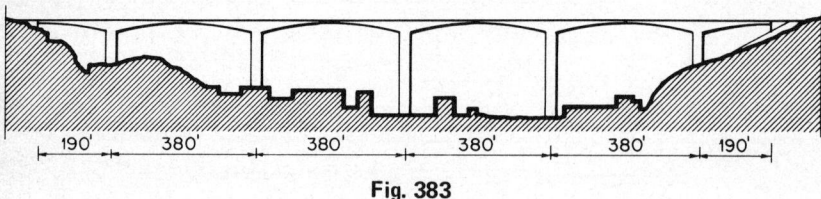

Fig. 383

The Nagoya-Genkai Bridge in Japan, with a main span of about 578 ft and relatively short adjacent spans acting as counterweight, was also constructed by the free-cantilever method (Fig. 384).*

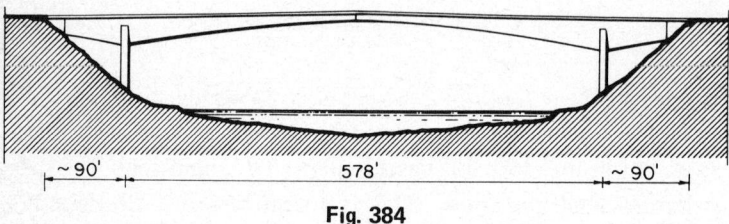

Fig. 384

For free-cantilever-method construction, the capital investment is rather small, no complex and high-capacity handling equipment is required, and because of the large number of repetitions of the same operations, skilled labor and overall man-hour requirement are relatively small. Due to this, the free-cantilever method has proved to be one of the most successful and safe methods in long-span bridge construction, and it has been used throughout the world.

In addition to the free-cantilever method, numerous other methods have been developed for constructing long-span bridges without expensive scaffolding. However, for economical reasons, they are of only limited value for shorter spans, whether poured in place or erected from precast units.

*Design and contractors: Sumitomo Construction Co., Ltd., Tokyo.

The methods developed by Campenon Bernard, Paris, Dyckerhoff & Widmann, Munich, and Polensky & Zollner, Frankfurt, are based on the same principle: a huge girder spanning from a previously finished end of the superstructure to the following pier, from which a movable form or carrier is suspended or supported, depending on the local conditions.

The Campenon Bernard system is suitable for the construction of prefabricated or poured-in-place bridges. The system as it is used in erecting prefabricated units is schematically illustrated in Fig. 385. The launching truss girder is supported on legs A and B atop the deck of a completed cantilever arm and on front leg C on a steel frame fastened to following pier. From this position, the center block D and a temporary steel tower at the top are erected (1). Then the launching girder over tower D is rolled forward (2) into position, and the next cycle of symmetric erection is carried out as indicated (3).

The sections, manufactured in a nearby casting yard, are commonly 10 to 12.5 ft long and the full width of the bridge. The section is

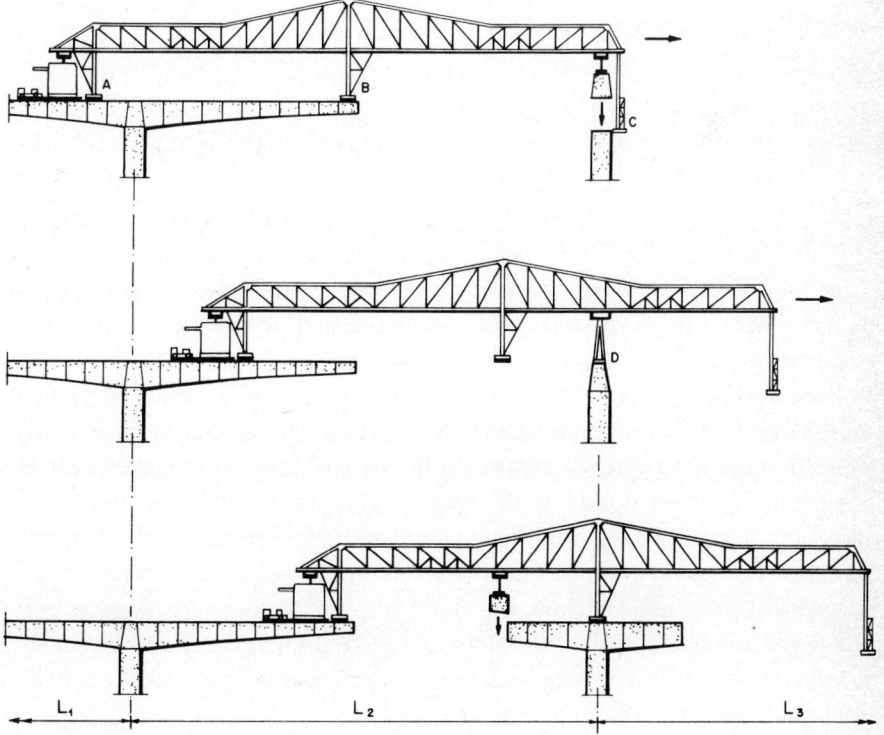

Fig. 385

transported by travel lift from the plant and along the completed part of the bridge to the front of the launching girder, from where it is picked up by the traveling-frame device. The traveling-frame device is suspended from the bottom chord of the girder, serving as a twin monorail. To the traveling frame is attached a spreader beam, which can be raised, lowered, or rotated so that wide sections will pass between the supporting legs of the truss girder and can be lowered into proper position. The sections are fastened into place with epoxy and prestressing tendons.

The piers are commonly slip-formed. However, they can also be precast and erected by the same launching truss girder.

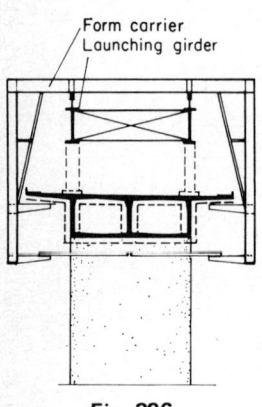

Fig. 386

The same type of launching girder and sequence of operation are used to construct poured-in-place bridges. The launching girder in this case has to be a parallel-chord truss, so that the movable form carrier (Fig. 386) can run at the top chord of the truss and bypass the supporting legs of the truss. The traveling frame running on the lower chord serves for handling the reinforcing and delivering concrete.

The length of span depends on the length and carrying capacity of the launching girder. The initial cost of the overall system is considerable for spans up to 300 ft and, therefore, from a competitive aspect sets limits for its use.

Dyckerhoff & Widmann uses for construction of poured-in-place moderate-span bridges a movable scaffold carrier, which is schematically illustrated in Fig. 387. The movable scaffold carrier is a steel structure consisting of two longitudinal beams, one extended centering beam, and several transversal girders combined into a so-called *girder grillage*. It is supported on two points at the edge of the previously completed span and on one point on the following pier. The forms are suspended by steel rods during construction of the span. After the span is prestressed, the suspension rods are removed and the forms are laid down on arms which clasp the superstructure at the sides and below. Then the movable carrier is slid forward to the next span on the finished deck and the next column through the centering beam.

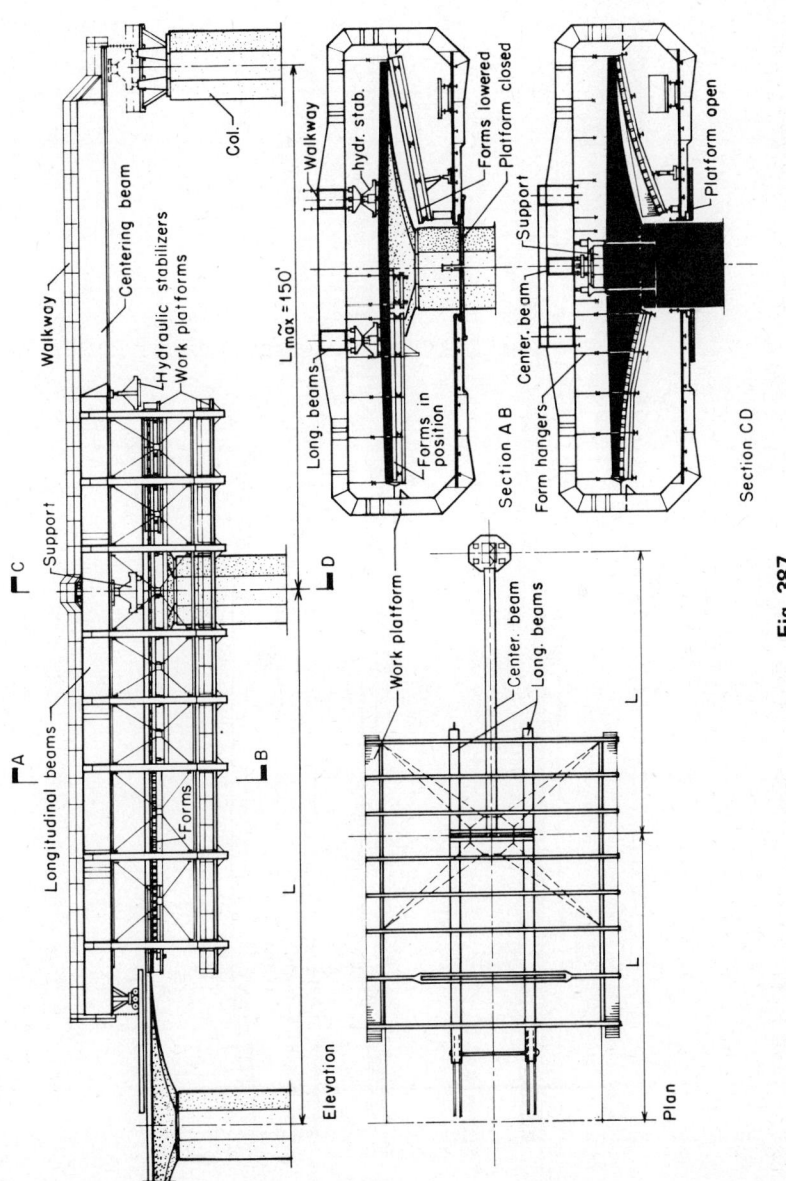

Fig. 387

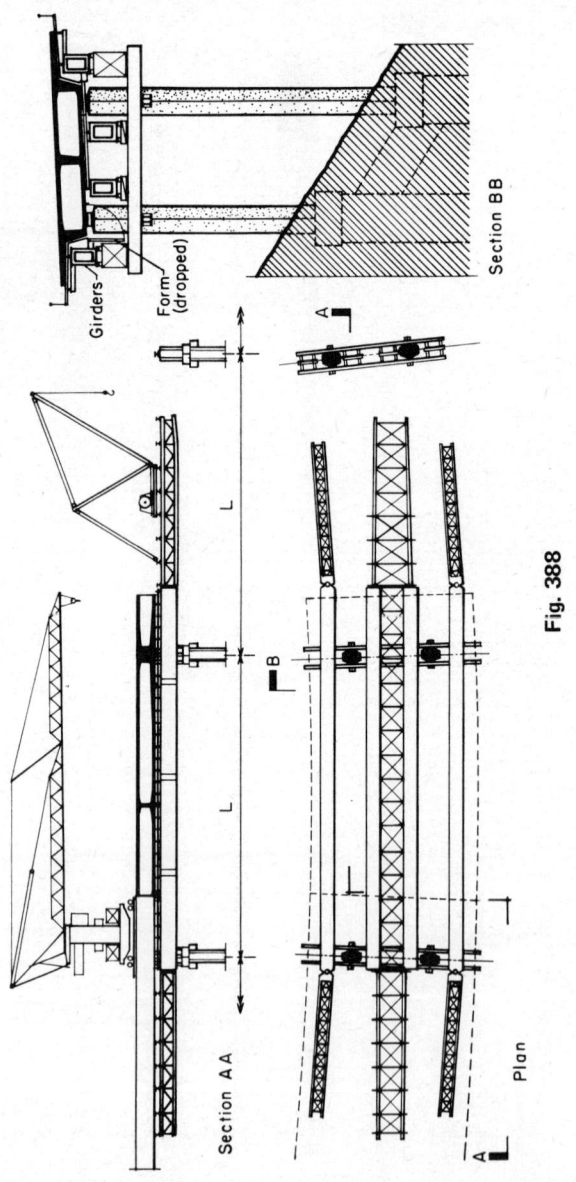

Fig. 388

The handling of reinforcing and the delivering of concrete are accomplished by means of the monorails fastened to the transversal beams of the grillage. The carrier is commonly overroofed so that the work can proceed without climatic interference.

The Polensky & Zollner system, known as *stepping formwork equipment*, is carried by stepping girders arranged beneath the superstructure supported by columns through stepping cross girders (Fig. 388). The equipment consists of two, three, or four stepping girders. The form panels are hinge-connected to the stepping girders and are adjustable to the curvature of the superstructure. For stripping and traveling, the stepping girders are lowered and the bottom form panels dropped to clear the piers. The inside forms are dismantled and reerected from span to span. The center girders are rigidly connected together but are detachable from the outer girders by bottom forms, thus allowing independent sliding forward. The stepping girders have truss extensions at both ends, making them approximately twice the span length, so that the equipment has stable support during the forward movement.

After the equipment has been moved to proper position, the stepping girders are hydraulically jacked into exact elevation and the forms are closed and readied for placing the reinforcement and tendons and for pouring the concrete.

Materials handling is carried out by a crane traveling on the finished deck slab. For erection of a cross girder on the next following pier a small derrick located on the extension trusses of the center girders is used.

For longer-span construction, Polensky & Zollner employs also a long launching girder (formwork equipment) similar to the one used by Campenon Bernard (Fig. 385) to carry the movable-form wagon. The system was successfully used for the construction of the Sieg Valley Viaduct, Eiserfeld, Germany, with maximum spans of 345 ft and a total length of 3,450 ft (Fig. 389). The operation proceeds symmetrically in

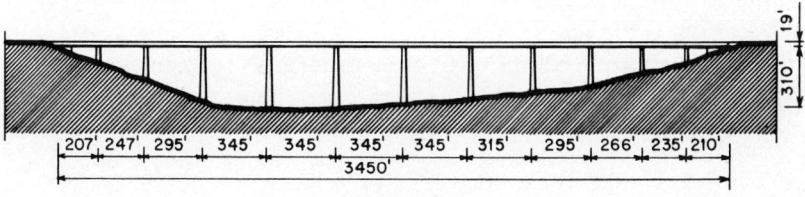

Fig. 389

both directions from one pier until the completed cantilever arm of the previous erection of the superstructure is reached. Then the launching girders are moved forward to the next span, etc. The length of the launching girder is approximately 1½ times that of the longest span. In the case of the Sieg Valley Viaduct, this is about 520 ft. The cross section of the girder and formwork equipment is given schematically in Fig. 390. The length of the section poured was approximately 40 ft.

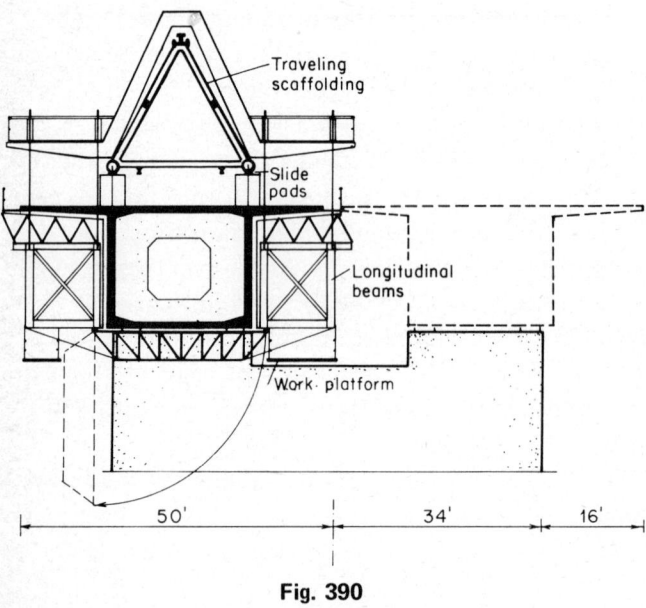

Fig. 390

For the Danish Great Belt Bridge, Polensky & Zollner proposed a prefabricated, prestressed continuous-beam design (Fig. 391),* with center span of about 985 ft and adjacent spans of 1,070 ft (total length, 4.5 miles), executed by the use of about 1,150-ft-long formwork equipment similar to that used for the Sieg Valley Viaduct. The equipment is designed wide enough to allow the erection of both superstructures at the same time.

Under construction conditions, the structural system consists of statically determined free cantilevers, which will be prestressed to continuous beams for live-load condition.

*One among the four first prizes in the International Competition for the Great Belt Bridge (155 entries, 1967).

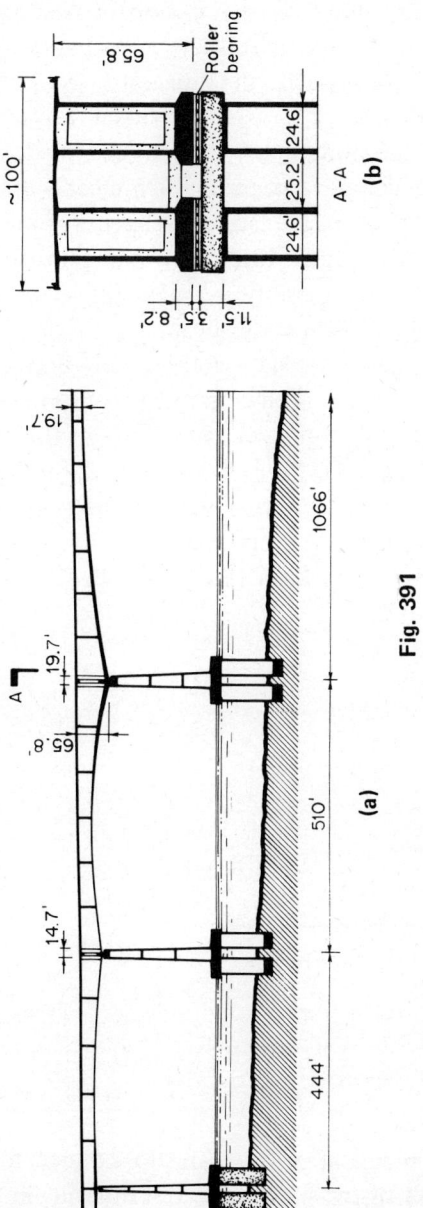

Fig. 391

A quite different type of prestressed bridge for considerable span, up to 1,000 ft or even more, is the diagonal-cable bridge (cable-stayed bridge) shown in Fig. 392. Statically the diagonal-cable bridge is a prestressed bridge in which the diagonal cables are independent structural elements arranged outside the superstructure of the bridge but acting similarly to the nonbonded tendons confined inside the girders of the superstructure. The advantages of such a cable arrangement over the typical nonbonded tendons are larger lever arm for prestressing force and self-prestressing by dead load. Under live-load condition, the prestressing force does not increase notably because the cable structure is comparatively more flexible than the beam structure. However, the interaction of the cable and live load is considerably improved if the diagonal cables are encased in prestressed concrete because the ratio f/E (strain) is smaller for concrete than for steel. Advantage of this fact has been taken in the Polcevera Bridge (Fig. 393),* with a maximum span of 689 ft. For the same reason and to increase bond, prestressed concrete bars have been used successfully as reinforcing in critical areas to control elongation and crack formation (see also the discussion of K systems, Sec. 49).

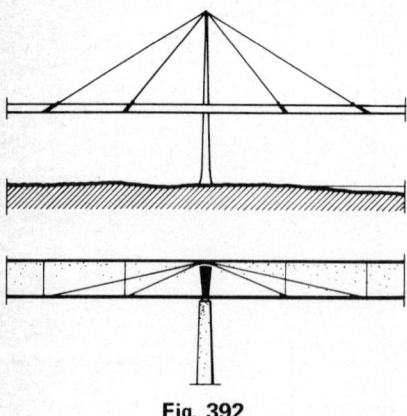

Fig. 392

The cables, for practical reasons, are nonbonded to allow adjustment during construction of the superstructure. Bonding in this particular case does not have appreciable advantage, except for protection of the cable against possible corrosion, because the cables are subjected to uniform

*Designer: Riccardo Morandi; Genoa, 1966. A similar bridge (main span, 940 ft) designed by Professor Morandi (1971) is under construction in Libya.

stress along their entire length. The number of diagonal cables depends on the length of span.

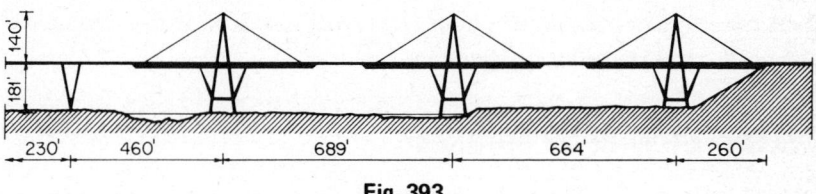

Fig. 393

Aesthetically the system is acceptable, provided that the pylons and their connections with the superstructure are simple. Trestle-type pylons and supports composed of several inclined elements create confusion and distorted views and therefore should be avoided.

For the Great Belt Bridge, Dyckerhoff & Widmann proposed a diagonal-cable bridge* having spans up to 1,150 ft. The deck is 145 ft and carries six lanes of traffic and a double-track railway located in the middle of the deck between two rows of the diagonal cables. The bridge deck is a solid prestressed concrete slab having an average thickness of 2 ft. The design is shown in Fig. 394.

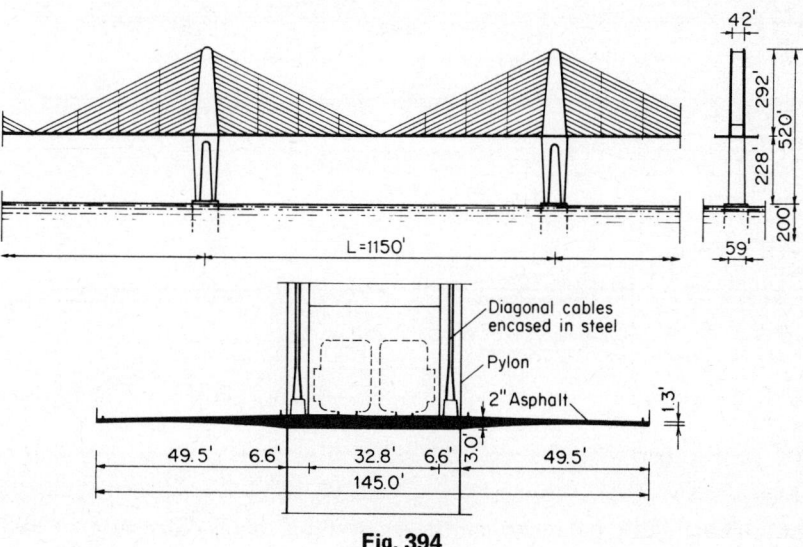

Fig. 394

*One among six second prizes in the International Competition for the Great Belt Bridge, 1967.

596 Structural Systems and Methods of Construction

Structurally the bridge consists of a series of independent balanced cantilevered systems, each carried by single pylon. The overall system is jointed together by shear hinge at midspan. The great advantages of this system are its simplicity and economy; the deck slab is rigidly connected with pylons, and no roller bearings are required.

72. SUSPENSION-BEAM BRIDGES

A unique bridge-design entry in the Great Belt Competition is illustrated in Fig. 395.* Statically it can be classified as a *suspension-beam system*. The bridge carries six traffic lanes and a double-track railway. Overall width of the deck is 141 ft. The bottom slab of the huge box is a box in itself, having constant depth of 13.1 ft and serving as the compressive chord for the static system. The depth of the huge box at support is about 93.4 ft and at midspan is 16.4 ft. The top slab of the box ends 450 ft from the center of support to provide clearance for the railway. It serves as tension chord for the system and has a varying width, 7.6 ft at the center of the pier to about 35 ft where it ends. The bottom slab has a constant width of about 41 ft.

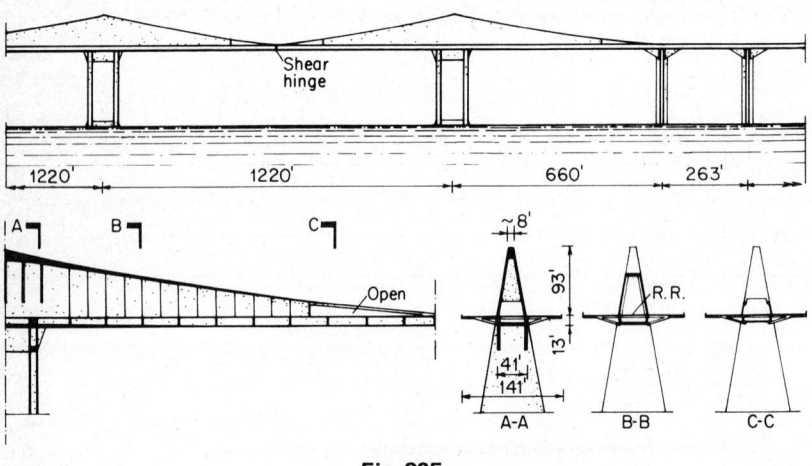

Fig. 395

The independent, balanced cantilever systems, each consisting of two cantilevers, project symmetrically from the 92-ft heavy pier for about

*Designer: Herbert Schambeck, Germany; second prize.

564 ft. Thus the overall span is 1,220 ft. The free cantilevers are jointed by shear hinge at midspan.

The railway runs on the deck slab inside the 41-ft-wide box. The traffic lanes at the same level are arranged on a cantilevered deck slab each projecting from the box about 50 ft.

The bridge is most economically constructed by free-cantilever method in about 25-ft sections.

The most fascinating long-span bridge type is the Finsterwalder stress-ribbon bridge. Statically the system is based on the most economical carrying action, that is, suspension action. It is an extremely simple design, in which all elements of the conventional suspension bridge are eliminated or, more precisely, replaced by a single structural element—the stress ribbon—which follows the grade of the roadway.

The suspended ribbon of prestressed concrete is anchored in the abutments and is supported at the intermediate supports, which are provided with relatively long cantilevers in order to obtain satisfactory grade. The bridge is schematically illustrated in Fig. 396.

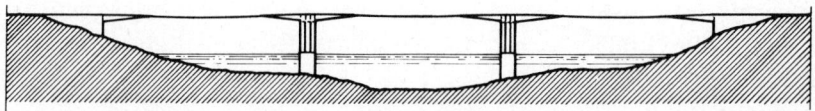

Fig. 396

The construction of the bridge proceeds, after the cantilevered piers are completed, by placing the required number of tendons piece by piece and stressing them so that the desired funicular curve, to meet the grade requirements, will be obtained under full-load condition. The formwork is hung from the tendons. The stress ribbon, about 10 in. thick, is heavily reinforced in the transversal direction at both the top and the bottom so that it can resist the torsion and bending moments resulting from unsymmetric live loading. The concreting starts in the middle of the spans and proceeds symmetrically toward the supports without interruption. After the concrete is properly cured and hardened, the forms and preloading, if required, are removed. The over-prestress in the stress ribbon due to the removed weight must be adequate to control and maintain the tensile stresses due to live load and shrinkage below the tensile strength of the concrete.

The counterweight, necessary to resist the thrust of the stress ribbon under favorable soil conditions, is about twice the horizontal thrust.

About a 2-in. thickness of asphalt wearing surface on the stress ribbon is considered adequate.

This bridge type makes it possible to compete successfully with steel bridges from spans longer than 650 ft up to 1,500 ft and even longer, provided that the topographic and soil conditions are favorable.

73. ARCH AND ARCH-BEAM BRIDGES

The economical span limit for prestressed beam bridges is reached with spans of approximately 600 ft. Beyond this span limit, the solution is forced, and still longer spans are justified only by special local conditions. For span ranges of 600 to approximately 1,200 ft, the diagonal-cable bridges, suspension-beam bridges, and stress-ribbon bridges are dominating at present.

Arch bridges, which structurally also belong in this span category, have lost or are losing their natural place in the long-span bridge field, in spite of their aesthetic superiority over the other bridge types and even though their carrying action is based on the most efficient use of concrete. The most significant, most recent, and longest concrete arch bridge built is the Gladesville Bridge across the Paramatta River in Sidney harbor, Australia (Fig. 397).*

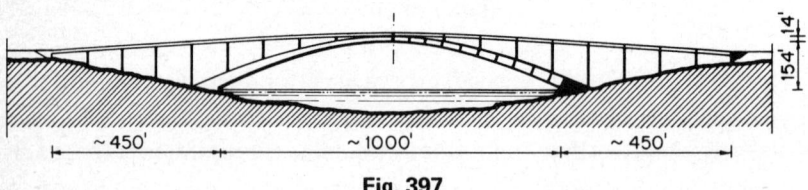

Fig. 397

The superstructure has approximately 1,000-ft spans. The bridge carries six lanes and 6-ft sidewalks at each side, for a total width of 84 ft. The fixed arch consists of four interconnected precast box sections, each 20 ft wide and about 9 ft long. The depth of the arch is about 23 ft at abutments and 14 ft at the crown. Abutments are of mass concrete cast into the sandstone riverbanks.

*Designer: G. Maunsell & Partners, London and Melbourne, 1964. Contractors: Reed & Mallik Ltd., Salisbury, England, and Stuart Bros. Pty. Ltd., Sydney, Australia.

The four arch ribs are erected one at a time on steel falsework. The 3-in. (approximately) joints between the arch ribs and the 2-ft-wide transversal ribs, arranged in intervals of five box sections, are cast in place. The transversal ribs are prestressed to obtain unity. The stripping of falsework of the individual arch ribs was accomplished by jacking the ribs at quarter points and raising them about 8 in. at the crown.

The superstructure consists of eight prestressed, prefabricated I beams, each 6 ft deep and 100 ft long. The deck slab is poured in place. The method of construction of this bridge is the best available at present, but when compared with contemporary long-span beam-bridge construction, it is expensive and time-requiring. Also, the prevailing arch-bridge systems involve practically two unrelated bridge designs: arch design and beam design for the superstructure carried by the arch, which obviously adds to the expense. In addition, the structural analysis of long-span arches is rather complicated for most bridge engineers. This all explains why arches are seldom seriously considered, even for locations most favorable for arch bridges.

To make arch bridges more economical, radical changes in design as well as construction methods must be developed and adapted. As far as design is concerned, the arch and superstructure can be combined to some degree into a unified structural system in which both participate in the overall carrying action. Structurally such a combination of beam and arch action is identified as *arch beam*.

One of the first engineers who used the arch-beam bridge in simple form was Robert Maillart. Maillart designed a series of arch-beam bridges in Switzerland. The most interesting and longest span among them is the Schaffhausen Bridge over the Rhine (1935), illustrated in Fig. 398.

The system consists of a shallow polygonal arch, with a thickness able to take only normal forces and too flexible to resist notably any bending, and a relatively stiff beam resisting the bending moments. The two carrying elements are tied to unity by thin vertical transversal bearing walls. The rather stiff beam acts also as a stabilizer for the slender arch, and the arch, on the other hand, offers elastic support to the beam. Thus all the elements the system is composed of participate in a most efficient way in the carrying action of the bridge.

The spacing of vertical walls is controlled by the buckling stability and acceptable shape of the thin polygonal arch. The most suitable cross section for the stiffening beam is a one- or two-chamber box or T sections with cantilevers. At the crown the arch and beam are joined and

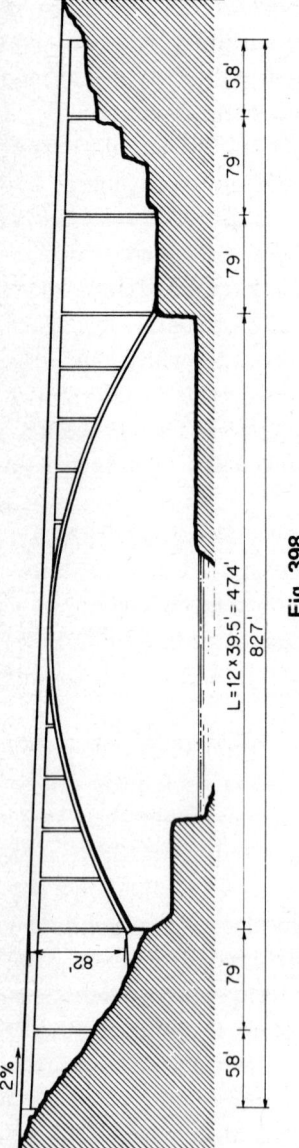

Fig. 398

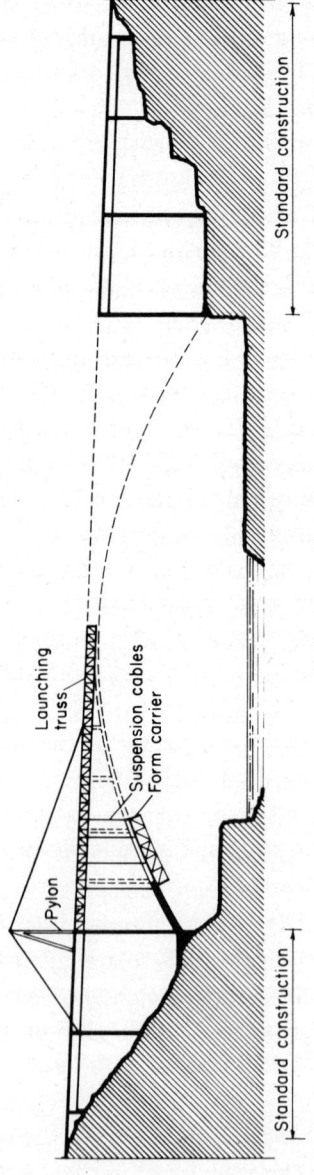

Fig. 399

form a combined rigid section of the system. To express the true functions of the two carrying elements—beam and arch—it is advisable to design the arch slightly wider than the box section of the beam so that the arch runs its course uninterrupted from abutment to abutment.

The construction of such a relatively light arch-beam bridge is simple and economical. First, the thin arch is constructed upon a suspended formwork. Then, after the concrete is hardened, the superstructure, supported by the arch, can be constructed. Construction of the arch proceeds in steps, as illustrated in Fig. 399. The form carrier for the arch can be suspended simply and economically from a prestressed launching truss, which can be cast in and which serves as reinforcing for the stiffening beam (Fig. 359). For long spans use can be made, if required, of pylons and auxiliary supports at the crown, whichever is most economical and practical. The form carrier for construction of the superstructure (stiffening beam and deck slab) will be supported by the transversal walls and launching truss.

For single-arch spans, the launching truss, after it has been used, would commonly be to large extent wasted. Pouring it in not only allows it to be made full use of but also permits the tendons to be assembled inside the truss prior to their use (see Fig. 359) in an exact and economical way. Some of the tendons can be prestressed gradually to help the truss carry the load it is subjected to while carrying the formwork and its own dead load.

If larger spacing of the walls or columns supporting the superstructure is desired, ribs must be provided or a box section used to increase the stability of the arch, as indicated in Fig. 400. For relatively wide bridges and long spans, more than two ribs or chamber box for the arch or even twin arches will be required.

To control temperature stresses and simplify construction by free-cantilever method, the three-hinged arch is structurally the most appropriate and economical. However, a two-hinged arch-beam system, at least for live-load condition, is possible and in some cases preferable.

A long-span arch-beam design of this type over open water is illustrated in Fig. 400.* The eight-lane superhighway in the east-west direction has to cross two channels with branch-off bridges to the south at the location of a small island between the channels (Fig. 371). The minimum clearance for seagoing boats is 85 ft above both channels. The east bank

*Study by the author for the Baltimore Inner Harbor Bridge, 1967.

602 Structural Systems and Methods of Construction

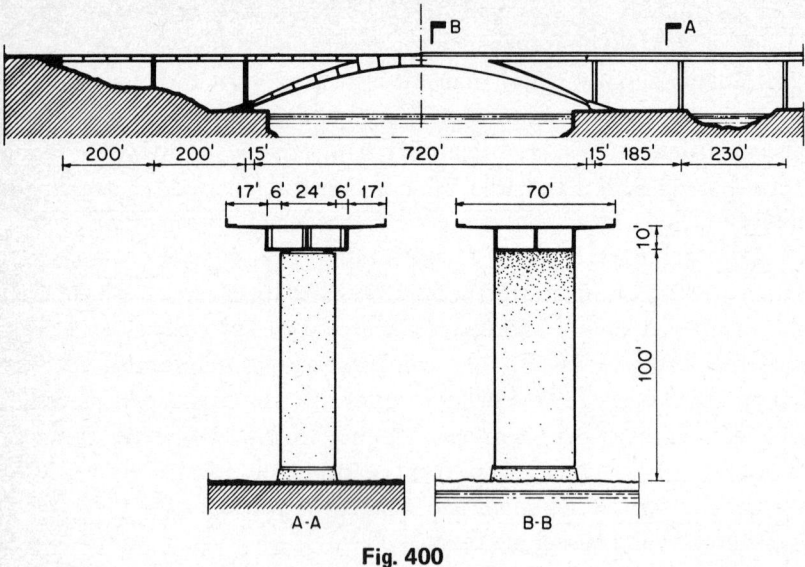

Fig. 400

of the twin bridge rises sharply up to 70 ft from the ordnance datum. The west side at the intersection is a level, populated area. The minimum clear span from the east bank to the small island is approximately 700 ft; for the curved branch-off bridges, about 200 ft is required to clear the small channel crossings and the west-east roadways.

The arch of such a bridge can be erected in a manner similar to that described above (Fig. 399) by the use of launching trusses which will be cast in. The construction of the arch starts, after the abutments and adjacent spans are finished up to the expansion joints, by erecting launching trusses and pylons. At the crown the trusses must be supported by barges until pylons and diagonal cables are erected. The forms for arches are suspended from the launching trusses. The work must be carried out simultaneously from both abutments. After the arch is finished, the beam (superstructure) can be constructed, starting symmetrically from the crown. The forms for the beam are also suspended from the trusses.

A similar arch-beam design is shown in Fig. 401.* The minimum clearance for navigation is 115 × 300 ft. By the use of an 800-ft main span and 400-ft adjacent spans, only two piers in deep water and two piers in rather shallow water are required. To reduce the arch span and to

---

*Study by the author for a bridge across the Delaware River, 1968.

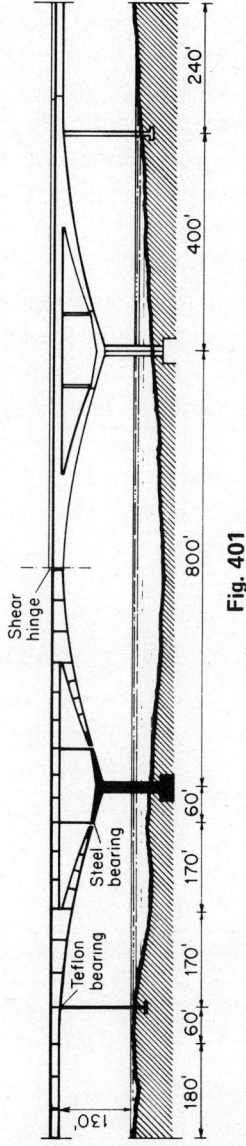

Fig. 401

simplify the free-cantilever construction, the main piers are cantilevered at each side symmetrically about 60 ft. The arch halves, supported by the cantilevers, are identical and symmetrical and are tied by prestressing at the superstructure level to stable, balanced structural unity for dead-load condition. Due to this, the units supported by each cantilevered pier can be constructed independently. The construction of the arch halves starts, after the cantilevered piers and the pier and part of the superstructure supported by cantilevers are finished, by erecting the pylon and launching trusses. Since the launching truss can be used several times, they will not be of the cast-in type. The construction sequence for the arches and superstructure is similar to that described above.

To keep the size of the launching trusses and pylons within manageable limits and to simplify the construction, the box section of the arch can be cast in two separate pours: first the bottom slab and then, after the concrete is hardened, the ribs and top slab. In this way, the bottom slab will help to support the upper part of the box section. Also, auxiliary support at the crown for the same purpose can be used, provided that the soil condition is favorable and that it does not interfere with river traffic.

18

MISCELLANEOUS STRUCTURES

74. WATER TOWERS

The stereotyped steel water towers decorating the skylines of communities small and large are, regardless of how economical they may be, uninspiring ugly landmarks. This applies also to exposed water tanks on the roofs of high buildings. Besides performing its function, a well-designed water tower can be an aesthetic asset for any community, but this has been widely overlooked by professionals as well as by authorities responsible for the design and approval.

However, since prestressing techniques have advanced, concrete water towers, designed mainly in European countries (especially in Finland, Germany, and England), are works of contemporary art. The structural analysis and design of prestressed concrete water towers are rather complex and constitute a real challenge to engineers because the system is statically highly indeterminate, subjected to variable loading conditions (seismic forces, wind, water pressure, and prestressing) and disuniform temperature changes. In addition, severe crack formation has to be controlled and concrete has to be durable and watertight.

Commonly the water tank consists of two or three chambers which can be separately emptied for cleaning. In severe climatic conditions, the external walls of the tanks must be insulated against freezing. Keeping the variation of hydraulic pressure within acceptable limits controls the height as well as the shape of tanks. The capacities of the tanks vary widely, depending on the size of the community, peak-hour supply requirements, etc.

A typical layout for a prestressed concrete water tower with two chambers is shown in Fig. 402. The shaft of the tower consists commonly of one or two cylinders. The inside cylinder serves to support the fall, rise, and overflow piping and to carry the stairs, elevators, etc. The outside cylinder supports mainly the outside conical and inside cylindrical tanks. The outside cylinder of higher towers can be reinforced by tapering legs for stability.

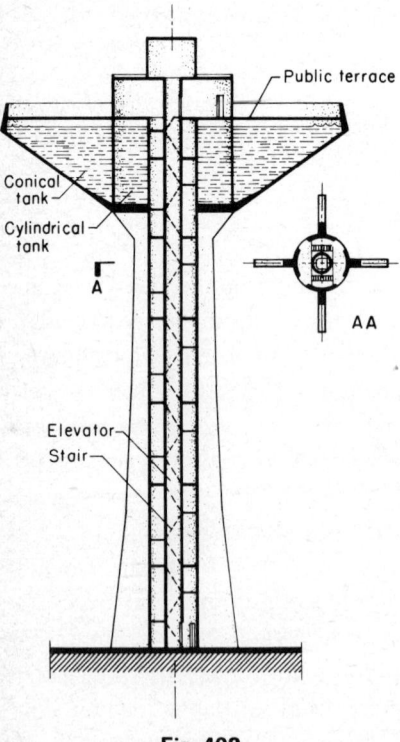

Fig. 402

In many communities, the water tower serves also as an observation tower and is opened to the public to view the surrounding countryside. In such cases, and also for relatively high towers, an elevator is a necessity. It is located inside the inner cylinder of the shaft or outside the water tower as an addition (Fig. 406). The shafts are commonly slip-formed, and the slip-form work platform serves to support the forms for tanks.

Finland, having a profound architectural tradition, has used the monumentality of water towers for enriching their communities. In recent years, a series of water towers with excellent aesthetic qualities has been built throughout Finland. The cross section of the Lauttasaari water tower is shown in Fig. 403.* It has a capacity of about 1.2 million gal. The two chambers of the tank can be emptied for cleaning. The conical shells of tank and roof are protected against severe weather by a heat-insulating outer ribbed shell, which is not a part of the tank.

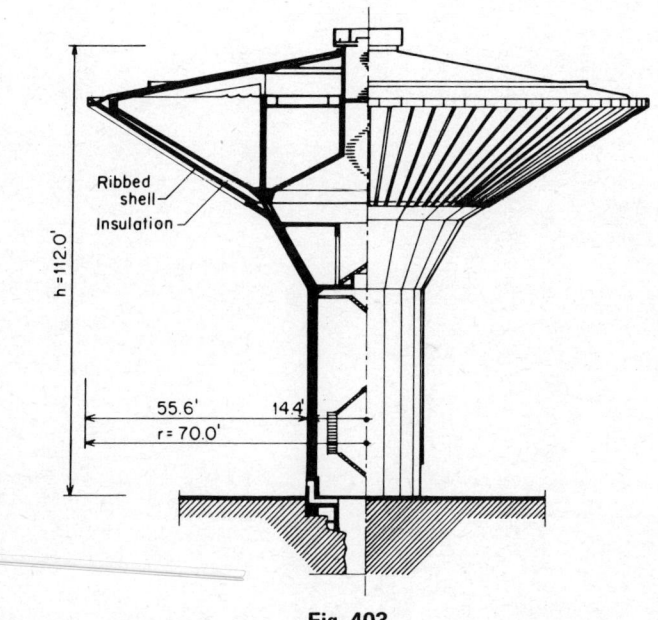

Fig. 403

*Design: Paavo Simula et Co., consulting engineers, Helsinki. Contractors: Yleinen Insinööritoimisto Oy and A-Betoni Oy, Helsinki, 1958.

608 Structural Systems and Methods of Construction

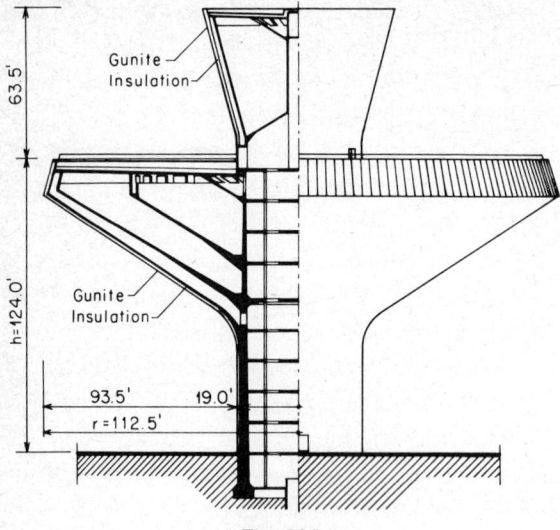

Fig. 404

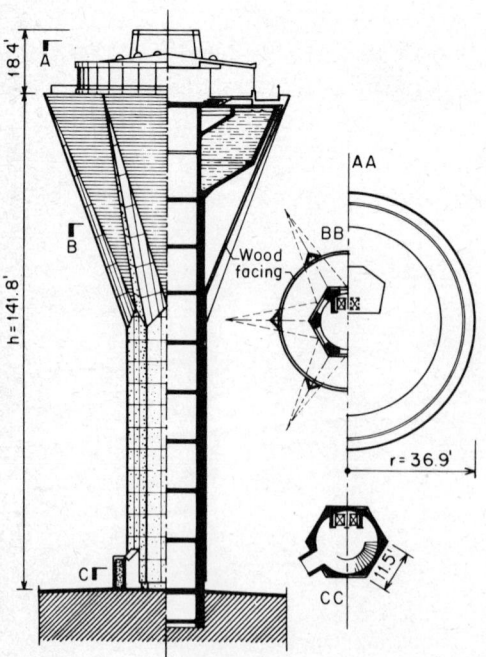

Fig. 405

Miscellaneous Structures 609

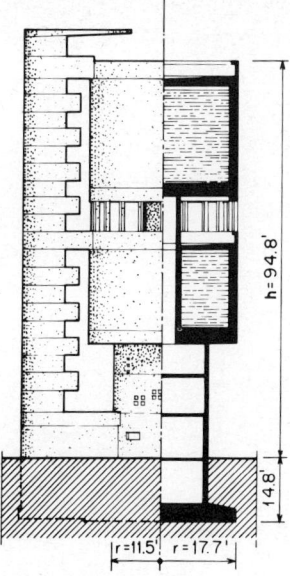

Fig. 406

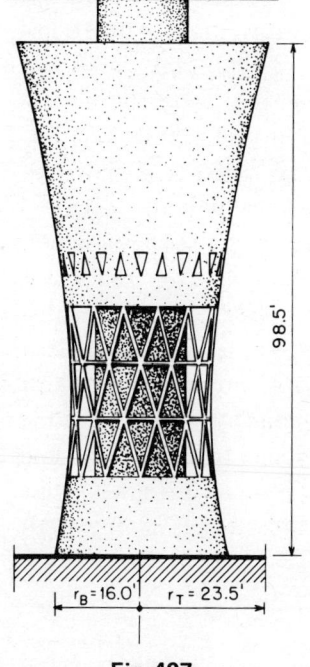

Fig. 407

The Herttoniemi water tower, near Helsinki, illustrated in Fig. 404,* has a capacity of about 3.6 million gal. The tank consists of three chambers. The outside walls are heavily insulated, and the insulation is protected by a layer of gunite.

Some of the Finnish water tanks have wooden casing—for example, the Raisio, Tornio, Karhula, and Lahti water towers—which includes the heat insulation. The Lahti water tower, illustrated in Fig. 405,† is a two-chamber tank with a capacity of 265,000 gal. It has two elevators, which serve also the observation terrace, about 142 ft over the grade. The wooden casing is supported by six concrete ribs which include the heat insulation. Heavy insulation is of the utmost importance in Finland because of the severe climate. Most Finnish water towers were constructed with the aid of rigid scaffolding.

German water towers, by local ordinance, commonly serve both for water storage and as observation towers. The Backnang tower, illustrated in Fig. 406,‡ consists of two tanks in a cylindrical building supported on a 23-ft-diameter cylindrical shaft, which houses offices and equipment. The upper-tank capacity is 140,000 gal, and that of the lower tank is 100,000 gal. The outside

---

*Design: Paavo Simula et Co., consulting engineers, Helsinki. Contractors: Siltaja Satama Oy, Helsinki, 1965.

†Design: Paavo Simula et Co. Architect: Reino Koivula. Contractors: Boman et Katajamäki, Lahti, and Pentti Kaista et Co., consulting engineers, Helsinki.

‡Design: F. Erdle and H. Kaiser, Stuttgart, Germany.

610 Structural Systems and Methods of Construction

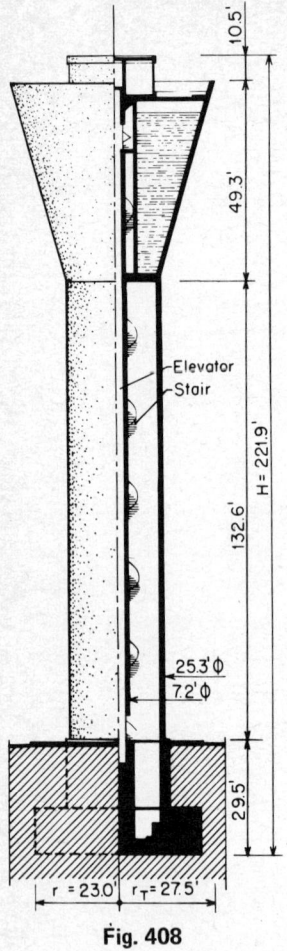

Fig. 408

diameter of the tanks is about 35.4 ft, and the total height of the tower is 94.8 ft. The stair is designed outside the tower.

Moglingen's water-observation tower (Fig. 407)* is a 98.5-ft-high hyperboloid with a uniform wall thickness of 9¼ in. The base diameter is 32 ft, and the top about 47 ft. The capacity of the tank is approximately 650,000 gal. The observation terrace is overroofed by an umbrella cantilevered from the inside cylinder.

The Basel, Switzerland, water tower, most recently built, is illustrated in Fig. 408.† The conical tank capacity is about 290,000 gal, and the total height of the tower is 222 ft (about 190 ft above ground level). The construction of this tower is unique. First, the inside cylinder (elevator shaft) was slip-formed; then the tank was constructed at the ground level, resting on a bracket mounted on the elevator shaft. The 10-in.-thick outside wall of the tank was constructed by guniting the concrete on external form. The cylindrical internal wall was slip-formed. The total weight of the tank (including pipes and equipment installed) that had to be lifted to the required height from the top of the elevator shaft was approximately 715 tons (metric). Then the outer cylinder, which supports the tank, was slip-formed, and the tank was lowered into place. The top of the cylinder was provided with about ¼-in.-thick neoprene bearing to ensure uniform placing without stress concentration. Since the relatively slender elevator shaft was not strong enough to resist high wind moments during the lifting operation, three wire ropes (120° apart) were attached to the tank for safety and to keep the stresses in the shaft within allowable limits.

*Design: Richard Kesseler, Stuttgart.
†Design and contractors: Preiswerk & Cie. AG, Ingenieure ETH/SIA, Basel.

Contemporary water tanks are prestressed horizontally and vertically to eliminate crack formation. The frictional losses in prestressing force depend on the total angular change α between the jacking ends, the wobble k, and the frictional coefficient μ [Eq. (341)]. To reduce these losses and to eliminate expensive grouting operations, the most recently developed teflon-covered rods or strands can be used. The frictional coefficient of these tendons is rather small ($\mu < 0.1$, $k = 0$), so that longer tendons can be used without considerable frictional loss.

For water towers located in seismic and high-wind areas, the legs also have to be prestressed. To obtain required ductility, sufficient elongation length must be provided for tendons and the legs at the base must have neoprene bearings.

75. SURFACE TANKS

Tanks of this type are generally located on hilltops or mountainsides and have large storage capacities. To accommodate large volumes of water requires large tank sizes. Thus diameters up to 130 ft and storage heights up to 50 ft and even more are not unusual. A standard cylindrical prestressed tank is illustrated in Fig. 409. The tank is overroofed by a dome. The floor is rather thin and flexible to allow moderate settlement without cracking. The foundation of the wall is relatively heavy to ensure a uniform distribution of the load imposed by the weight of the structure. There are considerable loading and rigidity differences between the foundation and floor slab. Commonly the floor slab is poured last. For these reasons, a joint between the foundation and slab

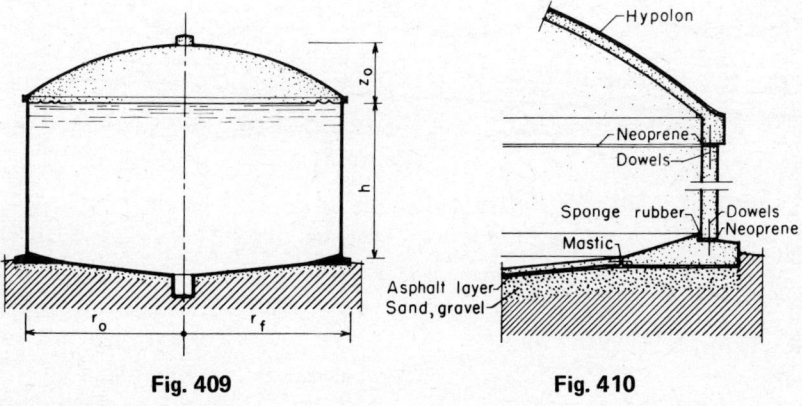

Fig. 409 Fig. 410

612 Structural Systems and Methods of Construction

should be provided to prevent cracking. The cylinder wall can best be connected to the foundation and dome by hinged neoprene joint. The recommended connections are illustrated in Fig. 410.

The wall of the tank is commonly poured in place in sections. The joints between sections usually have a neoprene interlay to ensure watertightness. The wall is horizontally prestressed by a self-propelled machine which winds the cold-drawn high-strength small-diameter wires (f_{UL} = 260,000 psi, d = 0.196 to 0.296) around the wall and dome ring in a continuous operation, stressing and spacing it within accurate limits. The wires are coated by guniting for protection against corrosion and to establish bond between concrete and wires (see Sec. 46). The dome is most economically constructed by guniting methods.

Aesthetically, these tanks are rather attractive. However, the appearance and quality of the tanks can be considerably improved by the use of prestressing tendons placed inside the wall and anchored in the outside pilasters, as indicated in Fig. 411. When the tank wall is constructed in sections, the joints between the sections are placed at the center of the pilasters. The roof of the tank can be of precast elements. This requires a center compression ring and an outside tension ring. In cases in which a center column is acceptable, no rings are required.

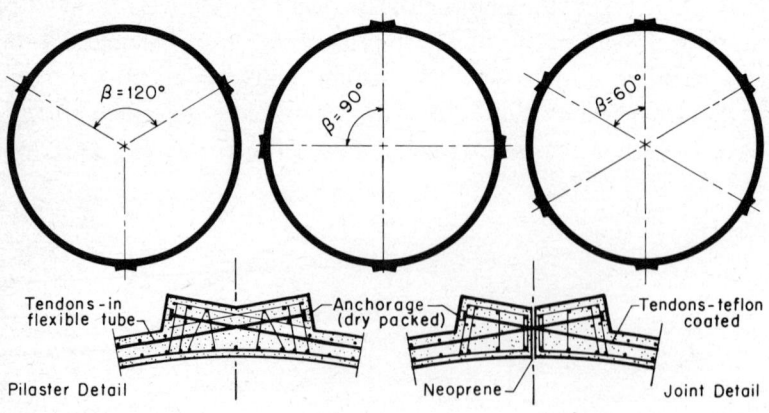

Fig. 411

Gasoline Tanks

Neither conventional reinforced-concrete tanks nor conventionally designed prestressed tanks meet the requirements for the storage of gasoline or other light liquids, even though crackless concrete is achieved. Because the specific weight of gasoline is approximately 0.75 that of water, the former percolates readily through the concrete walls if there are any defects in the tank lining. This disadvantage can be eliminated by using water insulation for prestressed concrete gasoline tanks. In this design (Fig. 412), two concentric cylindrical shells are used, separated from each other by a narrow space filled with water. The pressure on the water side of the inner wall is greater than that on the gasoline side because of the greater specific weight of water. Therefore, gasoline cannot percolate outward through the concrete walls. Any water which might percolate through the inner shell settles to the bottom and there merges with the water normally kept in the bottom of gasoline tanks to pick up sludge deposits. Such inward leakage is not a disadvantage in this type of design.

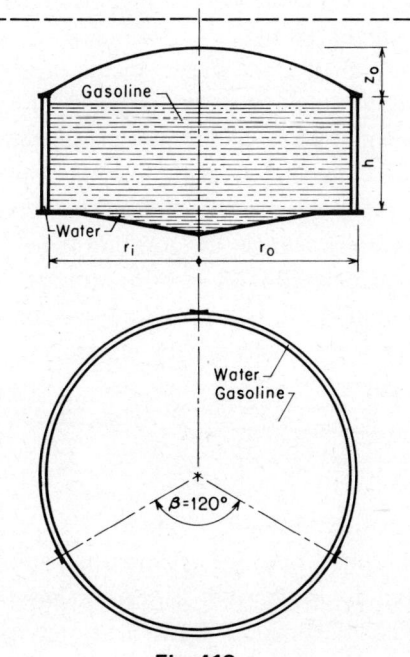

Fig. 412

Since the interior shell is always under compression $[\Delta p_y = (\gamma_W - \gamma_G)z]$, no horizontal prestressing is required. However, when a wall of constant thickness with fixed bottom edge is used, vertical prestressing may be required to eliminate flexural stresses in vertical direction in the wall. When the shell wall is designed to be simply supported, with its thickness increasing toward the bottom so that $f_{cT} =$ constant, the vertical prestressing is commonly not required. If properly designed, the interior shell can be made very thin.

Since the space between the two shells is always filled with water, the primary loading on the exterior shell is constant and is not affected by variations in the level of gasoline in the tank. Thus, if proper allowances are made for shrinkage, creep, and changes in temperature, the exterior shell of a tank above ground can be designed with horizontal prestressing only.

In underground tanks around which the earth has been properly backfilled, the earth pressure prestresses the exterior shell. In such cases, even the horizontal prestressing can be omitted, or at least greatly reduced. Changes in temperature are of minor importance in an underground tank, and since both walls of such a tank are continuously wet, the shrinkage of the concrete is greatly reduced.

In tanks above ground, the outer shell and water space provide sufficient insulation to reduce the daily changes in temperature of the inner shell to the point of insignificance. Where necessary, steam pipes can be placed in the water space to prevent freezing during winter months. In addition to the advantages mentioned above, the water between the two shells of this tank greatly reduces evaporation losses of gasoline and the fire hazard normally associated with gasoline storage.

Slip forms can be used in the simultaneous erection of the two shells. The level of water in the space between the shells can be controlled by any one of several automatic devices currently on the market.

76. SILOS

Silos are used for storage of grain, flour, cement, clinker, sugar, salt, granular chemicals, etc. Some of these materials are hygroscopic, and in contact with moist air, cohesion builds up to such a degree that the material can be loosened only by mechanical means. Raw sugar, salt, and some granular chemicals belong in this category. Cement and flour are also hygroscopic, but in relatively dry storage the cohesion is weak and the material can be easily loosened by compressed air so that it flows out of the cells by gravity.

Nonhygroscopic or lightly hygroscopic materials are stored most economically in relatively high cellular silos whose individual cells are emptied by gravity flow. Hygroscopic materials are stored in large-space silos where the material can be broken loose mechanically (by clamshells, bulldozers, etc.) before being taken out by appropriate conveyer systems.

Also, some of the highly hygroscopic materials (sugar, salt, etc.) can be aggressive to concrete, so that protective coating for inside surfaces of the walls is required unless special cement and dense, closed surfaces are used.

Assuming that the vertical pressure in the silo cell is uniformly distributed (Fig. 413) at any level z, the equilibrium requires that

$$dp_z A + \mu p_y U dz - \gamma A dz = 0 \tag{577}$$

where dp_z = increase of vertical pressure in interval dz
p_y = horizontal pressure on silo walls
μ = frictional coefficient between wall and material stored
γ = unit weight of material
A = area of silo cell
U = perimeter of silo cell

Further, representing the natural slope of the material by ϕ and applying Coulomb's theory,

$$k = \frac{p_y}{p_z} = \tan^2\left(45 - \frac{\phi}{2}\right) \tag{578}$$

the differential equation (577) gives the following exponential functions for horizontal and vertical pressures (Fig. 413):

$$p_y = \frac{\gamma A}{\mu U}(1 - e^{-z/z_0})$$

$$p_z = \frac{\gamma A}{k\mu U}(1 - e^{-z/z_0})$$

$$z_0 = \frac{1}{k\mu}\frac{A}{U}$$

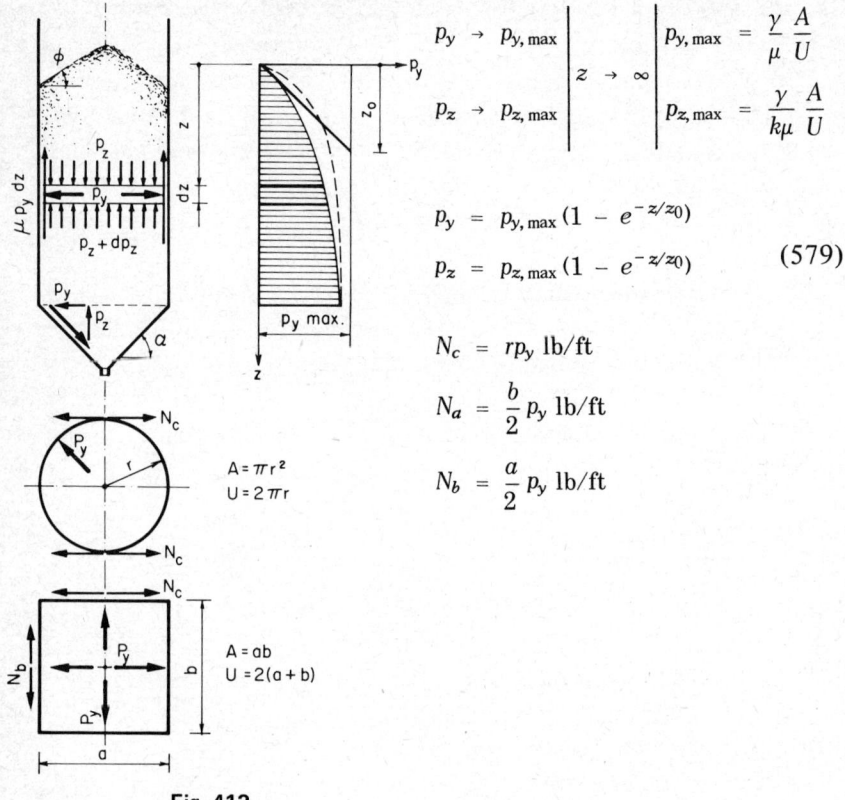

Fig. 413

These pressures are correct only for full filling of the cell. As the measurements indicate, the horizontal pressure at the upper part of the fill can be up to 20 percent higher than computed by the above formulas. This phenomenon can be explained by the arching of the material resulting in thrust upon the walls. Also, the vertical pressure at the bottom of the cell is notably increased when the arching collapses and the material falls. Arching most often occurs in cement and flour silos.

A 10,000-ton grain silo with 18 round cells is illustrated in Fig. 414.* The silo is provided with a drier and equipment for infected-grain treatment accommodated in a separated square cell between the cells' battery and the elevator tower. In the elevator tower are located a scale, cleaner, and dust collector and also three small cells for weighing and

*Designed for India, 1954.

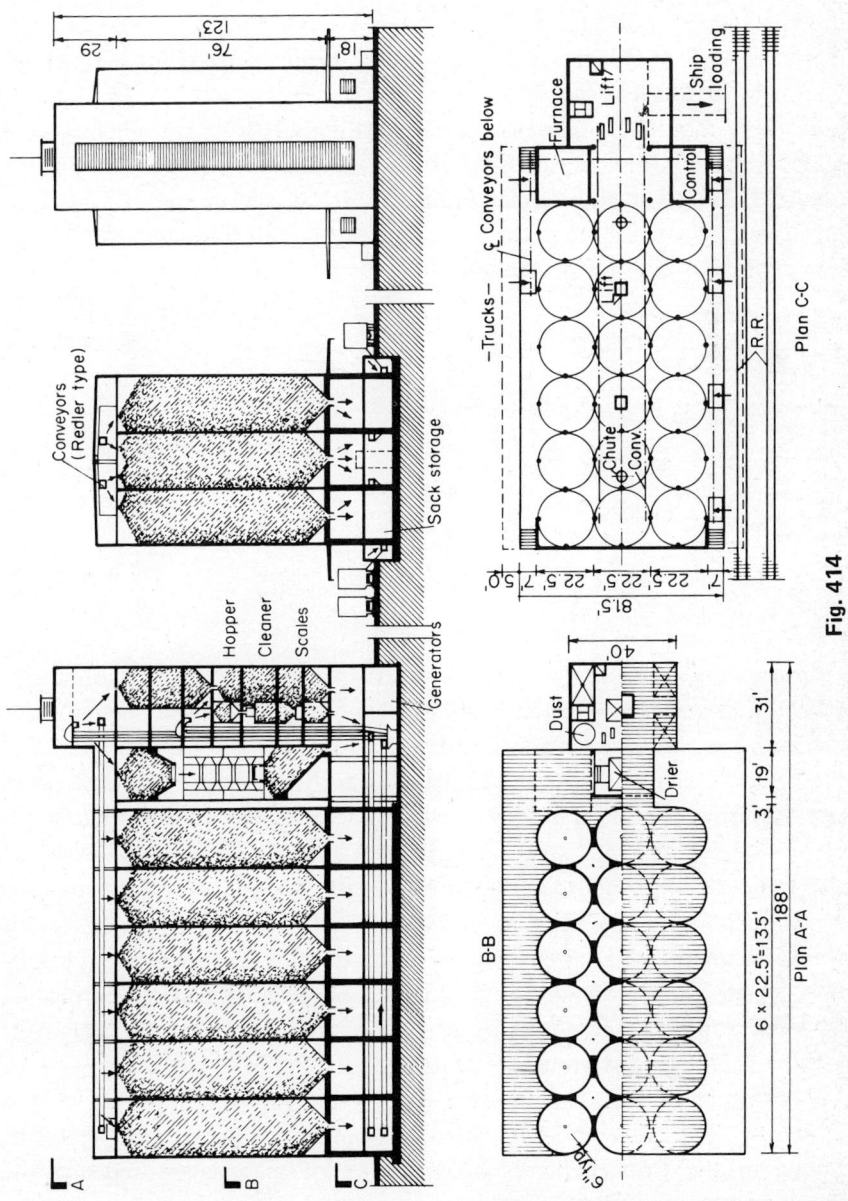

Fig. 414

shipping. The basement and upper floor below the cells are used for sacking and sack storage. For ventilation purposes, the sidewalls of the upper level are omitted.

Since the horizontal pressure p_y for a typical-size grain silo cell is relatively small ($\gamma \simeq 45$ lb/ft^3, $\mu = 0.45$, $\phi = 26°$, $k \simeq 0.39$), the tension force and moments in walls can be easily balanced by conventional reinforcing, so that prestressing for cylindrical cells is commonly not required. However, for square cells (Fig. 415), the moments as well as the tensile force in high grain silos are considerable and crack formation is unavoidable without prestressing.

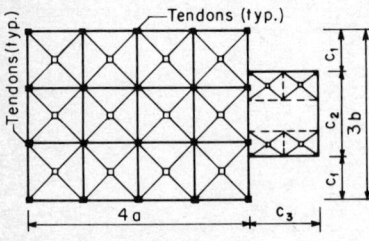

Fig. 415

Prestressing of straight walls is rather simple and is more economical than the use of conventional reinforcement only. Since the normal force and moments can be reversed, centric prestressing by rods or strands is most economical and proper. The use of a bottom slab instead of cones for slip-form operation is more economical and practical. It has to be computed for own weight as a flat slab. For live load it hangs from the silo walls and is computed as a two-way slab. Due to this, no prestressing is generally required. The recommended minimum wall thickness for slip-form operation is about 6 in. For lesser thickness, horizontal cracks may occur during placing of concrete because the frictional forces between concrete and slip form can be balanced only by the weight of the freshly placed concrete.

For storage of materials in sacks or coarse-grain bulk material, such as clinker, a large-space silo (Fig. 416)* is appropriate and economical. The design of the floor depends on the type of material stored and the handling system used.

*Designed for Kuksaing Co., Seoul, S. Korea, for rice storage.

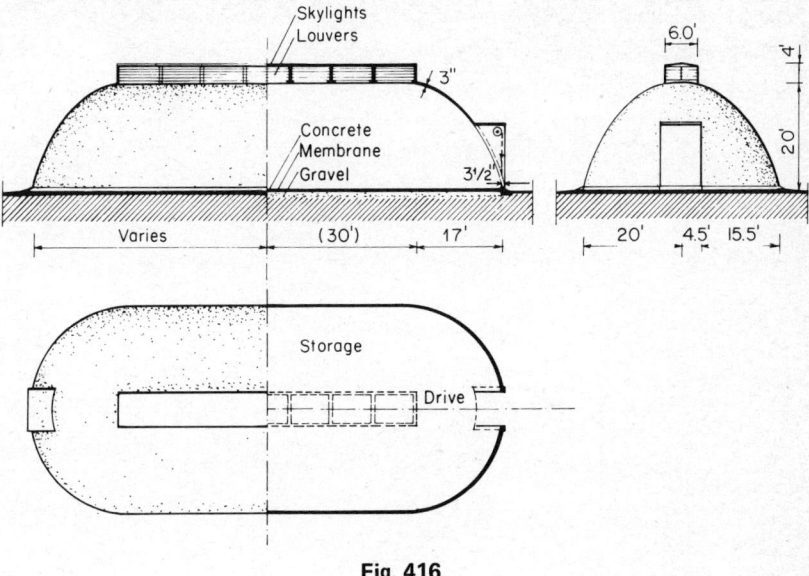

Fig. 416

The construction of such a silo is rather simple. First, the foundation and floor are constructed; then the lower frame for skylights or conveyer is erected on the scaffolding. As the curvature of the parabolic shell is equal throughout the shell, heavy reinforcing rods bent to match the curvature are welded to the dowels projecting from the foundation and skylight frame. The rods are tied horizontally by welding small-size bars to the rods. The steel grid is covered from inside with chicken wire. Concrete is applied by gunite. The minimum thickness of the shell is 3 in. at the top and 3½ in. at the foundation. When the shell must also carry the conveyer, and to resist the lateral pressure of stored materials, the wall thickness must be slightly increased to control the tensile stresses in the shell.

Silos for storage of cement, clinker, etc., usually consist of large-diameter single cylinders arranged either as independent units or in a battery. The reason for this is mainly to avoid the high flexural stresses due to unsymmetric loading and the temperature change. Stored cement commonly has high temperature, up to 150°F or even higher, which results in considerable expansion in the cell walls. In the battery-type arrangement, if the cylinders are rigidly connected, the deformations are restrained, which leads to considerable bending moments and crack formation in walls. Even in single-cylinder cells the loading due to arching can be unsymmetric, and with fixed bottom edge the flexural stresses cannot be balanced without cracking, unless two-way prestressing is applied.

A battery-type cement silo is illustrated in Fig. 417.

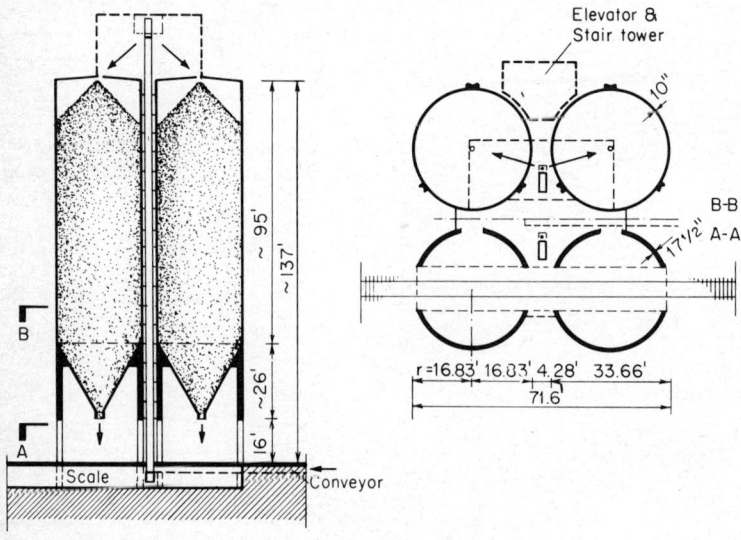

Fig. 417

The four structurally independent cylinders have a diameter of 32 ft, a storage height of approximately 100 ft, and a wall thickness of 10 in. The clear space between cylinders is determined so that they can be slip-formed independently. When no separate stair-elevator tower is provided, stairs and elevator can be located in the space between the cylinders.

The hoppers may be of steel or concrete (Fig. 417). Steel hoppers are preferred mainly for easy fastening of slide gates, air pipes, vibrators, etc., and for increased vibration efficiency. The hopper is fastened by dowels into a ring beam. Equal shear in the dowels can be achieved by coating the lower dowels with teflon or rubber-based paint. Depending on the supporting condition, the ring beam is subjected to torsion in addition to the radial moment. The radial moment is balanced by arch actions: compression in the top part and tension in the lower part.

The cells are filled by conveyors, by pumps, or by chuting directly from the elevator. The cells are commonly emptied directly to railroad cars or trucks. In this case, scale pits must be constructed below the trucks. Also, a belt conveyor below the cell hoppers and extra loading cells are used. The clinker silos are mostly emptied by belt conveyors.

Since the horizontal pressure (for cement: $\gamma = 95 \text{ lb/ft}^3$, $\mu \simeq 0.40$, $\phi \simeq 40°$) in such silos is relatively high and moments are developed in the vertical direction due to arching thrust, the walls commonly must be prestressed horizontally and vertically.

Silos used for the storage of sugar, Glauber's salts, etc., must also have structurally independent large-diameter cylinders, similar to those in cement silos. Since raw sugar is hygroscopic and can be rather sticky if not properly dried before storage, considerable cohesion and arching can be expected. Due to this, the cell walls are subjected to unsymmetric loading during emptying and have to resist considerably higher horizontal pressure than in the fully filled condition. Therefore, to avoid cracking, the walls must be prestressed horizontally and vertically about 30 percent higher than required to balance the pressure given by the formulas and computed vertical bending stresses. Sugar is not aggressive to concrete when it is properly cured; therefore, no protective coating is required.

A raw-sugar silo, about 76 ft in diameter and with about 95-ft storage height, is illustrated in Fig. 418.* In order to reduce the formwork, the walls in this particular case were designed to be constructed in sections and prestressed to unity by cables. The tensile stresses due to vertical moments were balanced by conventional reinforcing. For loading sugar into the silo, an elevator or inclined belt conveyor can be used. Generally the elevators are more economical. The main entrance to the silos was designed between the cylinders. To keep the entrances open, conveyors are provided inside the silos. Due to considerable arching, protective movable umbrellas are commonly used to protect the workers during the emptying operation.

Highly hygroscopic materials, such as salt and some chemicals, which in longer storage time turn practically solid and have to be broken loose for handling even by the use of explosives, and especially those which are

---

*Designed for Kuksaing Co., Seoul, S. Korea; a similar one was designed for Central Lafayette, Puerto Rico.

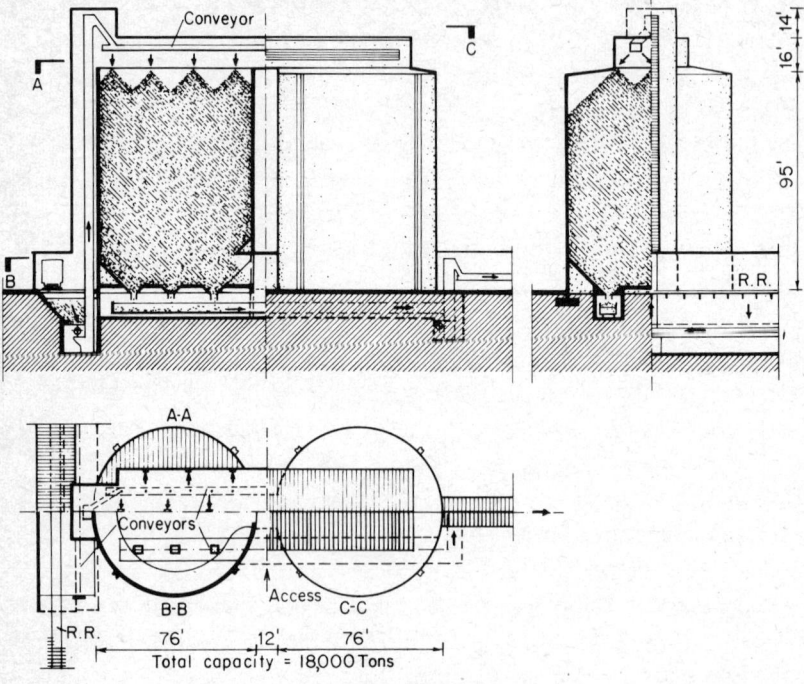

Fig. 418

aggressive toward concrete are best stored in large-space silos, as illustrated in Fig. 419.

The pylons are of prestressed concrete, and the walls between them are of parabolic face-brick arches. The pylons at the top level are tied horizontally to unity in the wall direction by prestressed concrete beams. Transversally, they can be combined with roof beams into frames or

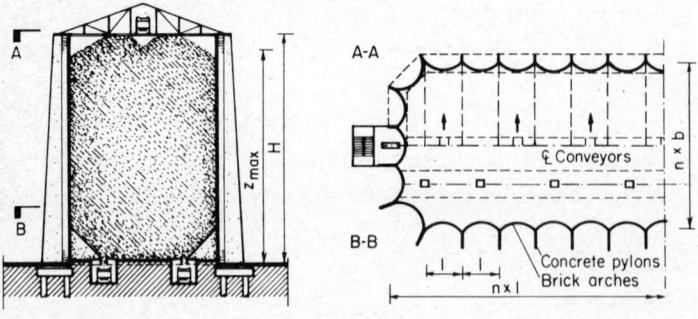

Fig. 419

designed as vertical beams tied at the top through roof trusses and fixed or simply supported at the bottom. The relatively high thrusts (top and bottom) are best balanced by prestressing cables embedded in concrete. Which of those systems is appropriate depends on the material stored, local conditions, and above all, economical considerations. This applies also for the handling system (elevator, inclined belt conveyor, etc.) to be used.

Pressure upon the walls is commonly computed by Coulomb's theory (for salt: $\gamma = 75$ lb/ft^3, $\mu \simeq 0.30$, $\phi \simeq 40°$). The pressure is approximately linear down to $0.75\, z_{max}$, and from there it remains almost constant to $z = z_{max}$.

77. OBSERVATION AND TRANSMISSION TOWERS

Such towers have a considerable height in order to obtain a large range of optical view and the most efficient transmission. At present, one of the highest of these towers is the Vienna tower, with a total height of 855 ft above grade (Fig. 421). The high-quality materials and theories available make it structurally feasible to build even far higher towers. The limits for the height are set only by construction difficulties and economy. Both of these, however, can be overcome by new, unique construction methods. One such construction technique already available is the slip-form method. In favorable climatic conditions, an average casting height per 24 hr can be as much as 16 ft; thus a 1,000-ft tower would require theoretically only about 65 casting days and practically not more than 3 months to construct. In principle, the slip-form method is not new; it has been used for about 80 years. Since that time it has developed into a very efficient and economical construction method. One of the most advanced slip-form techniques available at present is the *Prometo slip-form system*, illustrated in Fig. 420, developed by Swedish engineers.

The system is completely mechanized and automatically controlled so as to achieve uniform lifting of the form at all points. The leveling is also automatically controlled, by a so-called *distance setter*. The hydraulic jacks are designed to climb along plane bars about 1¼ in. in diameter. The jack has relatively long toothed climb jaws of high-grade steel which firmly and securely grip the climb rod. The safe lifting capacity of the 1¼ in. ϕ rods is about 6 tons of vertical load. The jacks are connected with each other and with an electrically operated oil pump so that they

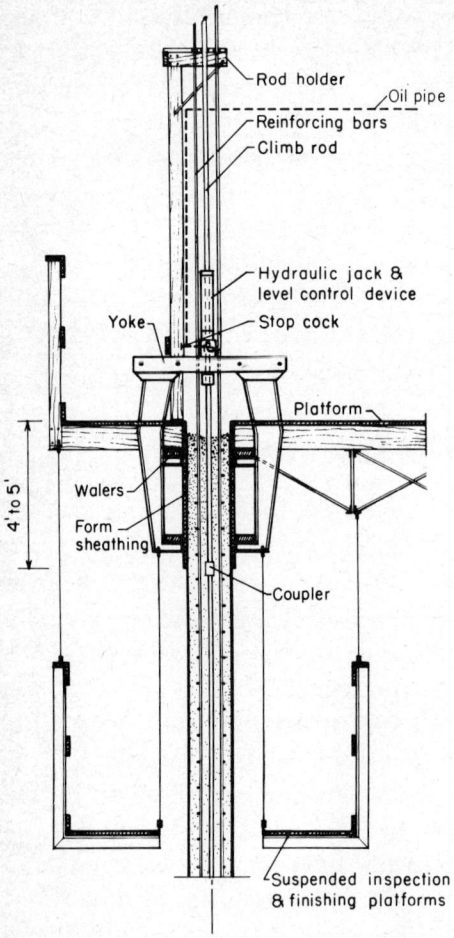

Fig. 420

all raise the form at the same time up one step, a distance of approximately 1 in. The impulse time of the pump is adjustable to any desired lifting speed.

The forms, about 4 to 5 ft high, are supported by a steel yoke through walings of the forms. Commonly the forms and yokes are of universal type, which can be quickly assembled and dismantled and are adjustable for straight as well as curved walls. The slip form is also adjustable for tapered or sloping walls and surfaces. The high wind force upon the forms and not yet hardened concrete is transferred through extended yokes to more hardened concrete. The yokes are arranged in locations, such as corners, where the frictional forces are highest. The horizontal wind-force influence upon climbing is counteracted by regulation of the stopcock. The form surfaces in contact with concrete can have stainless-steel or water-repellent neoprene lining to reduce wearing and to obtain high-quality concrete surface. The working platform is attached to the yokes and is commonly designed strong enough to support a climbing crane used for materials handling.

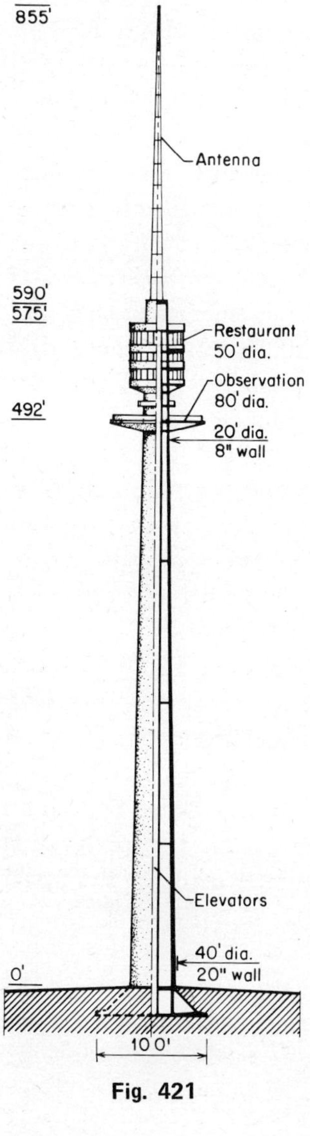

The quality of slip-formed concrete is relatively high due to the troweling action and vibration associated with this type of operation.

The slip-formed Vienna tower is illustrated in Fig. 421.* The main cylindrical shaft tapers from a 40-ft diameter at the base to a 20-ft diameter at the 500-ft level. The wall thickness at the base is 20 in. and at the top is 8 in. Above the top of the main shaft are an about 85-ft-diameter observation terrace and a smaller platform one level higher for a children's playground. Above these platforms is a three-story, 50-ft-diameter, revolving structure accommodating restaurants, terraces, etc. At the top of the concrete structure is a 260-ft steel antenna. Thus the total height of the tower is 855 ft. It is founded on about 25-ft-deep circular-conical slab resting on thoroughly compacted ground. The tower is aesthetically very attractive and considerably more economical than those constructed by stepwise, conventional methods.

Fig. 421

*Designer: Robert Krapfenbauer. Architect: Hannes Lintl. Both of Vienna, 1963.

The Stuttgart observation-transmission tower is about 530 ft high and topped with an approximately 170-ft-high antenna. Thus the total height of the tower is about 700 ft. It is one of the first towers of such a height (1956). The Stuttgart tower is illustrated in Fig. 422.* The circular tower shaft is tapered from about 33 ft at the foundation to 16.5 ft at 450 ft from grade. The respective wall thicknesses are 12 in. and 7.5 in. From the 450-ft level upward, four floors are cantilevered from the shaft up to the 491-ft elevation, where the main observation platform is located. The diameter of the cantilevered floors increases from 20 ft to a maximum of 48 ft for the observation platform. A smaller observation terrace, diameter about 31.5 ft, is arranged at elevation 500 ft. Above this level, the shaft is used for utilities. The two floors beneath the main observation platform are occupied by a restaurant and transmission facilities.

Inside the tower shaft are two elevators and an emergency stair. For support of the elevator and stair and for stability of the shaft, rigid platform-column systems are arranged approximately 33 ft apart. The tower is supported through a conical shell by a circular prestressed girder about 90 ft in diameter, 27 ft below the grade.

The Belgrade tower, illustrated in Fig. 423,† differs structurally from the Stuttgart and Vienna towers. The triangular shaft, with 24-ft-long sides, is supported by a rather heavy tripod pin-connected to the foundation bearing on rock about 6 ft below the grade. The three free legs merge into the corners of the triangular shaft and continue up to elevation 335 ft, providing stiffness to the shaft. The shaft is topped with a five-story structure 115 ft high, which houses transmission equipment, utilities, a restaurant, and an observation terrace. The antenna is 205 ft high. Thus the total height of the tower is 655 ft.

From the bottom frame, tying the legs, hangs a massive box and a lower observation terrace approachable over a bridge. Up to elevation 120 ft, the shaft was constructed by conventional method; the upper part was slip formed.

---

*Designer: Fritz Leonhardt. Architect: E. Heinle. Both of Stuttgart, 1955.
†Designer: M. Krstic. Architects: U. Bogunovic and S. Janic. All from Belgrade, Yugoslavia.

Miscellaneous Structures 627

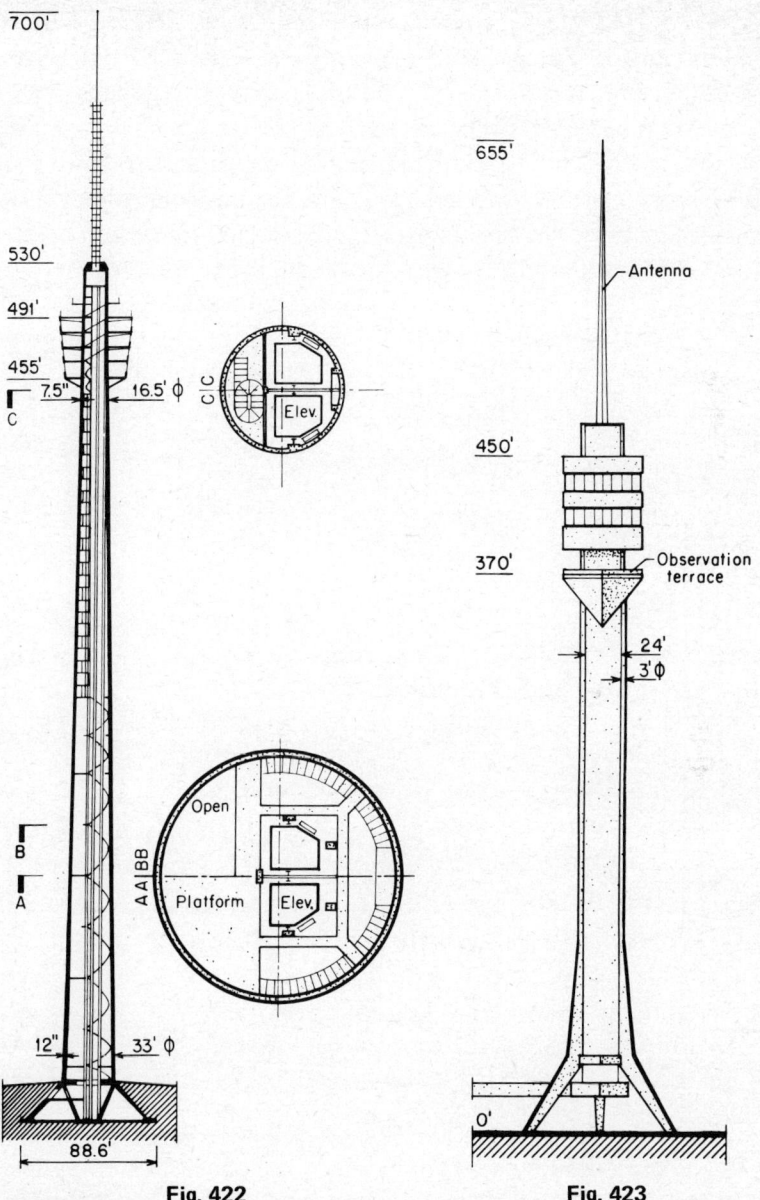

Fig. 422

Fig. 423

The Niagara Falls, Ontario, observation tower, illustrated in Fig. 424,* also has a triangular shape. The three legs are hollow trapezoids, diminishing in sections up to the 380-ft level. The legs support a three-story observation platform and a restaurant. The roof of the machine room, above the 100-ft-diameter galleries, is 450 ft above the grade, and a spire atop the roof is 505 ft. The dining room rotates. The tower is served by three outside elevators riding on rails mounted outside the stairwell between the interior ends of the legs. The elevator cars are glass-fronted.

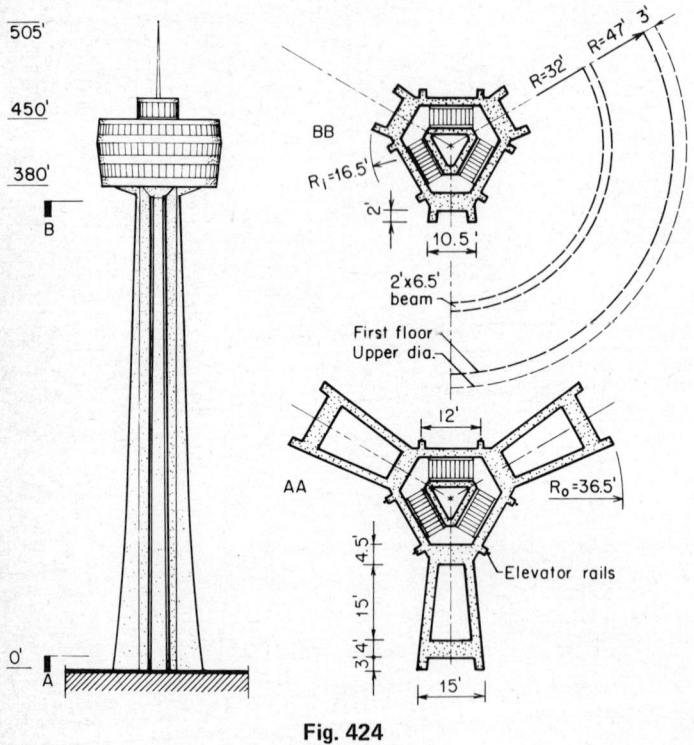

Fig. 424

The tower is supported by a 6-ft-deep reinforced concrete mat on sand and silt 35 ft below the grade. The tower shaft and sloping legs were cast by the Prometo slip-form systems.‡ The slip-form work platform and the

*Design: Farkas, Barron & Jablonsky, structural engineers, Toronto, Ontario. Architect: Bregman & Hamann, Toronto. Contractor: Pigott Construction Co., Hamilton, Ontario.
‡Slip-form design: B. M. Heede Canada, Ltd., Montreal.

crane mounted on it (total weight, approximately 273 tons) were raised by 81 climbing rods and jacks.

The forms for the first floor were assembled on steel frames, completed at the ground level, and jacked from the top of the triangular shaft to the 380-ft elevation. The first-floor structural system consists of reinforced-concrete radial beams cantilevered from the shaft carrying a 6.5-ft-deep and 2-ft-wide circular beam, about 64 ft in diameter, and a tapered folded-plate deck cantilevering beyond the ring beam. The floor was designed to carry the steel frame supporting the upper floor and roof.

The rather complicated shape of the tower and the tapering and sloping surfaces of the legs created a difficult construction, involving many problems for designers and contractors. But the finished work resulted in a most pleasing tower of its kind.

78. POWER-LINE TOWERS, TELEPHONE AND LIGHTING POLES

The existing, and especially the secondary, power-line towers and telephone poles are ugly and are least conformable to the twentieth-century art form. For this purpose, prestressed towers and poles would be the most suitable, economical, and aesthetically pleasing, as well as structurally sound. Prestressed towers and poles can be designed to satisfy any requirement for strength, height, etc. To obtain lasting durability, sharp corners must be avoided and ample coverage of the conventional reinforcing as well as the prestressing tendons must be provided. Minimum coverage for conventional reinforcing should be 1½ in.; for teflon-covered tendons, at least 2 in.; and for grouted tendons, 3 in. Crossarms are also of prestressed concrete or, for telephone poles, of corrosion-resistant aluminum. To secure the stability of the towers, the joints between the crossarms and pole are filled with epoxy. The joints between foundation and pole are filled up to about 1 ft with grout, and the top part is dry-packed. The high-strength concrete mix (up to 7,500 psi) must be designed with special care so that the concrete is weatherproof even in severe climatic conditions. The surfaces of the poles should be treated with concentrated silicon (two applications).

The recommended types of high-voltage power-line towers and poles are illustrated in Fig. 425. Since these poles are mass-produced, universal-type molds are used, so that any size of pole can be cast in the same mold, which is modified simply by adding sections and installing pilot liners at the top and bottom parts of the mold. The molds are placed on vibrotables or shock tables for consolidation of concrete.

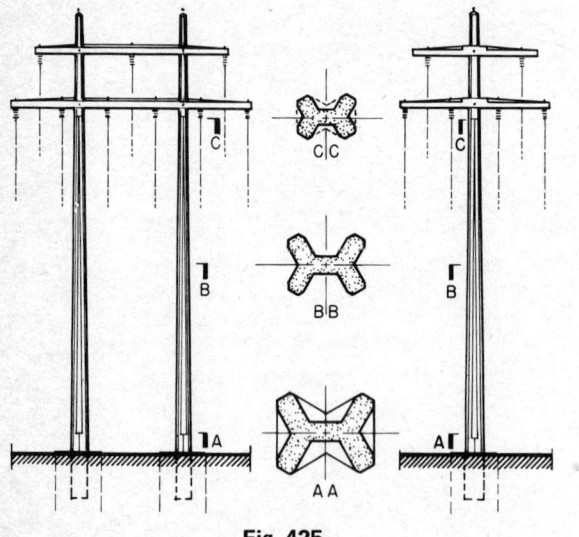

Fig. 425

The recommended reinforcing and posttensioning tendons as well as the forms are given in Fig. 426. To control the proper concrete coverage and obtain rigidity of cages, the reinforcing must be welded. The poles and crossarms can be simply erected by helicopter crane.

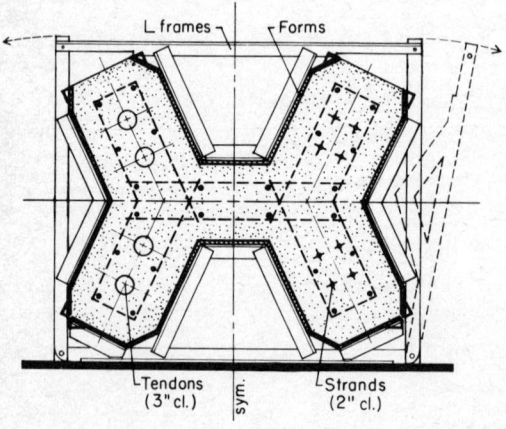

Fig. 426

The height of telephone poles is seldom more than 45 ft. The poles are provided with relatively short crossarms. Since these poles, even in hurricane areas, are subjected to rather small loading, the recommended pole type and cross section are as illustrated in Fig. 427.

Lighting poles are subjected to wind pressure only. The recommended cross section of the poles is octagonal, with a hole at the center for wiring and for cable for the lamp raising and lowering device. A door is provided at the bottom part above the grade to allow access to the wires and cables. The minimum dimensions of the poles are 5 in. at the top and about 6 in. at the bottom. The lamp arms are made either of stainless steel or of corrosion-resistant aluminum. Such poles have a very pleasing appearance and do not require any maintenance (Fig. 428).

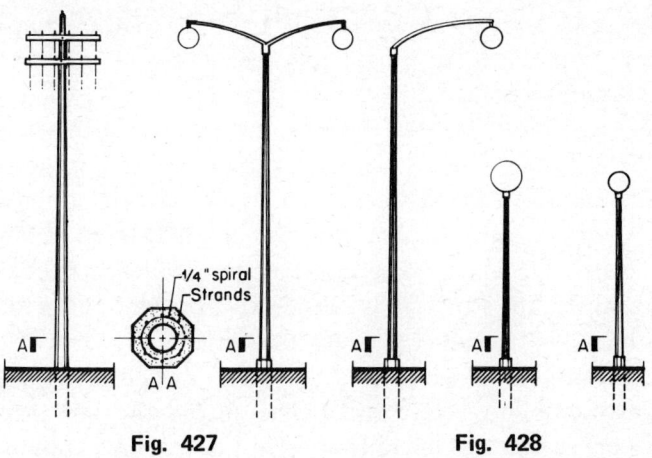

Fig. 427 Fig. 428

79. WIND AND EARTHQUAKE FORCES

The statical computations and stress analyses of towers and poles subjected to wind and earthquake loading constitute a rather complex problem. Both of these loadings raise a vibration in the structure, resulting in increased static loading. The magnitude of the forces and their duration are a function of time which can be only roughly estimated. The wind-pressure intensity depends on the weight and velocity of the air and on the shape of the structure [Eq. (245)]. The maximum and extreme wind velocities vary from location to location. They are highest in coastal areas, on hilltops, and in hurricane and tornado areas. The maximum mean wind velocity increases with the elevation and decreases downward to the ground, depending on the nature and roughness of the terrain. The mean wind velocity for a certain location is given in weather bureau records.

The velocities of extreme gusts are highly turbulent, depending on and caused mainly by terrain characteristics and shape of the structure. They are highest at a certain height from the ground level and gradually decrease upward. Due to this nature of gusts it cannot be expected that the entire height of a structure will be subjected to the gust force. The gusts are of short duration, and velocities as high as 150 mph have been recorded. Also, the pressure distributions in a steady wind may cause vibrations which can build up to larger periodic forces when the frequency of wind (ω) coincides with the natural frequency of the structure (ω_n). The gust wind force commonly controls the structural design and stresses, and the steady maximum mean velocity of wind force controls the stability of the structure.

Theoretically there is no difference between wind and earthquake analyses because both of these forces are similar in nature; that is, both are a series of erratic impulses. The dynamic responses of the structure to these impulses are motion and acceleration, the magnitude of which depends on the elasticity, mass, mass distribution, sequence of the impulse, etc. For analysis, the structure is considered to be stationary for both loadings (Fig. 429). In wind loading, the impulses are applied directly to the structure, and in earthquake, they are applied indirectly by ground displacement and acceleration. The forces the vibrating structure is subjected to can be determined from the dynamic equilibrium conditions (D'Alembert's principle) expressed by the following differential equation:

$$\frac{W}{g}\frac{d^2u}{dt^2} + c\frac{du}{dt} + ku = F_0 \sin\omega t \tag{580}$$

where u = displacement
$W/g = m$ = mass
W = weight of vibrating structure
g = acceleration due to gravity
c = coefficient of damping
k = spring coefficient
F_0 = disturbing force
ω = circular frequency of disturbing force

The ω values of the disturbing force are known or assumed. For example, if the duration of a gust or earthquake is 2 sec, then

$$\omega = \frac{\pi}{2t} = \frac{3.14}{2 \times 2} = 0.785$$

and the period (581)

$$T = \frac{2\pi}{\omega} = \frac{6.28}{0.785} = 8 \text{ sec}$$

The natural frequency of a structure is computed from the differential equation of harmonic vibration:

$$EI_z \frac{d^4 u_{(z)}}{dt^4} - \omega_n^2 m u_{(z)} = 0 \qquad (582)$$

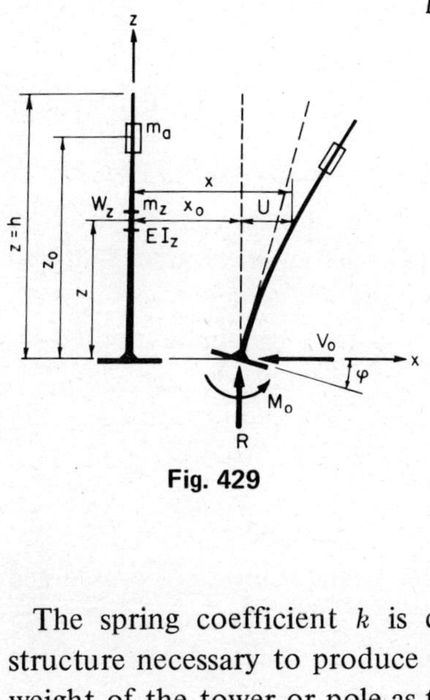

$$\omega_n^2 = \frac{EI_{(z)} \dfrac{d^4 u_{(z)}}{dt^4}}{m u_{(z)}} = \frac{w_{(z)}}{m u_{(z)}} \qquad w_{(z)} = k u_{(z)}$$

$$= \frac{k}{m}$$

$$\omega_n = \sqrt{\frac{k}{m}} \qquad m = \frac{\sum_0^h w_z}{g} \qquad g = 32.2 \text{ ft/sec}^2$$

$$T_n = \frac{2\pi}{\omega_n}$$

Fig. 429

The spring coefficient k is defined as the force or loading of the structure necessary to produce a deflection equal to unity. Applying the weight of the tower or pole as the horizontal loading and computing the moment and deflection (Fig. 430), the spring constant is

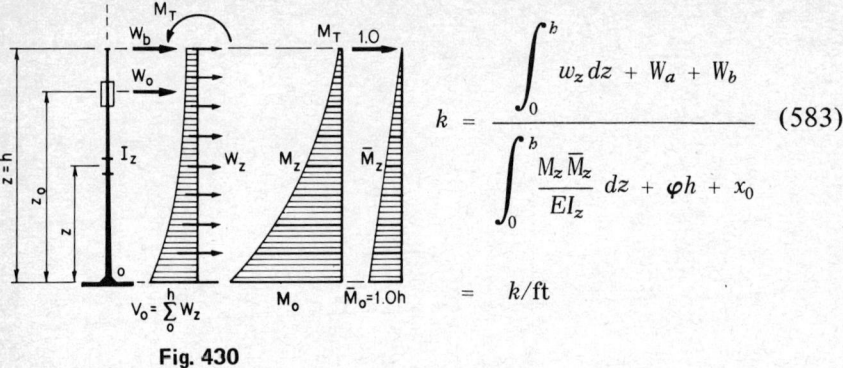

$$k = \frac{\int_0^b w_z \, dz + W_a + W_b}{\int_0^b \frac{M_z \overline{M}_z}{EI_z} \, dz + \varphi h + x_0} \quad (583)$$

$$= k/\text{ft}$$

Fig. 430

where φ is the rotation and x_0 is the horizontal displacement of the foundation due to the base moment M_0 and base shear V_0. The φ x_0 values are

$$\varphi = \frac{M_0}{C_m}$$
$$x_0 = \frac{V_0}{C_v} \quad (584)$$

The C_m and C_v are dependent on the flexibility of the foundation and soil characteristics. Substituting the M_0 by a vertical load couple $R = M_0/2e$, the deflection of the foundation (δ_e) can be computed by Eqs. (117) to (125) (Fig. 431):

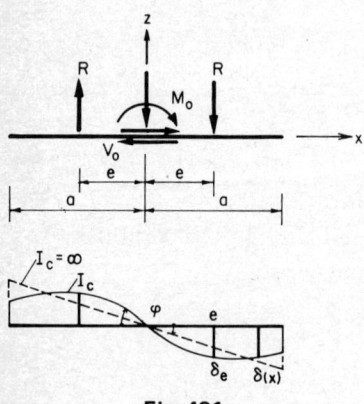

Fig. 431

$$\varphi = \frac{\delta_e}{e} \quad C_m = \frac{M_0}{\varphi}$$

If the foundation can be assumed rigid ($I = \infty$), the deflection of foundation due to M_0 is linear and the rotation

$$\frac{d\delta x}{dx} = \frac{M_0}{I_A k_s} = \varphi$$
$$C_m = I_A k_s \quad (585)$$

where I_A is the areal moment of inertia of the foundation and k_s is the modulus of soil reaction (ksf/ft).

The C_v value is a fraction of the C_m value and for normal cases has little influence upon deflection.

The damping coefficient c for reinforced concrete is rather small (up to 0.04) and about twice that for average soil. Using the notation (Timoshenko)

$$\omega_d^2 = \omega_n^2 - \left(\frac{c}{2m}\right)^2 \tag{586}$$

the damped vibration has a period

$$T_{nd} = \frac{2\pi}{\omega_d} = 2\pi\sqrt{\omega_n^2 - (c/2m)^2} \tag{587}$$

As can be seen, the influence of damping upon the period, for even a relatively large c value, is insignificant and can be disregarded. Thus,

$$T_n \simeq \frac{2\pi}{\omega} = 2\pi\sqrt{m/k} \tag{588}$$

The above given Rayleigh's method is mostly used for computing the natural circular frequency ω_n or period T_n of the fundamental mode. However, this method does not generally give accurate values of the mode shape, especially for variable mass distribution and moment of inertia. The F. H. Schroeder method,* which will be given below, allows one to determine the natural frequency of the fundamental mode more accurately in a simple operation from Eq. (582) by the use of Green's function (Fig. 432). It expresses the deflection of the cantilever at s

---

*Bauingenieur, no. 6, 1970.

which is caused by the dummy loads 1.0 applied at ξ and is

$$G(s,\xi) = \int_0^b \frac{M_1(z,\xi)\,\overline{M}_1(z,s)}{EI_{(z)}}\,dz + \frac{1.0\xi}{C_m}s + \frac{1.0}{C_v} \tag{589}$$

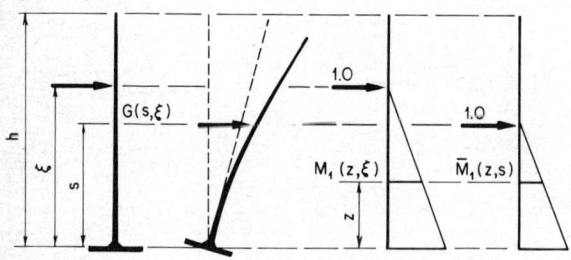

Fig. 432

The integral equation of the differential equation (582) is

$$u_{(z)} = \omega_n^2 \int_0^b G_{(z,\xi)}\,m_{(\xi)}\,u_{(\xi)}\,d\xi \tag{590}$$

Multiplying both sides of this integral equation with $u_{k(s)}$, $m_{(s)}$ and integrating over the cantilever, we obtain

$$\int_0^b u_{(s)}\,u_{k(s)}\,m_{(s)}\,ds = \omega_n^2 \int_0^b [G_{(s,\xi)}\,m_{(\xi)}\,u_{(\xi)}\,d\xi]\,m_{(s)}\,u_{k(s)}\,ds \tag{591}$$

$$m = \frac{w}{g} \quad k = 1, 2, \ldots, n$$

Assuming that the fundamental mode follows the function $u_{(s)} = \sum_{i=1}^{i=n} a_i u_{i(s)}$, we obtain a homogeneous equation system with n unknown coefficients a_i:

$$a_1(u_{ik} - \omega_n^2\,\overline{u}_{ik}) + a_2(u_{2k} - \omega_n^2\,\overline{u}_{2k}) + \cdots + a_n(u_{nk} - \omega_n^2\,\overline{u}_{nk}) = 0 \tag{592}$$

where

$$u_{ik} = \int_0^b m_{(s)} u_{i(s)} u_{k(s)} \, ds = \int_0^b m_{(z)} u_{i(z)} u_{k(z)} \, dz$$

$$\bar{u}_{ik} = \int_0^b \left[\int_0^b G_{(s,\xi)} m_{(\xi)} u_{i(\xi)} \, d\xi \right] m_{(s)} u_{k(s)} \, ds \tag{593}$$

Because of symmetry, $u_{ik} = u_{ki}$, $\bar{u}_{ik} = \bar{u}_{ki}$.

Introducing Eq. (589) into Eq. (593), we obtain

$$\bar{u}_{ik} = \int_0^b \int_0^b \int_0^b \frac{M_{1(z,\xi)} m_{(\xi)} u_{i(\xi)} \bar{M}_{1(z,s)} m_{(s)} u_{k(s)} d\xi \, ds \, dz}{EI_{(z)}}$$

$$+ \int_0^b \int_0^b \left(\frac{\xi s}{C_m} + \frac{1.0}{C_v} \right) m_{(\xi)} u_{i(\xi)} m_{(s)} u_{k(s)} \, d\xi \, ds \tag{594}$$

Carrying out the integration with respect to ξ and s and denoting

$$\int_0^b M_{1(z,\xi)} m_{(\xi)} u_{i(\xi)} \, d\xi = M_{i(z)}$$

$$\int_0^b \bar{M}_{1(z,s)} m_{(s)} u_{k(s)} \, ds = M_{k(z)}$$

$$\int_0^b \xi m_{(\xi)} u_{i(\xi)} \, d\xi = M_{0i}$$

$$\int_0^b s m_{(s)} u_{k(s)} \, ds = M_{0k} \tag{595}$$

$$\int_0^b m_{(\xi)} u_{i(\xi)} \, d\xi = V_{0i}$$

$$\int_0^b m_{(s)} u_{k(s)} \, ds = V_{0k} \qquad \text{(595)}$$
$$\text{(Cont.)}$$

Thus,

$$\overline{u}_{ik} = \int_0^b \frac{M_{i(z)} M_{k(z)}}{EI_{(z)}} \, dz + \frac{M_{0i} M_{0k}}{C_m} + \frac{V_{0i} V_{0k}}{C_v} \qquad (596)$$

In most cases, even with a very poor first assumption for the fundamental mode deflection, the natural frequency is sufficiently accurate for practical use, even if only one term of Eq. (592) is used. Assuming $u_{(z)} = a_1 u_{1(z)}$ and taking $u_{1(z)}$ proportional to z, in accordance with Eq. (592),

$$a_1(u_{11} - \omega_n^2 \overline{u}_{11}) = 0 \qquad k = 1$$

$$\omega_n^2 = \frac{u_{11}}{\overline{u}_{11}}$$

$$= \frac{\int_0^b m_{(z)} z^2 \, dz + m_\alpha z_\alpha^2 + m_b z_b^2}{\int_0^b \frac{M_1^2}{EI_{(z)}} \, dz + \frac{M_{01}^2}{C_m} + \frac{V_{01}^2}{C_v}} \qquad (597)$$

The moment M_1 is obtained by considering $m_{(z)} z$, $m_\alpha z_\alpha$, and $m_b z_b$ as ideal loading of the tower and V_{01} and $M_{01} = M_{1(z=0)}$ as the base shear and moment, respectively, for this loading, as illustrated in Fig. 433. The integrals in Eq. (597) are best solved by the use of Simpson's rule [Eq. (112)].

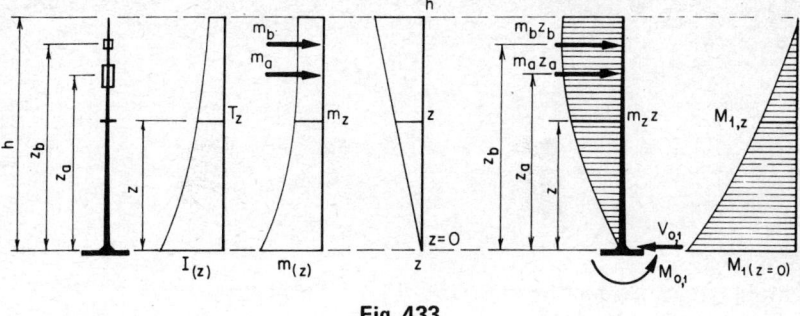

Fig. 433

Since the disturbing impulses are rather random and erratic, the related vibration-amplitude time curvature differs from harmonic vibration (Fig. 434). In accordance with Rausch,* denoting the degree of suddenness of the impulse by

$$\alpha = \frac{\omega}{\omega_n} \quad \omega_n = \frac{\omega}{\alpha}$$

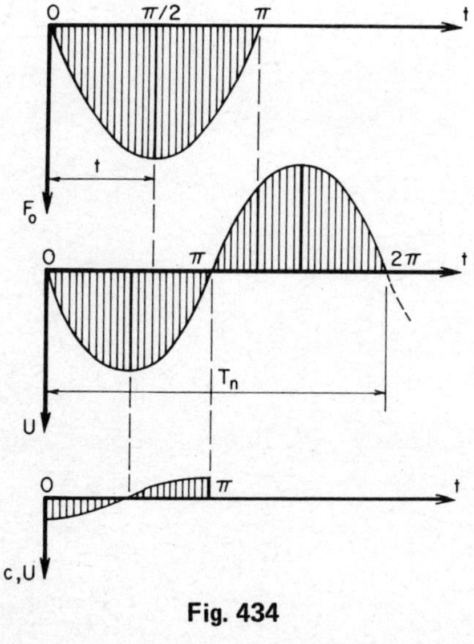

Fig. 434

*E. Rausch, "Maschinenfundamente und andere dynamische Bauaufgaben," VDI-Verlag, Berlin, 1936.

the amplitude of the forced vibration from 0 to π is

$$u = u_0 \frac{1}{1 - \alpha^2} [\sin(\omega t) - \alpha \sin(\omega_n t)]$$

$$\frac{du}{dt} = u_0 \frac{\alpha \omega_n}{1 - \alpha^2} [\cos(\omega t) - \cos(\omega_n t)]$$

(598)

and from π to 2π (free vibration) is

$$u = u_1 \sin(\omega_n t - \varphi)$$

$$\frac{du}{dt} = u_1 \omega_n \cos(\omega_n t + \varphi)$$

(599)

where $\omega = \dfrac{\pi}{2t}$ = circular frequency of disturbing impulse ($T = 4t$)

u_0 = assumed deflection due to F_0 as static force
u_1 = maximum deflection or amplitude of vibration due to dynamic impulse

At time $\omega t = \pi$, the amplitudes and velocities of both waves must be equal; thus,

$$u = -u_0 \frac{\alpha}{1 - \alpha^2} \sin\frac{\pi}{\alpha} = u_1 \sin\left(\frac{\pi}{2} + \varphi\right)$$

$$\frac{du}{dt} = -u_0 \frac{\alpha}{1 - \alpha^2} \left(1 + \cos\frac{\pi}{2}\right) = u_1 \cos\left(\frac{\pi}{2} + \varphi\right)$$

(600)

Substituting $u_1 = v u_0$, we can determine the v, φ values from these two equations:

$$\varphi = -\frac{\pi}{2\alpha} \quad \text{phase difference}$$

(601)

and

$$v = \frac{u_1}{u_0} = \frac{\alpha}{\alpha^2 - 1} 2 \cos\frac{\pi}{2\alpha}$$

(602)

The v is called the *magnification factor* due to dynamic condition. It is, for harmonic vibration,

$$v_b = \frac{1}{1 - \dfrac{\omega^2}{\omega_n^2}} = \frac{1}{1 - \alpha^2} \tag{603}$$

The amplitude of the free vibration [Eq. (599)] gradually decreases with the time due to the damping by a factor $e^{-ct/2m}$ and also diminishes after every cycle ($n = 1/T_n$) by constant ratio

$$\frac{e^{-ct/2m}}{1} \tag{604}$$

until $v = 0$.

When now a second disturbing impulse, of the same magnitude as the first, occurs at a time of n cycles ($t = nT_n$), the resulting maximum amplitude will be

$$u + u_n = v_r u$$
$$v_r = \frac{u + u_n}{u} \tag{605}$$

The v_r is commonly called the *repetition factor*. The equivalent static force or loading is thus

$$F = v v_r F_0 \sin(\omega_n t) \quad \omega_n t = \frac{\pi}{2};\ \sin(\omega_n t) = 1$$
$$F_{max} = v v_r F_0 \tag{606}$$

Wind Loading

The steady wind pressure, in accordance with the U.S. Weather Bureau, is approximately

$$p_a \simeq 0.0033\, v^2\ psf \tag{607}$$

in which v is the velocity of wind in miles per hour [see Eq. (245)].

The effective static wind pressure depends not only on the magnification and repetition factors but also on the shape of the structure. The shape factor ζ can be determined by model testing in a wind tunnel. It is smallest for streamlined cross sections. For cylindrical towers and poles, it is about 0.50, and for cross sections with large drag, it can be higher than 1.5. Thus the effective wind pressure for a design is

$$w_w = (p_a + vv_r p_G)\zeta \ psf \tag{608}$$

where p_a is the steady wind pressure and $(p_G = (p_{max} - p_a)$ is the gust wind pressure.

Earthquake Loading

In earthquake analysis, the disturbing force F_0 is applied not directly to the structure but through the ground motion. Thus the mass of the structure must first be accelerated and the inertia modified to the dynamic effect of the applied load. For this, the dynamic equilibrium condition is expressed by the following differential equation (Fig. 429):

$$m\frac{d^2x}{dt^2} + c\frac{du}{dt} + ku = 0 \tag{609}$$

Considering that

$$\frac{d^2x}{dt^2} = \frac{d^2x_0}{dt^2} + \frac{d^2u}{dt^2}$$

the differential equation (609) is transformed into the same form as Eq. (580):

$$m\frac{d^2u}{dt^2} + c\frac{du}{dt} + ku = -m\frac{d^2x_0}{dt^2} \tag{610}$$

If the ground motion were a simple harmonic vibration with a period $T = 2\pi/\omega$ and $c = 0$, the effective earthquake force would be

$$F_0 \sin(\omega t) = -m\frac{d^2x_0}{dt^2}\sin(\omega t) \tag{611}$$

and the maximum base shear [Eq. (599)] would be

$$V_{max} = \sum_0^b m_{(z)} \frac{1}{1 - T_n^2/T^2} \frac{d^2 x_0}{dt} \quad v_b = \frac{1}{1 - T_n^2/T^2}$$
$$= v_b \frac{\Sigma w_{(z)}}{g} \frac{d^2 x_0}{dt^2} \tag{612}$$

The ground acceleration $(d^2 x_0/dt^2)$ is an unknown and thus has to be assumed.

Very often it is more convenient to express the response of the structure to a specific ground motion in terms of spectral value of the velocity S_v (ft/sec) relative to the ground, as follows:

$$V_{max} = \frac{\Sigma w_{(z)}}{g} \omega_n S_v \tag{613}$$

where, for undamped structure,

$$S_v = v_b \omega_n \frac{d^2 x_0}{dt^2}$$
$$v_b = \frac{1}{1 - \omega^2/\omega_n^2} \tag{614}$$

and for damped structure,

$$S_{vd} = v_{vd} \omega_n \frac{d^2 x_0}{dt^2}$$
$$v_{vd} = \sqrt{\frac{1 + 4\beta^2 \alpha^2}{(1 - \alpha^2)^2 + 4\beta^2 \alpha^2}} \quad \beta = \frac{c}{2m\omega_n}; \quad \alpha = \frac{\omega^2}{\omega_n^2} \tag{615}$$

The impulses and velocities recorded and evaluated for actual earthquakes are irregular, erratic, and not harmonic. Therefore, if the spectral curves for a certain location are available, the given maximum S_v values should be used in Eq. (613). Also, the maximum base shear V_{max} can be computed in the same way as outlined for gust wind loading [Eqs. (598) and (606)].

Structural Systems and Methods of Construction

The base shear V_{max} is distributed over the height of a structure in accordance with the following formula:

$$F_z = \frac{V_{max} w_z z}{\sum_0^b w_z z} \tag{616}$$

Commonly it is assumed that wind and earthquake loadings do not occur simultaneously; therefore, the loading which produces the higher values is used for design.

19

WHY STRUCTURES COLLAPSE OR FAIL

Contemporary concrete structures are complex, slender, exposed, and used for architectural expression. Structural systems are highly statically indeterminate, and spans up to 300 to 500 ft are not uncommon. Such structures require up-to-date advanced engineering, efficient use of materials, sophisticated construction methods, and high-quality performance. When these requirements are not met, and considering the high stresses commonly involved, the possibilities of failure are greater than for ordinary structures. But seldom does an exposed structure under normal conditions collapse suddenly without due warning, provided there is nothing seriously wrong. The first indication that a structure is suffering is crack formation. The second indication is severe cracking associated with undue deflections. If no attention and help are offered to such a structure already in agony, most certainly it will collapse in the course of time.

There is a good, recently published book available on this subject: "Construction Failures," by Dr. Jacob Feld. In this book Dr. Feld thoroughly and extensively analyzes the causes of construction failures. Due to this, in the following only the common reasons for failures of most advanced structures will be discussed.

80. DESIGN

Any design in reinforced or prestressed concrete is based on theories, facts, and certain assumptions tested against reality and determined to be true or at least within acceptable accuracy. Experience proves that a structure designed in accordance with accepted theories and executed properly performs as predicted by computations. In cases in which the design has deviated in large degree from the certified theories and facts or in which the assumptions have been too arbitrary and unrealistic, some kind of deficiency has almost always been evidenced.

The advanced theories on reinforced and prestressed concrete are rather complex, and their interpretation and application require extensive mathematical knowledge not commonly possessed by average structural engineers. Due to this, in practice very often the so-called "exact" theories have been replaced with simplified ones or with empirical formulas. The commonly given justification for this is that the exact theories have been based on ideal conditions and materials and include uncertain material coefficients, such as modulus of elasticity, moment of inertia, and Poisson's ratio, which may vary considerably. To a certain degree this is true, but to counteract these uncertainties, one can always use limiting values between which the unknown material coefficients will lie and within which a structure will be qualitatively safe in terms of strength as well as deformation.

The fact that concrete is not a perfect solid homogeneous material and as such is subjected to capillary action and plastic deformations has very often been overlooked in design. The same applies for temperature changes. Due to this inadequacy in design, considerable stress concentration occurs, especially in long-span structures, resulting commonly in extensive crack formation and even collapse of the structure. In practice, to counteract the change of length and also to avoid complicated structural analysis, statically determinate systems are used and expansion joints provided. But this remedy is associated with serious deficiencies such as reduced economy, increased deflections, and even more extensive crack formation because the expansion joint seldom performs as required and most of all because the statically determinate structure lacks the ability of stress redistribution due to some type of structural weakness or unbalance in design. Therefore, as a conclusion, the indeterminate systems should be preferred to determinate ones. Change of length and rotation are best counteracted by designing the system as elastic as possible so that displacement and rotation can take place without exceeding the safe stress limits. In cases in which this is not possible or in which extensive ductility of the structural system is required, elastically controlled joints between the elements (neoprene interlays and increased elongation length for reinforcing and prestressing tendons) should be provided to absorb the energy and to reduce the effect of dynamic forces.

When determining the allowable stresses for steel, it should be kept in mind that the modulus of elasticity of steel for all grades is practically equal; therefore, allowable stresses for reinforcing steel should be kept below the acceptable crack-width limit of the concrete ($\triangle < 0.01$ in.). For prestressing steel, the stresses should not exceed $0.75\ f'_s$ even temporarily because the technical creep of high-carbon steel may start at $0.8\ f'_s$ and will proceed even at a lower stress level until failure (see Sec. 45).

The connections of precast elements should be designed rigid, but ample elongation length should be provided for welding plates to allow a certain amount of elongation and rotation so that the tensile strength of concrete will not be exceeded.

The most common failures in posttensioned concrete have occurred at the anchorage of tendons. The reasons for this have always been inadequate thickness of anchor plates, resulting in stress concentration, and not considering the tension in concrete perpendicular to the prestressing stresses.

In pretensioned elements, the strands are commonly flame-cut too suddenly, causing dynamic shock, loss of bond, and splitting of concrete. If no conventional reinforcing is provided and too small supporting length is used, it may lead to sudden collapse, especially at low temperature, even when neoprene seating is provided. Also, very often the numerous reinforcing bars are overlapped or stopped in the same section. This most certainly leads to crack formation because of sudden change in moment of inertia, shrinkage, and plastic flow. To prevent cracking, the overlapping and stopping must be staggered symmetrically. The bending of compressive reinforcing, to provide space for overlapping, causes considerable tension in concrete and reduces the effectiveness of such bars, mainly because of plastic flow. It can be easily proved that in the course of time compressive steel is stressed to yield strength; in this state, even straight bars can buckle if not properly tied, but bent bars, regardless of the ties, most certainly will be displaced from their original position, cracking the concrete and destroying the bond.

Large-scale cracking and deflection in prestressed poured-in-place multispan beams and frames have occurred because the resistance of columns or walls to the prestressing force has not been considered in design, which results in considerable reduction in prestressing force, especially where rather stiff columns or walls are involved.

Considerable internal stresses occur in concrete when heavy reinforcing bars are designed in many layers in narrow spacing, so that there is lack of tie between the inside and outside concrete. The physical characteristics of concretes separated by steel bars differ considerably due to overvibration, honeycombs, and most of all, segregation. Such a condition is extremely dangerous for compressive zones and may cause separation of the steel from concrete and collapse of the entire structural system.

Finally, structural drawings are seldom completed extensively and properly. For advanced designs, it is the rule that all structural reinforcing bars and prestressing tendons be indicated on drawings and that an adequate number of larger-scale cross sections be provided. Also, significant details should be worked out to the fullest extent. Turning over this work to reinforcing-bar detailers and to precasting and prestressing tendon manufacturers and providing only tables for the amount of reinforcing in beams and columns commonly lead to misinterpretation of the designer's intentions and to serious errors. Besides, it puts an unjustified burden on the contractor and manufacturers to prepare the shop drawings. On the other hand, several checkings by the designer of the numerous shop drawings require in most cases more time than is involved in the preparation of complete structural drawings, so that shop drawings are not even required, except some details for clarification. Also, the technique of preparing structural drawings leaves much to be desired. It is a sound rule that every drawing must be independent, and self-sufficient, including all elements concerned, and that references to other drawings should be kept to a minimum. Reference sketches and drawing numbers of adjoining elements should be indicated on each drawing for clarification. If more than one sheet is required for an element, all such sheets should be designated with the same number (for example: No. 10, 10A, 10B, . . .).

The philosophy of specification writing should be based on the principle that the designer should specify only what is significant and different from codes and should give only the special data the design is based upon, such as quality of concrete, strength, and type of tendons, and that the method of obtaining the required results is the contractor's duty and responsibility. Only in this way can the contractor's experience be utilized to the fullest extent and can the best construction results be obtained qualitatively as well as economically. Designers, and especially specification writers, commonly lack adequate construction experience and are not able to determine how the work can be best performed. Very often the specifications do not specify what is necessary and important in a particular design but instead only rewrite the code. Due to this, the contractor usually does not study the large volumes of specifications; therefore, it is essential to spell out the most important requirements on the drawings concerned.

81. CONSTRUCTION

By construction a design becomes reality, that is, a structure. These two phases are interrelated and control equally the quality of the structure; perfect design and poor construction, or vice versa, commonly result in deficiencies and failures. To obtain good results, both teams must work together, but very often the designer's part in the actual construction is entirely assumed by outsiders—the supervisor-inspection team.

Construction firms in general have become in recent years more management-finance organizations than professional construction engineers. This trend has already advanced so far that in many quarters construction is no longer considered qualified to belong in any branch of engineering. For this, government building departments, architects, and engineers are principally to be blamed, because if every performance of the contractor is specified in detail, professional engineers are not required in a construction firm. In addition, the financing of projects has been made extremely difficult for any contractor, so that projects can be handled only by highly specialized financial organizations. No wonder one finds construction firms which do not have a single professional engineer on the staff! There still are highly qualified construction firms, but all things considered, there is less possibility that they will be the low bidders on an advanced project. Considering the situation as it is, one should not be surprised that deficiencies and failures occur. However, in an advanced project, major failures are rare in comparison with those in ordinary projects, mainly because the designers, contractors, subcontractors, inspectors, and even workers take their duties more seriously and are more careful in their performance. Furthermore, the structurally less qualified firms generally use subcontractors who are quite qualified to carry out the work in their specialized field. The minor mishaps which occasionally occur but can be easily repaired cannot be considered in the true sense as failures.

Most serious structural failures occur during construction. The reasons for failure may lie in the design, in the method of construction, or in both. Any structure must be designed safe not only after completion but also during construction. The designer, contractor, and supervisor-inspection team should be fully aware of the fact that, depending on the structural system and construction method used, the highest stresses, no matter how temporary, may occur during construction. Overlooking this fact, both in design and in construction, can have disastrous results. After completion, a structure seldom fails without due warning. Serious failures at a later date are caused mostly by ignorance and negligence—for example, poor grouting, leaving water pockets in metal tubing, crack formation along the tendons. Poor grouting causes corrosion of tendons, which leads to failure in the course of time. Water pockets seldom occur if grouting distances are not extensive, grout mix is proper, and grouting is started from the lowest point of the tendons. Before grout is forced into tubing, the tubing must be properly washed with water, and if the grout does not flow out from the far end, it must be blown out immediately and the mix modified. Crack formation along flexible metal tubing indicates insufficient curing, resulting in early shrinkage, inadequate spacing and coverage for tubing, or too sharp curvature and flattening of the tendons. In prefabricated elements, in addition, the fact has been ignored that heat travels almost 100 times faster along steel than along concrete; if the ends of the tubes are not closed properly during steam curing, considerable temperature differences between tubing and concrete occur, resulting in expansion in tubing and high tensile stresses in concrete, which the concrete is unable to resist in its early phase of hardening. Such cracking can be extensive in columns in which, as is often the case, light plane tubing is used in order to ease erection. To prevent cracking, such tubing must be covered from the outside with soft compressible material to allow expansion and free shrinkage. For the same reason, all edges of anchor and welding plates should be covered, preferably by sponge-rubber tape. Also, small-size reinforcing at the ends around the anchor plates should be provided to prevent breaks at the ends of precast units. Not providing distribution reinforcing behind the anchor plates is the usual reason for the frequent anchorage failures of the posttensioning tendons, especially when large cables are used.

82. SUPERVISION AND INSPECTION

The main responsibility and purpose of general supervision are to make sure that the superintendent and foremen of the contractor, subcontractors, and inspectors interpret the drawings properly, understand the structure, and know their duties and responsibilities and, also, to help them solve technical problems arising during construction. To be able to do this, the engineer in charge of the supervision must know the project thoroughly and must have adequate construction experience. Most suitable would be the structural designer or his personal representative, a qualified professional engineer. Commonly, but especially in government works, this is not the case. Supervision is usually carried out by the architect or outside technicians, seldom by qualified engineers. The designing structural engineer visits the construction site only when something has happened and he is asked to solve special problems. Experience has shown that the supervision performed by architects and outside technicians, even on ordinary reinforced-concrete structures, is valueless. Due to this fact, possible errors are not discovered at the right time, causing unwarranted delays and failures for which nobody accepts responsibility.

83. RESPONSIBILITIES

Divided and not clearly defined responsibilities are the most common reasons for failures. If every member of the team knows exactly what his responsibilities are, he will be much more careful in carrying out his duties, because it is human nature to protect one's own interests. Practically, responsibilities shared with others mean avoidance of personal obligations; this is most pronounced in committee works, which present the lowest degree of responsibility.

It is just and logical that the designer should be responsible for the design as a whole and the contractor for quality of execution of the work. The supervision-inspection team, as the authority on the site, should have the sole responsibility for the work being carried out in accordance with the design drawings and specifications.

Nobody can expect the contractor to be a specialist, especially in advanced design, which involves complex theories. How can it possibly be expected that the contractor, who is mainly in a money-making business, will have the time to check the drawings for their correctness, omissions, and errors? On the other hand, it can also not be expected that the designer will inspect every step of the contractor's performance, especially when the supervision is carried out by others. The supervision-inspection team is charged with the quality of contractor's performance and is mainly responsible for deficiencies and failures, provided that the design is correct. In many failure cases, the supervision-inspection team tries to avoid responsibility by asserting that the contractor did not carry out the team's orders for correcting the deviations from design or the poor workmanship. Such an excuse only emphasizes the inadequacy of the team in allowing the contractor to proceed until failure occurs instead of stopping the work immediately, leaving the site, and notifying the designers. Such a case clearly demonstrates that the supervision-inspection team is valueless and should not be used at all and that, instead, the contractor should be solely responsible for the quality of his work. Experience shows that in most cases of failure, the professional competence and performance of the inspectors have been inadequate. On the other hand, when the contractor has been made responsible for the quality of his work, failures seldom have occurred.

Most deficiencies occur because of the poor quality and incompleteness of the structural drawings required for execution of the work. The numerous shop drawings, usually prepared by subcontractors and submitted by the contractor to the designers for checking and approval, are seldom qualitatively acceptable and are commonly checked inadequately. Designers try to protect themselves for this inadequacy by adding a well-known paragraph in the specifications: "Any errors, omissions, etc., in shop drawings do not release the contractor from responsbility and prevent the architect from insisting upon making all work right. ..." If now the inspector checks the work against such erroneous shop drawings and not against the structural drawings—provided they are complete—failures occur. In this particular case, no doubt, whoever checked and approved the shop drawings is at least partly responsible.

Very often in advanced designs, a structural engineering firm or engineer is commissioned as a specialist to design the structure or a part of it, but the obligation for making the finished drawings, checking shop drawings, and general supervision is kept for the overall designer. In this case, the overall designer should also carry full responsibility for his performance.

Adopting this philosophy of definite responsibilities, as outlined above, one can be sure that the number of deficiencies will be reduced considerably and major failures avoided.

APPENDIX

TABLE A-1 Elementary Transcendental Functions

x	x, deg	e^x	e^{-x}	$\sinh x$	$\cosh x$	$\tanh x$	$\sin x$	$\cos x$	$\tan x$
0.00	0.00	1.00000	1.00000	0.00000	1.00000	0.00000	0.00000	1.00000	0.00000
0.01	0.58	1.01010	0.99005	0.01000	1.00005	0.01000	0.01000	0.99995	0.01000
0.02	1.15	1.02020	0.98020	0.02000	1.00020	0.02000	0.02000	0.99980	0.02000
0.03	1.72	1.03050	0.97045	0.03000	1.00045	0.02999	0.03000	0.99955	0.03001
0.04	2.28	1.04080	0.96079	0.04001	1.00080	0.03998	0.03999	0.99920	0.04002
0.05	2.87	1.05130	0.95123	0.05002	1.00125	0.04996	0.04998	0.99875	0.05004
0.06	3.43	1.06180	0.94176	0.06004	1.00180	0.05993	0.05996	0.99820	0.06007
0.07	4.00	1.07250	0.93239	0.07006	1.00245	0.06989	0.06994	0.99755	0.07011
0.08	4.58	1.08330	0.92312	0.08009	1.00320	0.07983	0.07991	0.99680	0.08017
0.09	5.17	1.09420	0.91393	0.09012	1.00405	0.08976	0.08988	0.99595	0.09024
0.1	5.73	1.10517	0.90484	0.10017	1.00500	0.09967	0.09983	0.99500	0.10033
0.2	11.46	1.22140	0.81873	0.20134	1.02007	0.19738	0.19867	0.98007	0.20271
0.3	17.19	1.34986	0.74082	0.30452	1.04534	0.29131	0.29552	0.95534	0.30934
0.4	22.92	1.49182	0.67032	0.41075	1.08107	0.37995	0.38942	0.92106	0.42279
0.5	28.65	1.64872	0.60653	0.52110	1.12763	0.46212	0.47943	0.87758	0.54630
0.6	34.38	1.82212	0.54881	0.63665	1.18547	0.53705	0.56464	0.82534	0.68414
0.7	40.11	2.01375	0.49659	0.75858	1.25517	0.60437	0.64422	0.76484	0.84229
0.8	45.84	2.22554	0.44933	0.88811	1.33743	0.66404	0.71736	0.69671	1.02964
0.9	51.57	2.45960	0.40657	1.02652	1.43309	0.71630	0.78333	0.62161	1.26016
1.0	57.30	2.71828	0.36788	1.17520	1.54308	0.76159	0.84147	0.54030	1.55741
1.1	63.03	3.00417	0.33287	1.33565	1.66852	0.80050	0.89121	0.45360	1.96476
1.2	68.75	3.32012	0.30119	1.50946	1.81066	0.83365	0.93204	0.36236	2.57215
1.3	74.48	3.66930	0.27253	1.69838	1.97091	0.86172	0.96356	0.26750	3.60210
1.4	80.21	4.05520	0.24660	1.90430	2.15090	0.88535	0.98545	0.16997	5.79789
1.5	85.94	4.48169	0.22313	2.12928	2.35241	0.90515	0.99749	+0.07074	+14.10142

1.6	91.67	4.95303	0.20190	2.37557	2.57746	0.92167	0.99957	−0.02920	−34.23254
1.7	97.40	5.47395	0.18268	2.64563	2.82832	0.93541	0.99166	−0.12884	−7.69660
1.8	103.13	6.04965	0.16530	2.94217	3.10747	0.94681	0.97385	−0.22720	−4.28626
1.9	108.86	6.68589	0.14957	3.26816	3.41773	0.95624	0.94630	−0.32329	−2.92710
2.0	114.59	7.38906	0.13534	3.62686	3.76220	0.96403	0.90930	−0.41615	−2.18504
2.1	120.32	8.16617	0.12246	4.02186	4.14431	0.97045	0.86321	−0.50485	−1.70985
2.2	126.05	9.02501	0.11080	4.45711	4.56791	0.97574	0.80850	−0.58850	−1.37382
2.3	131.78	9.97418	0.10026	4.93696	5.03722	0.98010	0.74571	−0.66628	−1.11921
2.4	137.51	11.02318	0.09072	5.46623	5.55695	0.98367	0.67546	−0.73739	−0.91601
2.5	143.24	12.18249	0.08208	6.05020	6.13229	0.98661	0.59847	−0.80114	−0.74702
2.6	148.97	13.46374	0.07427	6.69473	6.76901	0.98903	0.51550	−0.85689	−0.60160
2.7	154.70	14.87973	0.06721	7.40626	7.47347	0.99101	0.42738	−0.90407	−0.47273
2.8	160.43	16.44465	0.06081	8.19192	8.25273	0.99263	0.33499	−0.94222	−0.35553
2.9	166.16	18.17415	0.05502	9.05956	9.11458	0.99396	0.23925	−0.97096	−0.24641
3.0	171.89	20.08554	0.04979	10.01787	10.06766	0.99505	0.14112	−0.98999	−0.14255
3.1	177.62	22.19795	0.04505	11.07645	11.12150	0.99595	+0.04158	−0.99914	−0.04162
3.2	183.35	24.53253	0.04076	12.24588	12.28665	0.99668	−0.05837	−0.99829	+0.05847
3.3	189.08	27.11264	0.03688	13.53788	13.57476	0.99728	−0.15775	−0.98748	0.15975
3.4	194.81	29.96410	0.03337	14.96536	14.99874	0.99777	−0.25554	−0.96680	0.26442
3.5	200.54	33.11545	0.03020	16.54263	16.57282	0.99818	−0.35078	−0.93646	0.37470
3.6	206.26	36.59823	0.02732	18.28546	18.31278	0.99851	−0.44252	−0.89676	0.49347
3.7	211.99	40.44730	0.02472	20.21129	20.23601	0.99878	−0.52984	−0.84810	0.62473
3.8	217.72	44.70118	0.02237	22.33941	22.36178	0.99900	−0.61186	−0.79097	0.77356
3.9	223.45	49.40245	0.02024	24.69110	24.71135	0.99918	−0.68777	−0.72593	0.94742
4.0	229.18	54.59815	0.01832	27.28992	27.30823	0.99933	−0.75680	−0.65364	1.15782
4.1	234.91	60.34029	0.01657	30.16186	30.17843	0.99945	−0.81828	−0.57482	1.42353
4.2	240.64	66.68633	0.01500	33.33567	33.35066	0.99955	−0.87158	−0.49026	1.77778
4.3	246.37	73.69979	0.01357	36.84311	36.85668	0.99963	−0.91617	−0.40080	2.28585
4.4	252.10	81.45087	0.01228	40.71930	40.73157	0.99970	−0.95160	−0.30733	3.09632

TABLE A-1 Elementary Transcendental Functions (Cont.)

x	x, deg	e^x	e^{-x}	$\sinh x$	$\cosh x$	$\tanh x$	$\sin x$	$\cos x$	$\tan x$
4.5	257.83	90.01713	0.01111	45.00301	45.01412	0.99975	−0.97753	−0.21080	4.63733
4.6	263.56	99.48432	0.01005	49.73713	49.74718	0.99980	−0.99369	−0.11215	8.86018
4.7	269.29	109.9472	0.00910	54.96904	54.97813	0.99983	−0.99992	−0.01239	+80.71280
4.8	275.02	121.5104	0.00823	60.75109	60.75932	0.99986	−0.99616	+0.08750	−11.38487
4.9	280.75	134.2898	0.00745	67.14117	67.14861	0.99989	−0.98245	0.18651	−5.26749
5.0	286.48	148.4132	0.00674	74.20321	74.20995	0.99991	−0.95892	0.28366	−3.38052
5.1	292.21	164.0219	0.00610	82.00791	82.01400	0.99993	−0.92581	0.37798	−2.44939
5.2	297.94	181.2722	0.00552	90.63336	90.63888	0.99994	−0.88345	0.46852	−1.88564
5.3	303.67	200.3368	0.00499	100.1659	100.1709	0.99995	−0.83227	0.55437	−1.50128
5.4	309.40	221.4064	0.00452	110.7009	110.7055	0.99996	−0.77276	0.63469	−1.21754
5.5	315.13	244.6919	0.00409	122.3439	122.3480	0.99997	−0.70554	0.70867	−0.99558
5.6	320.86	270.4264	0.00370	135.2114	135.2150	0.99997	−0.63127	0.77557	−0.81394
5.7	326.59	298.8674	0.00335	149.4320	149.4354	0.99998	−0.55069	0.83471	−0.65973
5.8	332.32	330.2996	0.00303	165.1483	165.1513	0.99998	−0.46460	0.88552	−0.52467
5.9	338.05	365.0375	0.00274	182.5174	182.5201	0.99999	−0.37388	0.92748	−0.40311
6.0	343.77	403.4288	0.00248	201.7132	201.7156	0.99999	−0.27942	0.96017	−0.29101
6.1	349.50	445.8578	0.00224	222.9278	222.9300	0.99999	−0.18216	0.98327	−0.18526
6.2	355.23	492.7490	0.00203	246.3735	246.3755	0.99999	−0.08309	0.99654	−0.08338
6.3	360.96	544.5719	0.00184	272.2850	272.2869	0.99999	+0.01681	0.99986	+0.01682

TABLE A-2 Minimum Ultimate Strength of Uncoated Wire

Nominal diameter, in.	Sectional area, in.^{2}	Minimum ultimate tensile strength	
		psi	kips
0.276	0.0598	225,000	13.445
0.250	0.0491	230,000	11.293
0.196	0.0304	240,000	7.288
0.192	0.0289	240,000	6.936
0.160	0.0201	250,000	5.025
0.128	0.0129	260,000	3.354
0.120	0.0113	260,000	2.938
0.104	0.0085	270,000	2.295
0.080	0.0050	280,000	1.400

Approximate modulus of elasticity: 29×10^6 psi.

TABLE A-3 Minimum Breaking Strength of Uncoated Seven-wire Strand

Nominal strand diameter, in.	Approximate steel area, in.^{2}	Minimum breaking strength, kips	Approximate weight of strand per 1,000 ft lb
1/4	0.036	9.0	122
5/16	0.058	14.5	198
3/8	0.080	20.0	274
7/16	0.109	27.0	373
1/2	0.144	36.0	494

Approximate modulus of elasticity: 27×10^6 psi.

TABLE A-4 Minimum Breaking Strength of Uncoated Seven-wire (270 K) Strands

Nominal strand diameter, in.	Approximate steel area, in.^{2}	Minimum breaking strength, kips	Approximate weight of strand per 1,000 ft lb
3/8	0.085	23.0	292
7/16	0.117	31.0	400
1/2	0.153	41.3	525
0.600	0.215	51.0	740

Approximate modulus of elasticity: 28×10^6 psi.

TABLE A-5 Roebling Galvanized Prestressing Strands

Nominal strand diameter, in.	Approximate steel area, in.^{2}	Minimum breaking strength, kips	Approximate weight of strand per ft lb
0.600	0.215	46.0	0.737
0.835	0.409	86.0	1.412
1	0.577	122.0	2.000
1 1/8	0.751	156.0	2.610
1 1/4	0.931	192.0	3.220
1 3/8	1.120	232.0	3.890
1 1/2	1.360	276.0	4.700
1 9/16	1.480	300.0	5.110
1 5/8	1.600	324.0	5.520
1 11/16	1.730	352.0	5.980

Approximate modulus of elasticity: 24×10^6 psi.

TABLE A-6 Galvanized Bridge Rope 6 x 7 and 6 x 19 with Wire Core

Nominal rope diameter, in.	Approximate steel area, in.^{2}	Minimum breaking strength, kips	Approximate weight of rope per ft lb
1	0.471	91.4	1.67
1 1/8	0.596	115.6	2.11
1 1/4	0.745	144.4	2.64
1 3/8	0.906	175.6	3.21
1 1/2	1.076	208.0	3.82
1 5/8	1.270	246.0	4.51
1 3/4	1.470	286.0	5.24
1 7/8	1.690	328.0	6.03
2	1.920	372.0	6.85
2 1/8	2.170	420.0	7.73

Approximate apparent moduli of elasticity: 6 x 7, 16×10^6 psi; 6 x 19, 15×10^6 psi; prestressed, 20×10^6 psi.

TABLE A-7 Plain SS Bars

Nominal bar diameter, in.	Approximate steel area, in.^{2}	Minimum breaking strength, kips		Approximate weight of bars per foot, lb
		Regular	Special	
3/4	0.442	64.1	70.7	1.50
7/8	0.601	87.1	96.2	2.04
1	0.785	113.8	125.6	2.67
1 1/8	0.994	144.1	159.0	3.38
1 1/4	1.227	177.9	196.3	4.17
1 3/8	1.485	215.3	237.6	5.05

Approximate modulus of elasticity: 30×10^6 psi.

TABLE A-8 Parallel Wire Cables

Number of wires ($d = 0.250$)	Steel area, in.^{2}	Load, kips				Weight, lb/100 ft	Tendon ϕ, in.
		$0.6f_s'$	$0.75f_s'$	$0.8f_s'$	f_s'		
1	0.04909	7.0	8.8	9.4	11.8	16.7	3/8
6	0.29454	42.5	53.0	56.5	70.7	100.0	3/4
8	0.39272	56.5	70.7	75.4	94.2	134.0	7/8
12	0.58908	84.8	106.0	113.1	141.4	200.0	1
14	0.68726	99.0	123.7	132.0	165.0	234.0	1 1/4
18	0.88362	127.2	159.0	169.7	212.1	301.0	1 1/4
20	0.98180	141.4	176.7	188.5	235.6	334.0	1 1/4
24	1.17816	169.7	212.0	226.2	282.8	401.0	1 1/2
27	1.32541	190.9	238.6	254.5	318.2	451.0	1 1/2
28	1.37452	197.9	247.4	263.9	330.0	468.0	1 1/2
30	1.47270	212.0	265.0	282.8	353.5	501.0	1 3/4
36	1.76724	254.5	318.0	339.3	424.2	601.0	2
38	1.86542	268.6	335.7	358.7	447.7	634.0	2
40	1.96360	282.8	353.3	377.0	471.3	667.0	2
46	2.25814	325.2	406.5	433.6	542.0	767.0	2 1/4

TABLE A-9 Strand Cables

Tendon type	No. and size of strands	Load, kips				Weight, lb/100 ft	Tendon ϕ, in.
		$0.6f_s'$	$0.7f_s'$	$0.8f_s'$	f_s'		
S1-5	1 × 0.5	24.8	28.9	33.0	41.3	54	3/4
S1-6	1 × 0.6	35.2	41.0	46.9	58.6	74	7/8
S3-5	3 × 0.5	74.4	86.7	99.0	123.9	162	1 3/8
S3-6	3 × 0.6	105.6	123.0	140.7	175.8	222	1 1/2
S4-6	4 × 0.6	140.8	164.0	187.6	234.4	296	1 5/8
S5-6	5 × 0.6	176.0	205.0	234.5	293.0	370	1 3/4
S6-6	6 × 0.6	211.2	246.0	281.4	351.6	444	1 7/8
S7-6	7 × 0.6	246.4	287.0	328.3	410.2	518	2
S12-5	12 × 0.5	297.6	346.8	396.0	495.6	648	2 1/2
S12-6	12 × 0.6	422.4	492.0	562.8	703.2	888	2 7/8
S19-5	19 × 0.5	471.2	549.1	627.0	784.7	1,026	3
S19-6	19 × 0.6	668.8	779.0	891.1	1,113.4	1,406	3 1/4
S28-5	28 × 0.5	694.4	809.2	924.0	1,156.4	1,512	3 1/2

TABLE A-10 Standard Sizes of Reinforcing Bars

No.	Size, in. Deformed	Area, in.^{2}	Weight, lb/lin ft	Perimeter, in.
¼	—	0.05	0.167	0.785
3	3/8	0.11	0.376	1.178
4	1/2	0.20	0.668	1.571
5	5/8	0.31	1.043	1.963
6	3/4	0.44	1.502	2.356
7	7/8	0.60	2.044	2.749
8	1	0.79	2.670	3.142
9	1	1.00	3.400	4.000
10	1 1/8	1.27	4.303	4.500
11	1 1/4	1.56	5.313	5.000

TABLE A-11 Common Styles of Welded Wire Fabric (One-way Types)

Style designation	Spacing of wires, in.		Size of wires, W & M gage		Sectional area, in.^{2}/ft		Weight, lb/100 ft^2
	L	T	L	T	L	T	
2 × 16-0/6	2	16	0	6	0.443	0.022	163
2 × 16-1/7	2	16	1	7	0.377	0.018	140
2 × 16-2/8	2	16	2	8	0.325	0.015	119
2 × 16-3/8	2	16	3	8	0.280	0.015	104
2 × 16-4/9	2	16	4	9	0.239	0.013	89
3 × 16-2/8	3	16	2	8	0.216	0.015	83
2 × 16-5/10	2	16	5	10	0.202	0.011	75
3 × 16-3/8	3	16	3	8	0.187	0.015	72
2 × 16-6/10	2	16	6	10	0.174	0.011	65
3 × 16-4/9	3	16	4	9	0.159	0.013	61
2 × 16-7/11	2	16	7	11	0.148	0.009	55
4 × 16-3/8	4	16	3	8	0.140	0.015	56
4 × 16-4/9	4	16	4	9	0.120	0.013	48
4 × 16-5/10	4	16	5	10	0.101	0.011	40
4 × 16-6/10	4	16	6	10	0.087	0.011	35
4 × 16-7/11	4	16	7	11	0.074	0.009	30
4 × 12-8/12	4	12	8	12	0.062	0.009	26
4 × 12-9/12	4	12	9	12	0.052	0.009	22
4 × 12-10/12	4	12	10	12	0.043	0.009	19
4 × 12-11/12	4	12	11	12	0.034	0.009	15
4 × 12-12/12	4	12	12	12	0.026	0.009	13

L, longitudinal; T, transverse.

TABLE A-12 Common Styles of Welded Wire Fabric (Two-way Types)

Style designation	Spacing of wires, in.		Size of wires, W & M gage		Sectional area, $in.^2/ft$		Weight, $lb/100\ ft^2$
	L	T	L	T	L	T	
2 x 2-10/10	2	2	10	10	0.086	0.086	60
2 x 2-12/12	2	2	12	12	0.052	0.052	37
2 x 2-14/14	2	2	14	14	0.030	0.030	21
3 x 3-10/10	3	3	10	10	0.057	0.057	41
3 x 3-12/12	3	3	12	12	0.035	0.035	25
3 x 3-14/14	3	3	14	14	0.020	0.020	14
4 x 4-4/4	4	4	4	4	0.120	0.120	85
4 x 4-6/6	4	4	6	6	0.087	0.087	62
4 x 4-8/8	4	4	8	8	0.062	0.062	44
4 x 4-10/10	4	4	10	10	0.043	0.043	31
4 x 4-12/12	4	4	12	12	0.026	0.026	19
6 x 6-0/0	6	6	0	0	0.148	0.148	107
6 x 6-2/2	6	6	2	2	0.108	0.108	78
6 x 6-4/4	6	6	4	4	0.080	0.080	58
6 x 6-6/6	6	6	6	6	0.058	0.058	42
6 x 6-8/8	6	6	8	8	0.041	0.041	30
6 x 6-10/10	6	6	10	10	0.029	0.029	21

L, longitudinal; T, transverse.

TABLE A-13 Engineering Conversion Factors

Multiply	By	To obtain
Inches	2.54001	Centimeters
Inches	2.54001×10^{-2}	Meters
Inches	25.4001	Millimeters
Feet	30.4801	Centimeters
Feet	0.304801	Meters
Feet	304.801	Millimeters
Yards	0.914402	Meters
Square inches	645.163	Square millimeters
Square inches	6.45163	Square centimeters
Square feet	0.0929034	Square meters
Square feet	9.29034×10^{-6}	Hectares
Square feet	929.034	Square centimeters
Square yards	0.83613	Square meters
Cubic inches	16.38716	Cubic centimeters
Cubic feet	2.8317×10^{4}	Cubic centimeters
Cubic feet	2.8317×10^{-2}	Cubic meters
Cubic yards	0.764559	Cubic meters
Gallons, U.S.	0.13368	Cubic feet
Gallons, U.S.	231.	Cubic inches
Gallons, U.S.	3.78543	Liters
Gallons, British Imperial	1.20091	Gallons, U.S.
Pounds, avoirdupois	453.592	Grams, metric
Pounds, avoirdupois	0.453592	Kilograms
Pounds, avoirdupois	4.53592×10^{-4}	Tons, metric
Kips	453.592	Kilograms
Tons, short	907.185	Kilograms
Tons, long	1,016.05	Kilograms
Cubic feet, water	28.317	Kilograms
Pounds per foot	1.48816	Kilograms per meter
Pounds per square foot (psf)	4.88241	Kilograms per square meter
Pounds per square inch (psi)	7.031×10^{-2}	Kilograms per square centimeter
Pounds per square inch (psi)	7.031×10^{-4}	Kilograms per square millimeter
Pounds per cubic foot	16.0184	Kilograms per cubic meter
Foot-pound (ft-lb)	0.1383	Kilogram meters

INDEX

A-Betoni Oy., 607
Aas-Jakobsen, A., 306
Acceleration, gravity, 633
ACI (American Concrete Institute) "Manual of Concrete Practice," 358
Admixtures, 362, 373
Aggregates:
  thermal expansion of, 359, 361, 362
  water absorption of, 359–362, 368
Air, relative humidity, 393, 394
Air entrainment, 362, 371
Air voids, 371, 393
Airy's stress function, 93
Alloying elements, 402
Anchorage, 521, 577, 612, 647
Antisymmetric and symmetric loading, 40, 41, 161, 162, 165, 173
Arena in Raleigh, North Carolina, 513
Arrangement of tendons:
  beams, 448, 451, 454, 456, 459, 462, 467, 472, 488
  shells, 305, 408, 492, 510, 516, 519
  trusses, 471, 485
ASCE Manuals of Engineering Practice, 306
ASCE Proceedings, 270
Atomic hydrogen corrosion of steel, 406
Auberlen, Friedemann, 275, 277

Backnang water and observation tower, 609
Bagon, Arnold, Bureau d'Études, 585
Base shear, 465, 634, 638, 643
Basel water tower, 610
Bearings:
  neoprene (*see* Neoprene bearings)
  steel, 500, 580
  stresses for, 99
  two-point, 467, 573
Belgrade tower, 626
Bendorf Bridge, 580
Beyer, Kurt, 138, 213, 239

Bisagno Bridge, 585
Bleich, F., 31
Bogunovic, U., 626
Boman et Katajamäki, 609
Bond stresses, 55, 56, 61, 89
Boundary conditions, 33, 36, 248, 252, 278, 303, 322
Boundary disturbance, 238, 250, 294
Boussiron Enterprises, 499
Breaking moment, 88
Breaking strength, 84
Bregman & Hamann, 628
Buckling, 181, 186, 272, 340, 344, 498, 648

Cable reactions, 134, 136, 232, 316
Cables and strands, wires for, 401, 403
  (*See also* Appendix)
Campenon Bernard, 587, 591
Cantilever construction method, 580, 583
Capillary constant, 393
Capillary forces, 394
Cauchy-Riemann equations, 333, 336
Cement:
  chemical heat, 351, 352
  color, 362, 366, 387
  constituents and chemical composition, 350–353
  fineness, 355, 367
  flush setting, 381
  gel volume, 355
  initial and final set, 357
  Klein cements, 354
  Lossier cements, 354
  magnesium and sulfur expansion, 353
  retarders, 362
  specific surface, 355, 367
  specific weight, 357
  voids volume, 357
Change in redundants, 80, 82
Column stations, 548, 555
Column strip, 171, 481

Columns:
  corner, 483
  fork-type, 473, 511
  prestressed, 432, 435, 453
  tapered, 453, 473, 494
Compression ring, 510
Concrete:
  amount of cement in, 369, 372
  cracks limit, 400
  durability of, 366, 369, 370
  efflorescence on, 353
  fine-coarse aggregate ratio, 369
  modulus of elasticity, 389, 392
  modulus of shear, 392
  thermal properties of, 392
  two-layer, 424, 467
  voids volume, 349
  water-cement ratio, 368, 380
  watertightness of, 366, 376
Connections:
  elastically controlled, 447, 448, 452, 467, 536, 573
  rigid, 430, 462, 480
Corrosion, 400, 405, 612
Corrugated-type arches, 479
Coulomb's theory, 615
Cracks, 375, 385, 400, 647, 650
Critical stresses, 84, 341, 404
Czernin, Wolfgang, 350

D'Alembert's principle, 632
Danish Great Belt bridges, 592, 595
David, Barott, Boulva, 535
Decking, 522
Deformation, $f(t)$, 67, 108, 394, 397
Depth-span ratio (*see* Span-depth ratio)
Difference equations, 30, 31, 207, 334, 336
Dischinger, Franz, 68, 131, 186, 258, 344, 426
Disturbing impulses, 632, 639
Drainage, 493, 507, 517, 520
Ductility, 401, 403, 408, 453, 468, 532, 611
Dyckerhoff & Widmann, 570, 580, 587, 588, 595

Economy, 14, 417, 482, 487, 502, 542, 548

Effective depth, 20, 76, 182, 468
Elastic limit:
  for concrete, 390
  for steel, 400, 403
Electrochemical corrosion of steel, 406
End-means relationship, 14
Engineering conversion factors (*see* Appendix)
Equations of equilibrium, 17, 240, 256, 262, 295, 308
Erdle, F., 609
Esquillan, Nicholas, 499
Eugema, S. C. E., 585
Expansion joints, 452, 457, 465, 467
Extrusion device, 419, 421

Factor of safety (*see* Safety factor)
Failures of structures, 85, 645
Farkas, Barron & Jablonsky, 628
Feld, Jacob, 646
Film water, 359
Finsterwalder, Ulrich, 559, 579, 597
Fire resistance, 359, 372, 404
Flexicore, 422
Floors:
  two-skin grid, 480, 488
  waffle-type, 480
Flügge, W., 275, 342
Fluttering of suspension roofs, 518
Forces:
  earthquake, 631, 642
  frictional, 232, 447
  internal, 20, 53, 239
  prestressing, 63, 64, 319, 323
  wind, 167, 461, 631
Form carrier, 588, 592
Form-true prestressing, 443, 481, 554
Forms, 507, 522, 575
Freezing, 370
Frequency:
  natural and circular, 633, 638, 640
  of vibrators, 378, 380
Freyssinet, E., 367, 393
Frictional coefficients, 411, 611, 618, 621, 623
Frictional losses, 232, 611
Frost action, 371
Funicular center line, 498

Gladesville Bridge, 598
Gravity acceleration, 633
Greased and wrapped tendons, 79, 448, 506, 521
Green's function, 635
Grid system, two-skin, 480, 488
Ground motion, 632, 648
Grout, 374
Guniting, 373, 503, 517, 610, 650
  (*See also* Shotcrete)

Habitat '67, 535
Harmonic vibration, 633
Heat conductivity, 359, 360, 362, 392, 412
Heede, B. M., Ltd., 628
Heinle, E., 626
Herttoniemi water tower, 609
Hollow-core slabs and planks, 421
Hooke's law, 66, 237

Igneous rocks, 361
Impresa Moviter, S.p.A., 585
Inflection points, 449, 451, 470
Influence of plastic flow and shrinkage, 66, 82, 131
Influence lines, 25
Insulation, 507, 523, 609, 613
Integration formulas, 103
Internal stresses, 53, 54, 394
Intrusion aid, 375

Jacking, hydraulic, 181, 521
Janic, S., 626

K system, 429, 546
Kahn, Louis I., 504, 518, 549, 555
Kaiser, H., 609
Kaista, Pentti, et Co., 609
Kesseler, Richard, 610
Kimbell Art Museum, 504
Klimov, Boris, 332
Klumb, Henry, 503
Koivula, Reino, 609
Komendant, A. E., 63, 270, 543

Krapfenbauer, Robert, 625
Krstic, M., 626
Kuksaing Co., 618, 621

Laboratory towers, Philadelphia, 485
Lahti water tower, 609
Laplace equation, 333
Launching truss, 587, 591
Lauttasaari water tower, 607
Leonhardt, Fritz, 109, 626
Libyan bridge, 594
Lintl, Hannes, 625
Lohmer, Gerd, 570
Lundgren method, 307, 318

Magnification factor, 641
Maillart, Robert, 559, 599
Marcus, H., 35, 39
Maunsell, G., & Partners, 598
Melan, E., 31
Membrane stresses, 238
Membrane theory, 239, 273
Merchants Refrigeration Building, 467
Middendorf, K. H., 99
Mix design, 366, 370, 372, 373
Modulus of elasticity, 66, 390, 403, 409
  apparent, 67
Moglingen's water-observation tower, 610
Mohr's theory, 389
Moment of inertia, 54
Moment transfer, 70, 436
Moments, static, 53, 55, 309, 318
Morandi, Riccardo, 594

Nagoya-Genkai Bridge, 586
Neoprene bearings, 408, 452, 478, 500, 562, 610
  compressive strength of, 410
  heat conductivity of, 412
  modulus of elasticity, 409
  plates, 445, 452, 483, 531
Nervi bridge, 586
Niagara Falls observation tower, 628
Nibelungen Bridge, 580
Noise propagation, 408, 487, 532, 536

Observation towers, 607, 609, 610, 623
Olivetti Factory Building, 492

Particular integrals, 241, 249, 264
PCI Journal, 99
Pier-beam relationship, 561
Pigott Construction Co., 628
Plastic flow, 66, 75, 434
Poisson's ratio, 30, 99, 263, 369, 426
Polcevera Bridge, 594
Polensky & Zollner, 587, 591, 592
Powers, T. C., 371
Prefabricated arches, 476
Prefabricated bridge elements, 563
Prefabricated houses, 524
Preiswerk & Cie. AG, Ingenieure ETH/ SIA, 610
Preload Company, 327
Pressure line, 111, 125, 137, 236
Prestress, loss in, 75, 78
Prestressing, form-true, 443, 481, 554
Prestressing force, 64, 77, 302, 310, 313, 319
Prestressing steel, 401
 (*See also* Appendix)
Principal system, 21, 27, 42, 120, 126, 153, 156, 254, 288, 322
Prometo slip-form system, 623, 628
Pumpcrete, 373

Rausch, E., 639
Rayleigh's method, 635
Reed & Mallik Ltd., 598
Reinforcing steel, 399
 (*See also* Appendix)
Relative humidity, 393, 394
Repetition factor, 641
Ridge loading, 259
Rochester church, 495
Rŏs equation, 390
Rosman, Riko, 205

Safdie, Moshe, 535
Safety factor, 88, 343, 345, 487, 506
Salk Laboratory, 464, 466

Saul's rule, 382
Scaling, 372
Schaffhausen Bridge, 599
Schambeck, Herbert, 596
Schiersteiner Bridge, 570
Schroeder method, 635
Secondary stresses, 182
Sediment stones, 361, 362
Severud, Elstad, and Krueger, 513
Shear hinge, 582
Shear modulus, 392
Shear strength, 389, 411
Shelley system, 530, 534
Shotcrete, 373
Shrinkage, 66, 372, 393, 394, 397
Sieg Valley viaduct, 591, 592
Silicon treatment, 387, 504
Silo pressure, 265, 615
Siltaja Satama Oy, 609
Simpson's rule, 101
Simula, Paavo, 607, 609
Slabs, 418, 423
Slip form, 551, 555, 614, 620
Slip-form system, Prometo, 623, 628
Soil pressure, 104, 265
 reaction, 104, 634
Span-depth ratio, 182, 419, 422, 423, 428, 429, 431, 443, 463, 471, 479, 559, 561, 580, 585
Span-rise ratio, 114, 126
Specific weight, 359–361
Spectral value, 643
Spring coefficient, 633, 634
Stability, 340, 439, 498, 504
Static moments, 53, 55, 309, 318
Steel:
 corrosion, 400, 404, 405
 modulus of elasticity, 401, 404
 prestressing, 401, 405
 reinforcing, 399
 strain hardening, 402
 thermal properties, 401, 404
 yield strength, 400, 403, 404
Stiffness:
 flexural, 20, 70
 relative, 21, 68, 153, 161, 172
 strain, 20, 75, 76
 torsional, 50
Stress transfer, 69, 70, 433–436

Stuart Bros. Pty. Ltd., 598
Stuttgart observation tower, 626
Sumitomo Construction Co., 586
Support settlement, 22, 83, 102, 130
Symmetric and antisymmetric loadings,
    40, 41, 161, 162, 165, 173

Temperature change, 100, 118, 130,
    159, 266
Tension ring, 251, 521
Thermal expansion, 359, 362, 372
Thermal properties, 392
Timoshenko, S., 239, 635
Torsion, 49, 511, 542

Ultimate carrying capacity, 84
Ultimate moment, 88
Ultimate strength, 87, 388, 410

Venice Cultural Center, 518
Vibration:
    damped, 635
    damping factor, 632, 635, 639
    harmonic, 633
Vibrators, 378
Vienna tower, 623, 625
Viesville twin bridge, 585

Voids, 357, 370, 371, 377
Vreden, Verner, 213

Wall panels:
    curtain wall, 439
    load-bearing, 439
Water:
    capillary, 355, 356
    chemically bound, 356
    film, 359
    free, 356
    gel, 356
Water interface, 453, 505
Water reducing agents, 362
Watertightness, 366, 370, 376
Weber, 50
Wind force, gust, 632
Wind pressure, 243, 265, 641
Wind velocity, 632

Yield stresses, 400, 403
Yielding supports, 177, 277, 293, 536
Yitzhaki, David, 277
Yleinen Insinööritoimisto Oy., 607

Zorzi, Silvano, Studio Tecnico, 585

TA
681
K66

JAN 30 1974